IET ENERGY ENGINEERING SERIES 227

Processing and Manufacturing of Electrodes for Lithium-Ion Batteries

Other volumes in this series:

Volume 1	**Power Circuit Breaker Theory and Design** C.H. Flurscheim (Editor)
Volume 4	**Industrial Microwave Heating** A.C. Metaxas and R.J. Meredith
Volume 7	**Insulators for High Voltages** J.S.T. Looms
Volume 8	**Variable Frequency AC Motor Drive Systems** D. Finney
Volume 10	**SF_6 Switchgear** H.M. Ryan and G.R. Jones
Volume 11	**Conduction and Induction Heating** E.J. Davies
Volume 13	**Statistical Techniques for High Voltage Engineering** W. Hauschild and W. Mosch
Volume 14	**Uninterruptible Power Supplies** J. Platts and J.D. St Aubyn (Editors)
Volume 15	**Digital Protection for Power Systems** A.T. Johns and S.K. Salman
Volume 16	**Electricity Economics and Planning** T.W. Berrie
Volume 18	**Vacuum Switchgear** A. Greenwood
Volume 19	**Electrical Safety: A guide to causes and prevention of hazards** J. Maxwell Adams
Volume 21	**Electricity Distribution Network Design, 2nd Edition** E. Lakervi and E.J. Holmes
Volume 22	**Artificial Intelligence Techniques in Power Systems** K. Warwick, A.O. Ekwue and R. Aggarwal (Editors)
Volume 24	**Power System Commissioning and Maintenance Practice** K. Harker
Volume 25	**Engineers' Handbook of Industrial Microwave Heating** R.J. Meredith
Volume 26	**Small Electric Motors** H. Moczala *et al.*
Volume 27	**AC-DC Power System Analysis** J. Arrillaga and B.C. Smith
Volume 29	**High Voltage Direct Current Transmission, 2nd Edition** J. Arrillaga
Volume 30	**Flexible AC Transmission Systems (FACTS)** Y-H. Song (Editor)
Volume 31	**Embedded Generation** N. Jenkins *et al.*
Volume 32	**High Voltage Engineering and Testing, 2nd Edition** H.M. Ryan (Editor)
Volume 33	**Overvoltage Protection of Low-Voltage Systems, Revised Edition** P. Hasse
Volume 36	**Voltage Quality in Electrical Power Systems** J. Schlabbach *et al.*
Volume 37	**Electrical Steels for Rotating Machines** P. Beckley
Volume 38	**The Electric Car: Development and future of battery, hybrid and fuel-cell cars** M. Westbrook
Volume 39	**Power Systems Electromagnetic Transients Simulation** J. Arrillaga and N. Watson
Volume 40	**Advances in High Voltage Engineering** M. Haddad and D. Warne
Volume 41	**Electrical Operation of Electrostatic Precipitators** K. Parker
Volume 43	**Thermal Power Plant Simulation and Control** D. Flynn
Volume 44	**Economic Evaluation of Projects in the Electricity Supply Industry** H. Khatib
Volume 45	**Propulsion Systems for Hybrid Vehicles** J. Miller
Volume 46	**Distribution Switchgear** S. Stewart
Volume 47	**Protection of Electricity Distribution Networks, 2nd Edition** J. Gers and E. Holmes
Volume 48	**Wood Pole Overhead Lines** B. Wareing
Volume 49	**Electric Fuses, 3rd Edition** A. Wright and G. Newbery
Volume 50	**Wind Power Integration: Connection and system operational aspects** B. Fox *et al.*
Volume 51	**Short Circuit Currents** J. Schlabbach
Volume 52	**Nuclear Power** J. Wood
Volume 53	**Condition Assessment of High Voltage Insulation in Power System Equipment** R.E. James and Q. Su
Volume 55	**Local Energy: Distributed generation of heat and power** J. Wood
Volume 56	**Condition Monitoring of Rotating Electrical Machines** P. Tavner, L. Ran, J. Penman and H. Sedding
Volume 57	**The Control Techniques Drives and Controls Handbook, 2nd Edition** B. Drury
Volume 58	**Lightning Protection** V. Cooray (Editor)
Volume 59	**Ultracapacitor Applications** J.M. Miller
Volume 62	**Lightning Electromagnetics** V. Cooray

Volume 63 **Energy Storage for Power Systems, 2nd Edition** A. Ter-Gazarian
Volume 65 **Protection of Electricity Distribution Networks, 3rd Edition** J. Gers
Volume 66 **High Voltage Engineering Testing, 3rd Edition** H. Ryan (Editor)
Volume 67 **Multicore Simulation of Power System Transients** F.M. Uriate
Volume 68 **Distribution System Analysis and Automation** J. Gers
Volume 69 **The Lightening Flash, 2nd Edition** V. Cooray (Editor)
Volume 70 **Economic Evaluation of Projects in the Electricity Supply Industry, 3rd Edition** H. Khatib
Volume 72 **Control Circuits in Power Electronics: Practical issues in design and implementation** M. Castilla (Editor)
Volume 73 **Wide Area Monitoring, Protection and Control Systems: The enabler for smarter grids** A. Vaccaro and A. Zobaa (Editors)
Volume 74 **Power Electronic Converters and Systems: Frontiers and applications** A.M. Trzynadlowski (Editor)
Volume 75 **Power Distribution Automation** B. Das (Editor)
Volume 76 **Power System Stability: Modelling, analysis and control** A.A. Sallam and B. Om P. Malik
Volume 78 **Numerical Analysis of Power System Transients and Dynamics** A. Ametani (Editor)
Volume 79 **Vehicle-to-Grid: Linking electric vehicles to the smart grid** J. Lu and J. Hossain (Editors)
Volume 81 **Cyber-Physical-Social Systems and Constructs in Electric Power Engineering** S. Suryanarayanan, R. Roche and T.M. Hansen (Editors)
Volume 82 **Periodic Control of Power Electronic Converters** F. Blaabjerg, K. Zhou, D. Wang and Y. Yang
Volume 86 **Advances in Power System Modelling, Control and Stability Analysis** F. Milano (Editor)
Volume 87 **Cogeneration: Technologies, optimisation and implementation** C. A. Frangopoulos (Editor)
Volume 88 **Smarter Energy: From smart metering to the smart grid** H. Sun, N. Hatziargyriou, H.V. Poor, L. Carpanini and M.A. Sánchez Fornié (Editors)
Volume 89 **Hydrogen Production, Separation and Purification for Energy** A. Basile, F. Dalena, J. Tong and T.N. Veziroğlu (Editors)
Volume 90 **Clean Energy Microgrids** S. Obara and J. Morel (Editors)
Volume 91 **Fuzzy Logic Control in Energy Systems with Design Applications in MATLAB®/Simulink®** İ. H. Altaş
Volume 92 **Power Quality in Future Electrical Power Systems** A.F. Zobaa and S.H.E.A. Aleem (Editors)
Volume 93 **Cogeneration and District Energy Systems: Modelling, analysis and optimization** M.A. Rosen and S. Koohi-Fayegh
Volume 94 **Introduction to the Smart Grid: Concepts, technologies and evolution** S.K. Salman
Volume 95 **Communication, Control and Security Challenges for the Smart Grid** S.M. Muyeen and S. Rahman (Editors)
Volume 96 **Industrial Power Systems with Distributed and Embedded Generation** R. Belu
Volume 97 **Synchronized Phasor Measurements for Smart Grids** M.J.B. Reddy and D.K. Mohanta (Editors)
Volume 98 **Large Scale Grid Integration of Renewable Energy Sources** A. Moreno-Munoz (Editor)
Volume 100 **Modeling and Dynamic Behaviour of Hydropower Plants** N. Kishor and J. Fraile-Ardanuy (Editors)
Volume 101 **Methane and Hydrogen for Energy Storage** R. Carriveau and D. S-K. Ting
Volume 104 **Power Transformer Condition Monitoring and Diagnosis** A. Abu-Siada (Editor)
Volume 106 **Surface Passivation of Industrial Crystalline Silicon Solar Cells** J. John (Editor)
Volume 107 **Bifacial Photovoltaics: Technology, applications and economics** J. Libal and R. Kopecek (Editors)

Volume 108 **Fault Diagnosis of Induction Motors** J. Faiz, V. Ghorbanian and G. Joksimović
Volume 109 **Cooling of Rotating Electrical Machines: Fundamentals, modelling, testing and design** D. Staton, E. Chong, S. Pickering and A. Boglietti
Volume 110 **High Voltage Power Network Construction** K. Harker
Volume 111 **Energy Storage at Different Voltage Levels: Technology, integration, and market aspects** A.F. Zobaa, P.F. Ribeiro, S.H.A. Aleem and S.N. Afifi (Editors)
Volume 112 **Wireless Power Transfer: Theory, technology and application** N. Shinohara
Volume 114 **Lightning-Induced Effects in Electrical and Telecommunication Systems** Y. Baba and V.A. Rakov
Volume 115 **DC Distribution Systems and Microgrids** T. Dragičević, F.Blaabjerg and P. Wheeler
Volume 116 **Modelling and Simulation of HVDC Transmission** M. Han (Editor)
Volume 117 **Structural Control and Fault Detection of Wind Turbine Systems** H.R. Karimi
Volume 118 **Modelling and Simulation of Complex Power Systems** A. Monti and A. Benigni
Volume 119 **Thermal Power Plant Control and Instrumentation: The control of boilers and HRSGs, 2nd Edition** D. Lindsley, J. Grist and D. Parker
Volume 120 **Fault Diagnosis for Robust Inverter Power Drives** A. Ginart (Editor)
Volume 121 **Monitoring and Control using Synchrophasors in Power Systems with Renewables** I. Kamwa and C. Lu (Editors)
Volume 123 **Power Systems Electromagnetic Transients Simulation, 2nd Edition** N. Watson and J. Arrillaga
Volume 124 **Power Market Transformation** B. Murray
Volume 125 **Wind Energy Modeling and Simulation Volume 1: Atmosphere and plant** P.Veers (Editor)
Volume 126 **Diagnosis and Fault Tolerance of Electrical Machines, Power Electronics and Drives** A.J.M. Cardoso
Volume 127 **Lightning Electromagnetics Volume 1: Return stroke modelling and electromagnetic radiation 2nd Edition** V. Cooray, F. Rachidi, M. Rubinstein (Editors)
Volume 127 **Lightning Electromagnetics Volume 2: Electrical processes and effects, 2nd Edition** V. Cooray, F. Rachidi, and M. Rubinstein (Editors)
Volume 128 **Characterization of Wide Bandgap Power Semiconductor Devices** F. Wang, Z. Zhang and E.A. Jones
Volume 129 **Renewable Energy from the Oceans: From wave, tidal and gradient systems to offshore wind and solar** D. Coiro and T. Sant (Editors)
Volume 130 **Wind and Solar Based Energy Systems for Communities** R. Carriveau and D.S-K. Ting (Editors)
Volume 131 **Metaheuristic Optimization in Power Engineering** J. Radosavljević
Volume 132 **Power Line Communication Systems for Smart Grids** I.R.S. Casella and A. Anpalagan
Volume 134 **Hydrogen Passivation and Laser Doping for Silicon Solar Cells** B. Hallam and C. Chan (Editors)
Volume 139 **Variability, Scalability and Stability of Microgrids** S.M. Muyeen, S.M. Islam and F. Blaabjerg (Editors)
Volume 142 **Wind Turbine System Design: Volume 1: Nacelles, drive trains and verification** J. Wenske (Editor)
Volume 143 **Medium Voltage DC System Architectures** B. Grainger and R. D. Doncker (Editors)
Volume 145 **Condition Monitoring of Rotating Electrical Machines** P. Tavner, L. Ran, and C. Crabtree
Volume 146 **Energy Storage for Power Systems, 3rd Edition** A.G. Ter-Gazarian
Volume 147 **Distribution Systems Analysis and Automation, 2nd Edition** J. Gers
Volume 151 **SiC Power Module Design: Performance, robustness and reliability** A. Castellazzi and A. Irace (Editors)
Volume 152 **Power Electronic Devices: Applications, failure mechanisms and reliability** F. Iannuzzo (Editor)

Volume 153 **Signal Processing for Fault Detection and Diagnosis in Electric Machines and Systems** M. Benbouzid (Editor)
Volume 155 **Energy Generation and Efficiency Technologies for Green Residential Buildings** D. Ting and R. Carriveau (Editors)
Volume 156 **Lithium-Ion Batteries Enabled by Silicon Anodes** C. Ban and K. Xu (Editors)
Volume 157 **Electrical Steels, 2 Volumes** A. Moses, K. Jenkins, Philip Anderson and H. Stanbury
Volume 158 **Advanced Dielectric Materials for Electrostatic Capacitors** Q. Li (Editor)
Volume 159 **Transforming the Grid Towards Fully Renewable Energy** O. Probst, S. Castellanos and R. Palacios (Editors)
Volume 160 **Microgrids for Rural Areas: Research and case studies** R.K. Chauhan, K. Chauhan and S.N. Singh (Editors)
Volume 161 **Artificial Intelligence for Smarter Power Systems: Fuzzy logic and neural networks** M.G. Simoes
Volume 165 **Digital Protection for Power Systems, 2nd Edition** Salman K Salman
Volume 166 **Advanced Characterization of Thin Film Solar Cells** N. Haegel and M. Al-Jassim (Editors)
Volume 167 **Power Grids with Renewable Energy: Storage, integration and digitalization** A.A. Sallam and B. OM P. Malik
Volume 169 **Small Wind and Hydrokinetic Turbines** P. Clausen, J. Whale and D. Wood (Editors)
Volume 170 **Reliability of Power Electronics Converters for Solar Photovoltaic Applications** F. Blaabjerg, A.I. Haque, H. Wang, Z. Abdin Jaffery and Y. Yang (Editors)
Volume 171 **Utility-Scale Wind Turbines and Wind Farms** A. Vasel-Be-Hagh and D.S.-K. Ting
Volume 172 **Lighting interaction with Power Systems, 2 Volumes** A. Piantini (Editor)
Volume 174 **Silicon Solar Cell Metallization and Module Technology** T. Dullweber (Editor)
Volume 175 **n-Type Crystalline Silicon Photovoltaics: Technology, applications and economics** D. Munoz and R. Kopecek (Editors)
Volume 178 **Integrated Motor Drives** Xu Deng and B. Mecrow (Editors)
Volume 180 **Protection of Electricity Distribution Networks, 4th Edition** J. Gers and E. Holmes
Volume 182 **Surge Protection for Low Voltage Systems** A. Rousseau (Editor)
Volume 184 **Compressed Air Energy Storage: Types, systems and applications** D. Ting and J. Stagner
Volume 186 **Synchronous Reluctance Machines: Analysis, optimization and applications** N. Bianchi, C. Babetto and G. Bacco
Volume 190 **Synchrophasor Technology: Real-time operation of power networks** N. Kishor and S.R. Mohanty (Editors)
Volume 191 **Electric Fuses: Fundamentals and new applications, 4th Edition** N. Nurse, A. Wright and P.G. Newbery
Volume 193 **Overhead Electric Power Lines: Theory and practice** S. Chattopadhyay and A. Das
Volume 194 **Offshore Wind Power Reliability, Availability and Maintenance, 2nd Edition** P. Tavner
Volume 196 **Cyber Security for Microgrids** S. Sahoo, F. Blaajberg and T. Dragicevic
Volume 198 **Battery Management Systems and Inductive Balancing** A. Van den Bossche and A. Farzan Moghaddam
Volume 199 **Model Predictive Control for Microgrids: From power electronic converters to energy management** J. Hu, J.M. Guerrero and S. Islam
Volume 201 **Design, Control and Monitoring of Tidal Stream Turbine Systems** M. Benbouzid (Editor)
Volume 204 **Electromagnetic Transients in Large HV Cable Networks: Modeling and calculations** Ametani, Xue, Ohno and Khalilnezhad
Volume 208 **Nanogrids and Picogrids and Their Integration with Electric Vehicles** S. Chattopadhyay
Volume 210 **Superconducting Magnetic Energy Storage in Power Grids** M.H. Ali

Volume 211 **Blockchain Technology for Smart Grids: Implementation, management and security** H.L. Gururaj, K.V. Ravi, F. Flammini, H. Lin, B. Goutham, K.B.R. Sunil and C. Sivapragash
Volume 212 **Battery State Estimation: Methods and models** S. Wang
Volume 215 **Industrial Demand Response: Methods, best practices, case studies, and applications** H.H. Alhelou, A. Moreno-Muñoz and P. Siano (Editors)
Volume 213 **Wide Area Monitoring of Interconnected Power Systems, 2nd Edition** A.R. Messina
Volume 214 **Distributed Energy Storage in Urban Smart Grids** P.F. Ribeiro and R.S. Salles (Editors)
Volume 217 **Advances in Power System Modelling, Control and Stability Analysis, 2nd Edition** F. Milano (Editor)
Volume 225 **Fusion-Fission Hybrid Nuclear Reactors: For enhanced nuclear fuel utilization and radioactive waste reduction** W.M. Stacey
Volume 228 **Digital Technologies for Solar Photovoltaic Systems: From general to rural and remote installations** S. Motahhir (Editor)
Volume 238 **AI for Status Monitoring of Utility Scale Batteries** S. Wang, K. Liu, Y. Wang, D. Stroe, C. Fernandez and J.M. Guerrero
Volume 239 **Intelligent Control of Medium and High Power Converters** M. Bendaoud, Y. Maleh and S. Padmanaban (Editors)
Volume 905 **Power System Protection, 4 volumes**

Processing and Manufacturing of Electrodes for Lithium-Ion Batteries

Edited by
Jianlin Li and Congrui Jin

The Institution of Engineering and Technology

Published by The Institution of Engineering and Technology, London, United Kingdom

The Institution of Engineering and Technology is registered as a Charity in England & Wales (no. 211014) and Scotland (no. SC038698).

First published 2023

The Institution of Engineering and Technology
Futures Place
Kings Way, Stevenage
Hertfordshire SG1 2UA, United Kingdom

www.theiet.org

British Library Cataloguing in Publication Data
A catalogue record for this product is available from the British Library

ISBN 978-1-83953-669-4 (hardback)
ISBN 978-1-83953-670-0 (PDF)

Typeset in India by MPS Limited

Contents

About the editors

Jianlin Li is a senior R&D staff and group leader at Oak Ridge National Laboratory, USA. He is an inventor of 31 patents and patent applications, has authored more than 150 peer-reviewed journal articles, and given more than 130 presentations including 43 invited talks. His research interest lies in materials synthesis, electrode processing, and cell manufacturing for energy storage and conversion applications.

Congrui Jin is an assistant professor at the University of Nebraska – Lincoln, USA. Prior positions include postdoctoral fellow at the Oak Ridge National Laboratory and assistant professor at the State University of New York at Binghamton. She has authored more than 50 peer-reviewed journal publications. Her research interests encompass energy storage materials, additively manufactured materials, geological materials, and construction materials.

Chapter 1

Introduction

Jianlin Li[1] *and Congrui Jin*[2]

Lithium-ion battery (LIB) technology has achieved great success since being commercialized three decades ago. Production of LIBs reached 492 GWh in 2021 [1] and is projected to reach 2–3.5 TWh by 2030 [2]. The LIB market has increased simultaneously, which was reported worth $34.2 billion in 2020 and is to reach $87.5 billion in 2027 [3]. These come with substantial improvement in battery performance, more than a five-fold increase in volumetric energy density at the pack level since 2008 [4], and nearly a 98% reduction in production cost from $5,000/kWh in 1991 to $101/kWh in 2021 [5,6]. The improvement in energy density and reduction in cost are enabled by great progress in material science, engineering, manufacturing, and scale-up.

From the material perspective, transitioning from $LiCoO_2$ to high nickel, low cobalt cathode materials such as $LiNi_xMn_yCo_{1-x-y}O_2$, and introducing silicon into graphite anodes have improved the specific capacity of the active materials and thus the energy density of the cells. It also enables a significant reduction in production cost, as cathode material is the largest cost component in a LIB. Based on Ziegler *et al.*'s cost model [7], the increase in cathode capacity for a fixed volume contributes ~23% to the cost reduction, which results from concurrent improvement in the specific capacity of cathode materials and the volume ratio of active materials. The decline in cathode material cost via reducing cobalt content is also important, accounting for 18% of the cost decline. Compared to materials, engineering, manufacturing, and scale-up demonstrate a more significant contribution to cost reduction. Progress in engineering and manufacturing has improved the volume ratio of active materials, allowing packing more active materials in a fixed volume. The volumetric energy density at the pack level has increased by five folds since 2008 [4]. The majority of the improvement was ascribed to engineering and manufacturing, as the improvement from the volumetric energy density of active materials was less than 50% (i.e., NMC111 vs. NMC811). As a result, engineering and manufacturing are the main drivers for many major contributors to the cost decline, including the increase in cathode charge capacity for a fixed volume

[1]Division of Electrification and Energy Infrastructures, Oak Ridge National Laboratory, USA
[2]Department of Civil and Environmental Engineering, University of Nebraska – Lincoln, USA

(23%), increased charge capacity utilization (9.5%), reduced electrolyte mass to cathode charge capacity ratio, and lower N/P ratio. Scale-up (plant size) contributes to approximately 14% cost reduction.

It has been clearly demonstrated that both material development and manufacturing are crucial to improve energy density and reduce costs. However, the research activities in manufacturing are far less compared to materials discovery and development. This is mainly due to the fact that manufacturing is considered industrial research and there exists little peer-reviewed publication from the industry. Figure 1.1 compares the number of publications with the keywords of "lithium-ion batteries" and "manufacturing & lithium-ion batteries," respectively, based on the search results from Scopus. As shown in the figure, the publications related to manufacturing usually account for <1% of the total publications on LIB research. The lack of publications on processing and manufacturing could hinder the LIB development due to the following reasons. (1) There will be a gap between fundamental/applied research and commercialization. (2) The know-hows in engineering are not available to the general public and could make it very challenging for new companies to develop the system. And (3) the next-generation workforce is not trained with the skills and mindset to go to the industry.

Our motivation in editing this book is to provide comprehensive overviews of materials processing and manufacturing of LIBs and stimulate future efforts in these areas. Academia can play a critical role in manufacturing and scale-up. For example, academia could advance manufacturing science which will well prepare our next-generation workforce and facilitate novel manufacturing techniques. Academia can also lead the industry in exploring high-risk-and-high-reward manufacturing methods. Since LIBs are very complex and their manufacturing involves many processes, the industry may be hesitant to adopt innovative changes in

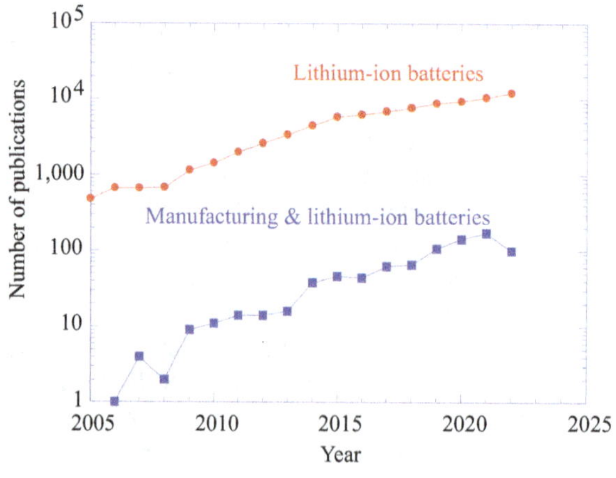

Figure 1.1 Comparison of the number of publications with keywords of LIBs and manufacturing and LIBs, respectively, from 2005 to 2022 based on the search results from Scopus

manufacturing process prior to being fully validated. For example, LIBs were previously manufactured without much consideration of recycling [8], but with increased mass production of LIBs, recycling the spent cells is becoming necessary, and thus it is highly desirable to design new materials and processes which can ease the recycling process and facilitate sustainable and green manufacturing.

References

[1] McKerracher, C. Electric Vehicle Outlook 2022. *Bloomberg New Energy Finance*, 2022. https://about.newenergyfinance.com/electric-vehicle-outlook/.

[2] IEA. Global EV Outlook 2022. *IEA*, 2022. https://www.iea.org/reports/gloval-ev-outlook-2022.

[3] Lithium-Ion Battery Market Size, Share & Trends Analysis Report by Product (LCO, LFP, NCA, LMO, LTO, Lithium Nickel Manganese Cobalt), by Application, by Region, and Segment Forecasts, 2020–2027, 2020.

[4] Li, J., Du, Z., Ruther, R. E., *et al.*, Toward low-cost, high-energy density, and high-power density lithium-ion batteries. *JOM* 2017, 69, 1484–1496. doi:10.1007/s11837-017-2404-9.

[5] Ziegler, M. S. and Trancik, J. E., Re-examining rates of lithium-ion battery technology improvement and cost decline. *Energy Environ. Sci.,* 2021, 14, 1639–1651.

[6] Kittner, N., Lill, F., and Kammen, D. M. Energy storage deployment and innovation for the clean energy transition. *Nat. Energy.* 2017, 2, 17125.

[7] Ziegler, M. S., Song, J., and Trancik, J. E., Determinants of lithium-ion battery technology cost decline. *Energy Environ. Sci.* 2021, 14, 6074–6098.

[8] Li, J., Lu, Y., Yang, T., Ge, D., Wood, D. L., and Li, Z., Water-based electrode manufacturing and direct recycling of lithium-ion battery electrodes – a green and sustainable manufacturing system. *iScience*, 2020, 23(5), 101081.

Chapter 2

Materials for lithium-ion batteries

Hanlei Zhang[1]

This chapter introduces materials for the cathode, anode, and electrolyte of Li-ion batteries (LIBs), which make up the structural and chemical foundations for an electrochemical battery cell. In this chapter, we will discuss the development and application of battery materials ever since the invention of LIBs in 1970s. We further investigate the advantages and disadvantages of varying battery materials from the electrochemical, structural, environmental, and safety perspectives. Afterwards, we summarize the trends for the development of next-generation battery materials to resolve the standing issues of current battery materials. Last but not the least, this chapter will present the practical design principles and limitations of battery materials, which are essential for the development and operation of battery cells.

2.1 Introduction

As schematically shown in Figure 2.1(a) [1], a standard Li-based electrochemical battery cell is composed of at least a cathode [2–4], an anode [5,6] and an electrolyte [7,8]. These components exhibit different electrochemical functionalities, and by working together they allow the battery cell to bear external loading. The cathode and the anode hold and release the Li^+ cations via their difference in chemical potentials, and the electrolyte functions for the transportation of Li^+ cations [9,10]. The different functionalities of the cathode, the anode and the electrolyte mean that they need to be made by materials with significantly different electrochemical properties. During the electrochemical discharge, Li^+ ions escape from the anode, pass through the electrolyte and eventually merge into the cathode, transforming the chemical energy stored in the battery cell into electric energy in the external circuit [9,11]. Reversibly, during the charging process, the external electric energy is transformed into the chemical energy of the battery cell, which is achieved by transporting the Li^+ cations from the cathode back into the anode [12]. The cyclic transportation of Li^+ cations between the cathode and the anode is the structural foundation of the electrochemical functionality of the Li-ion battery (LIB), which is

[1]Advanced Materials Characterization Laboratory, Materials Research Center, Missouri University of Science and Technology, USA

also termed as the "rocking-chair" mechanism [13,14]. To complete the rocking-chair mechanism, the cathode and anode materials should possess the capability to hold and release Li^+ cations at reasonable capacities and chemical potentials, and the electrolyte should be able to let the Li^+ cations pass at a good rate [12].

In the LIB cell, the combination of the physical and chemical properties of the cathode, the anode, and the electrolyte determines the overall electrochemical performance of the cell, such as the discharge capacity, the cutoff voltage, the rate

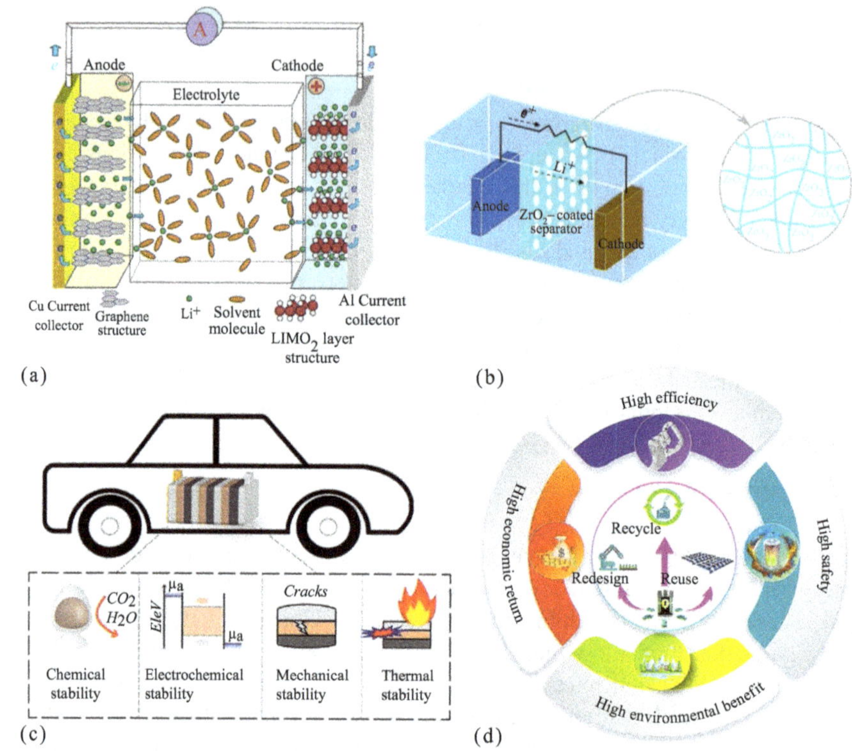

Figure 2.1 Components of the Li-based battery cell and the principal consideration for design and operation. (a) Schematic summarizing the fundamental configuration of the LIB cell. Reproduced with permission from Ref. [1]. Copyright 2004 American Chemical Society. (b) Schematic exemplifying the structure of the separator for the LIB cell. Adapted with permission from Ref. [15]. Copyright 2017 American Chemical Society. (c) Diagram summarizing the major considerations for optimizing the materials of the LIB. Reproduced with permission from Ref. [16]. Copyright 2020 American Chemical Society. (d) Diagram summarizing the properties associated with the sustainability of the battery materials. Reproduced with permission from Ref. [17]. Copyright 2020 American Chemical Society.

capability, and the electrochemical/structural stability [18,19]. Therefore, developing comprehensive understandings and the associated strategies for the selection, combination and improvement of battery materials arises as a major challenge in optimizing the fabrication and operation of next-generation battery cells, which we will discuss in this chapter with details.

Beside the cathode, the anode, and the electrolyte, some battery cells also contain a separator to prevent the internal short circuit between the cathode and the anode [20, 21], as schematically shown in Figure 2.1(b). Some anodes surface can develop lithium dendrites during the electrochemical lithiation–delithiation [22], which expands in the battery cell and eventually connects the anode with the cathode [23,24], resulting in short circuiting inside the battery cell and finally fire and explosion. The separator can mitigate the kinetics of dendrite growth and thus postpone, if not prevent, the internal short circuiting between the cathode and the anode [25,26]. However, overgrowth of lithium dendrites can still spike through the separator and result in short circuiting, fire, and explosion [27].

Nowadays, the LIBs are widely utilized in portable electronics [28], grid energy storage [29], and most recently and excitingly, electric vehicles [30,31]. Accordingly, the selection, combination, and application of battery materials must consider the realistic conditions of operation [32]. The most important consideration is the electrochemical performance, including the discharge capacity, the discharge voltage, the rate capability, etc. [33]. Another critical aspect is the structural and electrochemical stabilities of the battery [34–36], which is determinant in elongating the service time and reducing the financial cost. Also, the safety of the LIBs needs serious attention, which usually calls for chemically and thermally stable materials that prevent fire or explosion even under extreme conditions of operation, such as high temperatures or high voltages [36,37]. The mechanical stability of the battery materials is another major consideration, which is essential in preventing the cracking of the battery and thus the associated electrochemical fade [38,39]. A diagram summarizing the considerations regarding the material selection for optimized electrochemical performances of LIBs is schematically presented in Figure 2.1(c) [16].

Besides the physical and chemical properties associated with the electrochemical performance of the battery material, another critical consideration for material selection is the influence of the battery on the environment, namely the sustainability of the battery materials and the associated recycling technologies, as schematically summarized in Figure 2.2(d) [17]. The battery should adopt materials that are easy to recycle with good efficiencies [40], meaning that a large percentage of the old cell materials can be collected. The financial cost should also be kept affordable to encourage the recycling. Additionally, the battery material should be nontoxic and safe to both personnel and the environment. It is also preferred to have the battery materials compatible with recycle processes that have little environmental impact, as well as feasible in the recycling process [41].

In the following sections of this chapter, we will give an introduction of the materials for the cathode, the anode, and the electrolyte, with practical examples from the literature. The prevailing families of battery materials will be presented, and their electrochemical performances will be evaluated and compared.

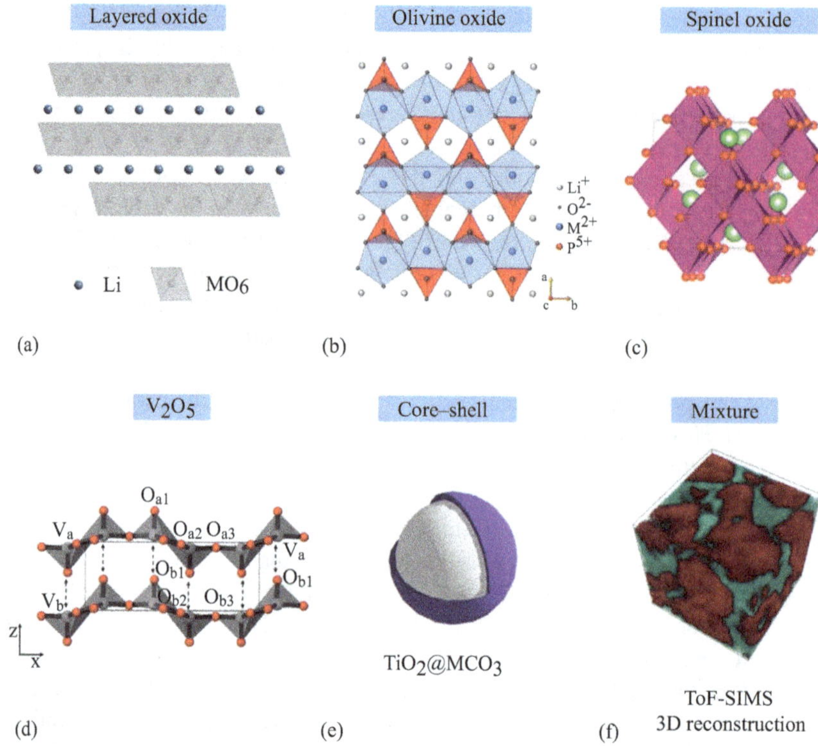

Figure 2.2 *Cathode materials for LIBs. (a) Schematic showing the $R\bar{3}m$ crystal structure of the layered oxides. Adapted with permission from Ref. [35]. Copyright 2018 American Chemical Society. (b) Schematic showing the crystal configuration of the olivine oxides. Reproduced with permission from Ref. [48]. Copyright 2021 American Chemical Society. (c) Schematic showing the $Fd\bar{3}m$ crystal configuration of the spinel oxides. Reproduced with permission from Ref. [49]. Copyright 2014 American Chemical Society. (d) Schematic showing the crystal configuration of the V_2O_5 oxide. Reproduced with permission from Ref. [50]. Copyright 2017 American Chemical Society. (e) Schematic showing the core–shell configuration of cathode particles. Reproduced with permission from Ref. [51]. Copyright 2020 American Chemical Society. (f) Three-dimensional reconstruction of a phosphate@NMC622 composite cathode by complete mixture of the two cathode materials. Reproduced with permission from Ref. [52]. Copyright 2011 American Chemical Society.*

2.2 Cathode materials

2.2.1 Layered oxides

Layered oxides are the most utilized cathode materials, which can be described with the chemical formula $LiMO_2$. M stands for transition metals such as Co, Mn, Ni, Al, V, and Zn, as well as their combinations [31,42]. By choosing the combination of transition metals, the electrochemical properties of layered oxides can be feasibly adjusted for the optimized performance [43,44]. The first layered oxide practically adopted as the LIB cathode was the $LiCoO_2$ oxide, which was proven to be electrochemically attractive by the pioneering work of Goodenough and Koichi Mizushima [45,46]. Ever since it invention in the 1980s, $LiCoO_2$ has been widely adopted in the modern electronics industry nowadays, with its outstanding electrochemical stability and satisfying, stable electrochemical performance [47]. Crystallographically, the layered oxide adopts the $R\overline{3}m$ crystal configuration, which is composed of alternating Li and Co slabs inserted in a face-centered-cubic (FCC) framework of O^{2-} anions, as schematically shown in Figure 2.2(a) [35]. The combination of Co^{+} and O^{2-} makes up a spatial framework where the Li^{+} cations can be feasibly inserted and extracted, namely the structural foundation of the electrochemical properties of $LiCoO_2$. This structural feature also applies to derivatives of $LiCoO_2$, as will be discussed in the following.

In the large family of layered oxides generated by varying the combination of transition metals, the most popular ones are $LiCoO_2$ [53], $LiNi_{1-x-y}Mn_xCo_yO_2$ (NMC) [54], $LiNi_{1-x-y}Co_xAl_yO_2$ (NCA) [35,55–57], Li- and Mn-rich oxides (LMR) [58] and Ni-rich oxides [59]. In general principles, higher Ni, Li, and Mn additions can increase the discharge capacity at the expense of the structural integrity and electrochemical stability [60,61]. On the other hand, Co and Al are stabilizers for maintaining the structural integrity, which helps the cathode to last during the electrochemical cycling, while these elements only yield medium discharge capacities [62,63].

The aforementioned layered cathodes exhibit significantly different advantages and disadvantages, thereby making the selection of cathode materials a critical task. For instance, the addition of Ni increases the discharge capacity [64], and the addition of Co helps with the structural stability [65]. Therefore, the coaddition of Ni and Co helps to make layered cathodes with both high performance and good cyclability. However, both Ni and Co are pricy elements, which thus increase the material cost of the cathode [66,67]. The LMR cathodes have both outstanding electrochemical performance and low costs [68], but they suffer from severe fade in discharge voltage due to their structural instability, majorly originated from the Jahn–Teller distortion [69,70]. The addition of Al^{3+} cations into the layered cathodes significantly increases the structural stability, but the Al^{3+} cations are electrochemically "dead" and do not contribute to the discharge capacity [11,71]. The reader is referred to the review articles by Manthiram [2] and Zhang *et al.* [72] for more discussions regarding the electrochemical performance and structural stabilities of layered oxide cathodes with different combinations of transition metals.

2.2.2 Olivine oxides

Beside the layered oxides, another popular family of cathode materials is the olivine oxides, which are lithium transition metal phosphate and can be described with the formula of $LiMPO_4$ (M = Fe^{3+}, Mn^{3+}, Co^{3+}, VO^{3+}, etc.) [73,74]. Compared with the layered oxides, the olivine oxides exhibit more distorted lattice configurations generated by the asymmetric electron configurations of transition metal cations [75], which exaggerates the Jahn–Teller effects and serves as a major source of structural degradations and failures [76,77]. A representative example of the olive cathodes is the $LiFePO_4$, which is being widely used in grid energy storage and electric vehicles with its low material cost and satisfying electrochemical performance, namely the satisfying cutoff voltage of ~4.0 V and the discharge capacity of ~170 mAh/g [78,79]. As schematically shown in Figure 2.2(b) [48], the olivine lattice only contains one-dimensional channels for the intercalation of Li^+ cations, which is the major reason for its slower ionic kinetics and lower discharge capacity, compared to the two-dimensional Li^+ channel network of the layered oxides (Figure 2.2(a)).

Another disadvantage of olivine cathodes is their poor electronic conductivities [80], which increases the internal resistance of the cathode and thus reduce the energy efficiency [81]. The conductivity of olivine cathodes can be improved by reducing the size of nanoparticles [82,83], adding a carbon coating layer [84,85] and adding high-conductivity, high-connectivity binding materials into the electrode [86,87]. Making nanosized particles also helps to improve the packing density of olivine cathodes [88], thereby overcoming another major issue limiting the wider applications of olivine cathodes. The volumetric energy density also affects the wider application of olivine cathodes. Due to the low Li:O ratio of olivine cathodes (1:4) [89], the olivine cathodes exhibit large lattices with low contents of redox Li^+, leading to their low volumetric energy densities. In other words, the low volumetric density is structurally intrinsic for olivine cathodes [76]. Also, the low discharge voltage of olivine cathodes also reduces the volumetric energy density. For example, the Fe^{2+}/Fe^{3+} redox couple only exhibits a discharge voltage of 3.4 V [90], which is much lower than the 4.3 V operation voltage of layered oxides [91]. This can be partially mitigated by the development of new redox couples with higher discharge voltages. For example, the V^{5+}/V^{4+} redox couple of the Li_2VOPO_4 cathode and its species exhibit a voltage of 4.0 V against the Li/Li^+ couple, which thus increases the energy density [73].

Although the low volumetric energy density limits the wider application of olivine cathodes on portable electronics and electric vehicles, they are outstandingly low in material cost for fabrication, which is based on the fact that no expensive transition metal is adopted [81,92]. By this mean, the olivine cathodes can be applied in scenarios that are not volume- or weight-sensitive, such as the storage of grid electric power [93,94]. Also, the electrochemical stability and safety of olivine cathodes are also outstanding, making them reliable candidates for electrochemical operations at elongated time or under extreme conditions [86,95].

2.2.3 Spinel oxides

Beside the prevailing layered and olivine oxides, there are also some novel families of oxides with unique lattice configurations and promising electrochemical properties. Among the large families of novel cathode materials, spinel oxides arise as shining stars, which are derived from the crystal configuration of the layered oxide and can be described with the chemical formula of LiM_2O_4 (M = Mn, Ni, Fe, Cr, Co, Cu, and their combinations) [55,96,97]. The spinel cathode has the same FCC oxygen framework as the layered oxides, with one-fourth of the 3a-sitting transition metal cations of the $R\bar{3}m$ layered configuration relocated onto the 4d sites of the $Fd\bar{3}m$ spinel structure [55,98]. The lattice configuration of the spinel oxides is schematically presented in Figure 2.2(c) [49]. Compared with the two-dimensional, plane-shaped Li^+ channels of the layered oxides, the spinel oxides exhibit a three-dimensional network for Li^+ intercalation [99,100], which increases the number of pathways for the Li^+ diffusion and is considered beneficial for promoting the electrochemical kinetics [101]. Structurally, the complex network of Li channels also increases their robustness against the structural degradations, which in turn promotes the electrochemical stability [102]. Similar as the olivine oxides, the spinel oxides also suffer from the Jahn–Teller distortion [103,104], which is a major source of lattice defects and the associated mechanical/electrochemical failures.

The first spinel cathode for LIBs was the $LiMn_2O_4$ oxide, which was developed by Thackeray *et al.* [105,106], showing an outstanding structural stability and satisfying discharge capacity. Regardless of the fact that the spinel cathode is not yet fully adopted in the industries of LIBs, it has proven its significant value serving as a protective coating layer of the core–shell cathode configuration [107,108], with the high-capacity, vulnerable cathode material underneath. The spinel coating drastically increases the electrochemical stability of the originally less stable core materials. Additionally, the usage of low-cost transition metals reduces the price of spinel cathodes [109,110]. Using $LiMn_2O_4$ as the prototype, derivative spinel oxide cathodes have been developed, such as $LiNi_{0.5}Mn_{0.5}O_4$ [111] and $Li_{1.05}M_{0.05}Mn_{1.9}O_4$ (M = Ni, Mg or Al) [112].

Besides the advantages of spinel cathodes, the shortcomings of the spinel oxides are obstacles in their way to wider applications. The spinel oxides are usually adopted in high-power systems, while their energy densities are usually low, which limits their applications in high-energy systems [113]. This is another reason that the spinel oxide is only adopted as the coating layer of the core–shell particles, rather than making into single-component battery cells, as mentioned above. So far, the spinel cathodes are largely restricted to the $LiMn_2O_4$ oxide, as the other derivative cathodes have been proven to exhibit intensive Jahn–Teller distortions and subject to severe structural degradations [114]. Even for the relatively stable $LiMn_2O_4$ oxide, it still suffers from dissolution of the Mn element and thus poisoning of the anode [2].

2.2.4 Vanadium oxides

Due to the multivalence nature of V, vanadium oxides such as V_2O_5 are also considered as promising cathode materials for LIBs [115,116]. Compared to the complex transition metal oxides, vanadium oxides are much easier to fabricate and manipulate using physical and chemical synthetic techniques [117,118], as well as being stable during electrochemical cycling [119,120]. A schematic summarizing the lattice configuration of the V_2O_5 oxide is presented in Figure 2.2(d) [50]. Compared to the layered, olivine and spinel oxides with most of the transition metal cations located on the octahedral sites of the FCC oxygen framework, the vanadium cations are located on the smaller tetrahedral sites of the oxygen framework, which results in significant lattice distortion [121].

Ever since its development in the 1970s and 1980s, the vanadium oxides have been actively tested as the cathode of LIBs [122], majorly due to its attractively low material cost. Regardless of the advantages of the vanadium oxide cathode, it suffers from restricted electrochemical kinetics and thus lower rate capabilities [123]. Its stability during the electrochemical cycling also needs improvements to reach a practical service life of a few hundred cycles [124].

2.2.5 Composite cathodes

Beside the single-component cathodes discussed above, another very important type of cathodes is the composite cathodes, which is composed of more than one active materials [125,126]. The development of composite cathodes is to combine the advantages of the composing materials [127,128], which is not easily achievable by tuning a single-component cathode. Also, the complex architectures of composite cathodes serve as another pathway for obtaining sophisticated combinations of electrochemical properties. Therefore, composite cathodes serve as a tunable synthesis strategy and thereby attract intensive research interest [129]. On the other hand, synthesizing composite cathodes is much more complicated than the traditional techniques such as chemical doping, chemical tunning or morphology manipulation [43], meaning that compositing can only be applied to high-value materials.

The most popular architecture of composite cathodes is the so-called core–shell configuration, as schematically shown in Figure 2.2(e) [51]. In this design, a thin layer of stabilizer material is coated on the surface of a spherical cathode particle, serving as a protective shell against the electrochemically induced structural degradations [130,131]. In direct contrast, the core region is composed of a material with good electrochemical properties but relatively low electrochemical stabilities [131,132], which is isolated from the degrading effects of the electrolyte by the protective surface. In other words, the core–shell configuration combines the cyclability of the surface shell and the outstanding electrochemical properties of the core.

Another prevailing configuration of composite cathodes is complete mixture of different cathode materials, as exemplified by the three-dimensional reconstruction in Figure 2.2(f) [52]. By mixing electrochemically stable and high-performance cathode materials, this design can also optimize the combination of electrochemical properties and stabilities [127,133], which is a similar design principle as the core–shell configuration in Figure 2.2(e). Compared to the core–shell configuration, the complete mixture allows the two composing materials to have better contacts with each other and thus results in better kinetic interactions, better overall conductivity and better structural integrity [134]. Additionally, the complete mixture of different materials is much more feasible to fabricate compared to the well-defined core–shell architecture [135], thereby reducing the cost of the cathode. However, the simple mixture lacks the well-defined architecture of the core–shell configuration, so it usually exhibits a less satisfactory electrochemical performance and reduced cyclability.

2.2.6 Comparison of cathode materials

Table 2.1 summarizes the commercialized cathode materials that have been applied on electronic vehicles, as well as their features, advantages and disadvantages. The data are adopted from the publications of Chen *et al.* [136], Catenaro *et al.* [137], Julien *et al.* [113], Want *et al.* [138], Blomgren [139], and Purwanto *et al.* [140]. In principle, $LiCoO_2$ is the most widely adopted cathode material, with satisfying electrochemical performance, high stabilities, and reliable safety against fire and explosion. However, the usage of the Co element greatly rises the price of the battery cell, thereby calling for urgent development of high-performance substitutes with lower costs, such as $LiFePO_4$, $LiNi_{0.80}Co_{0.15}Al_{0.05}O_2$ (NCA), lithium nickel manganese cobalt Oxide (NMC), and lithium manganese oxide (LMO).

Table 2.1 Comparison of commercialized cathode materials

Cathode	**Manufacturers**	**Gravitational energy density (Wh/kg)**	**Advantages**	**Issues**
$LiCoO_2$	Tesla, Panasonic, etc.	155	Good cyclability; satisfying electrochemical properties	High costs; moderate thermal stabilities
$LiFePO_4$	BYD, Tesla, etc.	160	Low costs; high power density; long cycle life	Relatively low energy density; poor conductivity
$LiNi_{0.80}Co_{0.15}Al_{0.05}O_2$ (NCA)	Tesla, Panasonic, etc.	256.6	Low costs; high energy density	Thermal and electrochemical instabilities
Lithium nickel manganese cobalt oxide (NMC)	Nissan, Tesla, Samsung, Bosch, etc.	200–260	Low costs; high energy density; tunable properties	Relatively low conductivity; electrochemical instabilities
Lithium manganese oxide (LMO)	Toshiba, Hitachi, LG Chem, Nissan, etc.	100–120	High power density; good electrochemical and thermal stabilities	Jahn–Teller distortions; relatively low energy density

The $LiFePO_4$ cathode has been recognized as a promising substitute for $LiCoO_2$, as it has comparable electrochemical performance and a significantly lower cost of materials by excluding the usage of expensive cobalt [141]. The NMC and NCA cathodes attract great interests from the manufacturing industry, as their discharge capacity can be as high as 260 Wh/kg, which drastically increases the energy density and thus the driving range of electric vehicles [142,143]. However, these two cathodes suffer from electrochemical and thermal degradations during their operation [144,145], which limits their cyclability and thus the overall durability of the battery cell. The LMO cathodes have demonstrated their stability during the electrochemical cycling, while they suffer from the Jahn–Teller distortions in lattice, which hurts their intercalation kinetics. Also, their discharge capacities are relatively low [103,146]. Therefore, the LMO materials are good candidates as the coating layer of the core–shell architectures, with a LMO coating layer in the surface and high-capacity materials underneath, which combines the electrochemical advantages of both materials [72].

2.3 Anode materials

2.3.1 Intercalation-based anodes

The intercalation-based materials are promising anodes for LIBs, which exhibit good structural stability and satisfying electrochemical performances [147,148]. Their high insertion kinetics of Li^+ cations promotes the electrochemical kinetics [149]. On the other hand, the intercalation-based anodes usually have a small number of sitting sites for Li^+ cations, which limits their electrochemical capacity [150]. This family of anodes has similar working mechanisms as the intercalation cathode materials, as we have discussed in Section 2.2. A typical intercalation-based anode is the graphite/graphene family, and the working mechanism is schematically shown in Figure 2.3(a) [151]. During the electrochemical charging, the Li^+ cations from the cathode are inserted into the interlayer spacings of the graphene anode, which migrate back into the cathode during the electrochemical discharging. Beside graphene, other intercalation-based anode materials include transition metal oxides [152,153], sulfides [154,155], carbides [156,157], silicide [158,159], and their composites [160,161].

Another representative intercalation-based anode is the $Li_4Ti_5O_{12}$ spinel oxide [164,165]. The $Li_4Ti_5O_{12}$ anode was originally developed as a cathode material, which was later determined to be more suitable for the anode as its discharge voltage is too low to be a cathode [166]. The advantages of the $Li_4Ti_5O_{12}$ anode include its outstanding electrochemical stability and safety [167,168]. However, the discharge voltage of $Li_4Ti_5O_{12}$ is a bit high for an anode material [169], which thus limits the capacity and energy density of the battery cell.

Currently, the mostly adopted intercalation anode is graphite and its composites and derivatives. Their low cost of fabrication [170], satisfying electrochemical performance [171], and good mechanical processability [172,173] are the major reasons for their wide applications in the LIBs. The limitations of the graphite

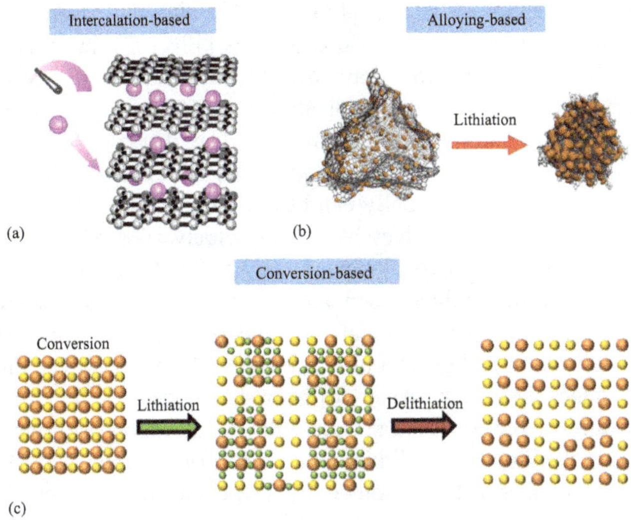

Figure 2.3 Anode materials of LIBs. (a) Schematic illustrating the intercalation of Li^+ cations into the interlayer spacings of graphene. Adapted with permission from Ref. [151]. Copyright 2020 American Chemical Society. (b) Schematic showing the alloying of Li into the α-Si anode. Adapted with permission from Ref. [162]. Copyright 2020 American Chemical Society. (c) Schematic showing the working mechanism of the conversion-based anodes. Adapted with permission from Ref. [163]. Creative Common CC BY license.

include its unstable SEI layers during the electrochemical cycling, which undermines the structural stability [174]. Also, its sluggish intercalation kinetics and low lithiation voltage against the cathode leads to deposition and growth of Li dendrites on the anode, which is a major reason for short circuiting and explosion [175]. To improve the structural stability and suppress the growth of Li dendrites, surface coating [176], and compositing [177] with stabilizers, balancing the properties of different anode materials.

2.3.2 Alloying-based anodes

Compared with the intercalation-based anodes, the alloying-based anodes have much higher capacities and lower costs [178,179]. However, this type of anodes suffers from significant volume changes and low conductivities [180,181], so they are usually mixed with other anode materials to mitigate their disadvantages. The alloying-based anodes functions via alloying between the anode materials and current-bearing Li^+ cations [182,183]. As an example, Figure 2.3(b) [162] schematically presents the functioning mechanism of an α-Si anode, which works via the alloying of the Si Cathode Materials Li couple. The common alloying-based anodes include Si, Ge, Sn, Sb, Bi, etc. [184,185]. The alloying-based anodes exhibit exceptional gravimetric and volumetric capacities [186,187], but their significant volume expansion and retraction upon the lithiation and delithiation processes remains as a major barrier for their application [188,189], which is a major source of mechanical and electromechanical failures. The development of composite materials with the capability to accommodate the significant volume change of alloying-based components internally can be a viable solution [190].

Currently, Si and its derivatives have been widely adopted as alloying-based anode materials. Si has a theoretical capacity of 3,590 $mAh\cdot g^{-1}$ [191], which is ten times larger than that of graphite-based anodes and thus attracts intensive research interests. Besides, the higher chemical potential of Si against the Li/Li^+ couple mitigates the deposition and growth of dendrites on the anode [192], although this feature also reduces the energy density of Si-based anodes. On the other hand, the amorphous nature of the Si-based anodes adopts tremendous volume expansion/retraction during the lithiation–delithiation process [193], which is attributed to the large volumetric number of Li^+ cations in Si. The cyclic volume changing leads to the formation of defects in the anode, such as fractures and nanovoids, which eventually lead to fatal structural degradations and reductions in the electrochemical performance [194]. Accordingly, intensive effort has been taken to prevent the electromechanical failures of Si-based anodes, such as the addition of mechanical stabilizers of polymers, carbon, metals, etc. into the anode [195,196]. Besides the simple mixtures, complicated composite architectures have also been proven to work effectively in mitigating the electromechanical failures. For example, introduction of tailored voids [197], constructing core–shell particles [198], and patterned silicon domains in separation [199] have been proven to maintain the structural integrity under hundreds or even over 1,000 electrochemical cycles, significantly improving the service life of Si-based anodes.

2.3.3 *Conversion-based anodes*

The conversion-based anodes include Mn-, Fe-, Co-, Ni-, and Cu-based transition metal oxides, sulfides, fluorides or selenides [182,200,201]. During the charging of the battery cell, the Li^+ cations from the cathode are inserted into the vacant lattice sites of the anode [202,203]. The inserted Li chemically combines with the anode metal and transform it into intermetallic compounds, as schematically shown in Figure 2.3(c) [163]. Compared with the intercalation-based and alloying-based anodes, the lithiation of the conversion-based anodes follows a much more ordered pattern and strict stoichiometry due to the stronger ionic bonds induced by Li^+ insertion [204,205], which is essential for the formation of Li–M intermetallic compounds.

Compared with the conventional intercalation-based anodes, the conversion-based anodes exhibits significantly improved discharge capacities and cyclabilities, which are thus considered as promising for the next-generation batteries [206]. However, similar as the alloying-based anodes, the conversion-based anodes also suffer from large volume changes during the electrochemical lithiation–delithiation

Table 2.2 Comparison of anode materials

Anode	Advantages	Issues
Carbonate	Good electronic conductivity; low costs; good processability	Low gravitational capacity; unstable SEI; sluggish intercalation kinetics
Alloys	High gravitational capacity (400–2,300 mAh·g^{-1})	Low electronic conductivity; dramatic volume changes
Silicon	Low costs; highest gravitational capacity (up to 3,600 mAh·g^{-1}); good safety	Dramatic volume changes; associated degradations
Transition metal oxide	Electrochemical stabilities	Low Coulombic efficiency; low electric conductivity

[207,208]. The disadvantage of conversion-based anodes can be mitigated by compositing with other anode materials with better structural stabilities [201,209].

2.3.4 Comparison of anode materials

Table 2.2 summarizes with advantages and disadvantages of representative anode materials being studied and applied, as adopted from the publications of Nzereogu *et al.* [210], Qi *et al.* [211], and Weiss *et al.* [174]. As discussed in the previous part of this section, the graphite is the most popular anode material due to its low material costs and satisfying electrochemical performance. Si and its derivatives arise as promising candidates for the next generation high-performance battery cells, major due to their highest theoretical gravitational capacity (up to 3,600 mAh·g^{-1}). However, the Si-based anodes suffer from significant volume changes induced by the electrochemical lithiation-delithiation (>400%), which is the major issue preventing their applications. The reader is referred to Section 3.2 for detailed discussions of this issue and the proposed solutions.

2.4 Electrolyte

Unlike the cathode and the anode, the electrolyte does not majorly contribute to the battery capacity. The functionality of the electrolyte focuses on guaranteeing smooth, uninterrupted transportation of Li^+ cations between the cathode and the anode [7,212]. In addition, electrolytes should have very low if at all electronic conductivity such that electrons will pass through external circuit and provide work to the load instead of forming an internal short circuit [213]. Other desirable properties of electrolytes include stability over a wide voltage window, low melting point, low viscosity, good wettability with other electrode components, etc. [214–216]. To serve this purpose, a large family of electrolytes have been developed with different electrochemical properties and functionalities, satisfying the different requirements of practical applications. In this section, we will discuss the major types of electrolytes utilized in LIBs, including the nonaqueous, aqueous, ionic

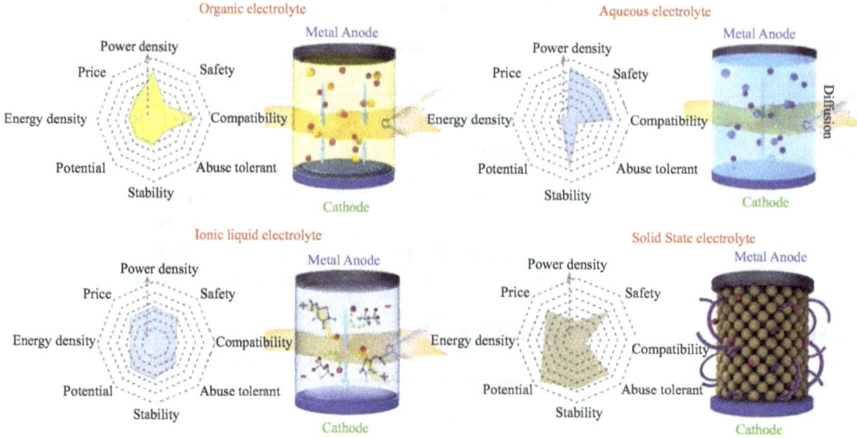

Figure 2.4 Electrolyte materials of LIBs. Schematic diagrams summarizing the advantages and disadvantages of different electrolyte materials for LIBs, including the electrochemical performance, compatibility, and stability. Reproduced with permission from Ref. [218]. Copyright 2020 Elsevier.

liquid, and solid-state electrolytes (SSEs) [217]. Together, the nonaqueous and aqueous electrolytes are also called liquid electrolytes. The advantages and disadvantages of different electrolytes are schematically summarized in Figure 2.4 [218]. The most important electrochemical property of the electrolyte is the ionic conductivity, while the stability, compatibility, and safety of electrolytes should also be taken into consideration for battery designs and operations. In this section, the properties of different electrolytes will be discussed and compared.

2.4.1 Nonaqueous electrolytes

Nonaqueous electrolytes are the most popular type utilized in commercial LIBs [219,220], with their satisfying ionic conductivity and electrochemical stabilities [221,222]. The most popular non-aqueous electrolytes are organic solutions of lithium hexafluorophosphate ($LiPF_6$) [223]. The most popular organic solvents include ethylene carbonate (EC) + dimethyl carbonate (DMC), propylene carbonate (PC), diethyl carbonate (DEC), and ethyl methyl carbonate (EMC) [1,224].

A major byproduct of the nonaqueous electrolytes during the electrochemical cycling is the solid–electrolyte interface (SEI), which is formed by the chemical and electrochemical interactions between the electrode and the electrolyte [225,226]. The SEI have both positive and negative influences on the battery performance. On the one hand, the SEI has a reduced conductivity of Li^+ cations compared to the pristine electrodes and electrolyte [227,228], thereby reducing the overall electrochemical kinetics of the battery cell. On the other hand, the SEI serves as a protective layer separating the electrodes from the harmful electrolytes, which mitigates the further structural degradations induced by the electrolyte–electrode interactions [229,230], thereby stabilizing the electrochemical performance of the battery cell. Particularly, the SEI layer is usually generated in the first few charge-discharge cycles due to the high surface activities of pristine electrodes, which is termed as "battery activation" [231,232].

2.4.2 *Aqueous electrolytes*

The common aqueous electrolytes usually contain a tuned amount of additives such as vinylene carbonate [233], lithium bis(trifluoromethylsulfonyl)imide [234], and disodium propane-1,3-disulfonate [235], which are used to introduce ionic conductivity into the electrolyte, as well as gaining a suitable stability [236]. Compared with the nonaqueous electrolytes, the aqueous correspondents are much less flammable and thus safer to work in large battery cells [237]. However, the ionic conductivity of the aqueous electrolytes is much lower compared with the nonaqueous ones [238], which greatly reduces the electrochemical kinetics of the battery cell. The aqueous electrolytes are also less electrochemically stable during long-term electrochemical cycling [239], which limits their applications.

2.4.3 *Ionic liquid electrolytes*

Ionic liquids are ionic compounds with low melting points (usually below 100°C), which thus show up as liquids at the room temperature [240]. Their good ionic conductivities and low electronic conductivities make them good candidates for the LIBs [241]. The impressive electrochemical safety of ionic liquids is another major advantage [242]. The low vapor pressure of ionic liquids also increases their chemical stability [243], by reducing the probability of ignition. The ionic liquids thus exhibit low flammability and are safe to be used in large battery cells under extreme conditions [244]. However, the ionic liquids also have drawbacks such as high prices and low flowabilities [245], which limits their applications as battery electrolytes.

2.4.4 *SSEs*

The SSEs arise as a novel, appealing family for LIBs, which include oxides, sulfides, hydrides, halides, thin films, and polymers [246,247]. The SSEs are usually ionic compounds or ion-containing compounds, which possesses the ability to transport Li^+ cations at satisfying kinetic rates [248,249]. The development of the SSEs was inspired by the safety concerns over the traditional liquid electrolytes [246], aiming at completely preventing the fire hazards and explosions resulting from the short circuiting and overcharging [250,251]. Also, the SSEs can act as the separator of electrodes [252,253], which simplifies the battery design.

The oxide, sulfide, and polymer SSEs are being intensively studied, due to their potential applications with optimized electrochemical performances [254]. Figure 2.5(a–c) [255] summarizes the advantages and disadvantages of the oxide, sulfide, and polymer SSEs. In principle, these electrolytes have relatively balanced electrochemical properties, making them good candidates for battery cells. Even

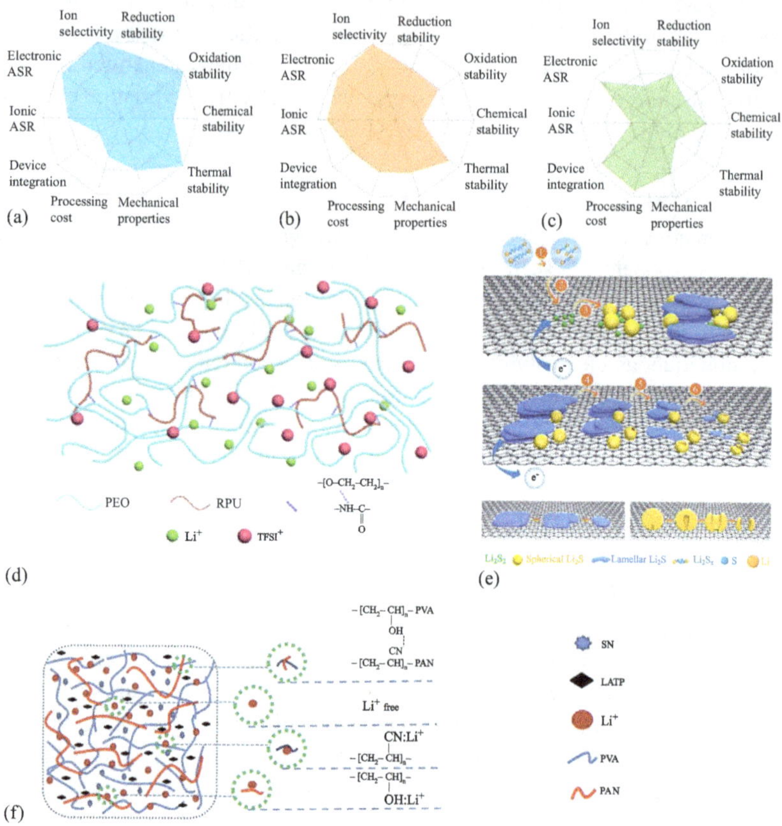

Figure 2.5 SSEs. (a–c) Diagrams summarizing the electrochemical properties of the oxide, sulfide, and polymer SSEs. Reproduced with permission from Ref. [255]. Copyright 2020 American Chemical Society. (d) Schematic showing the functioning mechanisms of the poly(ethylene oxide)-based solid-state electrolyte, as an example of oxide-based electrolytes. Reproduced with permission from Ref. [256]. Copyright 2020 American Chemical Society. (e) Schematic showing the working mechanisms of the Li_xS SSE, as an example of the sulfide-based electrolytes. Reproduced with permission from Ref. [257]. Copyright 2020 American Chemical Society. (f) Schematic showing the working mechanism of a polymer-based SSE. Adapted with permission from Ref. [258]. Copyright 2020 American Chemical Society.

though, compositing and trade-off are still important design strategies due to the different properties of electrolytes and their combinations.

Figure 2.5(d) [256] schematically presents the functioning mechanism of the poly(ethylene oxide)-based SSE. The amorphous segments of the oxide adopt Li^+

cations, and the transportation of Li^+ cations between neighboring segments enables the kinetic diffusion of lithium. Therefore, the distribution of the amorphous regions in the electrolyte along with their connectivity determines the rate capability of the oxide-based SSEs.

Figure 2.5(e) [257] schematically summarizes the functioning mechanisms of the Li_xS SSE. The long-chain polysulfides continuously interact with the lithium during electrochemical cycling, which can turn into nanoparticles and subsequently develop into cavities. This dynamic evolution of the polysulfides results in their loss of redistribution uniformity and loss of connectivity, which can reduce the conductivity of Li^+ cations and reduce the electrochemical kinetics.

Figure 2.5(f) [258] shows the working mechanism of the polymer-based SSEs. As can be seen, different ionic polymer chains are present in the electrolyte, which make up the basic motifs for the ionic conductivity of Li^+ cations. Similar as the oxide-based electrolytes, the interaction and connectivity of the Li^+ sites in the polymer-based electrolyte determine the rate capability.

The processing of SSE arises as a major challenge. For instance, compared with the liquidous electrolytes, the SSE adopts weaker mechanical bonding with the electrodes, which is majorly attributed to the lack of shape flexibility [259]. The disconnections between the SSE and the electrodes reduces overall ionic conductivity of the battery cell, thereby leading to the accumulation of resistance and even fire and explosion.

Also, the ionic conductivity of SSE is also a major issue. Unlike the liquidous electrolytes that are feasible for the transportation of multiple cations, the SSEs can only transport an individual type of cation, and the ionic kinetics are highly dependent on the structural nature of SSEs. A few approaches have been adopted to increase the ionic conductivity of SSEs, such as ionic substitution [260] and introduction of pre-existing mechanical strains [261].

Similar as the electrodes, the SSE can also suffer from mechanical degradations and the associated electrochemical fade, such as cracking and formation of nanovoids [262], which are not present in the liquidous electrolytes. The mechanical damages generate disconnections within the SSE, thereby reducing the ionic conductivity and increasing the resistance. A few approaches have been developed to mitigate the mechanical failures of SSE induced by the electrochemical cycling, such as increasing the electrochemical stabilities of the SSE material to prevent decomposition [263], adding a surface coating layer [264], compositing with strengthening fibers [265], and intentionally forming stable SEI layers [192].

2.4.5 Hybrid electrolytes

Hybrid electrolytes are electrolytes combined of two or more of the aforementioned electrolytes [266]. It can be a mixture of different ionic liquids [267], a mixture of different solid-state electrolytes [268], etc. Similar to the composite electrodes, the development of hybrid electrolytes is to optimize the electrochemical performance via combining different electrolytes.

Table 2.3 Comparison of anode materials

Anode	Advantages	Issues
High concentration solutions	Stable SEI formation; suppression of Li dendrites	High costs of Li salts
MOFs/COFs	Tailorable structures and properties	High manufacturing costs
Non-flammable or self-extinguishing liquid and gel	Good thermal stability; good ionic conductivity	Low stability towards reduction
Soggy sands	Good ionic conductivity	Complicated morphology control

2.4.6 Comparison of electrolyte materials

Table 2.3 compares the advantages and disadvantages of representative electrolyte materials with promising applications in the LIB industry, as adopted from the publications of Quartarone *et al.* [269,270]. Besides the intrinsic functionality of electrolytes, which largely relies on the ionic conductivity for the transportation of Li^+ cations between the electrodes [271], the next-generation electrolytes should have good stability to prevent decomposition-induced degradation and explosion [272], which are major limitations for their applications.

References

[1] Xu K. Nonaqueous liquid electrolytes for lithium-based rechargeable batteries. *Chem Rev.* 2004;104(10):4303–418. https://doi.org/10.1021/cr030203g

[2] Manthiram A. A reflection on lithium-ion battery cathode chemistry. *Nat Commun.* 2020;11(1):1550.

[3] Whittingham MS. Lithium batteries and cathode materials. *Chem Rev.* 2004;104(10):4271–302.

[4] Fergus JW. Recent developments in cathode materials for lithium ion batteries. *J Power Sources.* 2010;195(4):939–54.

[5] Goodenough JB and Kim Y. Challenges for rechargeable Li batteries. *Chem Mater.* 2010;22(3):587–603. https://doi.org/10.1021/cm901452z

[6] Agubra V and Fergus J. Lithium ion battery anode aging mechanisms. *Materials (Basel).* 2013;6(4):1310–25.

[7] Zhang SS. A review on electrolyte additives for lithium-ion batteries. *J Power Sources.* 2006;162(2):1379–94.

[8] Wang Q, Jiang L, Yu Y, and Sun J. Progress of enhancing the safety of lithium ion battery from the electrolyte aspect. *Nano Energy.* 2019;55:93–114.

[9] Yoshino A. The birth of the lithium-ion battery. *Angew Chemie, Int Ed.* 2012;51(24):5798–800.

[10] Horiba T. Lithium-ion battery systems. *Proc IEEE.* 2014;102(6):939–50.

[11] Manthiram A. An outlook on lithium ion battery technology. *ACS Cent Sci.* 2017;3(10):1063–9.

[12] Zuba M, Rana J, Siu C, *et al.* Operando XAS to illustrate the importance of electronic conductivity in vanadyl phosphate systems. *J Electrochem Soc.* 2021;168(5):50502.

[13] Guyomard D and Tarascon J-M. Rocking-chair or lithium-ion rechargeable lithium batteries. *Adv Mater.* 1994;6(5):408–12. https://doi.org/10.1002/adma.19940060516

[14] Megahed S and Scrosati B. Lithium-ion rechargeable batteries. *J Power Sources.* 1994;51(1):79–104. https://www.sciencedirect.com/science/article/pii/0378775394019568

[15] Pi J-K, Wu G-P, Yang H-C, Arges CG, and Xu Z-K. Separators with biomineralized zirconia coatings for enhanced thermo- and electro-performance of lithium-ion batteries. *ACS Appl Mater Interfaces.* 2017;9(26):21971–8. https://doi.org/10.1021/acsami.7b04505

[16] Chen R, Li Q, Yu X, Chen L, and Li H. Approaching practically accessible solid-state batteries: stability issues related to solid electrolytes and interfaces. *Chem Rev.* 2020;120(14):6820–77. https://doi.org/10.1021/acs.chemrev.9b00268

[17] Fan E, Li L, Wang Z, *et al.* Sustainable recycling technology for Li-ion batteries and beyond: challenges and future prospects. *Chem Rev.* 2020;120(14):7020–63. https://doi.org/10.1021/acs.chemrev.9b00535

[18] Spotnitz R. Simulation of capacity fade in lithium-ion batteries. *J Power Sources.* 2003;113(1):72–80. https://www.sciencedirect.com/science/article/pii/S0378775302004901

[19] Arora P, White RE, and Doyle M. Capacity fade mechanisms and side reactions in lithium-ion batteries. *J Electrochem Soc.* 1998;145(10):3647–67.

[20] Lagadec MF, Zahn R, and Wood V. Characterization and performance evaluation of lithium-ion battery separators. *Nat Energy.* 2019;4(1):16–25.

[21] Lee H, Yanilmaz M, Toprakci O, Fu K, and Zhang X. A review of recent developments in membrane separators for rechargeable lithium-ion batteries. *Energy Environ Sci.* 2014;7(12):3857–86.

[22] Li Y, Li Y, Pei A, *et al.* Atomic structure of sensitive battery materials and interfaces revealed by cryo–electron microscopy. *Science.* 2017;358(6362):506 LP–510. http://science.sciencemag.org/content/358/6362/506.abstract

[23] Selis LA and Seminario JM. Dendrite formation in silicon anodes of lithium-ion batteries. *RSC Adv.* 2018;8(10):5255–67.

[24] Park J, Jeong J, Lee Y, Oh M, Ryou M, and Lee YM. Micro-patterned lithium metal anodes with suppressed dendrite formation for post lithium-ion batteries. *Adv Mater Interfaces.* 2016;3(11):1600140.

[25] Zhao J, Chen D, Boateng B, *et al.* Atomic interlamellar ion path in polymeric separator enables long-life and dendrite-free anode in lithium ion batteries. *J Power Sources.* 2020;451:227773.

[26] Nie L, Li Y, Chen S, *et al.* Biofilm nanofiber-coated separators for dendrite-free lithium metal anode and ultrahigh-rate lithium batteries. *ACS Appl Mater Interfaces.* 2019;11(35):32373–80.

[27] Liu Z, Jiang Y, Hu Q, *et al.* Safer lithium-ion batteries from the separator aspect: development and future perspectives. *Energy Environ Mater.* 2021; 4(3):336–62. https://doi.org/10.1002/eem2.12129

[28] Scrosati B. Power sources for portable electronics and hybrid cars: lithium batteries and fuel cells. *Chem Rec.* 2005;5(5):286–97.

[29] Dunn B, Kamath H, and Tarascon J-M. Electrical energy storage for the grid: a battery of choices. *Science.* 2011;334(6058):928–35.

[30] Kai L, Yayuan L, Dingchang L, Allen P, and Yi C. Materials for lithium-ion battery safety. *Sci Adv.* 2022;4(6):eaas9820. https://doi.org/10.1126/sciadv.aas9820

[31] Zhang H, Omenya F, Whittingham MS, Wang C, and Zhou G. Formation of an anti-core–shell structure in layered oxide cathodes for Li-ion batteries. *ACS Energy Lett.* 2017;2(11):2598–606.

[32] Plaimer M, Breitfuß C, Sinz W, *et al.* Evaluating the trade-off between mechanical and electrochemical performance of separators for lithium-ion batteries: methodology and application. *J Power Sources.* 2016;306: 702–10.

[33] Liu Y, Li X, Guo H, *et al.* Effect of carbon nanotube on the electrochemical performance of C-LiFePO4/graphite battery. *J Power Sources.* 2008;184(2):522–6.

[34] Zhang H, May BM, Omenya F, Whittingham MS, Cabana J, and Zhou G. Layered oxide cathodes for Li-ion batteries: oxygen loss and vacancy evolution. *Chem Mater.* 2019;31(18):7790–8.

[35] Zhang H, May B, Serrano-Sevillano J, *et al.* Facet-dependent rock-salt reconstruction on the surface of layered oxide cathodes. *Chem Mater.* 30(3):692–9.

[36] Baba Y, Okada S, and Yamaki J. Thermal stability of LixCoO2 cathode for lithium ion battery. *Solid State Ionics.* 2002;148(3–4):311–6.

[37] Li Y, Yu L, Hu W, and Hu X. Thermotolerant separators for safe lithium-ion batteries under extreme conditions. *J Mater Chem A.* 2020;8(39):20294–317.

[38] Zhang H, Omenya F, Yan P, *et al.* Rock-salt growth-induced (003) cracking in a layered positive electrode for Li-ion batteries. *ACS Energy Lett.* 2017;2 (11):2607–15.

[39] Forouzan MM, Chao C-W, Bustamante D, Mazzeo BA, and Wheeler DR. Experiment and simulation of the fabrication process of lithium-ion battery cathodes for determining microstructure and mechanical properties. *J Power Sources.* 2016;312:172–83.

[40] Zheng X, Zhu Z, Lin X, *et al.* A mini-review on metal recycling from spent lithium ion batteries. *Engineering.* 2018;4(3):361–70.

[41] Li J, Lu Y, Yang T, Ge D, Wood III DL, and Li Z. Water-based electrode manufacturing and direct recycling of lithium-ion battery electrodes—a green and sustainable manufacturing system. *Iscience.* 2020;23(5):101081.

[42] Zhang H, Karki K, Huang Y, Whittingham MS, Stach EA, and Zhou G. Atomic insight into the layered/spinel phase transformation in charged LiNi0. 80Co0. 15Al0. 05O2 cathode particles. *J Phys Chem C.* 2017;121(3):1421–30.

[43] Sun Y-K, Myung S-T, Park B-C, Prakash J, Belharouak I, and Amine K. High-energy cathode material for long-life and safe lithium batteries. *Nat Mater*. 2009;8(4):320.

[44] Langdon J and Manthiram A. A perspective on single-crystal layered oxide cathodes for lithium-ion batteries. *Energy Storage Mater*. 2021;37:143–60. https://www.sciencedirect.com/science/article/pii/S2405829721000465

[45] Goodenough JB. Metallic oxides. *Prog Solid State Chem*. 1971;5:145–399. https://www.sciencedirect.com/science/article/pii/0079678671900185

[46] Goodenough JB, Mizushima K, and Takeda T. Solid-solution oxides for storage-battery electrodes. *Jpn J Appl Phys*. 1980;19(S3):305. http://dx.doi.org/10.7567/JJAPS.19S3.305

[47] Ozawa K. Lithium-ion rechargeable batteries with LiCoO2 and carbon electrodes: the LiCoO2/C system. *Solid State Ionics*. 1994;69(3–4):212–21.

[48] Ling J, Karuppiah C, Krishnan SG, *et al*. Phosphate polyanion materials as high-voltage lithium-ion battery cathode: a review. *Energy Fuels*. 2021;35(13):10428–50. https://doi.org/10.1021/acs.energyfuels.1c01102

[49] Wu Q, Liu Y, Johnson CS, Li Y, Dees DW, and Lu W. Insight into the structural evolution of a high-voltage spinel for lithium-ion batteries. *Chem Mater*. 2014;26(16):4750–6. https://doi.org/10.1021/cm5015288

[50] Kumar S, Qadir A, Maury F, and Bahlawane N. Visible thermochromism in vanadium pentoxide coatings. *ACS Appl Mater Interfaces*. 2017;9(25):21447–56. https://doi.org/10.1021/acsami.7b04484

[51] Cao J, Xie H, Lv F, *et al.* Stable Li2TiO3 shell–Li1.17Mn0.50Ni0.16Co0.17O2 core architecture based on an in-site synchronous lithiation method as a high rate performance and long cycling life lithium-ion battery cathode. *ACS Appl Energy Mater*. 2020;3(6):5462–71. https://doi.org/10.1021/acsaem.0c00446

[52] Neudeck S, Walther F, Bergfeldt T, *et al.* Molecular surface modification of NCM622 cathode material using organophosphates for improved Li-ion battery full-cells. *ACS Appl Mater Interfaces*. 2018;10(24):20487–98. https://doi.org/10.1021/acsami.8b04405

[53] Ruffo R, Wessells C, Huggins RA, and Cui Y. Electrochemical behavior of LiCoO2 as aqueous lithium-ion battery electrodes. *Electrochem Commun*. 2009;11(2):247–9.

[54] Li W, Lee S, and Manthiram A. High-nickel NMA: a cobalt-free alternative to NMC and NCA cathodes for lithium-ion batteries. *Adv Mater*. 2020;32(33):2002718.

[55] Zhang H, Karki K, Huang Y, Whittingham MS, Stach E, and Zhou G. Atomic insight into the layered/spinel phase transformation in charged $LiNi_{0.80}Co_{0.15}Al_{0.05}O_2$ cathode particles. *J Phys Chem C*. 121(3):1421–30.

[56] Ohzuku T, Ueda A, Nagayama M, Iwakoshi Y, and Komori H. Comparative study of LiCoO2, LiNi12Co12O2 and LiNiO2 for 4 volt secondary lithium cells. *Electrochim Acta*. 1993;38(9):1159–67.

[57] Yang F, Liu Y, Martha SK, *et al.* Nanoscale morphological and chemical changes of high voltage lithium–manganese rich NMC composite cathodes with cycling. *Nano Lett*. 2014;14(8):4334–41.

[58] Wu F, Li W, Chen L, *et al.* Renovating the electrode-electrolyte interphase for layered lithium-& manganese-rich oxides. *Energy Storage Mater.* 2020;28:383–92.

[59] Li H, Liu A, Zhang N, *et al.* An unavoidable challenge for Ni-rich positive electrode materials for lithium-ion batteries. *Chem Mater.* 2019;31(18): 7574–83.

[60] Li J, Li Y, Guo Y, Lv J, Yi W, and Ma P. A facile method to enhance electrochemical performance of high-nickel cathode material Li (Ni0.8Co0.1Mn0.1)O2 via Ti doping. *J Mater Sci Mater Electron.* 2018;29 (13):10702–8. https://doi.org/10.1007/s10854-018-9093-1

[61] Shan W, Huang S, Zhang H, and Hou X. Surface coating for high-nickel cathode materials to achieve excellent cycle performance at elevated temperatures. *J Alloys Compd.* 2021;862: 158022. https://www.sciencedirect.com/science/article/pii/S0925838820343863

[62] Li J, Li W, Wang S, Jarvis K, Yang J, and Manthiram A. Facilitating the operation of lithium-ion cells with high-nickel layered oxide cathodes with a small dose of aluminum. *Chem Mater.* 2018;30(9):3101–9. https://doi.org/10.1021/acs.chemmater.8b01077

[63] Kim U-H, Kuo L-Y, Kaghazchi P, Yoon CS, and Sun Y-K. Quaternary layered Ni-rich NCMA cathode for lithium-ion batteries. *ACS Energy Lett.* 2019;4(2):576–82.

[64] Kim Y, Park H, Warner JH, and Manthiram A. Unraveling the intricacies of residual lithium in high-Ni cathodes for lithium-ion batteries. *ACS Energy Lett.* 2021;6(3):941–8.

[65] Li M and Lu J. Cobalt in lithium-ion batteries. *Science.* 2020;367(6481): 979–80.

[66] Wentker M, Greenwood M, and Leker J. A bottom-up approach to lithium-ion battery cost modeling with a focus on cathode active materials. *Energies.* 2019;12(3):504.

[67] Ghatak K, Basu S, Das T, Sharma V, Kumar H, and Datta D. Effect of cobalt content on the electrochemical properties and structural stability of NCA type cathode materials. *Phys Chem Phys.* 2018;20(35): 22805–17.

[68] Yan P, Zheng J, Xiao J, Wang C-M, and Zhang J-G. Recent advances on the understanding of structural and composition evolution of LMR cathodes for Li-ion batteries. *Front Energy Res.* 2015;3:26.

[69] Xiao B and Sun X. Surface and subsurface reactions of lithium transition metal oxide cathode materials: an overview of the fundamental origins and remedying approaches. *Adv Energy Mater.* 2018;8(29):1802057.

[70] Zhu W, Wang Y, Liu D, *et al.* Application of operando X-ray diffractometry in various aspects of the investigations of lithium/sodium-ion batteries. *Energies.* 2018;11(11):2963.

[71] Li W, Liu X, Celio H, *et al.* Mn versus Al in layered oxide cathodes in lithium-ion batteries: a comprehensive evaluation on long-term cyclability. *Adv Energy Mater.* 2018;8(15):1703154.

[72] Zhang H, Liu H, Piper LFJ, Whittingham MS, and Zhou G. Oxygen loss in layered oxide cathodes for Li-ion batteries: mechanisms, effects, and mitigation. *Chem Rev.* 2022;122(6):5641–5681. https://doi.org/10.1021/acs.chemrev.1c00327

[73] Siu C, Seymour I, Britto S, *et al.* Enabling multi-electron reaction of ε-VOPO4 to reach theoretical capacity for lithium-ion batteries. *Chem Commun.* 2018;54:7802–7805.

[74] Wei S, Inoue S, Di Lecce D, Li Z, Tominaga Y, and Hassoun J. Towards a high-performance lithium-metal battery with glyme solution and an olivine cathode. *ChemElectroChem.* 2020;7(11):2376–88. https://doi.org/10.1002/celc.202000272

[75] Yu L, Cullin DW, Williamson JM, and Miller TA. High resolution laser spectroscopy of asymmetrically deuterated cyclopentadienyl radicals: a study of vibronic degeneracy resolution and Jahn–Teller distortion. *J Chem Phys.* 1993;98(4):2682–98.

[76] Yamada A, Hosoya M, Chung S-C, *et al.* Olivine-type cathodes: achievements and problems. *J Power Sources.* 2003;119–121:232–8. https://www.sciencedirect.com/science/article/pii/S0378775303002398

[77] Yamada A and Chung S-C. Crystal chemistry of the Olivine-Type Li (Mn_yFe_{1-y}) PO_4 and $(Mn_y\ Fe_{1-y})$ PO_4 as possible 4 V cathode materials for lithium batteries. *J Electrochem Soc.* 2001;148(8):A960.

[78] Zaghib K, Dubé J, Dallaire A, *et al.* Enhanced thermal safety and high power performance of carbon-coated LiFePO4 olivine cathode for Li-ion batteries. *J Power Sources.* 2012;219:36–44. https://www.sciencedirect.com/science/article/pii/S0378775312008737

[79] Yamada A, Chung S-C, and Hinokuma K. Optimized LiFePO4 for lithium battery cathodes. *J Electrochem Soc.* 2001;148(3):A224.

[80] Li J, Fleetwood J, Hawley WB, and Kays W. From materials to cell: state-of-the-art and prospective technologies for lithium-ion battery electrode processing. *Chem Rev.* 2022;122(1):903–56.

[81] Liu X-H, Saito T, Doi T, Okada S, and Yamaki J. Electrochemical properties of rechargeable aqueous lithium ion batteries with an olivine-type cathode and a Nasicon-type anode. *J Power Sources.* 2009;189(1):706–10.

[82] Kim T-H, Park H-S, Lee M-H, Lee S-Y, and Song H-K. Restricted growth of LiMnPO4 nanoparticles evolved from a precursor seed. *J Power Sources.* 2012;210:1–6. https://www.sciencedirect.com/science/article/pii/S0378775312005009

[83] Kim D-H and Kim J. Synthesis of $LiFePO_4$ nanoparticles in polyol medium and their electrochemical properties. *Electrochem Solid-State Lett.* 2006;9(9):A439. http://dx.doi.org/10.1149/1.2218308

[84] Gong C, Xue Z, Wen S, Ye Y, and Xie X. Advanced carbon materials/olivine LiFePO4 composites cathode for lithium ion batteries. *J Power Sources.* 2016;318:93–112. https://www.sciencedirect.com/science/article/pii/S0378775316303494

[85] Chen J-M, Hsu C-H, Lin Y-R, Hsiao M-H, and Fey GT-K. High-power LiFePO4 cathode materials with a continuous nano carbon network for lithium-ion batteries. *J Power Sources.* 2008;184(2):498–502. https://www.sciencedirect.com/science/article/pii/S0378775308006484

[86] Yang L, Deng WW, Xu W, *et al.* Olivine $LiMn_xFe_{1-x}PO_4$ cathode materials for lithium ion batteries: restricted factors of rate performances. *J Mater Chem A [Internet].* 2021;9(25):14214–32. http://dx.doi.org/10.1039/D1TA01526E

[87] Huang S, Chen H, Chen M, *et al.* Design of conductive binders for LiFePO4 cathodes with long-term cycle life. *ACS Sustain Chem Eng.* 2021;9(39):13277–86.

[88] Paul BJ, Kang S-W, Gim J, *et al.* Nucleation and growth controlled polyol synthesis of size-focused nanocrystalline LiFePO4 cathode for high performance Li-ion batteries. *J Electrochem Soc.* 2014;161(9):A1468.

[89] Yang L, Deng WWW, Xu W, *et al.* Olivine $LiMn_xFe_{1-x}PO_4$ cathode materials for lithium ion batteries: restricted factors of rate performances. *J Mater Chem A.* 2021;9(25):14214–32. http://dx.doi.org/10.1039/D1TA01526E

[90] Tan Z, Wang X, and Zhou H. Highly energy density olivine cathode material synthesized by coprecipitation technique. *Electrochim Acta.* 2013;90:597–603.

[91] Xu J, Lin F, Doeff MM, and Tong W. A review of Ni-based layered oxides for rechargeable Li-ion batteries. *J Mater Chem A.* 2017;5(3):874–901.

[92] Zhu Y, Xu Y, Liu Y, Luo C, and Wang C. Comparison of electrochemical performances of olivine NaFePO 4 in sodium-ion batteries and olivine LiFePO 4 in lithium-ion batteries. *Nanoscale.* 2013;5(2):780–7.

[93] Whiteside A, Fisher CAJ, Parker SC, and Islam MS. Particle shapes and surface structures of olivine NaFePO 4 in comparison to LiFePO 4. *Phys Chem Chem Phys.* 2014;16(39):21788–94.

[94] Yang Z, Liu J, Baskaran S, Imhoff CH, and Holladay JD. Enabling renewable energy—and the future grid—with advanced electricity storage. *Jom.* 2010;62(9):14–23.

[95] Gwon H, Seo D, Kim S, Kim J, and Kang K. Combined first-principle calculations and experimental study on multi-component olivine cathode for lithium rechargeable batteries. *Adv Funct Mater.* 2009;19(20):3285–92.

[96] Wang R, Chen X, Huang Z, *et al.* Twin boundary defect engineering improves lithium-ion diffusion for fast-charging spinel cathode materials. *Nat Commun.* 2021;12(1):3085. https://doi.org/10.1038/s41467-021-23375-7

[97] Kocak T, Wu L, Wang J, Savaci U, Turan S, and Zhang X. The effect of vanadium doping on the cycling performance of LiNi0.5Mn1.5O4 spinel cathode for high voltage lithium-ion batteries. *J Electroanal Chem.* 2021;881:114926.

[98] Gummow RJ, Liles DC, and Thackeray MM. Lithium extraction from orthorhombic lithium manganese oxide and the phase transformation to spinel. *Mater Res Bull.* 1993;28(12):1249–56.

[99] Thackeray MM. Structural considerations of layered and spinel lithiated oxides for lithium ion batteries. *J Electrochem Soc.* 1995;142(8):2558–63.

[100] Uchiyama H, Hosono E, Zhou H, and Imai H. Three-dimensional architectures of spinel-type LiMn2O4 prepared from biomimetic porous carbonates and their application to a cathode for lithium-ion batteries. *J Mater Chem.* 2009;19(23):4012–6.

[101] Huang Y, Dong Y, Li S, *et al.* Lithium manganese spinel cathodes for lithium-ion batteries. *Adv Energy Mater*. 2021;11(2):2000997.

[102] Bhuvaneswari S, Varadaraju UV, Gopalan R, and Prakash R. Structural stability and superior electrochemical performance of Sc-doped LiMn2O4 spinel as cathode for lithium ion batteries. *Electrochim Acta.* 2019;301:342–51.

[103] Zuo C, Hu Z, Qi R, *et al.* Double the capacity of manganese spinel for lithium-ion storage by suppression of cooperative Jahn–Teller distortion. *Adv Energy Mater*. 2020;10(34):2000363.

[104] Yamada A, Tanaka M, Tanaka K, and Sekai K. Jahn–Teller instability in spinel Li–Mn–O. *J Power Sources.* 1999;81:73–8.

[105] Thackeray MM, David WIF, and Goodenough JB. Structural characterization of the lithiated iron oxides LixFe3O4 and LixFe2O3 ($0< x< 2$). *Mater Res Bull.* 1982;17(6):785–93.

[106] Thackeray MM, David WIF, Bruce PG, and Goodenough JB. Lithium insertion into manganese spinels. *Mater Res Bull.* 1983;18(4):461–72.

[107] Cho Y, Lee S, Lee Y, Hong T, and Cho J. Spinel-layered core-shell cathode materials for Li-ion batteries. *Adv Energy Mater.* 2011;1(5):821–8. https://doi.org/10.1002/aenm.201100239

[108] Chen M, Li W, Shen X, and Diao G. Fabrication of core–shell α-Fe2O3@ Li4Ti5O12 composite and its application in the lithium ion batteries. *ACS Appl Mater Interfaces.* 2014;6(6):4514–23. https://doi.org/10.1021/am500294m

[109] Kim J, Pieczonka NPW, and Yang L. Challenges and approaches for high-voltage spinel lithium-ion batteries. *ChemPhysChem.* 2014;15(10):1940–54.

[110] Lee S, Yoon G, Jeong M, Lee M, Kang K, and Cho J. Hierarchical surface atomic structure of a manganese-based spinel cathode for lithium-ion batteries. *Angew Chem*. 2015;127(4):1169–74.

[111] Luo Y, Lu T, Zhang Y, Yan L, Mao SS, and Xie J. Surface-segregated, high-voltage spinel lithium-ion battery cathode material LiNi0.5Mn1.5O_4 cathodes by aluminium doping with improved high-rate cyclability. *J Alloys Compd*. 2017;703:289–97.

[112] Lee K-S, Myung S-T, Bang HJ, Chung S, and Sun Y-K. Co-precipitation synthesis of sphericalLi1.05M0.05Mn1.9O4 (M=Ni, Mg, Al) spinel and its application for lithium secondary battery cathode. *Electrochim Acta.* 2007;52(16):5201–6. https://www.sciencedirect.com/science/article/pii/S0013468607002745

[113] Julien CM, Mauger A, Zaghib K, and Groult H. Comparative issues of cathode materials for Li-ion batteries. *Inorganics.* 2014;2(1):132–54.

[114] Lu J, Chang Y-L, Song B, *et al.* High energy spinel-structured cathode stabilized by layered materials for advanced lithium-ion batteries. *J Power Sources.* 2014;271:604–13.

[115] Bhandari C, Lambrecht WRL, and van Schilfgaarde M. Quasiparticle self-consistent GW calculations of the electronic band structure of bulk and monolayer ${\mathrm{V}}_{2}{\mathrm{O}}_{5}$. *Phys Rev B.* 2015;91 (12):125116. https://link.aps.org/doi/10.1103/PhysRevB.91.125116

[116] Wu J, Byrd I, Jin C, *et al.* Reinvigorating reverse-osmosis membrane technology to stabilize the V2O5 lithium-ion battery cathode. *ChemElectroChem.* 2017;4(5):1181–9. https://doi.org/10.1002/celc.201700102

[117] Zhu Y, Yang M, Huang Q, *et al.* V2O5 textile cathodes with high capacity and stability for flexible lithium-ion batteries. *Adv Mater.* 2020;32 (7):1906205. https://doi.org/10.1002/adma.201906205

[118] Zhang X-F, Wang K-X, Wei X, and Chen J-S. Carbon-coated V2O5 nanocrystals as high performance cathode material for lithium ion batteries. *Chem Mater.* 2011;23(24):5290–2. https://doi.org/10.1021/cm202812z

[119] Hao Y, Jiang Y, Zhao L, *et al.* Bimetallic antimony–vanadium oxide nanoparticles embedded in graphene for stable lithium and sodium storage. *ACS Appl Mater Interfaces.* 2021;13(18):21127–37.

[120] Wei Q, Tan S, Liu X, *et al.* Novel polygonal vanadium oxide nanoscrolls as stable cathode for lithium storage. *Adv Funct Mater.* 2015;25(12):1773–9.

[121] Suthirakun S, Jungthawan S, and Limpijumnong S. Effect of Sn-doping on behavior of Li-intercalation in V2O5 cathode materials of Li-ion batteries: a computational perspective. *J Phys Chem C.* 2018;122(11):5896–907.

[122] Aldebert P, Baffier N, Gharbi N, and Livage J. Layered structure of vanadium pentoxide gels. *Mater Res Bull.* 1981;16(6):669–76. https://www.sciencedirect.com/science/article/pii/002554088190266X

[123] Wang H-E, Zhao X, Yin K, *et al.* Superior pseudocapacitive lithium-ion storage in porous vanadium oxides@C heterostructure composite. *ACS Appl Mater Interfaces.* 2017;9(50):43665–73. https://doi.org/10.1021/acsami.7b13658

[124] Wang X, Zhang Z, Huang M, Feng J, Xiong S, and Xi B. In situ electrochemically activated vanadium oxide cathode for advanced aqueous Zn-ion batteries. *Nano Lett.* 2022;22(1):119–27. https://doi.org/10.1021/acs.nanolett.1c03409

[125] Kwak W-J, Lau KC, Shin C-D, Amine K, Curtiss LA, and Sun Y-K. A Mo2C/carbon nanotube composite cathode for lithium–oxygen batteries with high energy efficiency and long cycle life. *ACS Nano.* 2015;9 (4):4129–37.

[126] Wang L, Li Y, Wu J, *et al.* Synthesis mechanism and characterization of LiMn0. 5Fe0. 5PO4/C composite cathode material for lithium-ion batteries. *J Alloys Compd.* 2020;839:155653.

[127] Shellikeri A, Yturriaga S, Zheng JS, *et al.* Hybrid lithium-ion capacitor with LiFePO4/AC composite cathode – long term cycle life study, rate effect and charge sharing analysis. *J Power Sources.* 2018;392:285–95.

[128] Hwang J, Kim HM, Shin S, and Sun Y. Designing a high-performance lithium–sulfur batteries based on layered double hydroxides–carbon nanotubes composite cathode and a dual-functional graphene–polypropylene–Al2O3 separator. *Adv Funct Mater*. 2018;28(3):1704294.

[129] Kalybekkyzy S, Mentbayeva A, Kahraman MV, Zhang Y, and Bakenov Z. Flexible S/DPAN/KB nanofiber composite as binder-free cathodes for Li-S batteries. *J Electrochem Soc.* 2019;166(3):A5396.

[130] Helbig J, Beuse T, Siozios V, Placke T, Winter M, and Schmuch R. Li/Mn-rich cathode materials with low-cobalt content and core-shell particle design for high-energy lithium ion batteries. *J Electrochem Soc.* 2020;167 (6):60519.

[131] Wu Z, Ji S, Liu T, *et al.* Aligned Li+ tunnels in core–shell Li (Ni x Mn y Co z) O2@ LiFePO4 enhances its high voltage cycling stability as Li-ion battery cathode. *Nano Lett.* 2016;16(10):6357–63.

[132] Asakura D, Li CH, Mizuno Y, Okubo M, Zhou H, and Talham DR. Bimetallic cyanide-bridged coordination polymers as lithium ion cathode materials: core@ shell nanoparticles with enhanced cyclability. *J Am Chem Soc.* 2013;135(7):2793–9.

[133] Rahman MM, Wang J-Z, Zeng R, Wexler D, and Liu HK. LiFePO4–Fe2P–C composite cathode: an environmentally friendly promising electrode material for lithium-ion battery. *J Power Sources.* 2012;206:259–66.

[134] Rembelski D, Viricelle J-P, Combemale L, and Rieu M. Characterization and comparison of different cathode materials for SC-SOFC: LSM, BSCF, SSC, and LSCF. *Fuel Cells*. 2012;12(2):256–64.

[135] Chen L, Zhang M, and Wei W. Graphene-based composites as cathode materials for lithium ion batteries. *J Nanomater*. 2013;2013:2.

[136] Chen X, Shen W, Vo TT, Cao Z, and Kapoor A. An overview of lithium-ion batteries for electric vehicles. In: *2012 10th International Power & Energy Conference (IPEC)*. IEEE; 2012. pp. 230–5.

[137] Catenaro E, Rizzo DM, and Onori S. Framework for energy storage selection to design the next generation of electrified military vehicles. *Energy.* 2021;231:120695. https://www.sciencedirect.com/science/article/pii/S0360544221009439

[138] Wang Z, Yuan J, Zhu X, *et al.* Overcharge-to-thermal-runaway behavior and safety assessment of commercial lithium-ion cells with different cathode materials: a comparison study. *J Energy Chem.* 2021;55:484–98. https://www.sciencedirect.com/science/article/pii/S2095495620305234

[139] Blomgren GE. The development and future of lithium ion batteries. *J Electrochem Soc.* 2016;164(1):A5019–25. http://dx.doi.org/10.1149/2.0251701jes

[140] Purwanto A, Yudha CS, Ubaidillah U, Widiyandari H, Ogi T, and Haerudin H. NCA cathode material: synthesis methods and performance enhancement efforts. *Mater Res Express.* 2018;5(12):122001.

[141] Dupré N, Cuisinier M, Martin J, and Guyomard D. Interphase evolution at two promising electrode materials for Li-ion batteries: LiFePO4 and LiNi1/2Mn1/2O2. *ChemPhysChem.* 2014;15(10):1922–38.

[142] Pelegov DV and Pontes J. Main drivers of battery industry changes: electric vehicles—a market overview. *Batteries.* 2018;4(4):65.
[143] Andre D, Kim S-J, Lamp P, *et al.* Future generations of cathode materials: an automotive industry perspective. *J Mater Chem A.* 2015;3(13):6709–32.
[144] Zhang H, Omenya F, Whittingham MS, Wang C, and Zhou G. Structural degradations in the bulk of cathode particles for Li-ion batteries. *Microsc Microanal.* 2018;24(S1):1504–5. https://www.cambridge.org/core/article/structural-degradations-in-the-bulk-of-cathode-particles-for-liion-batteries/1B3528FFB98AB8785CD7F8E40FBCF4F6
[145] Li T, Yuan X-Z, Zhang L, Song D, Shi K, and Bock C. Degradation mechanisms and mitigation strategies of nickel-rich NMC-based lithium-ion batteries. *Electrochem Energy Rev.* 2020;3(1):43–80. https://doi.org/10.1007/s41918-019-00053-3
[146] Chae C, Park H, Kim D, Kim J, Oh E-S, and Lee JK. A Li-ion battery using LiMn2O4 cathode and MnOx/C anode. *J Power Sources.* 2013;244:214–21.
[147] Mahmood N, Tang T, and Hou Y. Nanostructured anode materials for lithium ion batteries: progress, challenge and perspective. *Adv Energy Mater.* 2016;6(17):1600374. https://doi.org/10.1002/aenm.201600374
[148] Kumar Sen U, Shaligram A, and Mitra S. Intercalation anode material for lithium ion battery based on molybdenum dioxide. *ACS Appl Mater Interfaces.* 2014;6(16):14311–9.
[149] Barsoukov E, Kim JH, Kim JH, Yoon CO, and Lee H. Kinetics of lithium intercalation into carbon anodes: in situ impedance investigation of thickness and potential dependence. *Solid State Ionics.* 1999;116(3–4):249–61.
[150] Kim T, Park J, Chang SK, Choi S, Ryu JH, and Song H. The current move of lithium ion batteries towards the next phase. *Adv Energy Mater.* 2012;2(7):860–72.
[151] Nijamudheen A, Sarbapalli D, Hui J, Rodríguez-López J, and Mendoza-Cortes JL. Impact of surface modification on the lithium, sodium, and potassium intercalation efficiency and capacity of few-layer graphene electrodes. *ACS Appl Mater Interfaces.* 2020;12(17):19393–401. https://doi.org/10.1021/acsami.9b23105
[152] Li S, Chen J, Gong X, Wang J, and Lee PS. Holey graphene-wrapped porous TiNb24O62 microparticles as high-performance intercalation pseudocapacitive anode materials for lithium-ion capacitors. *NPG Asia Mater.* 2018;10(5):406–16. https://doi.org/10.1038/s41427-018-0042-5
[153] Park JH, Choi WY, Lee S, Kim T-S, and Lee JW. Graphene intercalated free-standing carbon paper coated with MnO2 for anode materials of lithium ion batteries. *Electrochim Acta.* 2020;348:136310. https://www.sciencedirect.com/science/article/pii/S0013468620307027
[154] Lee S, Song M, Kim S, *et al.* High lithium storage properties in a manganese sulfide anode via an intercalation-cum-conversion reaction. *J Mater Chem A.* 2020;8(34):17537–49.
[155] Cupid DM, Rezqita A, Glibo A, *et al.* Understanding and modelling the thermodynamics and electrochemistry of lithiation of tin (IV) sulfide as an

anode active material for lithium ion batteries. *Electrochim Acta.* 2021; 375:137936.

[156] Hussain T, Farokh Niaei AH, Searles DJ, and Hankel M. Three-dimensional silicon carbide from siligraphene as a high capacity lithium ion battery anode material. *J Phys Chem C.* 2019;123(45):27295–304.

[157] Wang H, Zhang Y, Ang H, *et al.* A high-energy lithium-ion capacitor by integration of a 3D interconnected titanium carbide nanoparticle chain anode with a pyridine-derived porous nitrogen-doped carbon cathode. *Adv Funct Mater.* 2016;26(18):3082–93.

[158] Roberts GA, Cairns EJ, and Reimer JA. Magnesium silicide as a negative electrode material for lithium-ion batteries. *J Power Sources.* 2002;110 (2):424–9.

[159] Ogata K, Salager E, Kerr CJ, *et al.* Revealing lithium–silicide phase transformations in nano-structured silicon-based lithium ion batteries via in situ NMR spectroscopy. *Nat Commun.* 2014;5(1):1–11.

[160] Petnikota S, Teo KW, Chen L, *et al.* Exfoliated graphene oxide/MoO2 composites as anode materials in lithium-ion batteries: an insight into intercalation of Li and conversion mechanism of MoO2. *ACS Appl Mater Interfaces.* 2016;8(17):10884–96.

[161] Wang M, Yu X, Hou L, *et al.* 3D sandwich-shaped graphene-based nano-composite intercalated with double-shelled hollow MnCo2O4 spheres as anode materials for lithium-ion batteries. *Chem Eng J.* 2018;351:930–8.

[162] Ko M, Chae S, Jeong S, Oh P, and Cho J. Elastic a-silicon nanoparticle backboned graphene hybrid as a self-compacting anode for high-rate lithium ion batteries. *ACS Nano.* 2014;8(8):8591–9. https://doi.org/10.1021/nn503294z

[163] Lee B-S. A review of recent advancements in electrospun anode materials to improve rechargeable lithium battery performance. *Polymers.* 2020;12: 2035.

[164] Hao Y-J, Lai Q-Y, Lu J-Z, Wang H-L, Chen Y-D, and Ji X-Y. Synthesis and characterization of spinel Li4Ti5O12 anode material by oxalic acid-assisted sol–gel method. *J Power Sources.* 2006;158(2):1358–64.

[165] Zhang C, Zhang Y, Wang J, Wang D, He D, and Xia Y. Li4Ti5O12 prepared by a modified citric acid sol–gel method for lithium-ion battery. *J Power Sources.* 2013;236:118–25.

[166] Colbow KM, Dahn JR, and Haering RR. Structure and electrochemistry of the spinel oxides LiTi2O4 and Li43Ti53O4. *J Power Sources.* 1989;26(3):397–402. https://www.sciencedirect.com/science/article/pii/0378775389801521

[167] Yi T-F, Xie Y, Zhu Y-R, Zhu R-S, and Shen H. Structural and thermodynamic stability of Li4Ti5O12 anode material for lithium-ion battery. *J Power Sources.* 2013;222:448–54.

[168] Shi X, Yu S, Deng T, Zhang W, and Zheng W. Unlock the potential of Li4Ti5O12 for high-voltage/long-cycling-life and high-safety batteries: dual-ion architecture superior to lithium-ion storage. *J Energy Chem.* 2020;44:13–8.

[169] Thackeray MM and Amine K. Li4Ti5O12 spinel anodes. *Nat Energy*. 2021;6(6):683. https://doi.org/10.1038/s41560-021-00829-2

[170] Wu J, Cao Y, Zhao H, Mao J, and Guo Z. The critical role of carbon in marrying silicon and graphite anodes for high-energy lithium-ion batteries. *Carbon Energy*. 2019;1(1):57–76.

[171] Adams RA, Varma A, and Pol VG. Temperature dependent electrochemical performance of graphite anodes for K-ion and Li-ion batteries. *J Power Sources*. 2019;410:124–31.

[172] Gordon R, Orias R, and Willenbacher N. Effect of carboxymethyl cellulose on the flow behavior of lithium-ion battery anode slurries and the electrical as well as mechanical properties of corresponding dry layers. *J Mater Sci*. 2020;55(33):15867–81.

[173] Sabisch JEC, Anapolsky A, Liu G, and Minor AM. Evaluation of using prelithiated graphite from recycled Li-ion batteries for new LiB anodes. *Resour Conserv Recycl*. 2018;129:129–34.

[174] Weiss M, Ruess R, Kasnatscheew J, *et al.* Fast charging of lithium-ion batteries: a review of materials aspects. *Adv Energy Mater*. 2021;11 (33):2101126.

[175] Shen C, Hu G, Cheong L, Huang S, Zhang J, and Wang D. Direct observation of the growth of lithium dendrites on graphite anodes by operando EC-AFM. *Small Methods*. 2018;2(2):1700298.

[176] Lee M-L, Li YH, Liao S-C, Chen J-M, Yeh J-W, and Shih HC. Li4Ti5O12-coated graphite anode materials for lithium-ion batteries. *Electrochim Acta*. 2013;112:529–34.

[177] Bai A, Wang L, Li J, He X, Wang J, and Wang J. Composite of graphite/phosphorus as anode for lithium-ion batteries. *J Power Sources*. 2015;289:100–4.

[178] Yi X, Ge J, Zhou J, Zhou J, and Lu B. SbVO4 based high capacity potassium anode: a combination of conversion and alloying reactions. *Sci China Chem*. 2021;64(2):238–44.

[179] Edison E, Gogoi PK, Zheng Y, *et al.* Electrochemically induced amorphization and unique lithium and sodium storage pathways in FeSbO4 nanocrystals. *ACS Appl Mater Interfaces*. 2019;11(22):20082–90.

[180] Yuan Y, Yao W, Yurkiv V, *et al.* Beyond volume variation: anisotropic and protrusive lithiation in bismuth nanowire. *ACS Nano*. 2020;14(11):15669–77.

[181] Eshetu GG and Figgemeier E. Confronting the challenges of next-generation silicon anode-based lithium-ion batteries: role of designer electrolyte additives and polymeric binders. *ChemSusChem*. 2019;12(12):2515–39.

[182] Han Y, Huang G, and Xu S. Structural reorganization-based nanomaterials as anodes for lithium-ion batteries: design, preparation, and performance. *Small*. 2020;16(15):1902841. https://doi.org/10.1002/smll.201902841

[183] Bintang HM, Lee S, Shin S, *et al.* Stabilization effect of solid-electrolyte interphase by electrolyte engineering for advanced Li-ion batteries. *Chem Eng J*. 2021;424:130524. https://www.sciencedirect.com/science/article/pii/S1385894721021100

[184] Ding J, Zhou H, Zhang H, Tong L, and Mitlin D. Selenium impregnated monolithic carbons as free-standing cathodes for high volumetric energy lithium and sodium metal batteries. *Adv Energy Mater.* 2018;8(8):1701918. https://doi.org/10.1002/aenm.201701918

[185] Wang Y, Yu R, Luo T, *et al.* Solid solution of Bi and Sb for robust lithium storage enabled by consecutive alloying reaction. *Small.* 2021;17(38):2102915. https://doi.org/10.1002/smll.202102915

[186] Son Y, Kim N, Lee T, *et al.* Calendering-compatible macroporous architecture for silicon–graphite composite toward high-energy lithium-ion batteries. *Adv Mater.* 2020;32(37):2003286. https://doi.org/10.1002/adma.202003286

[187] Liu J, Yu L, Wu C, *et al.* New nanoconfined galvanic replacement synthesis of hollow Sb@C yolk–shell spheres constituting a stable anode for high-rate Li/Na-ion batteries. *Nano Lett.* 2017;17(3):2034–42. https://doi.org/10.1021/acs.nanolett.7b00083

[188] Wang S, Xiong P, Guo X, *et al.* A stable conversion and alloying anode for potassium-ion batteries: a combined strategy of encapsulation and confinement. *Adv Funct Mater.* 2020;30(27):2001588. https://doi.org/10.1002/adfm.202001588

[189] Qiu X, Wang B, Jing P, *et al.* Bismuth dots imbedded in ultralong nitrogen-doped carbon tubes for highly efficient lithium ion storage. *Inorg Chem Front.* 2020;7(24):4854–64. http://dx.doi.org/10.1039/D0QI01158D

[190] Guo S, Feng Y, Wang L, Jiang Y, Yu Y, and Hu X. Architectural engineering achieves high-performance alloying anodes for lithium and sodium ion batteries. *Small.* 2021;17(19):2005248. https://doi.org/10.1002/smll.202005248

[191] Liang B, Liu Y, and Xu Y. Silicon-based materials as high capacity anodes for next generation lithium ion batteries. *J Power Sources.* 2014;267:469–90.

[192] Angulakshmi N, Zhou Y, Suriyakumar S, *et al.* Microporous metal–organic framework (MOF)-based composite polymer electrolyte (CPE) mitigating lithium dendrite formation in all-solid-state-lithium batteries. *ACS Omega.* 2020;5(14):7885–94.

[193] Wu H and Cui Y. Designing nanostructured Si anodes for high energy lithium ion batteries. *Nano Today.* 2012;7(5):414–29.

[194] Ling M, Zhao H, Xiaoc X, *et al.* Low cost and environmentally benign crack-blocking structures for long life and high power Si electrodes in lithium ion batteries. *J Mater Chem A.* 2015;3(5):2036–42.

[195] Wang C, Wu H, Chen Z, McDowell MT, Cui Y, and Bao Z. Self-healing chemistry enables the stable operation of silicon microparticle anodes for high-energy lithium-ion batteries. *Nat Chem.* 2013;5(12):1042–8.

[196] Ko M, Chae S, and Cho J. Challenges in accommodating volume change of Si anodes for Li-ion batteries. *ChemElectroChem.* 2015;2(11):1645–51.

[197] Liu Y, Tai Z, Zhou T, *et al.* An all-integrated anode via interlinked chemical bonding between double-shelled–yolk-structured silicon and binder for lithium-ion batteries. *Adv Mater.* 2017;29(44):1703028.

[198] Qu J, Li H, Henry Jr JJ, *et al.* Self-aligned Cu–Si core–shell nanowire array as a high-performance anode for Li-ion batteries. *J Power Sources.* 2012;198:312–7.
[199] Su X, Wu Q, Li J, *et al.* Silicon-based nanomaterials for lithium-ion batteries: a review. *Adv Energy Mater.* 2014;4(1):1300882.
[200] Choi W, Choi YS, Kim H, *et al.* Evidence for the coexistence of polysulfide and conversion reactions in the lithium storage mechanism of MoS2 anode material. *Chem Mater.* 2021;33(6):1935–45. https://doi.org/10.1021/acs.chemmater.0c02992
[201] Nowroozi MA and Clemens O. Insights on the behavior of conversion-based anode materials for fluoride ion batteries by testing against an intercalation-based reference cathode. *ACS Appl Energy Mater.* 2018;1(11):6626–37. https://doi.org/10.1021/acsaem.8b01630
[202] Zhang X, Zhu Y, Bruck AM, *et al.* Understanding aggregation hindered Li-ion transport in transition metal oxide at mesoscale. *Energy Storage Mater.* 2019;19:439–45.
[203] Amiri A and Shahbazian-Yassar R. Recent progress of high-entropy materials for energy storage and conversion. *J Mater Chem A.* 2021;9(2):782–823.
[204] Veluri PS and Mitra S. Enhanced high rate performance of α-Fe2O3 nanotubes with alginate binder as a conversion anode. *RSC Adv.* 2013;3(35):15132–8. http://dx.doi.org/10.1039/C3RA41722K
[205] Sahoo A, Deka B, and Sharma Y. Incorporation of alloy-de-alloy phase with conversion based manganese oxide to enable high and stable capacity and density functional theory study of CdMn2O4. *J Electrochem Soc.* 2018;165(9):A1610–20. http://dx.doi.org/10.1149/2.0101809jes
[206] Liang L, Li J, Zhu M, Li Y, Chou S, and Li W. Cobalt chalcogenides/cobalt phosphides/cobaltates with hierarchical nanostructures for anode materials of lithium-ion batteries: improving the lithiation environment. *Small.* 2021;17(9):1903418. https://doi.org/10.1002/smll.201903418
[207] Ng B, Faegh E, Lateef S, Karakalos SG, and Mustain WE. Structure and chemistry of the solid electrolyte interphase (SEI) on a high capacity conversion-based anode: NiO. *J Mater Chem A.* 2021;9(1):523–37.
[208] Zhong Y, Yang M, Zhou X, Luo Y, Wei J, and Zhou Z. Orderly packed anodes for high-power lithium-ion batteries with super-long cycle life: rational design of MnCO3/large-area graphene composites. *Adv Mater.* 2015;27(5):806–12.
[209] Xie Y, Zhang W, Li H, *et al.* From lychee seeds to hierarchical Fe3O4/carbon composite anodes for lithium-ion batteries: a high additional value conversion-based self-assembly strategy. *Energy & Fuels.* 2022;36(9):5027–35. https://doi.org/10.1021/acs.energyfuels.2c00664
[210] Nzereogu PU, Omah AD, Ezema FI, Iwuoha EI, and Nwanya AC. Anode materials for lithium-ion batteries: a review. *Appl SurfSci Adv.* 2022;9:100233. https://www.sciencedirect.com/science/article/pii/S2666523922000253
[211] Qi W, Shapter JG, Wu Q, Yin T, Gao G, and Cui D. Nanostructured anode materials for lithium-ion batteries: principle, recent progress and future perspectives. *J Mater Chem A.* 2017;5(37):19521–40.

[212] Matsuda Y. Behavior of lithium/electrolyte interface in organic solutions. *J Power Sources*. 1993;43(1–3):1–7.

[213] Wang C, Fu KK, Dai J, *et al.* Inverted battery design as ion generator for interfacing with biosystems. *Nat Commun*. 2017;8(1):1–7.

[214] Li J, Daniel C, and Wood D. Materials processing for lithium-ion batteries. *J Power Sources*. 2011;196(5):2452–60.

[215] Xu K. Electrolytes and interphases in Li-ion batteries and beyond. *Chem Rev*. 2014;114(23):11503–618.

[216] Davoodabadi A, Li J, Liang Y, Wood III DL, Singler TJ, and Jin C. Analysis of electrolyte imbibition through lithium-ion battery electrodes. *J Power Sources*. 2019;424:193–203.

[217] Liu G, Sun Q, Li Q, Zhang J, and Ming J. Electrolyte issues in lithium–sulfur batteries: development, prospect, and challenges. *Energy & Fuels*. 2021;35(13):10405–27. https://doi.org/10.1021/acs.energyfuels.1c00990

[218] Pan Q, Gong D, and Tang Y. Recent progress and perspective on electrolytes for sodium/potassium-based devices. *Energy Storage Mater*. 2020;31:328–43. https://www.sciencedirect.com/science/article/pii/S2405829720302452

[219] Li Q, Chen J, Fan L, Kong X, and Lu Y. Progress in electrolytes for rechargeable Li-based batteries and beyond. *Green Energy Environ*. 2016;1(1):18–42. https://www.sciencedirect.com/science/article/pii/S2468025716300218

[220] Kanamura K. Anodic oxidation of nonaqueous electrolytes on cathode materials and current collectors for rechargeable lithium batteries. *J Power Sources*. 1999;81–82:123–9. https://www.sciencedirect.com/science/article/pii/S037877539900141X

[221] Jow TR, Ding MS, Xu K, *et al.* Nonaqueous electrolytes for wide-temperature-range operation of Li-ion cells. *J Power Sources*. 2003;119–21:343–8. https://www.sciencedirect.com/science/article/pii/S0378775303001538

[222] Zhang J-G, Xu W, Xiao J, Cao X, and Liu J. Lithium metal anodes with nonaqueous electrolytes. *Chem Rev*. 2020;120(24):13312–48.

[223] Kanamura K, Takezawa H, Shiraishi S, and Takehara Z. Chemical reaction of lithium surface during immersion in LiClO4 or LiPF6/DEC electrolyte. *J Electrochem Soc*. 1997;144(6):1900–6. http://dx.doi.org/10.1149/1.1837718

[224] Shu ZX, McMillan RS, Murray JJ, and Davidson IJ. Use of chloroethylene carbonate as an electrolyte solvent for a lithium ion battery containing a graphitic anode. *J Electrochem Soc*. 1995;142(9):L161.

[225] An SJ, Li J, Daniel C, Mohanty D, Nagpure S, and Wood III DL. The state of understanding of the lithium-ion-battery graphite solid electrolyte interphase (SEI) and its relationship to formation cycling. *Carbon N Y*. 2016;105:52–76.

[226] Balasubramanian M, Lee HS, Sun X, *et al.* Formation of SEI on cycled lithium-ion battery cathodes: soft X-ray absorption study. *Electrochem Solid-State Lett*. 2002;5(1):A22. http://dx.doi.org/10.1149/1.1423802

[227] Sankarasubramanian S and Krishnamurthy B. A capacity fade model for lithium-ion batteries including diffusion and kinetics. *Electrochim Acta*. 2012;70:248–54. https://www.sciencedirect.com/science/article/pii/S0013468612004021

[228] Hao F, Liu Z, Balbuena PB, and Mukherjee PP. Mesoscale elucidation of solid electrolyte interphase layer formation in Li-ion battery anode. *J Phys Chem C*. 2017;121(47):26233–40. https://doi.org/10.1021/acs.jpcc.7b09465

[229] Xu J, Hu Y, liu T, and Wu X. Improvement of cycle stability for high-voltage lithium-ion batteries by in-situ growth of SEI film on cathode. *Nano Energy*. 2014;5:67–73. https://www.sciencedirect.com/science/article/pii/S2211285514000196

[230] Zhao Y, Amirmaleki M, Sun Q, *et al.* Natural SEI-inspired dual-protective layers via atomic/molecular layer deposition for long-life metallic lithium anode. *Matter*. 2019;1(5):1215–31. https://www.sciencedirect.com/science/article/pii/S2590238519300724

[231] Erickson EM, Li W, Dolocan A, and Manthiram A. Insights into the cathode–electrolyte interphases of high-energy-density cathodes in lithium-ion batteries. *ACS Appl Mater Interfaces*. 2020;12(14):16451–61.

[232] Zheng H, Qu Q, Zhang L, Liu G, and Battaglia VS. Hard carbon: a promising lithium-ion battery anode for high temperature applications with ionic electrolyte. *RSC Adv*. 2012;2(11):4904–12.

[233] Stojković IB, Cvjetićanin ND, and Mentus S V. The improvement of the Li-ion insertion behaviour of Li1.05Cr0.10Mn1.85O4 in an aqueous medium upon addition of vinylene carbonate. *Electrochem Commun*. 2010;12 (3):371–3.

[234] Huang Q, Lee Y-Y, and Gurkan B. Pyrrolidinium ionic liquid electrolyte with bis (trifluoromethylsulfonyl) imide and bis (fluorosulfonyl) imide anions: lithium solvation and mobility, and performance in lithium metal–lithium iron phosphate batteries. *Ind Eng Chem Res*. 2019;58(50):22587–97.

[235] McElvain SM, Jelinek A, and Rorig K. Ethane-1, 2-and propane-1, 3-disulfonic acids and anhydrides. *J Am Chem Soc*. 1945;67(9):1578–81.

[236] Miyazaki K, Shimada T, Ito S, Yokoyama Y, Fukutsuka T, and Abe T. Enhanced resistance to oxidative decomposition of aqueous electrolytes for aqueous lithium-ion batteries. *Chem Commun*. 2016;52(28):4979–82. http://dx.doi.org/10.1039/C6CC00873A

[237] Ko S, Yamada Y, Miyazaki K, *et al.* Lithium-salt monohydrate melt: A stable electrolyte for aqueous lithium-ion batteries. *Electrochem Commun*. 2019;104:106488.

[238] Tron A, Jeong S, Park YD, and Mun J. Aqueous lithium-ion battery of nano-LiFePO4 with antifreezing agent of ethylene glycol for low-temperature operation. *ACS Sustain Chem Eng*. 2019;7(17):14531–8.

[239] Luo J-Y, Cui W-J, He P, and Xia Y-Y. Raising the cycling stability of aqueous lithium-ion batteries by eliminating oxygen in the electrolyte. *Nat Chem*. 2010;2(9):760–5.

[240] Binnemans K. Ionic liquid crystals. *Chem Rev*. 2005;105(11):4148–204.

[241] Armand M, Endres F, MacFarlane DR, Ohno H, and Scrosati B. Ionic-liquid materials for the electrochemical challenges of the future. *Mater Sustain Energy a Collect Peer-reviewed Res Rev Artic from Nat Publ Gr*. 2011;129–37.

[242] Guerfi A, Dontigny M, Charest P, *et al.* Improved electrolytes for Li-ion batteries: mixtures of ionic liquid and organic electrolyte with enhanced safety and electrochemical performance. *J Power Sources.* 2010;195(3):845–52.

[243] Zhao J, Jiang X-C, Li C-X, and Wang Z-H. Vapor pressure measurement for binary and ternary systems containing a phosphoric ionic liquid. *Fluid Phase Equilib.* 2006;247(1–2):190–8.

[244] Brutti S, Simonetti E, De Francesco M, *et al.* Ionic liquid electrolytes for high-voltage, lithium-ion batteries. *J Power Sources.* 2020;479:228791. https://www.sciencedirect.com/science/article/pii/S0378775320310958

[245] Eisinger RS and Keller GE. Process for CO2 capture using ionic liquid that exhibits phase change. *Energy & Fuels.* 2014;28(11):7070–8.

[246] Manthiram A, Yu X, and Wang S. Lithium battery chemistries enabled by solid-state electrolytes. *Nat Rev Mater.* 2017;2(4):16103. https://doi.org/10.1038/natrevmats.2016.103

[247] Zheng F, Kotobuki M, Song S, Lai MO, and Lu L. Review on solid electrolytes for all-solid-state lithium-ion batteries. *J Power Sources.* 2018;389:198–213. https://www.sciencedirect.com/science/article/pii/S0378775318303653

[248] Kim JG, Son B, Mukherjee S, *et al.* A review of lithium and non-lithium based solid state batteries. *J Power Sources.* 2015;282:299–322. https://www.sciencedirect.com/science/article/pii/S0378775315002773

[249] Ohta N, Takada K, Zhang L, Ma R, Osada M, and Sasaki T. Enhancement of the high-rate capability of solid-state lithium batteries by nanoscale interfacial modification. *Adv Mater.* 2006;18(17):2226–9.

[250] Huang W, Feng X, Han X, Zhang W, and Jiang F. Questions and answers relating to lithium-ion battery safety issues. *Cell Rep Phys Sci.* 2021;2(1):100285.

[251] Cao J, He R, and Kyu T. Fire retardant, superionic solid state polymer electrolyte membranes for lithium ion batteries. *Curr Opin Chem Eng.* 2017;15:68–75.

[252] Nykaza JR, Savage AM, Pan Q, *et al.* Polymerized ionic liquid diblock copolymer as solid-state electrolyte and separator in lithium-ion battery. *Polymer (Guildf).* 2016;101:311–8.

[253] Yersak T, Salvador JR, Schmidt RD, and Cai M. Hot pressed, fiber-reinforced (Li2S) 70 (P2S5) 30 solid-state electrolyte separators for Li metal batteries. *ACS Appl Energy Mater.* 2019;2(5):3523–31.

[254] Lu P, Ding F, Xu Z, Liu J, Liu X, and Xu Q. Study on (100-x)(70Li2S-30P2S5)-xLi2ZrO3 glass-ceramic electrolyte for all-solid-state lithium-ion batteries. *J Power Sources.* 2017;356:163–71.

[255] Chen L, Huang Y-F, Ma J, Ling H, Kang F, and He Y-B. Progress and perspective of all-solid-state lithium batteries with high performance at room temperature. *Energy Fuels.* 2020;34(11):13456–72. https://doi.org/10.1021/acs.energyfuels.0c02915

[256] Li W, Zhu X, Zhou N, *et al.* Helical polyurethane-initiated unique microphase separation architecture for highly efficient lithium transfer and

battery performance of a poly(ethylene oxide)-based all-solid-state electrolyte. *ACS Appl Energy Mater.* 2021;4(5):4772–85. https://doi.org/10.1021/acsaem.1c00358

[257] Lang S-Y, Xiao R-J, Gu L, Guo Y-G, Wen R, and Wan L-J. Interfacial mechanism in lithium–sulfur batteries: how salts mediate the structure evolution and dynamics. *J Am Chem Soc.* 2018;140(26):8147–55. https://doi.org/10.1021/jacs.8b02057

[258] Tran HK, Wu Y-S, Chien W-C, *et al.* Composite polymer electrolytes based on PVA/PAN for all-solid-state lithium metal batteries operated at room temperature. *ACS Appl Energy Mater.* 2020;3(11):11024–35. https://doi.org/10.1021/acsaem.0c02018

[259] Mahbub R, Huang K, Jensen Z, Hood ZD, Rupp JLM, and Olivetti EA. Text mining for processing conditions of solid-state battery electrolytes. *Electrochem Commun.* 2020;121:106860.

[260] Yang Z, Tang B, Xie Z, and Zhou Z. NASICON-type Na3Zr2Si2 PO12 solid-state electrolytes for sodium batteries. *ChemElectroChem.* 2021;8(6):1035–47.

[261] Liu S, Zhao Y, Li X, Yu J, Yan J, and Ding B. Solid-state lithium metal batteries with extended cycling enabled by dynamic adaptive solid-state interfaces. *Adv Mater.* 2021;33(12):2008084.

[262] Lewis JA, Tippens J, Cortes FJQ, and McDowell MT. Chemo-mechanical challenges in solid-state batteries. *Trends Chem.* 2019;1(9):845–57.

[263] Kim KJ, Balaish M, Wadaguchi M, Kong L, and Rupp JLM. Solid-state Li–metal batteries: challenges and horizons of oxide and sulfide solid electrolytes and their interfaces. *Adv Energy Mater.* 2021;11(1):2002689.

[264] Song Y, Yang L, Zhao W, *et al.* Revealing the short-circuiting mechanism of garnet-based solid-state electrolyte. *Adv Energy Mater.* 2019;9(21):1900671.

[265] Li Z, Zhang H, Sun X, and Yang Y. Mitigating interfacial instability in polymer electrolyte-based solid-state lithium metal batteries with 4 V cathodes. *ACS Energy Lett.* 2020;5(10):3244–53. https://doi.org/10.1021/acsenergylett.0c01465

[266] Liang J, Luo J, Sun Q, Yang X, Li R, and Sun X. Recent progress on solid-state hybrid electrolytes for solid-state lithium batteries. *Energy Storage Mater.* 2019;21:308–34. https://www.sciencedirect.com/science/article/pii/S2405829719308438

[267] Lu Y, Korf K, Kambe Y, Tu Z, and Archer LA. Ionic-liquid–nanoparticle hybrid electrolytes: applications in lithium metal batteries. *Angew Chemie Int Ed.* 2014;53(2):488–92. https://doi.org/10.1002/anie.201307137

[268] Liu X, Li X, Li H, and Wu HB. Recent progress of hybrid solid-state electrolytes for lithium batteries. *Chem Eur J.* 2018;24(69):18293–306.

[269] Quartarone E and Mustarelli P. Review—Emerging trends in the design of electrolytes for lithium and post-lithium batteries. *J Electrochem Soc.* 2020;167(5):50508. http://dx.doi.org/10.1149/1945-7111/ab63c4

[270] Balbuena PB. Electrolyte materials-Issues and challenges. In: *AIP Conference Proceedings*. American Institute of Physics; 2014. pp. 82–97.

[271] Meyer WH. Polymer electrolytes for lithium-ion batteries. *Adv Mater*. 1998;10(6):439–48.

[272] Arya A and Sharma AL. Polymer electrolytes for lithium ion batteries: a critical study. *Ionics (Kiel)*. 2017;23(3):497–540. https://doi.org/10.1007/s11581-016-1908-6

Chapter 3
Slurry preparation

W. Blake Hawley[1], William Kays[2] and Jianlin Li[3]

3.1 Introduction

As will be detailed throughout this book, the state-of-the-art lithium-ion battery (LIB) electrode manufacturing process consists of several interconnected steps. There are quality control checks strategically placed that correlate material properties during or after a particular step that provide details on the processability (i.e., compatibility with downstream manufacturing processes) and, sometimes, projected performance of the finished product. In this chapter, we will begin this exploration by starting with the first step in the state-of-the-art LIB process, which is preparation of the electrode slurry [1]. Alternative terms to "slurry," such as ink, paste, or (less commonly) dispersion, are sometimes used in academia or industry.

Electrode slurries are highly sensitive to many conditions in the mixing environment that must be accounted for prior to processing. These conditions include the mixing instrumentation, mixing protocol, and the ambient conditions. There are two types of slurry properties – independent and dependent. Independent properties are those that are determined prior to mixing while dependent properties are those that depend on the mixing and ambient conditions as well as the independent slurry properties. Independent properties include the actual slurry constituents selected, the ratio between slurry constituents, the size of the slurry constituents, and the total solid loading. The slurry constituents include: (1) the active material, which is responsible for the intercalation/de-intercalation of lithium or forming alloys with lithium during charge and discharge of the battery; (2) the conductive additive(s), which increase the electronic conductivity of the electrodes; (3) the binder(s), which provide adhesion to the current collector and cohesion between electrode components; and (4) any additional additives, which may be used to modify slurry and/or finished electrode properties. The terms "solid loading" and "solid content" may be used interchangeably in this chapter; both terms refer to the percentage of the total slurry mass (or volume) that is comprised of solid materials. The dependent properties of the electrode slurry are the

[1]Sparkz, Inc., USA
[2]R.W. Baron Process Equipment, Inc., USA
[3]Division of Electrification and Energy Infrastructures, Oak Ridge National Laboratory, USA

rheological properties, including the viscosity and viscoelasticity of the slurry, and the surface tension of the slurry.

Good slurry preparation requires one to have precise control over the mixing protocol, ambient conditions, and the independent properties. Knowledge of how these three factors affect the dependent properties is essential for one to properly interpret rheological data. We will work backwards by first discussing the dependent properties in Section 3.2. The independent properties will be introduced in Section 3.3 and the connection between the independent and dependent properties will be established. Ambient conditions will be considered in Section 3.4, while a larger discussion on mixing instrumentation and protocols is available in Section 3.5. Finally, the measurement of dependent properties on a laboratory and large-volume production scale will be reviewed in Section 3.6. The chapter is summarized in Section 3.7.

3.2 Dependent properties

As the name implies, the dependent slurry properties are those that are contingent on the mixing instrumentation and protocol, ambient conditions, and independent slurry properties. These dependent slurry properties include the viscosity, viscoelasticity, and surface tension of the slurry. By far, the most commonly measured dependent slurry property in academia and industry is viscosity, since it is relatively easily measured, interpreted, and gives significant information about the slurry. The dependent properties of the electrode slurry are a strong indicator of the processability of the slurry, though the electrochemical performance of the assembled cell is a far weaker function of the slurry's rheology. The rheological properties have a tremendous influence on the processability of the slurry since a viscous slurry will be far more resistant to flow and may clog the tubing or coating instruments. On the other hand, a low-viscosity slurry may be difficult to apply well to a moving substrate [1]. The rheological properties are a variable that determines the coating window of the slurry. The coating window refers to the sets of slot-die coater conditions (i.e., pump speed, line speed) that will produce a viable, defect-free electrode coating. These slot-die coater conditions, in turn, determine the thickness and thickness homogeneity of the coating, which will be discussed further in Chapter 4. Meanwhile, of all steps in electrode processing, slurry rheology is a relatively poor indicator of electrochemical performance since there are many downstream processes beyond slurry mixing. These downstream processes (coating, drying, calendaring) have more direct influence on the microstructure and performance of the electrode. This, of course, does not even consider the effect of the quality of the materials, nor the cell, module, and pack manufacturing steps on the performance of the battery.

3.2.1 Viscosity

Viscosity is a measure of a fluid's resistance to flow and is calculated according to (3.1):

$$\mu = \frac{\sigma}{\dot{\gamma}} \tag{3.1}$$

In (3.1), μ is the viscosity (in Pa·s), σ is the shear stress (in Pa), and γ is the shear rate (in s^{-1}). In many industries, viscosity is often reported as a single value, though it is better practice to report viscosity as a function of shear rate, henceforth referred to as a "viscosity curve" or "viscosity profile." Single-value viscosity figures are usually reported after measurement with a viscometer, which is an instrument capable of measuring the viscosity of a fluid at a single shear rate. A single value for viscosity is particularly meaningless when the shear rate at which it was measured is not reported. To generate a viscosity curve, a more advanced viscometer that is capable of multi-point measurements or a rheometer is required. More information on the instrumentation of viscometers and rheometers is available in Section 3.6.1.

To illustrate why a viscosity curve is preferred over a single value for viscosity (even if the shear rate is reported), consider the different forces experienced by a slurry. First, during mixing, a high amount of force is applied to sufficiently disperse the slurry constituents. Once mixed, the slurry is either left at rest or gently mixed to keep larger active material particles suspended. When the slurry is ready to be coated, it flows through tubing to reach the head of the coating instrument, where it is applied to the rolling substrate. After being coated, the slurry experiences essentially no shear as the coating is fed to the dryer for solvent removal. At each of these four phases, the slurry experiences high, mild-to-none, moderate, and no shear, respectively. It is impossible for one viscosity value to accurately describe how a slurry behaves at each of these stages! Therefore, a viscosity curve across a broad range of shear rates (ranging from about 10^{-3} to 10^4 s^{-1}) is more meaningful [1]. The shear rate involved in slot-die coating may be estimated according to (3.2):

$$\dot{\gamma} = \frac{6Q}{bh^2} \tag{3.2}$$

In (3.2), Q is the volumetric flow rate (in $m^3 \cdot s^{-1}$), b is the slot width (in m), and h is the height of the slot above the substrate (in m) [2].

Viscosity curves can be described using one of three descriptors, depending on the dependency of the viscosity on the shear rate (see Figure 3.1). If the viscosity is

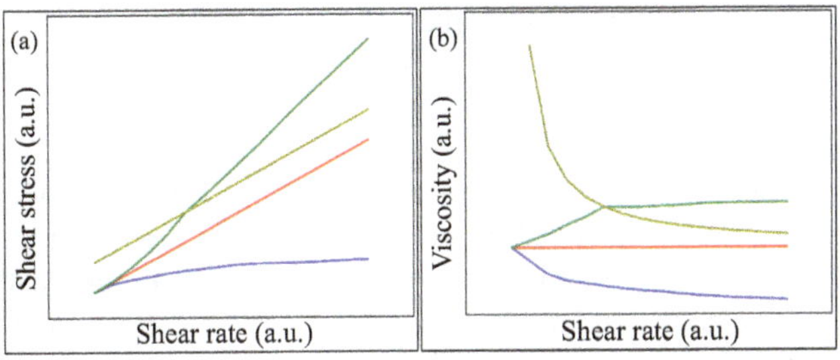

Figure 3.1 *(a) Stress vs. shear rate and (b) viscosity vs. shear rate plots for four kinds of fluids: Newtonian (red), shear thinning (blue), shear thickening (green), and Bingham (yellow). The units shown are arbitrary.*

unchanged regardless of the shear rate, the viscosity curve is a horizontal line, and the slurry is said to have Newtonian behavior. According to the definition of viscosity (see (3.1), Newtonian fluids have a shear stress that is a linear function of the shear rate [3]. If the viscosity decreases with increasing shear rate, the viscosity curve is downward sloping, and the slurry has pseudoplastic behavior. A more common term used in electrode slurry processing for pseudoplasticity is shear thinning. Shear-thinning fluids have a shear stress that increases at a slower rate than the shear rate is increasing [3]. Finally, if the viscosity increases with increasing shear rate, the viscosity curve is upward sloping and the slurry has dilatant, or shear thickening, behavior. Shear thickening fluids, the opposite of shear-thinning fluids, have a shear stress that increases at a faster rate than the shear rate is increasing [3]. In addition to the dependency of viscosity on shear rate, slurries can also be classified by whether or not they have a yield stress [3]. The yield stress is defined as the minimum stress needed to cause a fluid to flow. Fluids with a yield stress are classified as Bingham fluids or Bingham plastics. While most electrode slurries are technically Bingham pseudoplastics (i.e., slurries with a yield stress and shear-thinning behavior), this term is rarely used, and the slurries are usually called simply "shear thinning" with a mild yield stress often implied to exist.

The shear-thinning behavior of electrode slurries is a very desirable property. By definition, shear-thinning fluids have a high low-shear viscosity (LSV) and a low high-shear viscosity (HSV). The LSV describes the viscosity of the slurry during rest or under mild agitation after mixing and prior to coating as well as the viscosity of the slurry immediately after coating. A high resistance to flow after coating is preferred, since the slurry would ideally remain completely stationary after coating. Any flow from the slurry after coating and prior to drying could compromise edge quality. Conversely, the HSV describes the viscosity of the slurry during mixing and coating. A relatively low resistance to flow is needed during coating so that the slurry may be spread evenly onto the current collector [1]. The HSV is a major factor in determining the dry thickness of the coating. A slurry with high HSV will be more difficult to spread and will have a thickness closer to the gap set in the coating device, while a slurry with low HSV will spread more readily and have a lower thickness relative to the coating gap.

3.2.2 Viscoelasticity

Another dependent property of an electrode slurry is the viscoelasticity of the slurry. Materials that respond to perturbations strictly as a liquid are called viscous while materials that respond to perturbations strictly as a solid are called elastic. Specifically, a viscous material dissipates energy from a perturbation as heat, while an elastic material absorbs the energy [3]. Most fluids, including electrode slurries, cannot be considered purely viscous or elastic, hence the term viscoelastic. Particularly, viscoelastic materials dissipate some of the energy from a disturbance as heat and absorb the rest of that energy. The relative ratio between the energy dissipated and energy absorbed can determine how "liquid-like" or how "solid-like" a viscoelastic material is. The energy stored and dissipated are referred to as the storage (G') and loss (G'') moduli, respectively. The storage and loss moduli require sophisticated instrumentation and characterization to measure (see Section 3.6).

The storage and loss moduli can be measured as either a function of oscillation strain or angular frequency. When the moduli are measured as a function of oscillation strain, the angular frequency is held constant, and vice-versa. A test that is performed as a function of oscillation strain is referred to as an amplitude or strain sweep, while a test that is performed as a function of angular frequency is called a frequency sweep. For electrode slurries, amplitude sweeps are commonly performed between 0.01% and 10% strain, while frequency sweeps are often performed between 0.1 and 100 rad·s^{-1}, though lower frequencies are possible depending on how much time is allotted for the experiment.

The utility of measuring the viscoelasticity of an electrode slurry lies in acquiring a deeper understanding of the microstructure of the slurry. At rest, the ideal structure of a slurry is best described as a bridging flocculation network between particles. This bridging flocculation network is a state in which the active material secondary particles are not agglomerated and are connected by a long-chained binder network that "bridges" the gaps between particles (see Figure 3.2) [4]. When subject to sufficiently high shear rates, this bridging network breaks down.

The strength of the bridging flocculation network can be probed by an amplitude sweep [4]. At low strain, the storage modulus measured should be constant as strain increases if the proper structure is formed in the slurry. This region of constant storage modulus is referred to as the linear viscoelastic regime (LVR). Eventually, as the strain continues to increase, the storage modulus measured

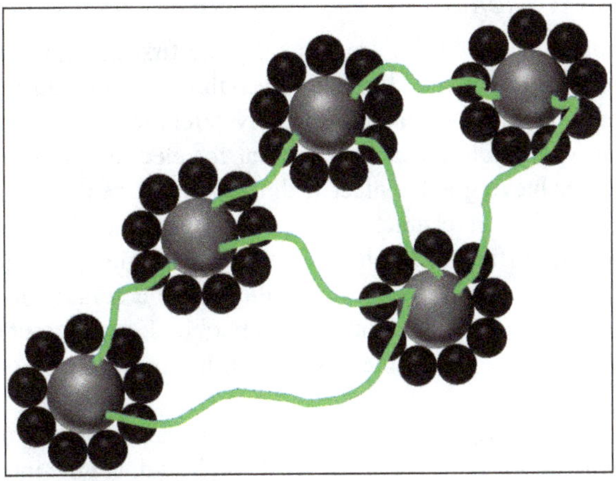

Figure 3.2 Cartoon of bridging flocculation network in an electrode slurry. The gray spheres represent active material particles, the black spheres represent conductive additive particles, and the green strands represent the binder.

decreases rapidly. At this point, the bridging flocculation network is beginning to be destroyed, and the slurry is no longer within its LVR. Quantitatively, there is not a formal standard for determining when the slurry is considered outside of its LVR, though energy values between 90% and 95% of the maximum storage modulus may be considered the point at which the structure is broken [5]. Determining the LVR is also important since the constant strain used in the frequency sweep must be within the LVR of the fluid for the data to be meaningful. Otherwise, the moduli measured do not probe the structure of the fluid, since it is destroyed at a strain beyond the LVR.

Both the amplitude and the frequency sweep are useful in determining the relative ratio between the storage and loss modulus. In both tests, when the storage modulus is greater than the loss modulus, the slurry is said to be more solid-like, and vice-versa. In amplitude sweeps, when the bridging flocculation network is formed, the storage modulus is typically greater than the loss modulus, even at strain values greater than the end of the LVR. At sufficiently high strain, though, the loss modulus is greater than the storage modulus, meaning the slurry behaves more like a liquid. In the frequency sweep, another indicator of the solid-like behavior of the slurry is the dependency of the storage modulus on frequency. When the storage modulus is barely dependent on frequency, the slurry behaves more solid-like. For both tests, the strain or frequency at which the loss modulus is greater than the storage modulus is called the sol–gel transition. Sometimes, the term “gel-like” is used for slurries whose storage and loss moduli are close in value for a given strain or frequency range. Gel-like slurries are considered a midway point between a watery, liquid state and an elastic, solid state.

3.2.3 Surface tension

Though the rheological properties are generally the first thought when discussing the dependent properties of the slurry, the wettability of the slurry is also a key property to note. The term “wettability” simply refers to the ability of a liquid to maintain contact with a solid. This is important for electrode processing since the liquid slurry must achieve good contact with the solid current collector during and after coating. Wettability is characterized by the contact angle formed between a liquid droplet on the surface of a solid. If the contact angle with water is less than 90°, the solid material is said to be hydrophilic, while a contact angle with water greater than 90° means a solid material is hydrophobic. If the current collector has a low contact angle with the slurry, then the wettability of the slurry onto the current collector is said to be good.

When a droplet of a liquid is added to a solid surface, there are three interfaces to consider: (1) the interface between the solid and liquid, (2) the interface between the solid and the surrounding atmosphere (gas), and (3) the interface between the liquid and the surrounding atmosphere. Young’s equation (3.3) correlates the surface free energy of these three interfaces to the contact angle between the liquid and the solid as follows:

$$\tau_{SG} - \tau_{SL} - \tau_{LG}\cos(\theta) = 0 \tag{3.3}$$

where τ is the interfacial free energy, the subscripts "*SG*," "*SL*," and "*LG*" indicate solid–gas interface, solid–liquid interface, and liquid–gas interface, respectively, and θ is the contact angle. Assuming a constant interfacial free energy between the solid and the gas and the liquid and the gas, hydrophilic surfaces (i.e., low θ) have a low solid–liquid interfacial free energy. Conversely, hydrophobic surfaces (i.e., high θ) have a high solid–liquid interfacial free energy. The solid–liquid interfacial free energy is a function of the surface tension of the slurry.

Like the viscoelastic properties, the wettability of a slurry is much less often characterized than the viscosity of a slurry, even at the academic and R&D levels. This is partially due to the collective knowledge of common binder and solvent pairings usually generating slurries with good wettability, which makes a measurement of the surface tension less necessary. Surface tension is also less commonly measured due to the time-consuming nature of the measurement; this is elaborated further in Section 3.6.5. Still, for novel slurry components, in particular solvents and binders, the importance of wettability of the slurry to the current collector is heightened due to less collective knowledge.

3.3 Independent properties

Most of the conditions influencing the dependent properties of an electrode slurry are the independent properties. The independent properties are those that relate to the slurry design that are a product of the properties of the slurry constituents used and the decisions made when developing the slurry recipe. Specifically, the size and the shape of the active material, conductive additive, and (if applicable) additive particles, the molecular weight (MW) of the binder, the ratio between the materials, and the total solid loading in the slurry may be considered independent properties. We will discuss the influence of each of these independent properties on the rheological properties of the slurry in the following subsections.

3.3.1 Particle size and shape and binder MW

LIB cathode active material particles follow a hierarchical structure, in which primary particles are aggregated into secondary particles. The primary particles usually have a diameter on the order of several hundred nanometers and the secondary particles are usually between 5 and 20 μm in diameter, though smaller secondary particles are possible in certain cases, such as for $LiFePO_4$ (LFP) particles. The primary particles are held together by strong, often electrostatic bonds [6], which are very difficult to break during mixing. Therefore, when one discusses the size of active material particles, it is often the case that one is really referring to the size of the secondary particles. What is important during mixing is to prevent the agglomeration of secondary particles into large agglomerates. Fortunately, these larger agglomerates (> 20 μm) are held together by relatively weak van der Waals forces [6] and can be dispersed using many different mixing media (see Section 3.5).

Assuming a constant mass fraction of hard, spherical particles, a slurry containing more large particles would be less viscous than a slurry containing more small particles across all shear rates. Hard spheres are a theoretical approximation that do not have any attractive or repulsive forces acting between them until they collide. Therefore, the only forces that can create shear stress in the slurry are the collisions between particles. If the total mass of particles and the density of all particles is equivalent, there will be significantly more particles when they are smaller than when they are larger. When the hard sphere assumption is removed, the attractive and repulsive forces that occur between particles are relatively more impactful at low shear rates; at high shear rates, the only meaningful force is the hydrodynamic force. Given the smaller particles have a larger surface area, they generally have greater attraction and repulsion forces between them, as well. In practice, this means that slurries that contain more carbon black (particles typically $\leq$ 100 nm in diameter) will be more viscous than those with less carbon black. Because of the vast difference in size and specific surface area between active material particles and conductive additive particles, electrode slurries are best approximated as a suspension with a bimodal particle size distribution in a polymer solution.

The shape of the particles is also a significant factor in the rheological profile of a slurry. Spherical particles are often assumed for simplicity in most steps of electrode processing and in electrochemical analysis, and particles that are roughly spherical are certainly used as active materials, particularly for cathodes. However, acicular, elliptical, and flake-like particles are also seen in both electrodes [7], and can complicate the interactions between components in a slurry beyond what is often assumed for spherical particles. These particle shapes can change as a function of slurry processing; for instance, aggressive milling conditions during mixing can break apart and distort primary and secondary particles into amorphous shapes. Additionally, the shape of particles changes as a function of calendaring after the electrode is dried; this consideration will be discussed further in Chapter 8.

As for the influence on rheology, the surface area of particles is largely dependent on the particle shape. The surface area impacts the amount of binding that can occur on the surface of a particle, which could lead to a strengthened or weakened network structure. Another consideration is the roughness of the particles; a rougher surface exposes additional surface area to allow physical or chemical adsorption to occur. The viscosity profile and viscoelastic properties, then, will reflect changes in these properties since they reflect the interactions within a slurry.

Finally, the binder MW has an influence on the rheological properties of a slurry. Longer binder chains, i.e., higher MW binders, will increase the degree of bridging in the soft solid structure [8,9]. As such, the slurry will have a higher LSV and yield stress since greater stress is needed to destroy the structure in the material. Higher MW binders will require an additional solvent for proper dispersion, meaning the maximum solid loading is slightly lower for those with high MW binders.

3.3.2 Ratio between materials and total solid loading

The ratio between materials in an electrode slurry, namely the active material, binder, and conductive additive, is perhaps the most important parameter of the slurry to optimize. Early academic research used rough mass percentages of 80–85% active material, 7.5–10% conductive additive, and 5–10% binder in electrode formulations [10–13]. This recipe is undesirable for manufacturing since the active material fraction is relatively low, leading to poor energy density. More recently, it is common to see studies using a 90 wt% active material, 5 wt% conductive additive, and 5 wt% binder recipe [5,14–17]. While this is still a lower active material mass fraction than what is typically seen in manufacturing [18], this is a more suitable baseline from which to develop electrode slurries. In recent years, studies that are focused on electrode manufacturing have increased the active mass fraction above 90 wt% [19–21].

There is also a stark difference in total solid loading that is seen academically and what is seen industrially. The solid loading in academic papers tends to be in the 30–50 wt% range, depending on the exact ingredients used, while industrially produced slurries have solid loadings >50 wt% [18]. There is an increasing number of academic papers using mixing equipment more analogous to industrial sized mixers, so it is becoming more common to see higher solid loadings in slurries in academic papers. Undoubtedly, this is a desirable trend that will assist in closing the gap between academia and industry [18,22].

From a cost perspective at manufacturing scale, it is desirable to increase the solid content of the slurry as much as possible. A high solid content means a low solvent content, meaning less factory floorspace and energy will be needed to dry out and recover the relatively small amount of solvent from the electrode [23]. Another advantage of this shorter drying time is a reduced possibility of binder and/or conductive additive migration, which refers to the diffusion of these components to the surface of the electrode during drying. This migration phenomenon is discussed further in Chapter 8. There are a couple of main causes for this difference in electrode composition and solid loading at lab-scale and production-scale. First, the physics of mixing in laboratory-scale mixers and larger-scale mixers are different, and more solvent may be required in smaller batches to achieve a reproducible slurry. Second, with the exception of papers dedicated to electrode processing, research tends to disregard the effect of slurry mixing in order to study another parameter of the electrode. While it is understandable to do this in order to minimize the number of variables in a particular study, it does lead to a gap between what is achievable in a lab and what is achievable at scale. For a particular technology, this gap must eventually be closed to lead to a commercial product.

With this in mind, how does one determine the best composition of an electrode? On the one hand, through an elaborate game of guess-and-check, one can land on a slurry composition that works well. On the other hand, with a deep understanding of the fundamentals of the interactions between electrode components, one could scheme a balanced ratio between components. In reality, an educated trial-and-error strategy is the most efficient approach.

The first principle of determining an electrode composition and thus the ratio between components is to maximize the active material fraction in the slurry. By doing so, the energy density is as high as possible at the electrode level, which is beneficial as many current and emerging applications demand higher energy density. The second principle is that it is more useful to design recipes around the volume fraction of the materials rather than the mass fraction of the materials. While most studies report on mass fraction due to the relative ease of determining mass instead of volume, the mass fraction provides no information on how much space the components will constitute in the slurry. For instance, LFP and $LiNi_xMn_yCo_{1-x-y}O_2$ (NMC) are among the most common cathode active materials, but LFP is significantly less dense and has a smaller particle size than NMC. Using the same formulation by mass for these two electrodes will lead to noticeably different interactions between the conductive additive, binder, and active material, since there will be greater LFP surface area than NMC surface area. This total surface area could also be influenced by particle shape, as explored in Section 3.3.1. The same can be said for different grades of carbon black and novel conductive additives, such as carbon nanotubes.

That said, the problem of slurry composition design can be simplified slightly to a question of ratio between the conductive additive and binder. It has been shown that a high conductive additive-to-binder ratio may lead to excessive conductive additive agglomeration and thus require a higher fraction of conductive additive and binder in the electrode to reduce impedance [24]. Though it may sound counterintuitive, it could be beneficial to have a relatively low conductive additive-to-binder ratio so that the overall fraction of the two components may be minimized.

An interesting approach that allows one to examine more relevant properties than simply the mass of each component is to optimize the ratio of binder to the combined surface area of conductive additive and active material [25]. Some binder coverage on the particles is necessary to provide adequate cohesion between electrode components, however binder strands block Li-ions and add unnecessary impedance to the electrode. Although they comprise a relatively small amount of the electrode, conductive additives have very high specific surface areas ($40–60\ m^2 \cdot g^{-1}$) as measured by Brunauer–Emmett–Teller analysis, especially compared to cathode active materials ($\leq 15\ m^2 \cdot g^{-1}$, often far less for oxide materials like NMC or synthetic graphite flakes). In addition to surface area, one should also consider the amount of carbon in the electrode slurry, as binders interact differently with carbon-rich surfaces as opposed to carbon-deficient surfaces.

The other primary consideration in slurry design is the total solid loading of the slurry. As mentioned previously, the goal is to maximize the solid loading without compromising the performance of the product or the compatibility of the slurry with downstream manufacturing processes. Ultimately, the interactions between and size of the slurry components will determine the achievable solid loading. At low solid loadings, the particles and binder strands have more spacing on average; as the solid loading increases, this average spacing is reduced and the interactions between components increases. These increased interactions lead to a higher slurry viscosity and a more gel-like (and eventually solid-like) slurry.

From the preceding discussion, all else being equal, a higher slurry LSV is beneficial to provide resistance from sedimentation. Indeed, slurries with higher solid loadings have been observed to have a higher LSV and be more stable again gravitational settling [26]. Due to the opacity of electrode slurries, examining the bridging flocculation network formed as a function of solid loading is extremely challenging, though studies have drawn conclusions from images of dried electrodes suggesting that a stronger network is formed in slurries with higher solid loadings [26]. Given that this observation is in line with the viscosity profile and viscoelastic properties of the slurry, and the increased interactions between components in a tighter, more-loaded volume, this is a logical conclusion. The achievable solid loading is also a function of the mixing protocol used, as will be examined in Section 3.5.2.

3.4 Influence of ambient conditions on dependent properties

In addition to dependent properties, which are selected by the manufacturer or researcher when developing a slurry recipe, the ambient conditions of the environment will have an influence on the independent properties of the slurry. In particular, the humidity and temperature of the environment can significantly alter the rheological profile of the slurry, though there are only a few studies on the topic.

Cathode slurries processed in *N*-methyl-2-pyrollidone (NMP) are best processed in a dry environment to minimize degradation of the cathode active material. This condition is true throughout much of the electrode manufacturing process, from material storage to cell assembly. A dry environment is especially important for Ni-rich layered cathode materials. When exposed to moisture, Ni^{3+} spontaneously reduces to Ni^{2+}, which eventually leads to decomposition products such as LiOH and Li_2CO_3 on the surface. These products may also remain on the surface from synthesis, as a stoichiometric excess of Li is needed to ensure a good, layered structure of the Ni-rich material. Several studies have shown that residual LiOH causes gelation of the slurry, specifically due to the polymerization of polyvinylidene fluoride [27–30]. A gelled slurry will be too viscous for coating, regardless of the mixing protocol, meaning that Ni-rich materials must be pretreated to remove residual LiOH and Li_2CO_3 from the surface [29].

The temperature of mixing can also have an influence on the rheology of the slurry. Predictably, at medium and high shear rates, increasing the temperature leads to a decrease in viscosity [5]. This is because the particles in the slurry have greater kinetic energy at high temperatures, leading to less resistance to flow. Interestingly, at low shear rate, the viscosity of the slurry increases with temperature since the bridging structure is more robust due to the formation of additional binder entanglements [5]. This correlates with an increase in the yield stress of the slurry, as more force is needed to disrupt the strengthened binder network at higher temperature [5]. Above a certain temperature, the LSV may decrease, likely due to the additional kinetic energy of the particles overcoming the strengthened binder network [5]. Increasing the temperature from 25°C to 50°C can lead to a roughly four-fold increase in the low-shear (0.01 s^{-1}) viscosity of an NMP-based cathode slurry, while the same temperature increase causes a roughly 30% decrease in the high-shear (1,000 s^{-1}) viscosity [5]. Of course, these values depend on the exact slurry recipes and mixing procedures. It is common for slurry mixers to have some method of temperature regulation, since high-energy mixers generate heat during the mixing process that must be considered.

Temperature also affects the surface tension of the slurry. The dependency of the surface tension of a liquid on temperature can be modeled using the Eötvös equation (see (3.4)):

$$\varsigma = \frac{k(T_{crit} - T)}{V^{2/3}} \tag{3.4}$$

In (3.4), ς is the surface tension of the slurry ($N{\cdot}m^{-1}$), k is the Eötvös constant ($2.1{\cdot}10^{-7}$ $J{\cdot}K^{-1}{\cdot}mol^{-2/3}$), T_{crit} is the critical temperature (K), T is the ambient temperature (K), and V is the molar volume ($m^3{\cdot}mol^{-1}$). The ratio between the viscous forces and the surface tension forces during flow is referred to as the dimensionless Capillary (Ca) number (see (3.5)) [31]. The Capillary number is also a function of the coating speed (v) and may be used to develop slot-die coating windows (see Chapter 4).

$$Ca = \frac{\mu v}{\varsigma} \tag{3.5}$$

3.5 Mixing instrumentation and protocols

3.5.1 Mixing instrumentation

At the pilot and industrial level, electrode slurries are commonly mixed in batches using double planetary mixers. This mixer works by stirring the slurry with a high-speed impeller that is dropped into the center of the slurry-filled mixing vessel. Simultaneously, a set of scrapers wipe the edges of the mixing vessel to clear any solids that may be sticking to the walls of the vessel and to redeposit them closer to the center of the mixer. The optimal design of the scrapers and impeller depends on the size of the mixing vessel and can change the mixing physics and, thus, the quality and properties of the slurry. Depending on the sizing of the mixers and output of the facility, the number of planetary mixers in a facility can vary significantly. A continuous mixing process is also possible, though is not as common at this time. Continuous mixing comes with varying advantages and disadvantages over the batch process and is discussed in detail in Chapter 9.

Alternative mixing methods have been investigated at the R&D level. One of the most common methods is hydrodynamic shear mixing, which consists of a high-speed impeller dipped into the slurry; this mixing mechanism mimics planetary mixing without the scrapers. Since R&D mixing occurs in smaller batches, wiping the edges of a smaller mixing container with a spatula at a few intervals during mixing is sufficient for the purposes of obtaining a few hundred mL of slurry, though insufficient for mapping an exact process to be used in manufacturing. Another method of preparing slurry is ball milling. Ball milling applies a consequentially higher amount of energy to the slurry than shear mixing in a short amount of time, making it a time-efficient method of breaking up agglomerates. However, ball milling can cause fracture in some active materials and binders and should be used with caution.

A third mixing technique to shear mixing and ball milling is ultrasonic mixing. In ultrasonic mixing, high-frequency acoustic waves are applied to a slurry and cause pressure fluctuations in the slurry. These pressure fluctuations can cause air bubbles to form when the oscillating pressure decreases below the vapor pressure. When the bubbles collapse, they impart significant energy to the slurry and serve as a means of mechanical stirring. This stirring is very energy-efficient and is effective at breaking up agglomerates in an electrode slurry [32]. However, the process has not been shown to be scalable since the intensity of bubble cavitation cannot be replicated throughout a mixer large enough to be of interest for manufacturing [18]. By reducing the frequency of the acoustic waves, the range of mixing can be increased at the expense of some mixing energy. However, since the energy of mixing in the ultrasonic range is sufficient, this trade-off is worthwhile for making the process more scalable. Acoustic mixing could be most effective in the frequency range of 58–62 Hz [33].

The three mixing mechanisms described in this section are compared in the spider chart in Figure 3.3. Each mixing mechanism was assigned an arbitrary score of 1–10 that was informed by data from the literature [32]. As the frequency is decreased from the ultrasonic range, the scalability score increases for acoustic mixing at a small trade-off from the other three categories. Unsurprisingly, the high

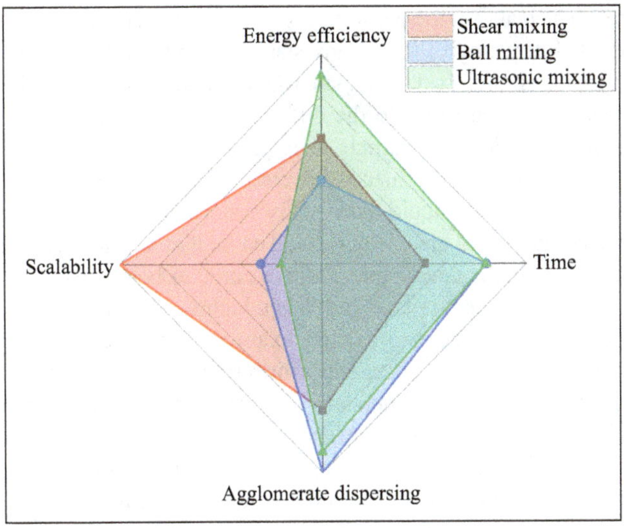

Figure 3.3 Spider chart comparing the three slurry mixing mechanisms discussed. Data from [32] informed decisions on energy efficiency and agglomerate dispersing.

scalability score of shear mixing (i.e., the mixing mechanism used in planetary mixers) enables its widespread use. It is also a method with adequate dispersing ability and energy efficiency, though it could improve in these categories. Also, shear mixing on the industrial scale lasts at least several hours and is followed by a critical degassing step that can require an additional 1–2 h. Slurry degassing is necessary to remove air bubbles from the slurry; if air bubbles are present in the slurry, these bubbles can burst after being coated onto the current collector and lead to blister defects that can harm product performance and consistency [34].

3.5.2 Mixing protocols

As mentioned in Section 3.3.1, the differences in properties between active material particles and conductive additive particles are significant. In any mixing protocol, the viscosity of the slurry is the highest when mixing begins (until more solids are added, if the protocol requires solid addition). Over time, the viscosity decreases logarithmically, with a steady-state viscosity being reached after the slurry is sufficiently mixed and the viscosity decreases little with time. In fact, mixing beyond the point at which the steady-state is reached in a planetary mixer is a waste of time, energy, and cost for the manufacturer [35], and can negatively impact the slurry if the heat generated from mixing is unaccounted [36].

The time required for a slurry to reach steady-state viscosity is a function of the order in which ingredients are added to the slurry. For example, when the conductive additive is added to the binder solution, held under vacuum, then the active material is added and mixed, the slurry requires about half the time to reach steady-state as a slurry in which the order of the conductive additive and active material addition are reversed [37]. The vacuum step allows the particles added to the binder solution in the first step to absorb liquid. Interestingly, the steady-state viscosity of the slurry with active material added first is slightly lower than when conductive additive is added first. These differences can be ascribed to the differences between the active material and conductive additive particles in particle size, particle size distribution, and specific surface area, all of which may influence solvent uptake [37]. For example, the active material particles may release liquid more easily than conductive additive particles due to their lower specific surface area, which could cause it to take longer to reach steady-state [37]. Additionally, the conductive additive particles being added in the vacuum step may lead to more solvent uptake and thus a higher viscosity.

The logical mixing sequence to first examine is the sequence in which all slurry components are added at once, henceforth referred to as one-step mixing (see Figure 3.4(a)). It is common for one-step mixing to include a pre-mixing sequence of binder solution, though this is not necessarily always the case. A common modification to one-step mixing is to dry mix the active material and conductive additive particles separately from the binder solution, then introduce the dry solids to the polymer solution, henceforth called the dry mixing sequence (see Figure 3.4 (b)). A third mixing procedure involves pre-making a concentrated binder solution, adding the active material and conductive additive particles, then stepwise adding additional solvent; this procedure will be called stepwise solvent addition (see

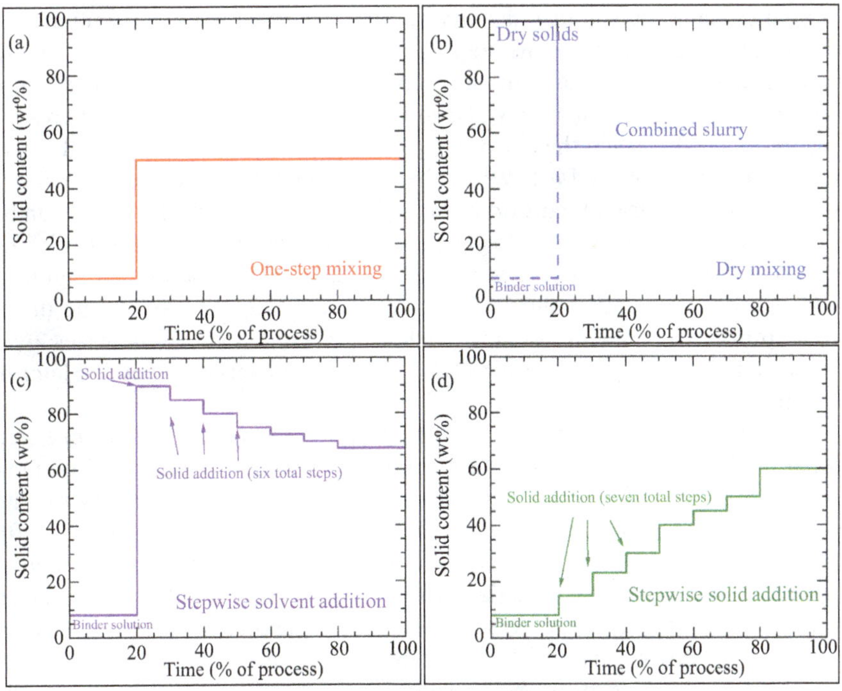

Figure 3.4 Slurry solid content development as a function of process time in the four mixing protocols discussed: (a) one-step mixing, (b) dry mixing, (c) stepwise solvent addition, and (d) stepwise solid addition. It should be noted that the values used for solid content, time, and number of solvent/solid addition steps in these processes are strictly for the purpose of process illustration and do not necessary represent an optimal process for slurry mixing. In other words, there is room to modify these parameters in these processes, though they will still follow the same scheme.

Figure 3.4(c)). The stepwise solvent addition can be modified to include stepwise addition of conductive additive and solvent to the concentrated binder solution, followed by stepwise addition of active material and solvent to the well-mixed conductive additive/binder slurry [20,38]. This mixing procedure will be called stepwise solid addition (see Figure 3.4(d)). These four mixing protocols are shown in Figure 3.4. The mixing protocols are illustrated by showing how the solid content changes as a function of the process mixing time in each protocol, though the protocols shown in Figure 3.4. It should be noted that the exact values of time, solid content, and, in the case of stepwise solvent addition and stepwise solid addition, number of addition steps, are only to be taken as an example. The exact values for time, solid content, and number of addition steps will vary based on the numerous factors discussed throughout this chapter.

Studies have shown that the dry mixing sequence causes a remarkable reduction in viscosity from one-step mixing across all shear rates [37,39,40]. The outcome of this is that, for the same slurry components and mass ratio between those slurry components, the solid loading of a slurry can be increased by as much as 34% when the dry mixing strategy is implemented instead of the one-step mixing strategy [40]. When the solid content is equal for a one-step mixed slurry and a dry mixed slurry, the dry mixed slurry has a LSV that can be up to two orders of magnitude lower than the one-step mixed slurry [40]. The stepwise solvent addition slurry is capable of a higher solid loading than the stepwise solid addition slurry. The solid loading decreases over time in the former process as solvent is added, while the solid loading increases over time since the majority of the solids (i.e., the active material) are added later in the process [20].

The microstructure of the slurry is also different when mixed via the one-step mixing or dry mixing processes. In a frequency sweep, one-step mixing commonly shows $G' > G''$ across a wide angular frequency range while dry mixing shows $G' < G''$ at moderate-to-high (>1 rad·s^{-1}) angular frequencies [40]. Slurries made with stepwise solvent addition have $G' < G''$ across a wide angular frequency range because the initial mixing is done in a highly viscous state, meaning a higher shear rate was needed during mixing which more readily destroyed the microstructure of the slurry [39]. Additionally, the absolute value of G' in the dry mixed slurry is about two orders of magnitude lower than for the one-step mixed slurry [40]. Slurries made with stepwise solid addition are more gel-like than slurries made from stepwise solvent addition, likely due to how the solid content changes over mixing time [20]. Sequential solvent addition has an advantage over sequential solid addition in that it allows for a greater final solid content, leading to a quicker drying process with less risk of binder and/or conductive additive migration. Meanwhile, the sequential solid addition has the advantage in viscoelasticity and LSV, leading to a slurry more stable against agglomeration. As with all aspects of electrode fabrication, it is up to the user to assess which trade-offs are to be made to provide the most value for their particular product needs.

Predictably, the impact of these vastly different rheological properties influences the electrode properties. One-step slurry mixing leads to poorer electrochemical performance in a dried cathode than the other three discussed strategies due to better dispersion of the cathode components [20,37,39,40]. When coated at the same shear rate, the dry mixed slurry has a dry thickness roughly 20% higher than that of the one-step mixed slurry [40]. This was a product of the higher solid loading enabled by the dry mixing strategy compared to the one-step mixed slurry [40]. Still, the dry mixed slurry and one-step mixed slurry had very similar viscosity profiles across a relatively wide shear rate range (0.1–200 s^{-1}), although there was a sizable difference in the electrochemical performance. This further supports the claim that while rheological properties are important to understand and measure, they alone are not reliable predictors of the performance of the finished electrode.

3.6 Characterization techniques

While there have been tremendous gains in our understanding of the slurry preparation process, including what factors influence the properties of the slurry and how to interpret the properties of the slurry, there are endless combinations of slurry constituents, ratios and properties of these constituents, solid loadings, and mixing protocols. Because of this, and as new materials are developed, it is important for academics and manufacturers to understand how slurries may be characterized.

3.6.1 Viscometer and rheometer configuration

Most viscometers and rheometers for electrode slurry characterization can be classified as rotational rheometers, in which a sample is placed between two surfaces, one of which is rotated and one of which remains stationary. The three classes of rotational viscometers and rheometers are parallel plates, cone-and-plate, and coaxial cylinders. Sometimes, parallel plate configurations are referred to as parallel disks. The coaxial cylinder configuration is also known as concentric cylinders and bob-and-cup. Across industries, parallel plate instruments are usually used for extremely viscous samples, while coaxial cylinder configurations are used for low-viscosity samples, with cone-and-plate instruments being a middle ground.

While only a few mL of sample are required for trials with parallel plate and cone-and-plate configurations, coaxial cylinders tend to require at least a few tens of mL of sample. Finally, in terms of keeping the instrument clean, parallel plate and cone-and-plate configurations are relatively simple, while the coaxial cylinders require more effort. This is because the diameter of the cup is typically low in order to keep the volume of sample low, which comes at the expense of making the instrument more challenging to clean. For electrode slurry characterization, coaxial cylinder configurations are used most often, with cone-and-plate configurations also sufficing in high solid loading cases. More detailed information on the three types of viscometers and rheometers may be found in more general texts about rheology [3]. Table 3.1 compares the important characteristics of the three configurations.

Table 3.1 A comparison of parallel plate, cone-and-plate, and coaxial cylinder configurations across key characteristics

Characteristics	Parallel plates	Cone-and-plate	Coaxial cylinders
Viscosity of samples	High	Medium	Low
Sample volume required*	Few mL	Few mL	Tens of mL
Cleaning and maintenance	Easy	Easy	Moderate

*Assuming no small sample adapter is used, which can reduce the sample required by an order of magnitude.

As mentioned in Section 3.2.1, viscometers may be capable of single-point or multi-point measurements. A single-point viscometer can apply a single rotation speed to a slurry and, knowing the geometry of the spindle and measuring the torque needed to turn the spindle submerged in the fluid, can determine the viscosity. Multi-point viscometers work similarly, except that they may be programmed to logarithmically sweep the rotational speed applied and measure the viscosity at multiple points. When discussing viscometers in this book, we will be referring specifically to multi-point viscometers, as single-point viscometers are not as useful in studying electrode slurries. Rheometers are a sophisticated type of multi-point viscometer, capable of a broader range of rotational speeds and able to perform the viscoelasticity measurements described in Section 3.2.2. It should be noted that rotational speeds (in rpm) can be converted to shear rates (in s^{-1}) depending on the properties of the spindle used in measuring viscosity and that the manufacturer should have a simple conversion factor for the user. Table 3.2 compares the characteristics of single-point viscometers, multi-point viscometers, and rheometers.

Although rheometers allow for the most comprehensive analysis, they can be 3–10 times as expensive as a viscometer and are more difficult and expensive to service given their more complex design. A quality multi-point viscometer can provide a good viscosity curve, although one that is more limited in shear rate range than one from a rheometer. A spindle designed for a low-viscosity sample, such as a binder solution, could apply a shear rate in the range of 0.1–100 s^{-1}, while one designed for a higher-viscosity sample, such as a slurry, could apply a shear rate in the range of 0.01–10 s^{-1}. Meanwhile, a rheometer could apply a shear rate range of 0.001–1,000 s^{-1} with a single spindle and provide a more complete viscosity curve.

Regardless of the instrument selected, having control over the temperature in a rheological measurement is very important, as temperature has been shown to have an effect on rheological properties [5]. The temperature of viscometers and

Table 3.2 A comparison of single-point viscometers, multi-point viscometers, and rheometers across key characteristics

Characteristic	Single-point viscometer	Multi-point viscometer	Rheometer
Range of shear rates	0.01–100 s^{-1}, one at a time	0.001–10,000 s^{-1}, can sweep between about four orders of magnitude of shear rate depending on instrument/spindle choice	0.001–10,000 s^{-1}, can sweep between about six orders of magnitude of shear rate
Capable of viscoelasticity measurements?	No	No	Yes
Cost*	\$500–\$3,000	\$5,000–\$20,000	\$50,000–\$150,000

*The cost metric is based on new units. The ranges are kept broad since there are many add-ons or features to instruments that could influence cost.

rheometers can be maintained by jacketing the sample cup and heating and cooling the sample with a circulating water bath or by using Peltier plates to thermoelectrically heat and cool the sample. While Peltier plates can heat and cool samples more rapidly than a water bath circulation system, they come at a higher cost. In addition to temperature, instruments may come outfitted with small sample adapters. These are useful, particularly at the R&D level, when slurry batches may be small.

3.6.2 Fitting and modeling viscosity data

There exist a few models to fit viscosity curves. The most general model is the simple power law model, as shown in (3.6):

$$\sigma = k\dot{\gamma}^n \tag{3.6}$$

In (3.6,) k is the consistency index and n is the dimensionless flow index. The consistency index is a function of viscosity; more viscous fluids will have a high consistency index and less viscous fluids will have a low consistency index. Essentially, the consistency index scales as viscosity increases for constant flow index [5]. Meanwhile, the flow index can be used to classify a fluid (see Section 3.2.1). An alternative definition of a Newtonian fluid is a fluid whose stress v. strain curve behaves according to (3.6) when $n = 1$. Similarly, a shear-thinning fluid behaves according to (3.6) with $n < 1$ and a shear thickening fluid behaves according to (3.6) with $n > 1$. However, as discussed in Section 3.2.1, most fluids have a yield stress, including electrode slurries. A simple modification to the power law index is to calculate the shear stress by adding the yield stress (σ_y) to the power law term (see (3.7)):

$$\sigma = \sigma_y + k\dot{\gamma}^n \tag{3.7}$$

Equation (3.7) is known as the Herschel–Bulkley equation or Herschel–Bulkley model. The yield stress can either be measured (see Section 3.6.4) or fit to a data set. The consistency index and flow index have the same meanings as in the power law model. These parameters, in addition to the yield stress, have been used to analyze the effect of temperature [5], solid loading [41,42], and storage conditions [43] on the rheology of a slurry.

After obtaining a viscosity curve from a viscometer or rheometer, there are a few reasons why one would want to fit that data to a model. The first is that the instrument may be limited in the shear rate that can be applied to the slurry and it would be useful to predict the viscosity at a higher or lower shear rate. Then, the viscosity of the slurry could be measured at the desired higher or lower shear rate, either validating the model or requiring a revision. Once a model is established for a given slurry recipe and mixing protocol, a viscometer that can only measure viscosity across a relatively narrow shear rate range is no longer restrictive.

Another reason one would be interested in modeling viscosity data is that fit parameters can summarize information about the curve. As seen in the two models presented, the flow index can provide a single value that describes the degree of shear thinning to the user. When repeating multiple viscosity tests on a multi-liter batch of slurry, it is easier for a data analyst to have single values to compare as opposed to examining multiple curves. When considering all the characterizations performed and data generated in electrode fabrication, simplifying the amount of data and the time needed to process and analyze that data is vital.

3.6.3 Rheological testing protocols

At the R&D level, it is useful to combine viscosity and viscoelastic testing of the same sample to have a complete rheological profile. However, running a flow ramp test (used to measure viscosity), followed by an amplitude sweep and a frequency sweep, is not advised for two reasons: (1) the slurry will be subject to hysteresis effects that cause measurements to be inaccurate, and (2) the strain selected for the frequency sweep must be within the LVR of the slurry. The latter issue is simple to resolve, since a small amount of sample may be used in a preliminary amplitude sweep to determine the LVR. Most slurries at the R&D level will be within the LVR at a strain of 0.1%, so this figure is generally advisable to use. For new formulations, however, a preliminary amplitude sweep is recommended.

Hysteresis effects, on the other hand, require special consideration. First, consider that there will be differences in how a sample is transferred from the mixer to the viscometer or rheometer. These differences include the time of transfer and ambient conditions and are impossible to totally normalize across multiple samples. To ensure that all slurries begin from a similar starting point, a brief (~1 min), gentle (~5 s^{-1}) pre-shear process can be used. The pre-shear process should also include time for the temperature of the slurry to acclimate to its new environment in the viscometer or rheometer, especially if the test is performed at a temperature less than or greater than room temperature.

Next, let us assume that after a pre-shear, a flow ramp test is performed. If the flow ramp test reaches a sufficiently high upper cutoff shear rate (several hundred s^{-1} or a couple thousand s^{-1}, for instance), the structure of the slurry will be completely destroyed. Therefore, performing an amplitude or frequency sweep immediately after the flow ramp test will not provide one with any information on the structure of the slurry, as is desired. To remedy this, soft perturbations may be applied to the slurry to rebuild its soft solid structure in a step called an oscillation time sweep. In an oscillation time sweep, an oscillation strain and an angular frequency are selected, and the rheometer acts on the sample according to these pre-determined figures. As with the frequency sweep, the oscillation strain selected must be within the LVR of the slurry, meaning a preliminary amplitude sweep may be necessary. As the oscillation time sweep proceeds, the structure will slowly be rebuilt. This is evidenced by a rising storage modulus over time, meaning the slurry is absorbing increasingly more energy as it is perturbed. Eventually, the storage modulus value will no longer increase with time, which is an indication that the slurry structure has been rebuilt. At this point, the oscillation time sweep may be ended, and the next test may be performed.

Oscillation time sweeps may be necessary to perform after flow ramp, amplitude sweep, and frequency sweep tests, until all characterization of the slurry has finished. Of course, the flow ramp test is far more destructive to the slurry than either the amplitude sweep or the frequency sweep, and the amplitude sweep is usually more destructive than the frequency sweep. The oscillation time sweeps will be longer for samples that are more damaged. If an incomplete structure rebuild occurs prior to an amplitude sweep, the storage modulus will actually increase as the strain is increased within the LVR, since these mild perturbations continue to rebuild the structure. An alternative to these hysteresis-combatting measures is to use a different sample of the slurry for the various tests one wishes to run. However, considering that the amount of slurry prepared at the R&D level is generally low, this could be difficult to realize in practice. This practice also introduces variation that could be caused by differing samples taken for each test. It is uncommon to see a rheology procedure written out to this level of detail in the literature, though some recent studies have increased the clarity of their rheological procedures [5,44].

Rheometers are sophisticated tools that provide a user with a wealth of information about the fluid being tested. However, in large-scale manufacturing, there are some issues with using rheometers. First, large vats of slurry can mix and store hundreds of liters of slurry, meaning that a 20 mL sample of that slurry tested in a concentric cylinder configuration is unlikely to be representative of the flow properties of the entire batch. Second, obtaining a small sample of slurry may require an interruption in the mixing process to retrieve. Third, though viscosity curves can be obtained in <10 min, oscillation time sweeps, amplitude sweeps, and frequency sweeps can cause the entire characterization to last longer than an hour. Fourth, for high-speed coating operations, shear rates in excess of 10,000 s^{-1} are possible [2]; most rheometers are not equipped to test at these high shear rates. Because of these limitations, such in-depth rheological analysis is scarcely seen in large-scale production [18].

3.6.4 Measuring yield stress

The yield stress of a fluid can either be estimated from fitting with a model, such as the Herschel–Bulkley model, or measured using a flow ramp test. As mentioned, flow ramp tests are used to produce shear stress vs. shear rate and viscosity vs. shear rate curves. The input parameters of a flow ramp test are the beginning shear rate, the ending shear rate, and the time for the test. The shear rate may be ramped linearly or as is done more often, logarithmically, between the shear rate limits for the given time. A wide range of shear rates is desirable to understand the material properties at various flow conditions. In particular, if the shear rate is sufficiently low ($<10^{-2}$ s^{-1}), it is possible to estimate the yield stress. In a double log shear stress vs. shear rate plot, there will exist a range of shear rates in which the shear stress is constant (or increases at a very low rate). The shear stress at shear rates outside of this range, either higher or lower, will be a much larger function of the shear rate. The average shear stress measured within this constant-stress region may be estimated as the yield stress (see Figure 3.5). This estimation is made based on the definition of yield stress, which is the minimum stress required to induce flow

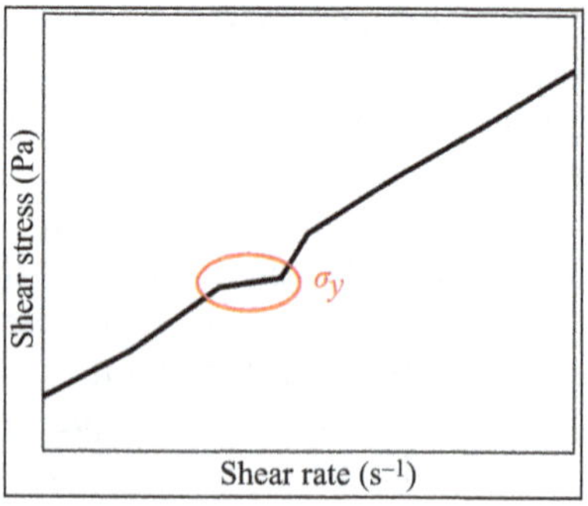

Figure 3.5 Yield stress estimation from a flow ramp test

in the slurry. Additional stress will not develop in the fluid until it begins flowing, causing the stress to be constant within a particular shear rate range.

Another method for determining yield stress is from an amplitude sweep. From the perspective of the fluid's microstructure, flow is induced once the network is broken down, which may be indicated by the end of the LVR. The yield stress may be calculated by multiplying the modulus at the end of the LVR by the critical strain [42]. There has been good agreement between the yield stress values determined from flow ramp and amplitude sweep tests with yield stress data fitted using models like the Herschel–Bulkley model being slightly different [42]. In addition, yield stress calculated from models is disconnected from a physical response from the material, unlike that from flow ramp and amplitude sweep experiments. While the latter is preferable, the constraints of the instrument used may require an approximation of the yield stress from a model (see Section 3.6.2).

3.6.5 Measuring wettability

In many scenarios, the easiest characterization of the wettability of a liquid to a solid is contact angle goniometry. In contact angle goniometry, either the surface tension of the liquid or the surface energy of the solid is to be determined with the other value being known for several test materials. First, a droplet of liquid is carefully dropped onto a surface. A light source and camera are used to capture an image of the shadow of the droplet immediately after it contacts the surface. Many contact angle goniometers come with software that can measure the contact angle on both sides of the droplet. It is important to repeat this measurement several times, as the contact angle may change within milliseconds of contacting the surface, which can lead to inaccurate measurements.

After collecting several measurements of a single liquid onto a single solid, the entity of unknown surface tension or energy remains while the entity of known surface tension or energy is replaced with another material of known surface tension or energy. For example, if the surface energy of a solid is to be determined, the liquid will be replaced with another liquid of known surface tension, and the contact angle will be measured between the solid and the new probe liquid. It is generally advised to use probe liquids with a variety of surface tensions, including those with different polar (σ_p) and dispersive (σ_d) components. Examples include water (σ = 72.8 mN·m^{-1}, σ_p = 46.4 mN·m^{-1}, σ_d = 26.4 mN·m^{-1}), diiodomethane (50.8 mN·m^{-1}, σ_p = 0.0 mN·m^{-1}, σ_d = 50.8 mN·m^{-1}), and ethylene glycol (σ = 47.7 mN·m^{-1}, σ_p = 21.3 mN·m^{-1}, σ_d = 26.4 mN·m^{-1}) [45]. The total surface tension (or energy) is calculated as the sum of the polar and dispersive components of the surface tension.

Once the contact angle between several probe materials and a static, unknown material have been collected, the Fowkes equation (see (3.8)) may be used to calculate the surface tension of the unknown material:

$$\left(\varsigma_L^D\right)^{0.5}\left(\varsigma_S^D\right)^{0.5} + \left(\varsigma_L^D\right)^{0.5}\left(\varsigma_S^P\right)^{0.5} = \frac{\varsigma_L(1+\cos\theta)}{2} \tag{3.8}$$

Continuing with the previous example, the surface energy of the solid will be known and the contact angle will be measured. The surface tension of the liquid can be re-written according to (3.9):

$$\varsigma_L = \varsigma_L^D + \varsigma_L^P \tag{3.9}$$

Therefore, the Fowkes equation will have two unknown variables and can be written in slope-intercept form. Using the data from several probe liquids, a trend can be drawn to calculate the dispersive and polar components of the surface energy of the solid. If probe solids are used and the surface tension of the liquid is unknown, then the same methodology applies, and the surface tension of the liquid can be calculated.

Previous studies have shown that the surface tension of the slurry is primarily a function of the solvent selected with the binder having some influence. The active material and conductive additive, then, have little impact on the wettability of the slurry to the current collector. Generally, water-based electrodes have poorer wettability to the current collector, given the significant difference in surface tension between NMP (40.79 mN·m^{-1}) and water (72.80 mN·m^{-1}) [46]. The challenges of water-processed cathodes are explored further in Chapter 10. It is worth mentioning that it could be challenging to measure the contact angle of slurry on current collector as the slurry is highly concentrated making it difficult to get a small droplet with a perfect round shape.

3.6.6 Agglomeration characterization

Electrode slurries are an opaque, black suspension that cannot easily be visually inspected; therefore, detecting particle agglomeration or sedimentation is nontrivial. Techniques such as dynamic light scattering, which are used in clear, dilute colloidal suspensions to measure particle size, do not work on electrode slurries. Agglomerated cathode particles that are not dispersed in the slurry stage have been linked to an inhomogeneous coating, causing inhomogeneous electrical field distribution, as well as by creating scratches in the electrode during coating [34]. Agglomerated conductive additive particles lead to regions of poor electrical connection [34].

Agglomerated particles can sometimes be seen in a dried coating by visual inspection or could be seen in a scanning electron microscopy image, though it would obviously be advantageous to detect them prior to coating to minimize waste. A simple instrument that can approximate the size of the largest agglomerates in a slurry is a Hegman gauge (sometimes also called a grind gage). To measure the agglomerates, a sample of slurry is placed at one end of the Hegman gauge and drawn down the length of the gauge using a flat edge. Down the length of the gauge are grooves that correspond to different particle sizes. Once the slurry is drawn down the gauge, the deposits of slurry within the grooves are inspected. The user will notice a groove with a peppery deposit which corresponds to the largest particle size evident within the sample.

3.7 Summary

As the first step in the electrode preparation process, proper slurry mixing is essential. Though mixing is a relatively inexpensive step, poorly mixed slurry can have downstream impacts that may not be detectable until much later in the process, leading to wasted time and material. Before mixing, slurry design is an important procedure, as the size, shape, and quantity of materials all influence how the slurry behaves. Another aspect of design is understanding how the slurry components interact with one another and optimizing the order in which they are added to the mixture and the intensity of mixing as ingredients are added. It is apparent that optimizing each of these parameters is necessary for low-cost, effective mixing. This optimization work is best done at the pilot-scale level with industrial equipment before being scaled up into large-scale manufacturing [47].

Unfortunately, due to the large volumes of slurry produced and the opacity of slurry, characterizing slurry at scale is very difficult. Generally, in-depth pilot-scale characterization is done to verify the reproducibility of a given slurry recipe and mixing protocol. These characterization methods are most commonly rheological techniques, which were discussed in depth in this chapter. Viscosity, which is a dependent property of the slurry, is particularly important due to the depth of information that can be obtained from a single curve. Additional properties, such as the viscoelasticity and wettability of the slurry, can provide information on the microstructure and eventual adhesion of the slurry, respectively, that are not as evident from viscosity data alone.

Though rheological properties provide significant information on the slurry, they are not a good predictor of electrochemical performance. There are too many downstream electrode and cell processing steps (coating, drying, calendaring, slitting, and winding, to name a few) to draw a correlation between viscosity and electrochemical performance. Still, slurry characterization is an essential tool in developing a robust, repeatable slurry recipe and mixing protocol at the pilot-scale for large-scale production of LIB electrodes.

Finally, as will be a theme throughout this book, there exist numerous complex trade-offs that teams must assess to design the best electrode slurry and mixing procedure for their needs. One recipe may be cost-effective but may hinder power density. One mixing process may allow for high solid content but may compromise the stability of the slurry and make it less reproducible. Modifying just one seemingly small aspect, such as an active material with a different particle size distribution, a binder with a different molecular weight, or using a mixer with different dimensions, can radically change how the components are distributed and the properties of the resulting electrode. While the literature presented here are useful in demonstrating design principles and showcasing how slurry data may be interpreted, it is up to the manufacturer to determine for their particular set of electrode ingredients and mixing instrumentation what recipe and process is best.

References

[1] Hawley, W.B. and J. Li, Electrode manufacturing for lithium-ion batteries – analysis of current and next generation processing. *Journal of Energy Storage*, 2019, 25, 100862.

[2] Reynolds, C.D., S.R. Peter, S.D. Hare, M.J.H. Simmons, and E. Kendrick, A review of metrology in lithium-ion electrode coating processes. *Materials and Design*, 2021, 209, 109971.

[3] Mewis, J. and N.J. Wagner, *Colloidal Suspension Rheology,* 2011. Cambridge: Cambridge University Press.

[4] Bauer, W. and D. Notzel, Rheological properties and stability of NMP based cathode slurries for lithium ion batteries. *Ceramics International*, 2014, 40, 4591–4598.

[5] Hawley, W.B. and J. Li, Beneficial rheological properties of lithium-ion battery cathode slurries from elevated mixing and coating temperatures. *Journal of Energy Storage*, 2019, 26, 100994.

[6] Kraytsberg, A. and Y. Ein-Eli, Conveying advanced Li-ion battery materials into practice: the impact of electrode slurry preparation skills. *Advanced Energy Materials*, 2016, 6, 1600655.

[7] Hawley, W.B., A. Parejiya, Y. Bai, H.M. Meyer III, D.L. Wood III, and J. Li, Lithium and transition metal dissolution due to aqueous processing in lithium-ion battery cathode active materials. *Journal of Power Sources*, 2020, 466, 228315.

[8] Swenson, J., M.V. Smalley, and H.L.M. Hatharasinghe, Mechanism and strength of polymer bridging flocculation. *Physical Review Letters*, 1998, 81 (26), 5840–5843.

[9] Chu, V., M.-Z. Tsai, Y.-R Chang, T.-J. Liu, and C. Tiu, Effects of the molecular weight and concentration of poly(vinyl alcohol) on slot die coating. *Journal of Applied Polymer Science*, 2010, 116, 654–662.

[10] Clemencon, A., A.T. Appapillai, S. Kumar, and Y. Shao-Horn, Atomic force microscopy studies of surface and dimensional changes in Li_xCoO_2 crystals during lithium de-intercalation. *Electrochimica Acta*, 2007, 52, 4572–4580.

[11] Thorat, I.V., D.E. Stephenson, and N.A. Zacharias, *et al.*, Quantifying tortuosity in porous Li-ion battery materials. *Journal of Power Sources*, 2009, 188, 592–600.

[12] Liu, H., Y. Yang, and J. Zhang, Investigation and improvement on the storage property of $LiNi_{0.8}Co_{0.2}O_2$ as a cathode material for lithium-ion batteries. *Journal of Power Sources*, 2006, 162, 644–650.

[13] Zhuang, G.V., G. Chen, S. Joongpyo, and X. Song, Li_2CO_3 in $LiNi_{0.8}Co_{0.15}Al_{0.05}O_2$ cathodes and its effects on capacity and power. *Journal of Power Sources*, 2004, 134, 293–297.

[14] Davoodabadi, A., J. Li, Y. Liang, *et al.*, Characterization of surface free energy of composite electrodes for lithium-ion batteries. *Journal of The Electrochemical Society*, 2018, 165(11), A2493–A2501.

[15] Mao, C., S.J. An, H.M. Meyer III, *et al.*, Balancing formation time and electrochemical performance of high energy lithium-ion batteries. *Journal of Power Sources*, 2018, 402, 107–115.

[16] Parikh, D., T. Christensen, C.-T. Hsieh, and J. Li, Elucidation of separator effect on energy density of Li-Ion batteries. *Journal of the Electrochemical Society*, 2019, 166(14), A3377–A3383.

[17] Parikh, D., T. Christensen, and J. Li, Correlating the influence of porosity, tortuosity, and mass loading on the energy density of $LiNi_{0.6}Mn_{0.2}Co_{0.2}O_2$ cathodes under extreme fast charging (XFC). *Journal of Power Sources*, 2020, 474, 228601.

[18] Li, J., J. Fleetwood, W.B. Hawley, and W. Kays, From materials to cell: state-of-the-art and prospective technologies for lithium-ion battery electrode processing. *Chemical Reviews*, 2022, 122, 903–956.

[19] Altvater, A., T. Heckmann, J.C. Eser, S. Spiegel, P. Scharfer, and W. Schabel, (N)IR-drying of lithium-ion battery electrodes - Influence of energy input on process speed and electrode adhesion. *Energy Technology*, 2022, 11, 2200785.

[20] Hoffmann, A., E.A. Heider, C. Dreer, C. Pfeifer, and M. Wohlfahrt-Mehrens, Influence of the mixing and dispersing process on the slurry properties and the microstructure and performance of ultra-thick cathodes for lithium-ion batteries. *Energy Technology*, 2022, 11, 2200484.

[21] Shodiev, A., M. Chouchane, M. Gaberscek, *et al.*, Deconvoluting the benefits of porosity distribution in layered electrodes on the electrochemical performance of Li-ion batteries. *Energy Storage Materials*, 2022, 47, 462–471.

[22] Ue, M., K. Sakaushi, and K. Uosaki, Basic knowledge in battery research bridging the gap between academia and industry. *Materials Horizons*, 2020, 7, 1937–1954.

[23] Wood, D.L., J.D. Quass, J. Li, S. Ahmed, D. Ventola, and C. Daniel, Technical and economic analysis of solvent-based lithium-ion electrode drying with water and NMP. *Drying Technology*, 2018, 36(2), 234–244.

[24] Liu, G., H. Zheng, S. Kim, *et al.*, Effects of various conductive additive and polymeric binder contents on the performance of a lithium-ion composite cathode. *Journal of the Electrochemical Society*, 2008, 155, A887–A892.

[25] Weichert, A., V. Goken, O. Fromm, T. Beuse, M. Winter, and M. Borner, Strategies for formulation optimization of composite positive electrodes for lithium ion batteries based on layered oxide, spinel, and olivine-type active materials. *Journal of Power Sources*, 2022, 551, 232179.

[26] Ouyang, L., Z. Wu, J. Wang, *et al.*, The effect of solid content on the rheological properties and microstructures of a Li-ion battery cathode slurry. *RSC Advances*, 2020, 10, 19360–19370.

[27] Cho, D.-H., C.-H. Jo, W. Cho, *et al.*, Effect of residual lithium compounds on layer Ni-rich $Li[Ni_{0.7}Mn_{0.3}]O_2$. *Journal of the Electrochemical Society*, 2014, 161, A920–A926.

[28] Sicklinger, J., M. Metzger, H. Beyer, D. Pritzl, and H. A. Gasteiger, Ambient storage derived surface contamination of NCM811 and NCM111: performance implications and mitigation strategies. *Journal of the Electrochemical Society*, 2019, 166, A2322–A2335.

[29] Liu, M., Y. Jiang, Y. Qin, Z. Feng, D. Wang, and B. Guo, Enhanced electrochemical performance of Ni-rich cathodes by neutralizing residual lithium with acid compounds. *ACS Applied Materials & Interfaces*, 2021, 13, 55072–55079.

[30] Bi, Y., Q. Li, R. Yi, and J. Xiao, To pave the way for large-scale electrode processing of moisture-sensitive Ni-rich cathodes. *Journal of the Electrochemical Society*, 2022, 169, 020521.

[31] Ding, X., J. Liu, and T.A.L. Harris, A review of the operating limits in slot die coating processes. *AIChE*, 2016. 62(7): p. 2508–2524.

[32] Schilde, C., C. Mages-Sauter, A. Kwade, and H.P. Schuchmann, Efficiency of different dispersing devices for dispersing nanosized silica and alumina. *Powder Technology*, 2011, 207, 353–361.

[33] Mixers, R.A. Resonance for power and efficiency: acoustics for performance and quality. https://resodynmixers.com/how-ram-mixes/ [accessed on 15 Oct 2022].

[34] Mohanty, D., E. Hockaday, J. Li, D.K. Hensley, C. Daniel, and D.L. Wood III, Effect of electrode manufacturing defects on electrochemical performance of lithium-ion batteries: cognizance of the battery failure sources. *Journal of Power Sources*, 2016, 312, 70–79.

[35] Wenzel, V., R.S. Moeller, and H. Nirschl, Influence of mixing technology and the potential to modify the morphological properties of materials used in the manufacture of lithium-ion batteries. *Energy Technology*, 2014, 2, 176–182.

[36] Adachi, N., M. Hashiba, and O. Sakurada, Rheological properties of slurries prepared using a planetary mixer. *Ceramics International*, 2004, 30, 1055–1058.

[37] Kim, K.M., W.S. Jeon, I.J. Chung, and S.H. Chang, Effect of mixing sequences on the electrode characteristics of lithium-ion rechargeable batteries. *Journal of Power Sources*, 1999, 83, 108–113.

[38] Kremer, L.S., A. Hoffmann, T. Danner, *et al.*, Manufacturing process for improved ultra-thick cathodes in high-energy lithium-ion batteries. *Energy Technology*, 2020, 8, 1900167.

[39] Lee, G.-W., J.H. Ryu, W. Han, K.H. Ahn, and S.M. Oh, Effect of slurry preparation process on electrochemical performances of $LiCoO_2$ composite electrode. *Journal of Power Sources*, 2010, 195, 6049–6054.

[40] Wang, M., D. Dang, A. Meyer, R. Arsenault, and Y.-T. Cheng, Effects of the mixing sequence on making lithium ion battery electrodes. *Journal of the Electrochemical Society*, 2020, 167, 100518.

[41] Ligneel, E., B. Lestriez, A. Hudhomme, and D. Guyomard, Effects of the solvent concentration (solid loading) on the processing and properties of the composite electrode. *Journal of The Electrochemical Society*, 2007, 154(3), A235–A241.

[42] Zhao, B., D. Yin, Y. Gao, and J. Ren, Concentration dependence of yield stress, thixotropy, and viscoelasticity rheological behavior of lithium-ion battery slurry. *Ceramics International*, 2022, 48, 19073–19080.

[43] Kwon, Y.I., J.D. Kim, and Y.S. Song, Agitation effect on the rheological behavior of lithium-ion battery slurries. *Journal of Electronic Materials*, 2015, 44(1), 475–481.

[44] Hawley, W.B., H.M. Meyer III, and J. Li, Enabling aqueous processing for $LiNi_{0.80}Co_{0.15}Al_{0.05}O_2$ (NCA)-based lithium-ion battery cathodes using polyacrylic acid. *Electrochimica Acta*, 2021, 380, 138203.

[45] Li, J., C. Rulison, J. Kiggans, C. Daniel, D.L. Wood, Superior performance of $LiFePO_4$ aqueous dispersions via corona treatment and surface energy optimization. *Journal of The Electrochemical Society*, 2012, 159(8), A1152–A1157.

[46] Du, Z., K. Rollag, J. Li, *et al.*, Enabling aqueous processing for crack-free thick electrodes. *Journal of Power Sources*, 2017, 354, 200–206.

[47] Keppeler, M., H.-Y. Tran, and W. Braunwarth, The role of pilot lines in bridging the gap between fundamental research and industrial production for lithium-ion battery cells relevant to sustainable electromobility: a review. *Energy Technology*, 2021, 9, 2100132.

Chapter 4
Slot die coating

William Kays[1], W. Blake Hawley[2] and Jianlin Li[3]

4.1 Introduction

Continuing with the concepts discussed in this text the next interconnected step in the lithium-ion battery (LIB) manufacturing process is electrode slurry application onto the metal foil current collector. This process is commonly referred to as a coating process; however, in practice, the current methods are founded and fundamentally very similar to legacy processes used in the printing, photographic film, and magnetic media industries [1]. Those industries have all seen a decline in market demand over the last few decades while the market demand for LIB has and continues to increase [2]. The state-of-the-art coating process has been and continues to be slot die coating, employing a variety of configurations depending on a host of process variables such as target electrode loading mg/cm^2, foil current collector thickness and quality, production plant layout, target coating width, and production machine speed range which is commonly referred to as the coating line speed or simply line speed.

The previous chapter described the necessary steps to prepare an electrode slurry with the appropriate properties both for downstream processing but also as a functional coating within the final cell assembly. In this chapter, we will pick up the process from slurry preparation/manufacturing and delve deeper into how the slurry is formed into a thin film using slot die coating techniques. The common term slot die is comprised of two separate words. A die as used in the context of a tool is a device for forming a material into a particular shape. In the case of slot die coating the word, die is used as the slurry is delivered to the tool under pressure at a controlled rate passing through precise flow geometries and channels to be formed into a thin layer with a uniform velocity across the entire width at the die exit. The exit of the die is the slot. It has a known width, thickness (commonly referred to as the slot gap), and overall length which is sometimes called the land or internal body land. The exit slot on the die is the most restrictive area of the slot die; it has the

[1]R.W. Baron Process Equipment, Inc., USA
[2]Sparkz, Inc., USA
[3]Division of Electrification and Energy Infrastructures, Oak Ridge National Laboratory, USA

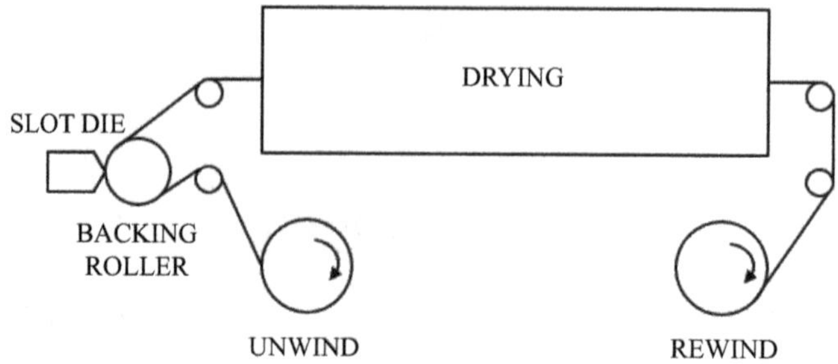

Figure 4.1 Diagram of simplified coating line

smallest cross-sectional open area and typically generates the majority of the pressure drop across the total slot die.

Slot die designs can differ from manufacturer to manufacturer. Many slot dies may look alike from the exterior but can be completely different internally, each specialized to process a unique fluid rheology. There are specialty manufacturers that focus exclusively on the design, fabrication, and finishing of slot dies. In many cases dies from industrial suppliers may share a lot of common attributes, meaning they may look similar, but there are differences in design standards, materials of construction, and finishing techniques. Outside of the offerings from the dedicated slot die manufacturers some larger, vertically integrated companies, may have developed their own unique die designs and manufacturing processes. This is less common in recent times, but there are companies still today that utilize their own designs and choose to keep them confidential as trade secret. For discussion purposes in this chapter, we will review more generalized slot die designs to illustrate their differences, use cases, and practical application limits.

Figure 4.1 depicts a simplified coating line diagram starting with the substrate or foil current collector unwind, moving through to the coating station with a slot die coating head position against a backing roller before going into the drying phase and finally the rewind of the dry coating. The layout and ancillary systems of a coating line may differ from machine to machine, but the basic functions shown here are fundamental.

4.2 Slot die coating system

Slot die coating methods and techniques can take on many different forms. Those in the industry will have a mental image of a slot die coating process by simply mentioning the words slot die. To provide a basis for this discussion, we will consider the two techniques most widely used for LIB electrode coating, as (1) slot die coating against a backing roller and (2) free-span or tensioned-web over slot die (TWOSD). Before getting into the details of different types of slot die coating configurations, it is important to understand the fundamentals of the slot die coating process as an interrelated system. The system can be broken down into discrete

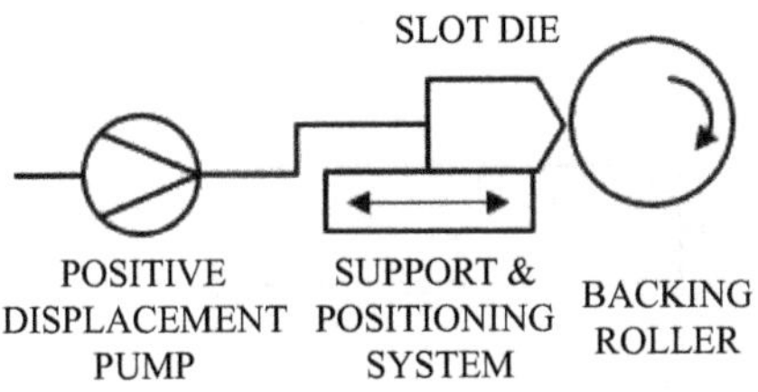

Figure 4.2 Diagram of slot die coating with backing roller

elements, which may be slightly different, depending on the specific slot die configuration being discussed. For a point of reference, we will start with the basic system configuration as slot die coating against a backing roller. The fundamental system elements are as follows and as depicted in Figure 4.2.

- Slot Die.
- Slot Die Support and Positioning System.
- Positive Displacement Fluid Delivery System.
- Precision Backing Roller.
- Process Inputs.

4.2.1 Slot die

A slot die, in the most basic form, consists of two pieces of metal, machined with a specific flow channel geometry and precision surface ground to be matched, creating an air/liquid-tight seal upon assembly. The seal surfaces are required to keep the coating fluid or in this case, LIB slurry contained within the flow cavity. Many slot die designs also utilize a body shim that is used to set the final slot exit gap and coating width. The electrode slurry is pumped to the slot die entrance at point. The entrance is designed with an adapter plate to transition from the process piping to the slot die flow channel. The cross-sectional area of the slot die entrance is typically the largest of the individual flow channel areas. The entrance of the slot die transitions into the throat. The throat carries the electrode slurry to the primary flow cavity or manifold. The primary flow cavity has one main function which is to provide bulk distribution of the electrode slurry out to the desired width. This cavity can have a wide range of shapes, but generally the design falls into one of two types. The first type is often referred to as an infinite cavity as it has a constant cross-sectional area at every point from the center to the end [3]. This type of cavity is better suitable for fluids that are more Newtonian. The second type of cavity is a "coat hanger" style that has a diminishing cross-sectional area [3]. This style has a change in cross-sectional area, being larger in the center and reducing by size out to the end of the die. The term coat hanger comes from the shape of the primary channel backline when looking down from above, refer to Figure 4.3. From the primary cavity, the electrode slurry then passes through the final slot gap. This section of the slot die has the smallest cross-sectional area which generates the highest pressure drop to perform the fine distribution.

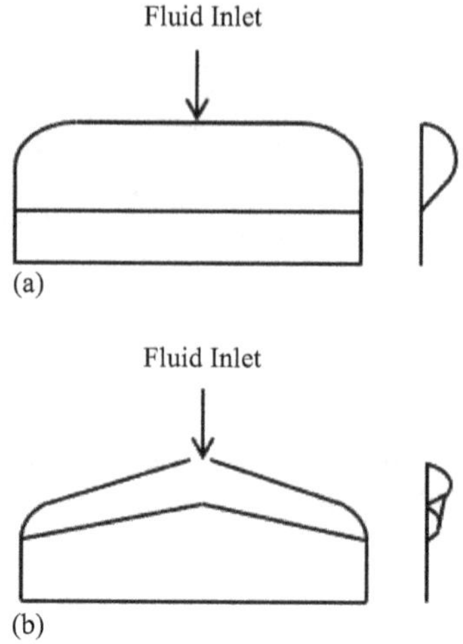

Figure 4.3 *(a) Slot die cavity, constant cross section, plan, and side view and (b) slot die cavity diminishing cross section, plan, and side view*

In theory, the pressure drop and corresponding fluid exit velocity should be uniform across the width of the final exit slot. Control of the final exit slot dimensions is accomplished by two methods. First, as in the case of a fixed lip slot die, the slot gap is set by the thickness of the die body shim selected during the slot die assembly. The slot die surfaces are precision surface ground and either a precision ground stainless steel body shim is used or a precise biaxially oriented polyethylene terephthalate (PET) shim is used. In either case, the three-piece assembly of the outer die halves and inner body shim are to fit together precisely to create an air and liquid-tight seal of the correct slot exit dimensions. As a fixed lip die the lip is fixed, meaning non-adjustable. If the resulting coat weight distribution does not fall within the desired tolerance, then the die must be taken apart so an alternate body shim of a different thickness can be selected. Then the process must be restarted to evaluate the result. This can be an iterative process but there are key indicators and other predictive tools that can be used to shorten the cycle time. For example, the first key indicator to consider is the total pressure drop through the slot die. If the pressure drop is too low, say only 1–2 psig, then the slot die is not being fully utilized, which typically results in heavier flow to the center section of the die. Conversely, a high-pressure drop can also have a negative impact. Depending on the fluid rheology there may be exceptions, but generally pressure drops greater than 50 psig are above the target range for a stable coating process. Slot dies operating at higher pressures can be more sensitive to process upsets. The most stable process with repeatable results will most often have a slot die pressure drop in the range of 10–30 psig [3]. This range is the starting point for process startup and development.

The second common type of die, opposed to a fixed lip die, is a flexible lip slot die. This type of slot die can also use a body shim to set the initial slot gap thickness, but it has a special feature machined into one of the die halves to allow the lip to flex, usually through the use of a series of threaded adjustment screws across the width. The advantage of a flexible lip slot die is when feedback from the process shows heavy coating in some areas and lighter coating in other areas the flexible lip can be adjusted to alter the actual slot gap at different points to redistribute the coating in real time to achieve the desired result. Through the use of online scanning gauge systems, this feedback can be sped up even further and can be automated through automatic adjustment systems. Figure 4.4 shows a schematic of both fixed lip and flexible lip slot die.

The choice to use a fixed lip versus a flexible lip slot die is not always clear. Sometimes the choice comes down to the history, experience, and knowledge of the facility that operates the coating process. Some facilities have more experience with one type of slot over the other and have built up a good history of what could be viewed as internal knowledge on how to get the best production out of the slot die tooling they have available. For this reason, there can be separate facilities working under the same corporate entity that employ different slot die systems to produce the same exact coated product. Over time each facility may have developed its operating procedures to fine tune its equipment and process to achieve its individual facility operational metrics.

Contrast that to other facilities that may use both fixed lip and flexible slot dies for different reasons. For example, when performing narrow width coating process development for a new product a fixed lip slot die may provide a quick turnaround on process experiments without the extra time needed to ensure the slot die lip adjustments are set correctly. Furthermore, as the coating width is reduced so is the need for slot die lip adjustments. One must keep in mind that when adjusting a flexible lip slot die it involves bending a beam; the narrower the coating width, the shorter the beam and the more difficult it is to adjust. Typical spacing on slot die lip adjusters is in the 25–40 mm range. It is not feasible to have tighter spacing than 25 mm as each adjuster does not act independently nor discretely. Rather lip adjustment screws work together in groups to apply pressure to the flexible lip to push, decreasing the slot gap or pull to increase the slot gap. Just as adjustments should be made using groups of adjusting screws, those adjustments should be balanced. In other words, adjustments made to close or reduce slot gap in one area should also be counter acted by adjustments to open the slot gap in another area. When looking at the coat weight profile either from an online gauging system or actual data taken from manual coat weight measurements, it is human nature to only look at the high spots and attempt to close the slot gap in those areas to reduce

(a) (b)

Figure 4.4 (a) Fixed lip slot die and (b) flexible lip slot die with die lip adjusting screws

the coat weight variation. This can work, but only for a period of time and series of adjustments. If all of the adjustments made only act to reduce the slot gap, then over time the total open cross-sectional area of the slot will be reduced, which will increase the total pressure drop through the exit slot. This may result in positive results initially, but at some point, the increase in pressure drop may cause the die to be unstable in which very small adjustments seem to have a big impact on the coating and the result of the adjustments may seem erratic. In a case such as this, the slot die needs to be reset to uninform nominal slot gap to restart the process. Then as feedback from the coat weight distribution is gathered balanced adjustments should be made. Small adjustments to reduce the slot gap in areas of higher coat weight should be followed by similar small adjustments to open the slot gap in areas of lower coat weight. Balanced adjustments will have a net effect to keep the overall pressure drop stable resulting in a more stable and predictable process throughout the production campaign. For those reasons, fixed lip slot dies may often be used for narrow width process development, while flexible lip slot dies may be used for wider width production of LIB electrode materials. Furthermore, potential changes in the electrode slurry rheology over time or from mix batch to mix batch may also make a flexible lip slot die a better choice, allowing for adjustments to be made over time.

4.2.2 Slot die support and positioning system

Procuring a precision slot die is not enough to ensure a robust, repeatable, and high quality coating result. To quote John W. Ryan, an influential figure in the extrusion and slot die industry, "[Slot] dies are static devices. They do not move or make noise." The spirit of that statement was intended to highlight how the slot die is only one element of the overall system and in this section, we will discuss how precise positioning and support of the slot die is equally as important as having a slot die manufactured to exact tolerances.

The slot die support and positioning system has two main functions as the name implies. First, it provides a stable base to support the slot die across the entire width. That seems like a non-issue, and it could be for smaller, narrow slot dies. However, as slot dies become wider to maximize operational efficiency it will be increasingly more important to ensure the slot die is fully supported across the entire width. Keep in mind that a slot die can be considered a beam. If it is only supported from the ends, then there will be some level of deflection in the center which may show up as an irregular coating pattern. For slot dies larger than around 500 mm multiple support pads and fastening locations may be required to maintain the slot die straightness. Slot die support systems are typically designed with precise locating features which allow the slot die to be removed and reinstalled while maintaining the positional accuracy. This helps to reduce setup errors during product changeover, which may require the removal of the slot die for cleaning.

Most slot die positioning systems will control to two positions—the retracted or off coat position and the extended or on coat position. In the retracted position, the slot die is moved away from the substrate; typically, a distance of 100–150 mm.

In this position general setup and minor cleaning of the slot die can be completed. Cleaning of the slot between coating stops and restarts can be done. Setup changes to the slot die can also be performed in this position. Although for more thorough cleaning and setup, the slot is usually removed from the machine and taken to another area or workshop.

During operation, the support and positioning system is used the set the slot die position relative to the substrate. That distance is commonly referred to as the coating gap and is critical to coating quality. The type of actuator used to drive the slot die to the desired coating position will vary by the design approach of the manufacturer. In most cases, the slot die position will be actuated by either a pneumatic or an electric system. In the case of pneumatic compressed air is used to actuate cylinders to extend moving the slot into position. The end point position is set by a mechanical stop with precise adjustment for modifying the coating gap during operation. For this type of system, the movement of the slot die can be considered digital as in extended or retracted. When extended electronic proximity sensors may also be used to confirm the slot die is in the on coat location, but there may not be a precise measurement of the actual position. The verification that the slot die is in the coating position may be a permissive condition needed to initiate the start of coating. It is common to have a series of permissive conditions written into the coating machine PLC program as verification prior to allowing the start of coating. Some of the permissive conditions could be: all safeties are in place and the emergency stop circuit is reset, the dryer is running and at the proper temperature, the condition in the drying chamber is below the lower flammable limit (LFL) and lower explosive limit (LEL), the coating line is running, meaning the web conveyance is turning, and the slot die is in the coating position. Those conditions must be met before the machine software will allow the operator to start the fluid delivery pump to supply coating to the slot die.

Some coating equipment manufacturers will use an electrically driven actuator that combines the slot die movement and precise positional control in one system. These types of systems use servo actuators with encoder feedback to monitor the slot die position. In most cases, the electronic systems are more repeatable for coating process setup, as manual gap control adjustments are not needed. However, the electronic systems require more upfront investment, both by way of hardware and controls programming. Later in Section 4.3.4, the concepts of intermittent coating using a slot die will be discussed. It is worth mentioning here that an electronic position control system is required to enable intermittent coating, allowing for precise movement of the slot die to facilitate quality coating start and stops.

Many slot die positioning systems also have another adjustment point that can impact the angle of the slot die. Generally, slot dies are positioned around a backing roller on the horizontal plane at the 9 o'clock position or set at some angle pointed up toward the backing roll. The actual application angle is less important as the coating fluid dynamic forces are dominant and for the most part, gravity can be considered negligible. It is common for slot die systems to be configured with the slot die pointed up toward the substrate or backing roller. This configuration helps to purge air out of the system during initial startup reducing scrap production. Irrespective of the application angle the other critical angle that is set by the positioning system is the slot die angle of attack. The angle of attack is a rotational adjustment centered around the die exit slot. A neutral angle of attack set to zero will have the die exit slot pointed directly at the center of the backing roller and exit lip faces parallel to the tangent point of the roller. From that reference point, the angle of attack can be adjusted positive and negative usually by as much as five degrees in either direction. A positive angle of attack tilts the rear of the die up causing the downstream lip face to tilt toward the backing roller thereby changing the degree of divergence in the coating gap. Since the center of the rotation is a point at the exit slot the coating gap should remain constant throughout changes in the angle of attack. A negative angle of attack is the opposite, tilting the back of the slot die down which in turn further increases the diverging area between the downstream lip face and the substrate. A negative angle of attack is rarely used. However, a positive angle of attack is almost always used. The angle of the lip face relative to the substrate can impact the appearance of the coating and overall coating quality. If a more neutral angle is used and the coating quality is not smooth, fully continuous, and free of any voids, then increasing the angle of attack can apply pressure from the downstream lip to help resolve those types of defects. Furthermore, within a certain range, this adjustment can be used to impact the final coating width. As the angle of attack is increased the pressure applied by the downstream lip causes the coating to spread wider. This is only true until a certain point. Eventually, with higher degrees of angle applied, the coating will stop spreading wider and start spilling upstream. In that situation, the slot die is no longer being operated as a pre-metered process. When upstream overspill occurs, the slot die lip acts like a knife scraping the access fluid off. The coating fluid will always follow the path of least resistance. In the case of upstream overspill, there is a higher pressure at the downstream lip edge, which prevents 100% of the coating from transferring to the substrate. The access spills upstream and becomes waste. To correct upstream overspill the angle of attack should be reduced. If that results in further coating defects like a rough appearance, coating voids, or horizontal barring, then the coating gap may need to be adjusted.

When balancing the parameters that impact coating quality, process engineers refer to the coating window for a given fluid or product. The coating window can be defined as the range of adjustment of the critical process parameters in which a quality coating can be produced [4,5].

- Exit slot gap.
- Coating gap.
- Angle of attack.
- Coating speed.
- Coating fluid surface tension versus substrate surface energy.

Coating process engineers must understand these parameters and how they are interrelated. During the development of a new coating or product methodical design of experiments (DOE) may need to be conducted to evaluate how the parameters are related and which combination has the biggest impact on coating quality for the given product. In the following, we will consider each parameter independently.

The exit slot gap impacts the pressure drop through the slot die which can be correlated to the coating cross-web distribution. The exit slot gap may not directly impact the wet coating quality, but poor cross-web coating distribution may show up as a pattern in the coating [3]. Areas with a higher relative flow rate may sufficiently fill the coating gap resulting in a smooth wet appears while areas with a lower flow rate may not fill the coating gap evenly resulting in a rough, "orange peel" appearance and/or voids in the coating.

The coating gap sets up the interface that allows for the application to the substrate as the fluid exits the slot die. The gap is set by the slot die positioning system using either manual adjustments or inputs to the control system through the coating line human–machine interface (HMI). For a smooth coating application, the coating gap must be set to a distance near the theoretical wet coating thickness. This may be a good starting point and can be increased or decreased as needed depending on the observed coating results. Too large of a coating gap and the wet coating appearance may go from smooth to rough with possible voids as described above. In this case, the volume of coating supplied to the application interface is not sufficient to maintain a stable upstream meniscus and downstream contact line. This could result in coating bead breakup as shown in Figure 4.5.

The angle of attack can be used to provide fine adjustments impacting the coating bead. As the angle of attack is adjusted from neutral to positive or negative, it changes the angle of the slot die lip face, which impacts the degree of divergence formed at the coating interface. Adjusting the angle of attack can shift the position of the upstream meniscus and downstream contact line. Adjustments made too far in either direction can lead to coating bead breakup, poor coating result, and possible upstream overspill.

Coating speed can impact the coating window for a given coat weight and fluid viscosity [6]. As the target coat weight and fluid viscosity decrease the coating window or upper limit for coating speed may also decrease. For electrode coating, the line speed may have less of an impact on the overall coating window as the dry/wet coat weight and relative slurry viscosities tend to be higher. If coating speed appears to be a limiting factor based on experimental results and process observations, then there are counter measures that may be employed. First, if the slot die application is coated against a backing roller, then adding a vacuum box may be a good option [3]. A vacuum box placed at the upstream side of the slot die lip alters the pressure around the coating interface. Pulling a vacuum on the upstream side creates a low-pressure zone, which can impact the size and position of the upstream meniscus. Typically, if the coating speed appears to be the limiting factor, then it is likely the upstream meniscus is approaching the die feed slot resulting in bead breakup. Adding vacuum can pull the upstream meniscus into a stable position to create a sealed coating bead. Second, if a thin electrode layer is required, then higher line speeds may be desired to increase the production rate while maintaining the mass balance transfer through the dryer. In other words when coating a thinner

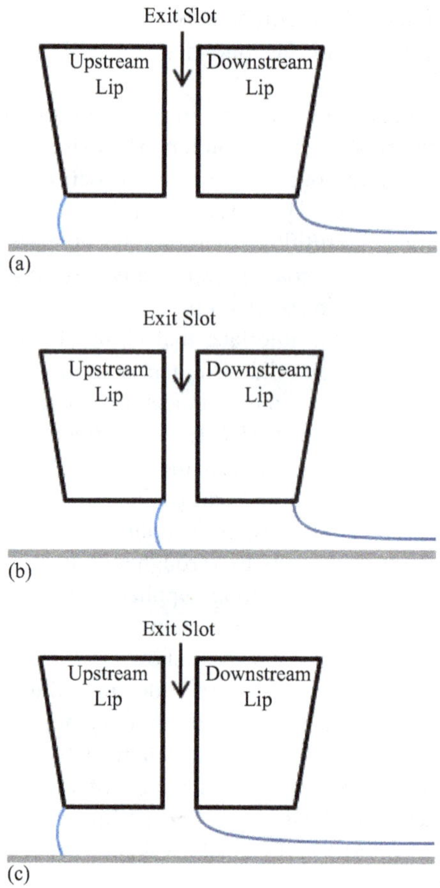

Figure 4.5 (a) Stable coating bead. (b) Coating bead breakup due to insufficient upstream meniscus. (c) Coating bead breakup due to insufficient downstream meniscus.

layer a higher line speed can be run to achieve the same solvent evaporation loading through the drying step. As coating thickness is reduced slot die positioning becomes more important. Maintaining very tight coating gaps without environmental contamination, leading to web breaks becomes increasingly difficult. This is precisely the use case for switching to a TWOSD coating configuration. TWOSD was developed for coating thin layers. The close contact and interaction of the slot die lip faces, substrate, and coating fluid allows for coating very thin wet layers. While traditional slot die against a backing roller may be able to coat wet layers down to around 25 μm, adding a vacuum box may reduce that further to around 15 μm. TWOSD can reduce that even further to produce a wet layer < 15 μm depending on the fluid properties, percent solids, and particle size.

Surface tension and surface energy also play a role. The units are the same for both, surface tension is often used when describing the coating fluid while surface energy is more commonly brought in the context when discussing the substrate. For the coating to properly wet out on the substrate the interfacial solid/liquid surface tension should be considered [6,7]. If the surface tension of the coating is higher than the substrate, then the coating will not wet the substrate properly. Interfacial surface tension can be evaluated using Young's equation (3.3), refer back to Chapter 3, Section 3.2.3 on surface tension.

For LIB electrode slurries changes to the chemistry may be limited based on performance requirements, which could rule out possible use of surfactants to modify the interfacial tension. Generally, solvent-based coatings will have lower surface tension and can more easily wet most substrates when compared to water-based chemistries [7]. If the chemistry is fixed, then the focus may shift to modifying the substrate. For example, some aluminum foil substrates may contain a very thin film of oil due to the roll forming process. The thin oil residue may be undetectable visually or even by touch, but it could be enough to prevent proper wetting. In a case like this pretreatment of the aluminum foil may be needed to clean the substrate thereby modifying the surface. Such treatment could be an online plasma system that is placed prior to the coating application point. The plasma treatment system can be used to clean, etch, and otherwise prepare the substrate to accept the coating more readily [8,9]. Pretreatment of the foil substrate not only promotes better wetting of the liquid slurry, but it may also enable better adhesion of the dry electrode layer [10] making the electrode more durable for postprocess handling, such as slitting, cell winding, or punching.

While slot die support and positioning system cannot directly impact the solid/liquid surface tension, it is important to remember precise positioning of the slot die is the key to setting up the correct physical proximity between the slot die and the substrate to enable coating bead formation. Furthermore, the support and positioning system adjustment points, positional accuracy, and repeatability are critical in enabling a robust coating process with repeatable results.

4.2.3 Positive displacement fluid delivery system

In this chapter, we have already discussed how the slot is a forming tool used to transform liquid stream into a thin-coated film applied to a substrate. We also discussed how slot die coating is a pre-metered coating method. As a pre-metered system, a metering device is needed to supply a uniform stream of coating fluid to the slot die inlet. This requires a positive displacement pump to meter the electrode slurry at a prescribed flow rate [3]. The term positive displacement is used because the type of pump that is required must provide a specific output per unit time. Pumps of this type can be set precisely to achieve a specific flow rate target. For LIB electrode coating, the type of pump most widely used in industry today is a progressive cavity type.

A progressive cavity pump (shown in Figure 4.6) is a type of positive displacement metering pump that has an internal helical rotor that rotates within an

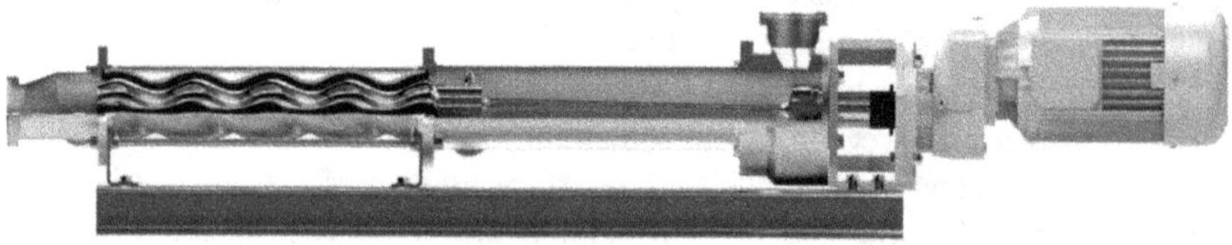

Figure 4.6 Image of progressive cavity pump, provided by SEEPEX Inc.

elastomeric stator with a mating dual helical pattern. As the rotor rotates, cavities within the stator helical pattern open and close on the suction side of the pump. Slurry is drawn into the open stator cavity and progresses through the stator helical geometry as the rotor continues to rotate. There are a few good reasons why this style of pump is used for LIB electrode coating. First, the pump technology is well known, robust, and relatively less expensive than other specialty pump technologies. Second, the progressing cavities allow for conveying slurries with a larger particle size. Other pumps, such as gear pumps and lobe pumps, have a rigid internal gear or lobe-shaped rotors. These internal components require a tight clearance between the gear or lobe and the housing to prevent internal slippage. That tight clearance can be problematic for fluids with larger particles as those particles will tend to get lodged between the internal housing and the major diameter of the rotating internal gear or lobe. The result is galling of the internal surfaces, increased slippage, and/or complete pump failure. In a progressive cavity pump, the larger particles in the slurry pass through the stator cavities. The rotor has an interference fit with the elastomeric stator which seals the cavities. There is no risk of galling of the stator, but over time there will be normal wear to the stator; and at some point the stator will need to be replaced to bring the pump capacity back to original condition. The last main reason why progressive cavity pumps are widely used for electrode coating is they have a very repeatable and linear output. Target flow rates can be precisely set by altering the pump speed. This can be done using simple proportional math. When evaluating the average coat weight one can simply use a proportional factor to calculate the corrected pump speed to achieve the desire coat weight. For example, if the target coat weight is 15 mg/cm^2 and the feedback from the process through online gauging or manual coat weight measurements shows the average actual coat weight is 12 mg/cm^2 at a pump speed of 50 RPM, then the following equation can be used:

$$\left(\frac{\text{Target coat wt.}}{\text{Actual coat wt.}}\right)\text{initial pump speed} = \text{new target pump speed} \tag{4.1}$$

Using (4.1) the resulting new target pump speed in this example is calculated to be 62.5 RPM to achieve the desired 15 mg/cm^2 coat weight target. This assumes that the coating line speed is at a fixed rate.

Another related point regarding the use of progressive cavity pumps for metering electrode slurry is this style of pump has been used for industrial and sanitary food processes in other market applications. That is important as many of the electrode coating operations look to leverage market experience from other applications to improve the effectiveness and overall quality of the coated electrode. To that end, lower cost industrial pumps can be used for process development and scale-up. Then with the process well defined the same type of pump can be used but upgraded to a sanitary style with polished internal surfaces which can allow for clean-in-place (CIP) systems to be used, enabling quick product changeover with very low risk of contamination from one product to the next. This will become more and more important as electrode and cell manufacturers need to optimize efficiency to further reduce manufacturing costs.

4.2.4 Precision backing roller

Slot die coating against a backing roller can be considered the most traditional method of applying a thin film fluid coating to a substrate. The term backing or back-up roller is used because the substrate wraps and is fully supported by a precise, typically driven, roller as shown in Figure 4.2. The backing rollers are most often larger in diameter when compared to other web conveyance idlers and directional rollers. Most idler and directional rollers are in the 76–152 mm range, usually non-driven and constructed of aluminum to keep the weight and inertia to a minimum. Backing rollers, on the other hand, are common in sizes from 203 to 355 mm in diameter. They are most often driven as previously mentioned and are constructed of industrial hard chrome plated steel, precision ground to very tight tolerances for total indicated runout, which leaves a mirror-like surface finish. Measured surface finishes could be in the range of 4 Ra (microinches) or 0.10 Ra (micrometers). That said the surface finish is not the most important specification. It is improved by the means of cylindrical surface grinding to achieve a very low total indicated runout of the exterior roller surface. Similar to slot dies, there are industrial companies that specialize in roller manufacture and have the skills, techniques, and proper equipment to produce very high-quality rollers. Typical runout requirements are in the 1–2.5 μm range. To achieve such a tight tolerance, it requires more than just proper tooling. In cases where the total indicated runout is measured down to 1 micron the bearings for the roller must be very high quality and matched. The roller is then surface ground in the bearings to eliminate or greatly reduce any tolerance buildup from the running bearings. Producing the roller is challenging enough, but then measuring the total indicated runout as verification can also be a challenge. Some manufacturers have developed their own measurement systems to better document roller inspection results. This was done primarily out of necessity as customers have become accustomed to getting inspection reports. Today there are a variety of roll measurement systems on the market. It is common for specialty roll manufacturers and customers alike to have similar measurement systems to perform independent quality checks. Such systems are available from PRO-MIC which can measure the profile of a given roller from end-to-end. The systems can be configured with Bluetooth data transfer to a remote PC. They can measure roll concentricity, perform sine cure roll shape analysis, and even roll temperature measurement if needed.

The backing roller has two main functions: (1) convey the substrate and (2) provide support by maintaining uniform tension around the full degree of wrap. This support of the substrate is why the backing roller needs to be manufactured precisely. If the roller were to have a more open tolerance with a larger amount of total indicated runout, that translates to variations in the substrate position, which impacts the distance between the slot die lip and the substrate. Simply as that distance varies it can translate into variations in the wet coating thickness that ultimately show up in the dry coating thickness and coat weight or loading. The distance between the slot die lip and the substrate is commonly referred to as the coating gap. It is very critical and must be held constant to achieve consistent coating results. More discussion on the coating gap is forthcoming.

The backing roller supports the substrate while the slot die applies the wet film as pre-metered from a positive displacement pump. It is important to note that slot die coating is a pre-metered process, unlike other roll coating processes in which there may be coating reservoirs or troughs that flood the gap between a knife or set of rollers and those gaps are what meter the coating. In the slot die process, a positive displacement pump is set to a prescribed output speed and flow rate to achieve the coat weight set point at the desired coating line speed. It is important that the flow rate through the slot die be consistent throughout the process run time as the output from the slot die and the velocity at the exit is a function of pressure drop. If the volumetric flow rate has variation, then the pressure drop across the slot die will also have variation which carries further to impart more variation at the exit velocity. This all impacts the final variation of the dry coating thickness and coat weight.

4.2.5 Process inputs

The last two elements of the system that impact coating quality are the process inputs, meaning the electrode slurry and foil substrate. The previous chapter has already discussed the electrode slurry preparation process and how highly sensitive the rheological properties of the final slurry can be impacted by the order of operations during the mixing steps [11–15]. There is no need to fully recap the points made in the previous chapter, but here we will stress the importance of the final slurry rheological properties and how they relate to the performance of the coating system and more specifically flow through the slot die coating head. While previous sections discussed the overall slot die design and differences in cavity types it is important to note the slot die can only process or form the coating fluid as it is delivered. The slot die cavity geometry and final slot gap are generally fixed based on a specific electrode slurry to be processed. If the final slurry rheology is found to be different from batch to batch, then the impact of how each of those slurry batches process through the slot die may also be different. In some cases, the differences may be minor and easily counteracted through a few adjustable parameters. In other cases, the differences may be significant enough to require major changes to the slot die setup or in an extreme case, a completely different slot die with a different flow cavity may be required. In an electrode manufacturing operation, there should be a good working relationship between the team that develops the LIB slurry chemistry and the team that develops the slot die coating process. Each team needs to understand the limits of its own area of responsibility and communicate those limits to the other team. Through pilot trials on slurry preparation and slot die coating both teams should work together to develop an optimized process. For a new electrode product this may require multiple process iterations to investigate how changes to the slot die setup and/or slurry chemistry impact the coating process results. This type of collaborative effort should reduce the process development time.

The foil substrate is also a critical input that can have an impact on the quality of the coating and ultimately the final electrode. Thin foil substrates can be difficult to process through web conveyance machinery, this is increasingly true if the foil is of a lesser quality. The condition of the thin foil and how it has been prepared, wound, packaged, and handled can all impact the quality before it is loaded on the coating machine [10]. Obvious issues like damage to the edges will cause stress risers that may result in a foil web break which contributes to machine downtime and production of scrap. Other more subtle issues like the consistency of the foil thickness and planarity of the foil may be more difficult to trace and measure but can also contribute to scrap production. Specifically, the planarity of the substrate can have a big impact on how well the foil processes through the entire coating line. The planarity of the foil substrate is most often associated with how it was produced. High-quality copper foils for use as current collectors in LIB are often produced via an electro-deposition (ED) method which provides a consistent thickness for a flat stable foil. Other foil production methods such as rolled foil, typically used to produce aluminum foil, may not yield as consistent of foil quality across the entire width as produced. For economies of scale, aluminum foil may be rolled and formed at a wider width to produce a jumbo or master roll. The master roll is then further processed to slit the width to the desired dimension for use as the input to the electrode coating process.

Unfortunately, through the aluminum foil production and slitting processes the planarity of the foil can be impacted. This becomes evident when the foil is processed on the electrode coating line. Foil substrates with poor planarity exhibit signs of a "baggy edge" which will require intervention to keep the foil running straight and true through the machine without wrinkling. The term baggy edge is used because the lack of planarity causes the foil to be tight on one side and very loose or slack on the other side. This becomes evident when observing how the foil passes between process rollers. With the foil held under tension between two fixed points through visual observation, it may be evident that one edge is taught, and the other edge is loose or baggy. Typically, all the process rollers on the coating line should be installed and measured to be parallel to each other and perpendicular to the centerline of the machine. This is the nominal reference position. To handle foil substrates that exhibit a baggy edge certain process rollers may need to be installed with a skew adjustment capability. Critical areas that may require rollers with skew adjustment are (1) in the unwind or foil payoff location to ensure the foil is brought into the web line without wrinkles. (2) Entering the coating section of the machine, again to ensure a flat, stable, and wrinkle-free substrate is presented to the slot die coating head(s). (3) At the exit of the dryer, it is usually automated to be what is called a steering roller. The steering roller at the end of the dryer helps to correct for any change in web alignment that may have taken place in the dryer and to reset the proper alignment heading into the rewind section of the machine. (4) In the rewind section, there may also be a roller with manual skew adjustment to help ensure uniform tension across the now coated foil, as it is wound for transfer to other downstream postprocessing. The use of rollers with skew adjustment can be very helpful to reduce scrap production by correcting non-uniform tension that likely may have caused a wrinkle in the foil substrate, which can not only lead to unusable coated material but also a higher frequency of web break further leading to unscheduled machine downtime. The rollers with skew adjustment allow for a trained coating line technician and/or process engineer to adjust the machine web line to better match the condition of the foil being processed. Each new roll of bare foil substrate may behave differently depending on how it was produced. The challenge for an electrode coating operation is to evaluate the condition of each roll of foil substrate and make the proper adjustments to ensure the foil passes through the complete system without defects. This means trained personnel must monitor the foil condition at the critical areas of the machine to make interventions through adjustments when needed.

4.3 Slot die coating techniques for LIB electrode production

4.3.1 Coating against a backing roller

In the previous section, the slot die process or system was broken down into basic elements. In this section, the elements are combined to better describe how the complete system functions in the configuration as slot die coating against a backing roller. Using a slot die as a direct coating on a substrate supported by a backing roller is an upgrade over other precursor roll coating methods. What makes slot die coating with a backing roller attractive from a processing standpoint is the pre-metered approach and direct application to the substrate. Other roll coating methods use more than one roller to meter and transfer the coating to the substrate. Those processes are generally run in an open system susceptible to contamination from the processing environment. They may also require changing out specific process rollers to produce coatings of different coat weights. Whereas the slot die coating process is a closed system, which better isolates the coating fluid from the working environment and allows for quick adjustments to the coat weight by adjusting the pre-metered fluid delivery rate.

Slot die coating against a backing roller is well suited for LIB electrode production. Considering the industry, in general, the rheological properties and target coat weights for most electrodes are similar. Meaning if you were to aggregate slurry rheology datasets you are likely to find they fall within a normal distribution with most of the rheology curves having a similar slope and within a similar viscosity range. The same may be true for electrode loading as the industry moves toward commonly adopted sizes for cells, physically there are limitations on electrode loading and thickness which may result in similar electrode loadings. The advantage of these similarities is that as knowledge is shared between researchers and processors, incremental process improvements are made resulting in a higher quality electrode produced at a lower cost. That said, as with any normal distribution there are outliers. Researchers and cell manufacturers continue to innovate and investigate new active materials and combinations of materials targeting a wider range of electrode loadings for different cell designs [16–18]. This can pose some significant challenges to slot die coating process development especially when attempting to coating both very thin and thick electrodes using the same equipment.

As is the case for higher energy cells, thick electrode layers are desired [16]. In the past typical electrode, loading may have been in the range of 10 mg/cm^2, with a wet coating thickness of around 140–190 μm depending on the slurry density and solids content. Today thick electrodes with loading in the 40–50 mg/cm^2 range are becoming increasingly more common. Electrodes in that range may require a wet coating thickness > 700 μm and perhaps even 1,000 μm, again depending on slurry density and solids content. As the wet coating thicknesses increase the slot die to substrate or coating gap must also increase. As coating process trials are carried out there will be a physical limit in which the coating gap will be too large, and the coating bead will become unstable with upstream overspill. If that limit is reached before the desired wet thickness and corresponding electrode loading is achieved, then modifications to the slurry and/or slot die lip geometry may be required. Increasing the slurry solids content can help as that impacts the coating application in two ways. First, an increase in solids content will decrease the wet thickness required to produce the same electrode loading. This change alone may be enough to allow for a stable coating bead. Second, increasing the solids content of the slurry also results in a shift of the rheology curve to yield a higher viscosity overall and more specifically an increase in the low shear viscosity [6,19]. The increase in the slurry low shear viscosity changes the fluid behavior at the coating interface [3,16]. At that point, the slurry is transitioning from a high shear zone of the slot gap to a lower shear zone of the coating gap followed by rest/recovery on the substrate. An increase in the low shear viscosity at this point can help enable a more stable coating bead. Based on that point alone, increasing the solids content sounds like it may be the best solution with no downside, but there are always tradeoffs. In this case, changes to the solids content may alter the fluid rheology to the point where it may not process well through the specific slot die cavity geometry in use. Modifications to the slot gap may be required if the shift in the rheology curve is enough to result in a higher than ideal pressure drop through the slot die for a given mass throughput rate. This is where work to define the coating window comes into play. It is worth restating the slot die coating process parameters are interrelated.

Similarly, for high-power cells with thin electrode layers, there can be challenges in achieving very low coat weights using a system originally setup for higher coat weights. Similar concerns may come regarding the slurry rheology, but the approach may be the opposite. As the desired coat weight decreases so does the resulting dry coating thickness; and if the solids content is constant, then the wet thickness also decreases. For slot die coating against a backing roller this means the coating gap must also decrease at which there is a finite limit. Generally, slot die positioning systems can control to a position within approximately 2 μm. The backing roller can have a total indicated runout (TIR) of approximately 2 μm and the foil substrate also has a tolerance on thickness. All these values impact the actual coating gap and as the coating gap is decreased the net impact of those variations increasing becomes a higher percentage of the target gap and resulting wet thickness. The practical limit on the wet coating thickness for slot die coating against a backing roller is somewhere in the 15–25 μm range. There are likely applications that can coat slightly thinner, but those may be limited to dissolve polymeric solutions with a higher degree of viscoelasticity not dispersions or slurries with larger undissolved solid particles. Next Section 4.3.2 will discuss the benefits of TWOSD coating and how it can enable achieve thin coated layers and low coat weights. For now, it is important to note the variation in the coating gap is a combination of a few different factors and that variation directly impacts the variation of the wet coating thickness and the resulting coat weight or electrode loading.

Aside from the obvious concern that variation in electrode loading is undesired and can contribute to under performance in the final cell, variation in loading can also impact the electrode while still in the production process. Referring to Figures 4.1 and 4.2, when using a slot die coating against a backing roll the foil substrate or current collector passes through the coating interface at the slot die application point. As the electrode needs to be coated on both sides of the current collector this requires a two-step coating process. For a coating machine designed to coat single sided as depicted in Figure 4.1, the current collector needs to be processed through the machine twice to achieve a double-sided coating. On the first pass, the base foil substrate is coated one side before drying and being rewound as a coated roll in process. That roll then needs to be brought back around to the machine to unwind for further processing to coat on the second side. In coating the second side, the first side dry coating thickness will now contribute to changes and variations in the coating gap. First to maintain a nominally similar coating gap for coating side two, the slot die position must be moved back by the same amount of the side one dry coating thickness. If this is not done, then the coating gap will be reduced and if all other parameters are constant, then the coating for side two may spread wider leading to an overall reduction in loading for that side and/or possibly an unstable coating bead. It is common to measure and calculate the average thickness of the first side dry coating and then move the slot die back by that amount to compensate prior coating side two. That accounts for the macro control in an effort to maintain the same coating gap for coating side two versus side one. In this coating configuration, the side one coating variation greatly impacts the side two coating variation.

Localized high and low spots from side one will change the coating gap in those areas and subsequently contribute to variation when coating side two. If the variations are small this may be less of an issue as long as the total composite (side one plus side two) coat weight is on target and within the total loading upper and lower control limits. Where this can have a negative impact when the side one coating contributes to defects when coating side two. Again, consider the case when targeting lower coat weights which require a decreased coating gap. The variation in the coating gap also now includes variation in the side one coating plus tolerances in the slot die positional accuracy, backing roller runout, and to a lesser degree thickness variation of the foil substrate. A tight coating gap combined with all those variations or tolerance stack up can result in a higher likelihood of particle streaks and defects. As the coating gap decreases and changes locally due to these variations it can be more susceptible to larger and sometimes foreign particles or agglomerates that can become lodged at the coating interface causing streaks or other defects which contribute to the production of scrap electrode material or in some cases the cause of the defect can put enough stress on the substrate to cause a web break which requires the machine to be shut down for cleaning and threading before restarting. Next Section 4.3.2 on TWOSD will further discuss how that coating configuration can mitigate effects of the side one coating on side two and enable simultaneous double-sided coating.

4.3.2 Free-span/TWOSD

With the basics of slot die coating against a backing roll covered already, we can introduce a new set of variables by removing the backing roller and converting the process to a slot die in free-span or TWOSD, as shown in Figure 4.7.

In this coating configuration, the precision of the backing roller is no long supporting the foil substrate. In free-span coating uniform tension of the substrate across the width is critical to coating quality and cross-web coat weight or loading control. Even if the slurry exiting the slot die is theoretically uniformly distributed, variation in foil tension will allow for coating redistribution and contribute to variation in coat weight uniformity. As discussed in a previous section, the planarity of the foil substrate plays an important role in the final electrode quality. In tensioned-web coating, the rollers, positioned before and after the slot die, are adjustable in and out to impact the web line position relative to the slot die lip face. Setting up the slot die position to just touch the foil substrate will not result in a smooth continuous coating. As shown in Figure 4.7, the slot die lip face must cross

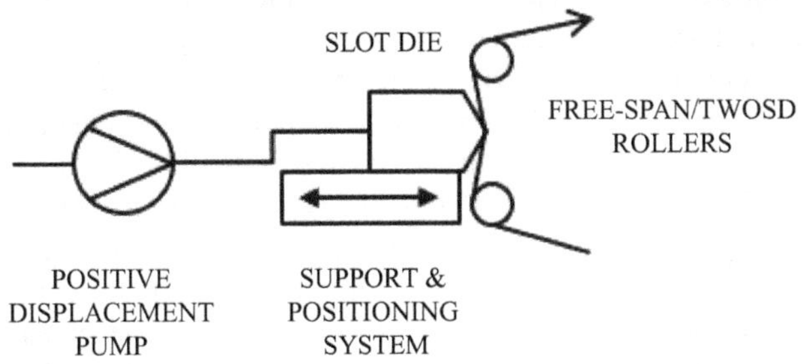

Figure 4.7 Diagram of slot die coating as free-span/TWOSD

the neutral web line created by the rollers. This distance can be referred to as the amount of indent. The specific amount of indent needed for a given product is dependent on the fluid properties (rheology and surface tension) and the desired coat weight. Generally, the lower the coat weight the greater the amount of indent and possibly the higher substrate tension in the coating section of the web line.

Substrate tension control within a coating line is usually broken up into zones. The number of zones may vary depending on the coating line design and number of elements. Some coating lines may have additional features such as vision inspection, calendering rollers, or slitting station. These areas may require separate tension control zones. For now, we will focus on the simplified coating line which can be broken up into three main tension control zones. First is the tension in the unwind section to bring the foil substrate into the machine usually at a lower tension. Second is the coating and drying section. This tension control zone is more critical as it impacts the coating quality and tracking of the coated foil substrate while passing through the dryer. The tension in the coating/drying section is usually higher than the unwind section. The third tension control zone is the rewind section. This section is isolated from the previous by way of a pull roll station usually using a rubber roll to steel roll nip point. The pull roll nip provides a mechanical means to separate the tension control zones. The rewind section tension needs to be separated as the actual tension required will vary as the coated roll accumulates and the diameter increase. Since the diameter of the wound roll increases over time, also does the surface speed. If the tension were to remain constant the internal stress of the wound roll will continue to build as the outer wraps start to crush the inner wraps. For this reason, a taper tension profile is applied through the controls program to compensate for the increase in surface speed, making for a more uniform roll as the diameter increases.

With a brief overview of tension control on the coating line, we can focus again on the TWOSD process. For the tensioned-web coating process substrate tension across the width is critical. If the foil substrate has a baggy edge as previously discussed, then the tension rollers can be adjusted to help compensate and provide a local uniform tension over the slot die lip. The adjustment of the rollers is independent on the left- and the right-hand side. With the slot on coat position fixed at a nominal amount of indent, each roller can be adjusted independently to impact the local tension over the slot die lip. If the foil substrate has a baggy edge, then the rollers can be adjusted to increase the amount of indent on that side to increase the local tension in that area and balance out the pressure drop at the coating application point. The secret to the TWOSD coating process is finding the right balance of slot die position or indent, overall substrate tension, and roller position to account for local tension as the substrate passes over the slot die lip.

With added complexity and the addition of new variables, one may ask why to choose TWOSD versus coating against a backing roller? The answer can be found by looking further at the needs of the coating line processor. First, TWOSD coating was originally developed to allow for thin coatings at faster line speeds. In coating against a backing roller, there is a fixed gap between the slot die and the substrate, which is supported by the backing roller. The physical coating gap distance can only be controlled to a certain degree. Capabilities on slot die position control are limited to around 2 μm, in addition to TIR on the backing roller, also around 2 μm. That combine control of around plus or minus 2 μm may seem very tight and it is. However, when trying to coat very thin wet layers, say 25 μm or less, the coating gap must be reduced to nearly that distance. As the coating gap is reduced the difference of a few microns becomes an increasingly larger percentage of the total distance. Coating very thin wet layers is susceptible to small variations in position control and the tight coating gap is more susceptible to particulate in the environment or agglomerates in the slurry which may become lodged at the coating interface, leading to streak defect or in some cases a web break which increases machine downtime. TWOSD solves this issue by allowing the substrate to float over the slot die lip by the hydraulic forces of the slurry exiting the slot die. Generally, any particulates in the environment and even agglomerates from the slurry will pass through the coating interface without issue. Those inclusions still can produce a spot defect that may have to be avoided or cut out of the final electrode, but the process can keep running with minimal risk of downtime caused by a web break in the coating area. Thin coatings for LIB components can be thin electrodes as is the case for cells designed for high power and conductive carbon primers that are applied to the foil current collector prior to the electrode coating. Conductive primers must be very thin coatings, usually less than 5 μm. To produce a thin primer coating less than 5 μm, the wet coating layer will likely be less than 25 μm depending on the solids content of the primer ink formulation.

The other main advantage of TWOSD coating is it enables dual-sided simultaneous coating. LIB electrode designs require coating on both sides of the foil current collector. This can be accomplished through a few methods. The first method and most common during process and product development is to produce the coating in a 2-step process using a traditional coating method such as slot die against a backing roller as previously covered. In this method, the foil current collector is fed through the coating machine two times. The first pass is to coat the electrode on one side, which is then dried, rewound, and brought back to the start of the machine to be reloaded on the unwind. As previously discussed, the slot die position must be adjusted to increase the coating gap to compensate for the dry electrode layer on the first side of the foil. The foil is then run through the machine for the second pass to coat side two. The multi-pass process is not ideal from a machine utilization standpoint, so operations for mass production will have a machine layout in a tandem coating configuration. Tandem coating essentially puts two coating and drying sections of a coating line together end-to-end. This allows the bare foil substrate to be unwound, coated on one side, via slot die against a backing roller, sent through the first dryer, and then passing through a second coating station to be coated on side two before passing through a second dryer. The tandem coating line increases machine productivity and output but requires nearly twice the floor space and two separate drying chambers which account for most of the utility consumption of the coating process. The third approach is to combine slot die coating against a backing roller with TWOSD coating to apply electrode slurry to both sides of the foil substrate at the same time prior to sending the coated foil through a flotation dryer, which is capable of drying both sides simultaneously.

Dual-sided simultaneous coating combined with a flotation dryer reduces the complexity and floor space needed for the coating machine while maintaining the productivity benefits of the tandem coating line configuration. The key to dual-sided simultaneous coating is the application of TWOSD and the use of a flotation dryer.

In this coating line configuration, the first side coating can and often is applied using slot die against a backing roller as that is more traditional; depicted in Figure 4.8 position a. Then, the second side coating application is done using a slot die positioned vertically up very near to the entrance of the dryer, shown in Figure 4.8 position b. The slot die in this position still relies on the same TWOSD coating principles in that there is control of the indent, die angle of attack, and substrate or web tension control. It is important to note the tension control zone in this area includes the side one coating, the side two coating, and the flotation dryer; with either side of the foil current collector containing a wet coating, there is no way to further split up the tension control zones. Typically, there is a pull roll station upstream of the coating station and downstream of the flotation dryer that is used to break up the coating line into separate tension control zones.

The TWOSD configuration for use in electrode coating and more specifically dual-sided simultaneous coating of electrodes is relatively new to the energy storage industry. While slot die coating has been around for several decades, the application of the technology in this manner has only come about within the last 10–15 years, as commercialization of Li-ion cells became a greater focus with specific cost reduction targets. Dual-sided simultaneous coating of electrodes using TWOSD and flotation drying technology requires a lot of process knowhow in the areas of slot die coating, flotation dryer technology, and sophisticated web handling of delicate foils. Building up experience in those areas takes time and the energy storage industry as a whole may still be in the early stages of the learning curve. Cell manufacturers or electrode processors looking to improve their operational metrics to increase productivity and reduce manufacturing costs should first consider slot die coating if they are not using that application method today. Second, if

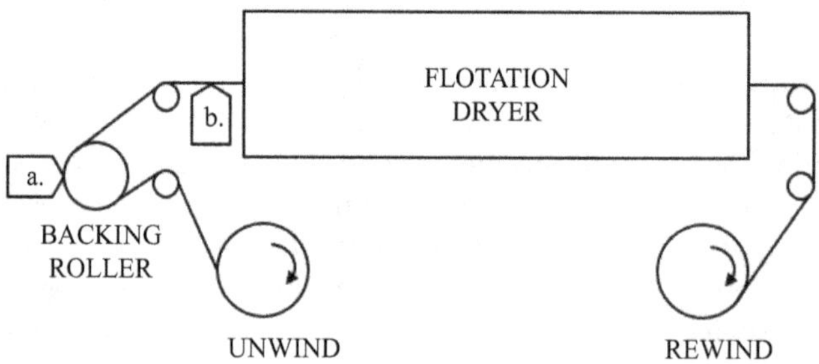

Figure 4.8 Diagram of coating line for dual-sided coating with (a) slot die coating against a backing roller and (b) slot die coating as TWOSD

they are using slot die, either in a multi-pass process or in a tandem coating line, then they should investigate dual-sided simultaneous coating utilizing tensioned web over slot die.

4.3.3 Lane/stripe coating

To take advantage of economies of scale [20], a wider electrode coating line should be more productive per square foot of facility floorspace than multiple narrower electrode coating lines. For this reason, and the need to reduce overall cell costs, for the mass adoption of electric vehicles there is a push to process foil current collectors at wider widths which requires electrode coating at wider widths. The fundamental scale-up of this process is nothing new and has been achieved in other industries as products are commercialized and markets mature. That said, coating wider widths can have challenges as previously discussed with regard to foil substrate quality and planarity along with slot die setup and adjustment. Assuming those challenges have been overcome the next logical question is what to do with wider width coating. Cell designs continue to evolve to require different electrode widths and lengths to be wound in the case of cylindrical cells or punched and stacked in the case of prismatic cells. In either cell type, there may be a need to have a coated electrode of a specific width, while still producing multiple narrow electrodes at the same time. To do this lane coating is required. As previously discussed, slot dies are generally used to create a coating of a specific width. The slot is machined with an overall maximum width and most often, a body shim is used to set the width corresponding to the desired coating width to produce one wide lane or stripe. This is considered full width coating and only leaves a narrow margin of uncoated foil substrate on the right and left edges. Full width coating can be later slit into individual electrodes of a specific width, but the slitting in the middle of the coated electrode will not result in an uncoated area. If an uncoated area is needed, then lane or stripe coating is required. With a slot die, this is achieved by modifying the internal body shim to create a mask or block off the slot in the areas where electrode coating is not desired.

Figure 4.9 depicts a slot die body shim cut for lane coating. In the desired coated areas, the shim is cut back to the slot die cavity, allowing electrode slurry to flow through to the exit slot. In the desired, uncoated areas, the body shim extends all the way through the internal slot gap to block off the exit slot in those areas. As a simple example, Figure 4.10 shows a three-lane coating pattern with two uncoated lanes. In this example, the resulting coated roll can be slit as needed to create discrete electrodes with an uncoated margin on either edge.

Lane coating on one side of the foil current collector is not the only requirement. As a coating on both sides of the current collector is needed the position and width of the coated lanes also becomes important. The individual lane width dimensions can be controlled by the body shim dimensions and the other slot die setup parameters. Control of the alignment of the individual lanes from one side of the coated foil to the other is a function of the slot die and foil alignment. Figure 4.10 shows a cross-sectional view of the coating lanes with some degree of

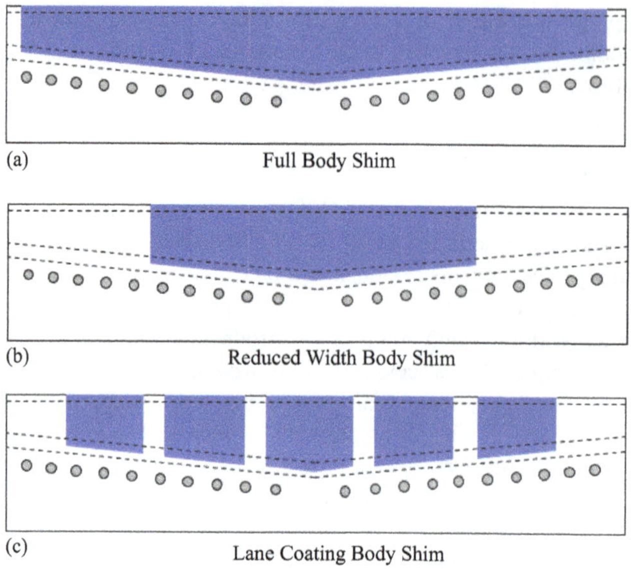

Figure 4.9 Slot die body shims for (a) full width coating, (b) narrow width coating, and (c) lane coating

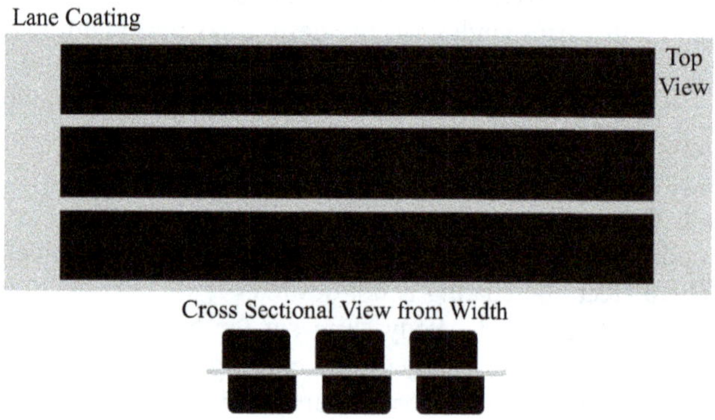

Figure 4.10 Diagram showing lane/stripe coating. Reprinted with permission from J. Li, J. Fleetwood, W. B. Hawley, W. Kays, From Materials to cell: state-of-the-art and prospective technologies for lithium-ion battery electrode processing, Chemical Reviews 122 (2022) 903–956 [3]. Copyright 2023 American Chemical Society.

misalignment between the side A coated lanes to the side B coated lanes [3]. To achieve precise alignment of the coated lanes on both sides of the foil some form of a measurement system combined with adjustment of the alignment must be employed. The measurement system could be as simple as a quality technician using a pin to mark the coated edge on one side of the foil, then measuring the difference in alignment on the opposite side of the foil. In this case, the feedback from the quality technician must be relayed to process technician to make adjustments to the slot die position to correct for misalignment. This process can be automated through a vision system that uses cameras to inspect the coated substrate, making real-time measurements of the coating widths, uncoated widths, and relative positions based on a datum which is typically the left or right edge of the base foil. Feedback from a vision system can be used to trigger the movement of an electric actuator connected to the slot die to make precision adjustments to the cross-web position, ensuring alignment of the coated lanes. Sophisticated systems like this do require significant investment and time to setup but may be required to achieve repeatable results to meet the electrode coating process operational metrics.

4.3.4 Intermittent/patch coating

Depending on the cell design, coating of discrete lanes may not be required. Some cylindrical cell designs require an uncoated gap to be used for tab welding. This requires a slightly different pattern of coating, referred to as intermittent, patch, or skip coating [3]. In this coating configuration, the coated pattern is not lanes or stripes running in the machine direction, but rather interrupted coating to create uncoated gaps across the width of the coating. Figure 4.11 depicts an example of intermittent coating looking from the top view as well as a cross-sectional side view.

Intermittent coating is not as straight forward of a process as lane coating. The slot die coating process is considered a continuous process. In the basic form, it is continuous across the width of the substrate (as in full width coating) and it is continuous in the machine direction along the length of the substrate. To create coated lengths and uncoated gaps for LIB electrodes requires augmentation of the

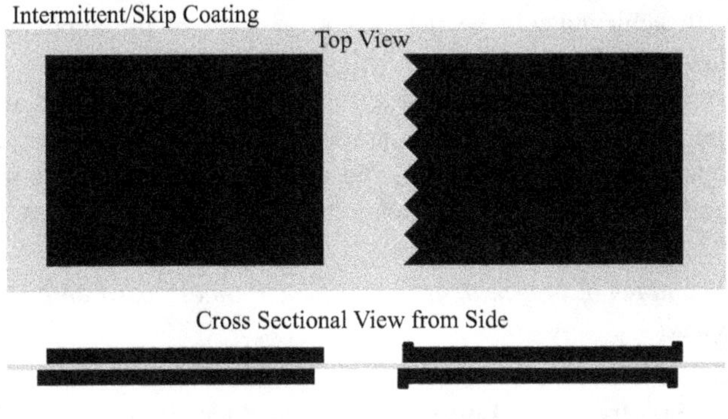

Figure 4.11 Diagram of intermittent/patch coating. Reprinted with permission from J. Li, J. Fleetwood, W. B. Hawley, W. Kays, From materials to cell: state-of-the-art and prospective technologies for lithium-ion battery electrode processing, Chemical Reviews 122 (2022) 903–956 [3]. Copyright 2023 American Chemical Society.

slot die process to be semi-continuous. To accomplish this goal there have been a variety of techniques, special tooling, and patented technology developed. Without debating specific techniques or technology, fundamentally, the coating must be interrupted via two main methods: (1) the slot die position relative to the substrate must be changed to break the coating meniscus and (2) the fluid flow must be stopped or redirected allowing the pressure in the slot die to reach atmospheric. If both of those actions take place, then the coating will be stopped. The relative success of the process is dependent on the timing of both of those actions to stop or interrupt the coating and then restart the coating at the coating line machine speed.

The coating line speed can increase the complexity of the intermittent coating process. Suffice to say that higher coating line speeds and shorter desired uncoated gaps can be increasingly more difficult to produce with quality edges on the starts/stops. Consider the following example, a coating process is developed at a slower speed. Let us assume 10 m/min coating line speed, 1,500 mm for the coated electrode length and 15 mm for the uncoated gap, which is often referred to as the mass-free zone. Based on the line speed the time factors for coating the electrode length and uncoated mass-free length are 9 s and 0.09 s, respectively. That means to create the mass-free zone, the coating gap must be increased to break the coating bead and the slot die pressure must be relieved to atmospheric and both reversed for coating restart all within 90 ms. This requires very sophisticated timing of the process variables. Now consider the same coated and uncoated lengths but increasing the line speed to something much faster, like 50 m per minute, as the industry works to increase productivity. Table 4.1 shows the impact on the cycle times.

The cycle times as shown in Table 4.1 highlight the complexity of trying to stop and restart the coating process within the time allotted. It is not feasible to both move the slot die position and drive the internal pressure drop to atmospheric and reverse the process all within 11.25 ms at 80 m per minute coating line speed. At higher line speeds, long uncoated mass-free zones will be required. The minimum uncoated length achieved will be dependent on a lot of the application-specific parameters including the intermittent coating methodology used.

For a given intermittent coating pattern, let us assume the coated and uncoated cycle times are within an acceptable range we then must address the quality of the starts and stops. Since the default slot die process is continuous it starts with a quality electrode already being applied to the foil substrate. When intermittent

Table 4.1. Examples of intermittent coating cycle times based on coating line speed

Coated length (mm)	Mass-free length (mm)	Line speed (m/min)	Coated length cycle time (ms)	Mass-free cycle time (ms)
1,500	15	10	9,000	90.00
1,500	15	50	1,800	18.00
1,500	15	80	1,125	11.25

coating control is initiated the first break in coating starts the semi-continuous process and ultimately sets up the start of the next coated patch. What is meant by this is the quality of the end of the preceding patch greatly impacts the quality of the start of the following patch. When working in the area of intermittent coating process development it is common to refer to the start of the coated patch as the "head" and the end of the coated patch as the "tail." A lot of experimental work goes into the development of the process parameters to impact the quality of the heads and tails. As mentioned, multiple times in this section, during the mass-free length the slot die internal pressure must be drawn down to atmospheric. This is important as excess pressure in the slot die during the mass-free cycle time will result in excess slurry accumulation on the slot die exit lip. This slurry then transfers to the foil substrate at the start of the next patch, creating a heavy coating or thick bead at the head. That scenario is an undesired negative result. The key to successful intermittent coating is the timing of the devices that impact the coating gap and slot die pressure. Most often the slot die pressure is relieved by adjusting valve(s) to redirect slurry flow back to the holding vessel. During this process, the fluid delivery pump remains running at the same speed to keep a constant flow rate through the process piping in preparation for restarting coating. The timing of diverting the flow from the slot die and movement of the slot die to break the coating bead are critical to impacting the quality of the coated patch heads and tails. The topic of intermittent coating process development is the subject of a lot of experimental and simulation work that continues to remain ongoing.

4.4 Summary

As the coating process is the primary step in the production of electrodes, the specific coating method chosen can have a large impact on the quality of the electrode and the productivity of the production operation. Battery producers and electrode processors alike have to balance operational metrics such as speed, productivity, and scrap rates across a full range of cell/electrode designs while maintaining a profitable operation. The production of the base electrode on the current collector is a critical step and it is important that the optimal coating method is selected to provide a wide range of flexibility while maintaining quality. While a lot of research and development for electrode chemistry is carried out using coating methods, such as a simple doctor blade, which may seem crude but effective at creating small sample coatings for evaluation. Once the electrode chemistry is defined and the process scale-up to mass production is required a more sophisticated coating system and method is required. The energy storage industry trend has been to leverage the experience and knowledge base of the slot die coating process and apply it to battery electrode production. Slot die coating has been utilized in other industries going back to the 1950s. Early examples may have included the coating of paper products using relatively crudely made slot dies. As metal cutting and machining technology advanced, more precise slot dies can be produced for specialty coated products, such as photographic film, magnetic media, transdermal medical coatings, and optical films for use in flat panel displays. Fortunately, there continue to be incremental improvements across the industry by way of slot die manufacturing/finishing techniques and coating system design with ancillary tooling. Unfortunately, simply selecting the slot die coating method and even procuring a high-quality slot die does not guarantee a high-quality electrode coating. As discussed in this chapter, there are a lot of interrelated variables that impact the quality of the coating and productivity of the process. Much of the knowledge and experience on electrode coating process development is kept as a trade secret and not shared openly. This contributes to the competitive nature of the industry but also may limit the rate of change on continuous improvement for slot process development.

For those new to the energy storage industry and slot die coating for electrode production the process can seem complex but also can be broken down into fundamentals. When planning an electrode coating process, first consider the production output requirements. What coat weight or target loadings are required? What coating widths and target speeds are required to meet production demand or forecasts? Answers to these questions will help guide the discussion on general sizing for the overall coating line which will result in a maximum machine width for the current collector and a maximum coating width. Once the slot die coating method is selected, it will be important to remember the slot die is part of a complete system comprised of the slot die, the slot die support and positioning system, the fluid delivery system, the precision backing roller, and the process inputs (electrode slurry and foil current collector). All those system elements should be reviewed in detail to ensure the individual components will work together for the planned production system. Many commercial companies who have already implemented electrode coating process systems may have a good handle on the interrelated process variables. For those needing assistance, it is recommended to survey the industry and seek out others with electrode coating process experience. In many cases, slot die manufacturing and coating equipment manufacturing companies may have the relevant experience and can help navigate key equipment decisions. They may also be able to assist with process development, troubleshooting, and training. A good working relationship between the equipment manufacturer(s) and the production/process engineering team can speed up the time to market for a new electrode product.

References

[1] Ding, X., J. Liu, and T.A.L. Harris, A review of the operating limits in slot die coating processes. *AIChE*, 2016. **62**(7): p. 2508–2524.

[2] BloombergNEF. *Global energy storage market to grow 15-fold by 2030*, 2022, 1/28/2023. https://about.bnef.com/blog/global-energy-storage-market-to-grow-15-fold-by-2030/.

[3] Li, J., *et al.*, From materials to cell: state-of-the-art and prospective technologies for lithium-ion battery electrode processing. *Chemical Reviews*, 2022. **122**: p. 903–956.

[4] Schmitt, M., *et al.*, Slot-die processing of lithium-ion battery electrodes – coating window characterization. *Chemical Engineering and Processing: Process Intensification*, 2013. **68**: p. 32–37.

[5] Schmitt, M., *et al.*, Analytical determination of process windows for bilayer slot die coating. *Journal of Coatings Technology and Research*, 2015. **12**(5): p. 877–887.

[6] Hawley, W.B. and J. Li, Beneficial rheological properties of lithium-ion battery cathode slurries from elevated mixing and coating temperatures. *Journal of Energy Storage*, 2019. 26: p. 100994.

[7] Du, Z., *et al.*, Enabling aqueous processing for crack-free thick electrodes. *Journal of Power Sources*, 2017. **354**: p. 200–206.

[8] Li, J., *et al.*, Superior performance of $LiFePO_4$ aqueous dispersions via corona treatment and surface energy optimization. *Journal of the Electrochemical Society*, 2012. **159**(8): p. A1152–A1157.

[9] Sahore, R., *et al.*, Performance of different water-based binder formulations for Ni-rich cathodes evaluated in $LiNi_{0.8}Mn_{0.1}Co_{0.1}O_2$//graphite pouch cells. *Journal of the Electrochemical Society*, 2022. **169**: p. 040567.

[10] Zhu, P., *et al.*, A review of current collectors for lithium-ion batteries. *Journal of Power Sources*, 2021. 485.

[11] Hoffmann, A., *et al.*, Influence of the mixing and dispersing process on the slurry properties and the microstructure and performance of ultra-thick cathodes for lithium-ion batteries. *Energy Technology*, 2022: p. 2200484.

[12] Kim, K.M., *et al.*, Effect of mixing sequences on the electrode characteristics of lithium-ion rechargeable batteries. *Journal of Power Sources*, 1999. **83**: p. 108–113.

[13] Lee, G.-W., *et al.*, Effect of slurry preparation process on electrochemical performances of $LiCoO_2$ composite electrode. *Journal of Power Sources*, 2010. **195**: p. 6049–6054.

[14] Kremer, L.S., *et al.*, Manufacturing process for improved ultra-thick cathodes in high-energy lithium-ion batteries. *Energy Technology*, 2020. 8: p. 1900167.

[15] Wang, M., *et al.*, Effects of the mixing sequence on making lithium ion battery electrodes. *Journal of the Electrochemical Society*, 2020. **167**: p. 100518.

[16] Hawley, W.B. and J. Li, Electrode manufacturing for lithium-ion batteries – analysis of current and next generation processing. *Journal of Energy Storage*, 2019. 25: p. 100862.

[17] Arnot, D.J., *et al.*, Thick electrode design for facile electron and ion transport: architectures, advanced characterization, and modeling. *Accounts of Materials Research*, 2022. 3: p. 472–483.

[18] Boyce, A.M., *et al.*, Design of scalable, next-generation thick electrodes: opportunities and challenges. *ACS Nano*, 2021. **15**: p. 18624–18632.

[19] Ouyang, L., *et al.*, The effect of solid content on the rheological properties and microstructures of a Li-ion battery cathode slurry. *RSC Advances*, 2020. **10**: p. 19360–19370.

[20] Mauler, L., F. Duffner, and J. Leker, Economies of scale in battery cell manufacturing: the impact of material and process innovations. *Applied Energy*, 2021. 286: p. 116499.

Chapter 5
Laser materials processing in manufacturing of lithium-ion batteries

Peter Smyrek[1] *and Wilhelm Pfleging*[1]

5.1 Introduction

Lithium-ion batteries (LIBs) were first commercially offered by Sony in 1991 [1]. Since then, the demand for maximum power and energy efficiency has increased rapidly. LIBs have become the dominating energy storage technology for electric vehicles (hybrid electric vehicles, plug-in hybrid electric vehicles, or fully electric vehicles) as well as for stationary electrochemical storage systems [2]. LIBs are characterized by their relatively high gravimetric energy and power density of 150 Wh/kg–260 Wh/kg and 340 W/kg–500 W/kg, respectively. For LIBs, the cost per kilowatt hour was USD 764 in 2009 and could be reduced to USD 111 by 2020 [3–5]. However, it should be noted that the costs and performance limits of current LIBs still are bottlenecks for electric vehicles and efficient operation in combination with solar and wind power plants. Efficient stationary storage devices are required to compensate for the intermittent properties of, for example, wind turbines and to smooth their load peak during power grid feeding.

In the field of LIB technology, recent research is concerned with increasing energy and power density by simultaneously lowering production and material costs. Inspired by the commercial success of lithium nickel manganese cobalt oxide ($Li(Ni_{1/3}Mn_{1/3}Co_{1/3})O_2$, NMC 111), new nickel-enriched layered oxides with various compositions ($Li(Ni_{1\text{-}x\text{-}y}Mn_xCo_y)O_2$, with $1\text{-}x\text{-}y > 0.5$) are currently under intense investigation for increasing the reversible capacity in lithium-ion cells. Contemporaneously, the manufacturing of electrodes with film thicknesses of about 320 μm could be realized on a laboratory scale. Here, the aim is to apply a reduced amount of inactive material and thus to realize an enhancement of energy density at the cell level. As calculated in [6] for NMC 111 batteries (vs. graphite), increasing the layer thickness from state-of-the-art 50 μm to values about 200 μm to 300 μm can contribute to an increase in energy density of more than 20–30% up to values of 250 Wh/kg and 580 Wh/L. However, for high-power application, the use of thick-film electrodes with thicknesses $\geq$ 100 μm is conditionally realizable. On

[1]Institute for Applied Materials – Applied Materials Physics, Karlsruhe Institute of Technology, Germany

the one hand, the mechanical integrity of the electrode layer is impaired, on the other hand, due to high-mass loading and areal capacity of > 20 mg/cm^2 and $>$ 4.0 mAh/cm^2, a diffusion overpotential is established, which reduces the available capacity in a lithium-ion cell at a certain C-rate. The term "C-Rate" is used to describe the applied power densities during battery operation. The C-rate indicates the current at which the battery is being discharged and/or charged, based on the practical capacity of the cell. For example, if the C-rate value is C/2 or 2C, then the battery will be fully charged or discharged within 2 h or 30 min, respectively.

An effective approach, which enables a battery operation with a simultaneous high energy and power density, is based on the so-called "3D-Battery-Concept," which was introduced by Long *et al.* [7] for micro-scaled batteries. Figure 5.1 presents a principle scheme of classical 2D and advanced 3D-Battery-Concept [8]. In comparison to the planar electrode arrangement, which represents the current state-of-the-art design (Figure 5.1(a)), the electrode geometry based on the 3D-Battery-Concept is arranged in a three-dimensional (3D) configuration (Figure 5.1 (b)). According to Long *et al.* [7], the general strategy of this approach is to fabricate a 3D electrode design including free-standing architectures that can maximize power and energy density while maintaining short lithium-ion transportation pathways. An increased active surface area is achieved, which shortens the transport pathways for lithium-ions in liquid electrolyte. It is worth mentioning that the active surface area is referred as the electrode area, which stands in direct contact with the free liquid electrolyte. Furthermore, the charge transfer resistance decreases and a reduction in material stress is expected due to volumetric expansion and contraction during lithiation and delithiation. Finally, the cycle life can be increased [6]. The approach of the 3D-Battery-Concept was mainly motivated for applications in micro- and nano-electromechanical systems [9]. Typically, process technologies such as photolithography, etching processes, or thin-film deposition were therefore introduced to produce 3D micro-batteries [10,11].

For classical composite electrodes, as they are nowadays manufactured and integrated in lithium-ion cells, the implementation of the 3D-Battery-Concept was demonstrated for the first time by the research group of Pfleging *et al.* [12–17] at

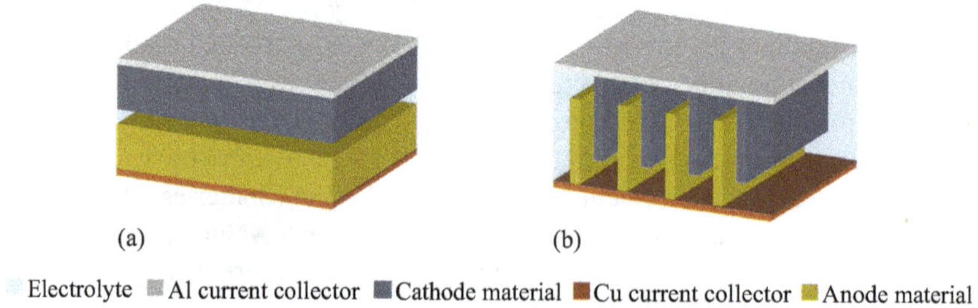

Figure 5.1 Schematic representation of the (a) 2D and (b) 3D-Battery-Concept for micro-scaled batteries. Reproduced from [8] (open access).

the Karlsruhe Institute of Technology (KIT). 3D electrode architectures were generated by laser-assisted processes to improve cycle stability, increase cell lifetime, and realize an accelerated electrolyte-filling process. The latter one can be referred to the laser structuring part of the electrode which finally leads to a spontaneous and homogeneous liquid electrolyte wetting [9,18,19]. Pfleging *et al.* [6,20–28] proved that the laser structuring process fulfills the goals of the 3D-Battery-Concept. Furthermore, the 3D-Battery-Concept can be transferred to different types of cathode materials (e.g., $LiFePO_4$, NMC 111, NMC 532, NMC 622, and NMC 811) with film thicknesses in the range from 50 μm to 250 μm. Pröll *et al.* [18] were able to show that the wettability of NMC 111 thick-film cathodes was significantly improved by adjusting the 3D-Battery-Concept. Due to the generated micro-capillary structures, spontaneous and homogenous wetting with liquid electrolyte was initiated. As a consequence, electrochemical data show that, compared to cells with unstructured electrodes, the operational lifetime increased significantly and an improved cycle stability at elevated C-rates was achieved. Rakebrandt *et al.* [25] and Zhu *et al.* [27] demonstrated for the first time that the process of laser structuring can also be transferred to ultra-thick film NMC electrodes with thicknesses of up to 250 μm. Depending on the electrode film thickness and applied 3D structure, an active mass loss of less than four percent could be achieved. Recently Pfleging's research group introduced laser-assisted structuring of thick-film electrodes, which are based on self-made water-based production route on cathode and anode side using NMC 622 and silicon-graphite composites, respectively [29–31]. Based on the mentioned pioneering work, several research groups are now heading for the process of laser structuring of battery materials [32–40]. Besides overcoming overpotential problems, one central goal hereby is to counteract fast charging issues of high-energy cells by applying the 3D battery concept. Review articles from Pfleging [6,22] and very recently from Li *et al.* [41] give quite a comprehensive overview of current international approaches in the laser processing of battery materials.

In general, the current and future requirements and parameters for next generation LIBs are shown in Figure 5.2, extracted from [42,43]. In addition to a gravimetric energy density of greater than 275 Wh/kg (at cell level), it is predicted that the cell-level power density will reach values of larger than 800 W/kg. In addition, the charging process should be able to be carried out within 10–15 min and the total operational lifetime at battery level for fully electric vehicles should be around 2,000 cycles [42].

Regarding the optimization of electrodes by laser structuring, it is essential to correlate the generated 3D architectures with electrochemical and chemical data. Depending on the electrochemical cycling, conclusions can be drawn about the degradation behavior of the composite materials. The assessment largely depends on the used characterization method. While *in situ* measurements enable direct monitoring of micro- and nano-scaled processes, *ex situ* techniques are effective when higher resolution is required during material characterization [44]. Characterization methods, such as energy-dispersive X-ray spectroscopy (EDX), X-ray photoelectron spectroscopy, Auger electron spectroscopy, or time-of-flight

	Energy density	Power density	Fast charging	Cell lifetime
2021 (today)	160 – 260 Wh/kg 450 – 730 Wh/l	340 – 500 W/kg ca. 1000 W/l	15 – 30 min (full electric vehicle, FEV)	≈ 5000 cycles (in combination with C–Anodes)
2030 (in future)	275 – 320 Wh/kg 750 – 900 Wh/l	800 –1200 W/kg ca. 2000 W/l	10 –15 min (full electric vehicle, FEV)	≈ 10000 cycles (in combination with C–Anodes)

Figure 5.2 Performance indicators (at cell level) for research and development of lithium-ion batteries extracted from data by Armand et al. [42] and the European Energy Storage Technology Development Roadmap [43].

secondary ion mass spectrometry, offer a high lateral resolution. However, they are less effective for volume analyses. Although EDX has a depth resolution in the range of a few microns, it cannot be used as an analysis method for detecting light elements such as lithium due to the beryllium window.

Laser-induced breakdown spectroscopy (LIBS) is a chemical analysis method that can be used to analyze electrode materials with a depth resolution of about 2 µm–8 µm. As an application of atomic emission spectroscopy, the elemental composition can be determined either qualitatively or quantitatively within a few milliseconds. Depending on the depth of field of the laser radiation, it is also possible to analyze film thicknesses of up to 500 µm without an adjustment of the laser focus position. With the selection of suitable measurement parameters, electrode thick layers can be completely recorded and examined in the range from a few microns up to pouch cell dimensions. LIBS has already been used successfully applied in the field of LIBs material characterization, for example, for solid electrolytes [45]. In addition, the binder distribution within the electrode could be qualitatively analyzed and described based on the chemical composition (also at interfaces) [6,46,47]. Other publications describe the use of LIBS on thick-film electrodes (anodes and cathodes) to investigate the lithium concentration distribution semi-quantitatively [32,48,49]. Recent publications show that LIBS is used for parameter optimization of laser structured NMC 532 cathodes. Depending on the structure geometry and depth, the lithium concentration and its distribution were investigated as a function of the applied C-rate [33]. The use of the LIBS method opens up new possibilities to fully characterize battery materials post mortem in order to be able to draw conclusions about the effects of the electrode design and degradation processes and their origin.

5.2 Materials and methods

In this chapter, concepts for ultrafast laser-assisted structuring of electrode materials are presented with suitable process sequences. Furthermore, the manufacturing process of thick-film NMC electrodes will be shortly described. Finally, the basic LIBS components for analyzing thick-film electrodes will be introduced.

5.2.1 Laser materials processing of battery materials

Laser materials processing systems, PS450-TO (Optec s.a., Frameries, Belgium) and MSV-203 (M-Solv Ltd., Oxford, UK), were equipped with different ultrafast laser sources providing medium laser average powers up to 35 W (Satsuma and Tangerine type, Amplitude Systèmes s.a., Bordeaux, France) and ultrafast laser systems with high average laser power of 300 W and 500 W (FX600 types, EdgeWave GmbH, Würselen, Germany). The processing systems are equipped with roll-to-roll technique and large field optical scanner to be able to process commercial and advanced electrode materials with a large footprint. The femtosecond lasers operate at fundamental wavelength of 1,030 nm or at its second harmonic (wavelength: 515 nm) with a pulse duration in the range of 280 fs (fs = 10^{-15} s) to 10 ps (ps = 10^{-12} s). Laser pulse repetition rates from 100 kHz up to 50 MHz can be applied for electrode cutting and structuring [34].

As an example, the PS450-TO system equipped with ultrafast fiber laser of type Tangerine will now be described in more detail. The ultrafast fiber laser emits laser radiation in the near-infrared range with a fundamental wavelength of 1,030 nm and enables pulse lengths in the range from 350 fs to 10 ps. The laser source is operated in the basic TEM_{00} mode and is specified with a diffraction index of $M^2 < 1.3$. The repetition rate f of the laser can be variably adjusted in the range from 1 Hz to 2 MHz. A maximum pulse energy of 175 μJ and a maximum average power of P=35 W can be applied. The optical beam path of the PS450-TO system is shown schematically in Figure 5.3(a).

Starting from the ultrashort pulsed (USP) laser beam source, the laser radiation with a wavelength of 1,030 nm is first guided into a frequency converter via a partially transparent deflection mirror. The laser wavelength can then be halved or

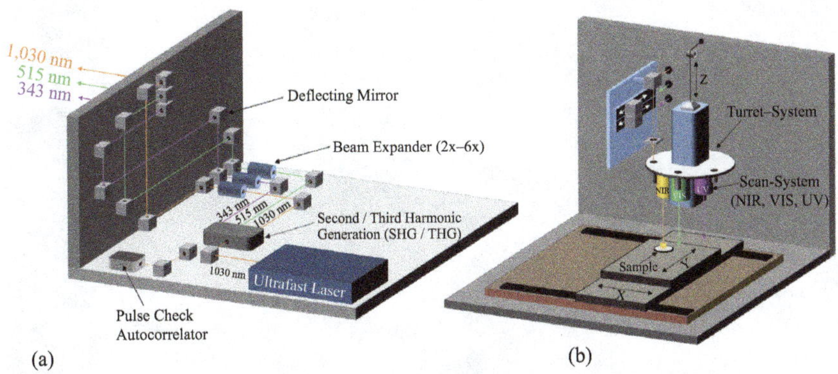

Figure 5.3 Schematic representation of the PS450-TO type laser processing system: arrangement of (a) the beam path and (b) the laser processing chamber. Reproduced from [8] (open access).

divided into three (515 nm or 343 nm). Depending on the process application or the degree of absorption of the material, the appropriate selection of the wavelength for laser-assisted processing of materials is crucial. A small part of the laser radiation is guided into an autocorrelator (PulsCheck, APE Angewandte Physik & Elektronik GmbH, Germany) via deflection mirrors. This is used to measure the laser pulse duration which can be adjusted in a range from 350 fs to 10 ps. For the laser radiation emerging from the frequency converter, a wavelength-specific deflection mirror was then used for transmission to a telescope. The electrodes were processed in a process chamber (Figure 5.3(b)) using a scanner system (RhothorTM Laser Deflection System, Newson Engineering BV, Overmere, Belgium), which has a deflection unit (Rhothor Active Smart Deflector, RTA) with a positioning accuracy of less than 8 μrad and a repeatability of less than 8 μrad. With integrated turret optics, it is also possible to carry out the laser-assisted structuring process with wavelengths in the NIR, VIS, and UV ranges without having to move the sample material. Depending on the used optics (F-Theta lens), the focus position on the sample surface is set via an axis that can be moved in the z-direction (accuracy: 0.5 μm, repeatability: ± 1 μm). This is determined using a microscope system, which has an optical zoom and is also attached to the turret optics. In addition to the scanner system, a motorized *X* and *Y* positioning table (Aerotech, Inc.) can be used, which is specified with a positioning accuracy of ± 1 μm and a repeatability of ± 0.5 μm. One suitable set of process parameters for laser structuring of NMC 111 electrodes is achieved for a wavelength of 515 nm, a pulse duration of 380 fs, and a repetition rate of 500 kHz. The composite layer is removed using the scanning method and using an F-Theta lens with a focal length of 100 mm. Depending on the layer thickness and the degree of porosity of the electrodes, the number of scan repetitions can be varied. Figure 5.4 shows examples of laser-structured electrodes with hole pattern and line pattern for cathode and anode, respectively.

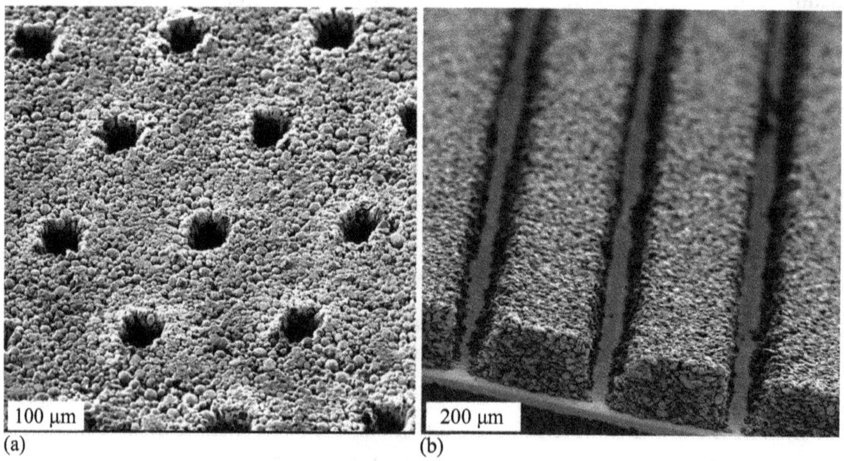

Figure 5.4 SEM images of ultrafast laser-structured electrodes: (a) hole pattern in NMC 811 cathode and (b) line pattern in graphite anode.

5.2.2 Battery materials for cell fabrication

Commercially available active materials were selected for electrode coating development via tape casting with a doctor blade (ZUA 2000.100, Proceq, Schwerzenbach, Switzerland) using a lab coater (MSK-AFA-L800-LD, MTI Corporation, Richmond, CA). The thickness of the electrode composite layers was controlled by varying the gap of the doctor blade. On cathode side NMC powders such as NMC 111 powder (MTI Corporation, USA), NMC 622 (BASF SE, Ludwigshafen, Germany), and NMC 811 (ECOPro BM, Chungcheongbuk-do, Korea) were used while on anode side active materials were applied such as flake-like graphite (Targray, Technology International Inc., Kirkland, QC, Canada) and/or nano-sized silicon particles (2W iTech, LLC, San Diego, CA). For slurry preparation, different types of binders were introduced such as NMP-based polyvinylidene fluoride (PVDF, MTI Corporation, USA), water-based carboxymethyl cellulose (CMC, CRT 2000PA, DoeWolff Cellulosic, Bomlitz, Germany), styrene-butadiene rubber (SBR) (MTI Corporation, USA), and water-based fluorine acrylic copolymer latex TRD 202A (JSR Micro NV, Leuven, Belgium). Finally, carbon black C-NERGY Super C65 (Imerys G & C Belgium, Willebroek, Belgium) or Timcal Super C65 (MTI Corporation, Richmond, CA, USA) were used as conductive additives. Detailed descriptions of the respective recipes for anodes and cathodes can be found in [30,35,36] and [31,37,38], respectively. The slurries were tape casted onto thin metallic aluminum (cathode) and copper (anode) current collector foils. Prior to laser structuring, the electrode was calendered using a hot rolling press machine (HR01 Hot Rolling Machine, MTI Corporation, Richmond, CA, USA). Finally, the manufactured electrodes reveal a porosity of 35% (cathode) and 42% (anode).

As an example of used electrode materials, Figure 5.5(a) shows a photograph of the wet film after the tape casting process has been carried out. The wet film was

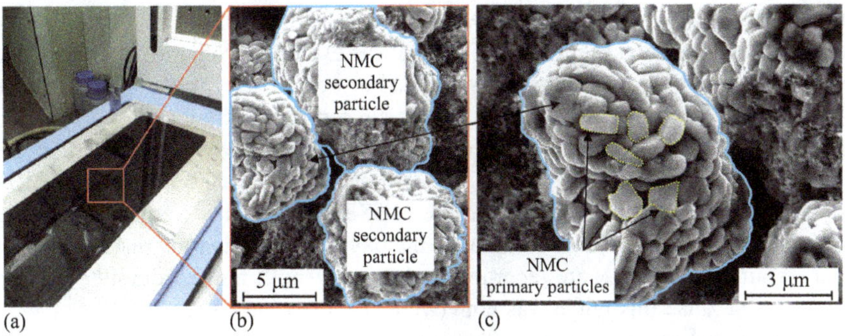

Figure 5.5 Fabrication and characterization of NMC 111 thick-film cathodes: (a) photograph of the electrode thick film (wet film) after the coating process (film casting process), (b) SEM image of the surface of an NMC thick-film cathode after the drying process. The blue-framed particles identify the NMC secondary particles, which consist of primary particles—shown in (c) with yellow dashed lines. Reproduced from [8] (open access).

dried at 90°C. The subsequent characterization of the produced electrodes was initially carried out by SEM investigations (Figure 5.5(b,c)). The NMC secondary particles, which are surrounded by binder (PVDF) and conductive carbon black (Timcal Super C65), are composed of individually irregular primary particles.

In a further processing step, the manufactured NMC thick-film cathodes were calendered. The calendering process is essential in the production of lithium-ion batteries. On the one hand, it is ensured that the particle density increases during the compression, and on the other hand, an improvement in the adhesion of the layer on the current conductor is guaranteed. Care must be taken to make sure that the desired porosity of the electrodes is adjusted.

5.2.3 Laser-induced breakdown spectroscopy (LIBS)

LIBS was used to determine lithium concentration profiles in thick-film electrodes. A passively mode-locked diode-pumped neodymium-doped yttrium-aluminum-garnet solid-state laser (Nd:YAG) with a fundamental wavelength of 1,064 nm was used to generate the plasma. The repetition rate can be variably adjusted in a range from 1 Hz to 100 Hz. The pulse energy and pulse duration are 3 mJ and 1.5 ns, respectively. The laser beam was guided via an optical fiber and subsequently focused onto the electrode surface using a focusing mirror with a focal length of 75 mm. The radiation emission of the plasma was then coupled in an optical fiber bundle and fed into two spectrometers arranged in parallel (type: AvaSpec-ULS2048XL, Avantes, Netherlands). The detector covers spectral ranges between 229–498 nm and 569–792 nm. The data were analyzed by SEC Viewer 1.9 (Secopta GmbH, Teltow, Germany). The measured intensity of the elements detected in the plasma was visualized as a function of the emission wavelength [39]. The identified emission lines were assigned with the help of the NIST database [40]. Due to the sensitivity of the LIBS method for light elements such as lithium, the two Li^{I} peaks at λ=610.4 nm and λ=670.8 nm are particularly pronounced. However, the Li^{I} peak at 670.8 nm was excluded from the calibration due to the observed self-absorption effect [50]. Further intense peaks, which can be assigned to the transition metals Ni, Mn, and Co in the excited state, can be also detected [39]. The oxygen contained in the NMC cathode was detected in the wavelength range from 777.2 nm to 777.5 nm but was neglected for the later LIBS calibration. The reason for this is that the LIBS measurements were carried out in ambient air and this can lead to values in the oxygen determination that are subject to uncertainty. No data were recorded in the wavelength range from 499.7 nm to 568.2 nm, because this range was not covered by the used detectors.

The measurement position on the sample surface was set with an axis system (X-direction) and with a positioning table that can be moved in the Y-direction. An additional camera attached to the axis system (Z-direction) was applied to adjust the focus position. In addition, the visual recognition system was used to determine the impact position of the first laser pulse with high accuracy ($\pm$5 μm) via an "offset" relative to the laser beam focusing optics. For a typical measurement area of 13 mm × 13 mm, a total number of 17,161 spectra per layer were recorded. Due to the effect of self-absorption at the wavelength of 670.8 nm, the Li^{I} emission wavelength of 610.4 nm was used for the analysis of lithium [39]. The calibration data were derived from LIBS and inductively coupled plasma optical emission spectroscopy (ICP-OES) as described in [39,45]. Lithium concentration and distribution related to C-rates were systematically investigated in structured and unstructured electrodes [30,35,39,45,51,52].

5.3 Lithium concentration profiles in unstructured NMC thick-film electrodes

The use of electrodes with high-mass loading is currently one of the most promising strategies to increase the volumetric/gravimetric energy density of the lithium-ion battery while using the well-established types of active materials. In the production of such high-energy electrodes, approaches are presented with the aim to reduce inactive cell components (e.g., current collector, separator) [6,53]. It could be shown in case of a 52 Ah battery (NMC 111 vs. graphite) that by increasing the cathode film thickness from 50 μm to 125 μm and 250 μm the specific energy density at cell level can be increased by 18% and 27%, respectively [6]. However, a further increase in layer thickness shows a saturation effect for the enhancement of volumetric and gravimetric energy density at the cell level. Taking into account that roll-to-roll processing of electrode material should be still possible, it can be assumed that the maximum electrode film thicknesses in the range of 200 μm–300 μm (for each side of the current collector) might be technically feasible.

Regarding low C-rates (< C/2), the thick-film approach might be still practicable. However, with the use of higher C-rates (> C/2) for cells with ultra-thick film electrodes, the electrochemical performance of the cell is severely impaired [54]. For example, Lee *et al.* [55] achieved less than 20% of the practical capacity at a C-rate of 2C for cathodes with a mass loading of about 25 mg/cm^2. Another aspect that poses challenges for the use of high-mass loaded electrodes is the wetting of electrodes with liquid electrolyte. The filling process becomes more complex and time consuming as the layer thickness increases, especially in the case of electrode configurations with low porosity [6,18,22]. In case of insufficient electrode wetting with liquid electrolyte dry areas with electrochemical inactivity are formed, especially in the deeper layers of the electrode which are close to the current collector. Those areas provide preferred starting points for degradation processes such as lithium plating. From LIBS analysis data, a model scheme description for NMC thick-film cathodes could be set up, which illustrates the possible degradation processes and respective locations within the cell. In a study published by Park *et al.* [56], NMC standard and high-energy electrodes were produced with an active mass loading of 20 mg/cm^2 and 28 mg/cm^2, which correspond to layer thicknesses of 70 μm and 100 μm, respectively. Depth profile measurements were carried out using LIBS in order to determine qualitatively the lithium distribution of NMC thick electrodes after manufacturing. The depth resolution that corresponds to the material removal per laser pulse during the LIBS measurement was in the range of 7 μm–10 μm. The relative lithium concentration was normalized to the intensity of the first measurement, each time recorded on the surface of the NMC thick-film electrode. The procedure was repeated for each data set of the depth profiles. The area of electrode exposed to the LIBS measurements was not specified. It becomes clear that the values of the lithium content for electrodes after manufacturing vary as a function of the electrode depth—starting from the electrode surface. For the standard electrode, the relative lithium content varies in the range from 0.95 to 1.125. The minimum value being detected after the second laser pulse. The maximum value of the lithium content is assigned to the seventh pulse, which corresponds to an electrode depth of 49 μm. The LIBS analysis on the high-mass loaded electrode also shows no significant change or deviation in the lithium concentration. The variation in the lithium content is almost in the same range, with a minimum and maximum of the lithium content after the fifth and tenth laser pulse, respectively. For the electrodes manufactured in [45], the averaged lithium concentration after the coating process was in the range between 0.99 ± 0.01 and 1.09 ± 0.02, which indicates a uniform particle distribution in the measured electrode. The relative lithium concentration does not provide any information about quantitative data that could indicate, for example, the over-lithiation of the cathode material. However, the study shows that LIBS as an analysis method in the field of battery development enables the detection of lithium concentration profiles. To investigate the lithium content in electrochemically cycled electrodes, Park *et al.* [56] assembled half-cells (vs. Li/Li^+) in coin cell design. The galvanostatic measurements were carried out using the constant-current method, with the LIBS analysis being carried out post mortem after cycle number 25th, 50th, 75th, and 100th.

The strong impact of the number of cycles, which significantly influences the lithium distribution both on the electrode surface and in the deeper regions of the electrode, could be shown. The inhomogeneous lithium distribution is clearly pronounced from a number of cycles of 50, in which the electrodes show an increase or decrease in lithium concentration in regions close to the current collector, depending on the process (charging or discharging). According to Park *et al.* [56], the relative lithium concentration nearby the current collector at the end of the charging cycle is significantly higher than on the electrode surface. It can be assumed that the charging process in the electrode near the interface between the composite material and the current collector was not completed. The increased lithium concentration is confirmed with an increase in the number of cycles and in particular after a number of cycles in the range between 75 and 100. In the discharged state, the electrode also shows an inhomogeneous distribution of the lithium content. However, compared to the charged state, the level of non-uniformity appears to be much less pronounced. Park *et al.* [56] explained this behavior with the fact that the discharging process begins with a charged state, which already contains a lower lithium content in the region close to the electrode surface. Furthermore, it was pointed out that during discharging the top layer achieves a much faster recovery of the lithium content, which indicates the significant inhomogeneity of the reaction in the electrode. The LIBS results obtained by Smyrek *et al.* [45] could confirm the results of Park *et al.* [56] in a first approximation and for selected positions in the electrode. Those results will be presented in the following chapter.

5.4 LIBS for evaluation of degradation processes in unstructured NMC thick-film electrodes

To investigate the influence of cell cycling regarding the lithium concentration and its distribution in the composite material, unstructured NMC 111 thick-film electrodes were assembled into half-cells. Post mortem LIBS was applied in order to determine the lithium concentration on the sample surface and in the entire electrode. In a first approach, the electrodes were measured after the formation process. For the NMC 111 cathode, the voltage range was set from 3.0 V to 4.2 V. The proportion of lithium-ions extracted from the cathode material was in the range of 50%–55%, so that the practical specific capacity corresponds to a value of around 150 mAh/g. The formation was performed with a C-rate of C/10 (charging and discharging) for three cycles. The cell was then disassembled; the cathode material was washed in dimethyl carbonate (DMC) and subsequently characterized by using LIBS. In general, it can be observed that the lithium concentration shows a homogeneous distribution after the formation process of the lithium-ion cell [45].

To evaluate the lithium concentration distribution in heavily degraded cathode materials, half-cells were examined which showed spontaneous cell failure behavior during battery operation [45]. At the beginning of the galvanostatic measurement, the cell was first cycled in a formation process with a low charging and discharging rate of C/10 for three cycles. A specific discharge capacity of 142 mAh/g could be measured. The cell was then subjected to a lifetime test, operated at a constant charge and discharge rate of C/2. The decrease in specific discharge capacity in the fourth cycle, which was initiated due to the higher current rate, can be clearly seen. The specific discharge capacity drops to a value of 122 mAh/g, which corresponds to 85.9% of the initial capacity. As the galvanostatic measurement continues, it becomes apparent that the specific discharge capacity of the cell decreases more rapidly from the 60th cycle and the cell finally fails in the 93rd cycle. The specific discharge capacity suddenly drops from 95 mAh/g to a value of 0 mAh/g. For the quantitative investigation of the lithium concentration distribution, the half-cell was disassembled, the NMC cathode was washed in DMC, and LIBS was performed post mortem.

The lithium concentration was averaged in the X-direction and in the range ΔX=0.3 mm, starting at an X value of X_0=4.7 mm (Figure 5.6). In the Y-direction and in the range from Y_0=7.5 mm to Y_1=11.0 mm, a measurement range (green marked in Figure 5.6(a)) was set to study the lithium concentration outside the detected lithium hot spot observed at location X=4.8 mm and Y=7.2 mm. The NMC thick-layer electrodes were initially charged with quite a low C-rate (C/50) so that a sufficiently uniform charge state could be set after complete lithium

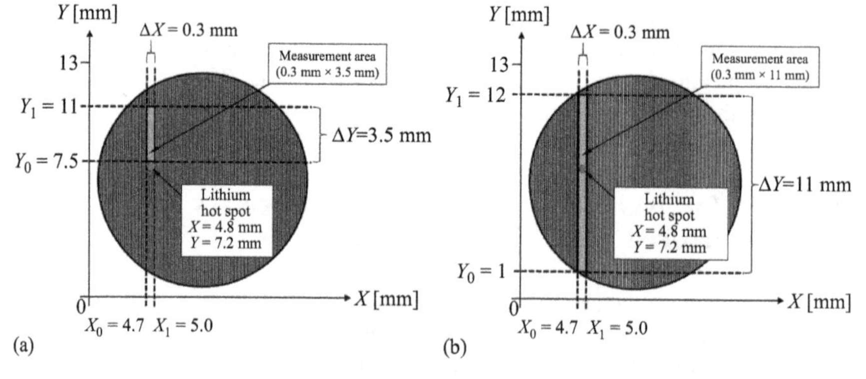

Figure 5.6 LIBS measurement ranges (marked in green) for investigating the lithium concentration distribution in unstructured and degraded NMC thick-film electrodes: (a) measurement range outside a detected lithium hot spot at (X=4.8 mm and Y=7.2 mm). (b) Measurement range covering the lithium hot spot. Reproduced from [8] (open access).

deintercalation. As expected, a similar lithium inhomogeneity was observed after full discharge. The result suggests that the electrode particles on the electrode surface that are in direct contact with the free electrolyte can be charged and discharged faster and more homogeneously than those particles that are placed in the area close to the current collector.

Figure 5.7(a) shows a range of the lithium concentration distribution for an NMC thick film, which showed spontaneous cell failure after 93 cycles. Hereby the measurement area was selected according to Figure 5.6(a), outside the detected lithium hot spot. Similar to Park *et al.* [56], a slow decrease of lithium concentration from the surface down to the current collector was detected. In the model scheme (Figure 5.7(b)), it is shown that the limited transport of lithium-ions within the electrode gradually leads to inhomogeneous reactions. The upper electrode regions are more affected during electrochemical cycling, suggesting that the ohmic loss and concentration polarization in the electrode are increasing. Accompanying side reactions and repeated charging and discharging of the electrode lead to a further increase in the impedance values against the transport of lithium-ions in the electrode. The result is an effectively higher current density in the top layer (Figure 5.7(a,b)), which means that the active particles exposed to a higher current density are more susceptible to electrochemical degradation. According to Park *et al.* [56] cracking is initiated within the particles, followed by SEI reformation at the exposed surfaces. The cracking and resulting surface products increase the complexity of the lithium-ion diffusion pathways, along with the increase in tortuosity. Overall, the proportion of active material that is still available for the electrochemical reactions is reduced. Finally, a considerable amount of the active material in the area close to the current collector can no longer

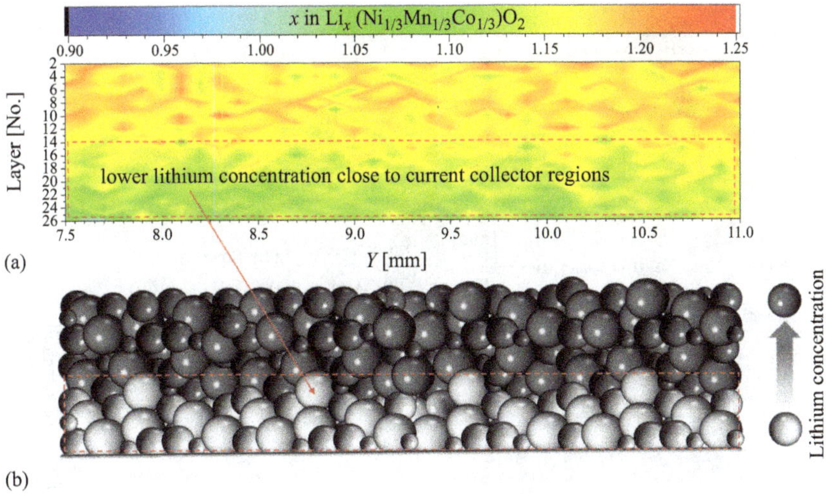

Figure 5.7 LIBS analysis (depth profile in cross-sectional view) of an NMC 111 thick-film cathode (measurement range shown in Figure 5.6(a)): (a) X*-direction averaged lithium concentration profile (quantitative) as a function of the* Y*-coordinate. (b) Model scheme of the lithium concentration distribution in the discharged state of the cell. Reproduced from [8] (open access).*

participate in the electrochemical reaction at all. On the other hand, the particles in the top layer of the electrode are exposed to a higher effective current density, with the risk of permanent damage over several cycles.

However, as shown in [45], the lithium mapping of an electrode surface shows in general strong deviations from a homogenous behavior. The local state-of-charge in general varies influenced by defects, electrode, and separator inhomogeneities, and external pressure variations on the cell. In case of separator inhomogeneities and their impact on lithium plating, this was entirely discussed by Liu *et al.* [57]. They found that external stresses lead to a local change in separator porosities which in turn induce a variation of current density and thus a local change in state-of-charge or even lithium plating can be detected. A variation in porosity on macro-scale is enforcing the formation of lithium hot spots during electrochemical cycling as shown in [45]. However, also for homogenous porosity cells show after cell failure the formation of lithium hot spots at the cathode side. The cross-sectional depth profile in Figure 5.8(a) shows clearly that a local increase in lithium concentration is penetrating through all layers down to the current collector. It is worth mentioning that a local increase in lithium concentration is also observed in an area close to the current collector. Possible reasons for this may indicate insufficient wetting with liquid electrolyte in deeper areas of the electrode, where dry areas provide starting points for cell degradation during cycling. However, an uneven pressure distribution during cell assembly could also lead to increased lithium enrichment.

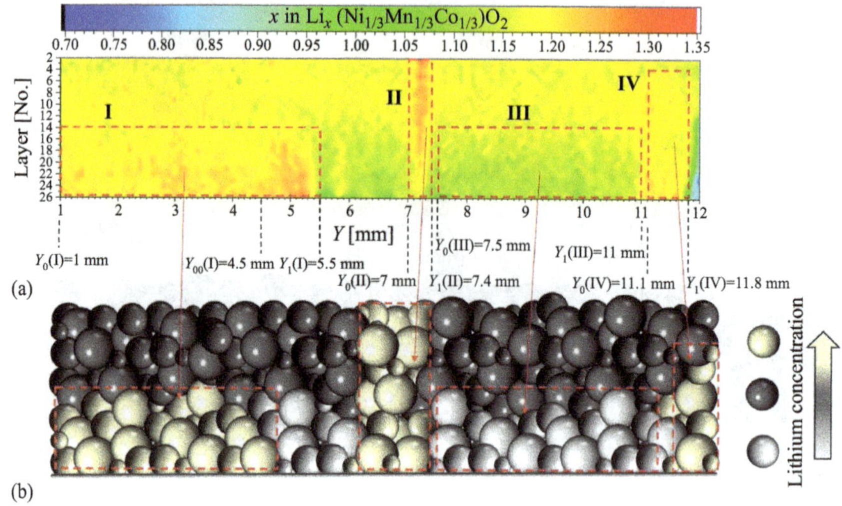

Figure 5.8 Post mortem LIBS investigation of unstructured NMC 111 thick-film electrode after spontaneous cell failing described in [45]: (a) 2D lithium distribution in cross-sectional view (measurement range shown in Figure 5.6(b)). (b) Model scheme derived from LIBS measurements to describe possible degradation processes. Reproduced from [8] (open access).

The zones of characteristic lithium concentration profiles are denoted by the "I" to "IV" in Figure 5.8(a). The lithium concentration significantly increases for layers 14–26 and especially close to the current collector. The lithium concentration is also most pronounced in the range from Y_{00}(I)=4.5 mm to Y_1(I)=5.5 mm. In layers 2–14, the lithium concentration is quite inhomogeneous. The intercalation and deintercalation rate of lithium-ions is reduced in electrode areas close to the current conductor due to the lithium-ion concentration gradient along the thickness of the electrode layer and a correlated increased diffusion overvoltage. In addition, dry areas of the electrode can accelerate cell degradation. For Y_0(II)=7.0 mm–Y_1(II) =7.4 mm (marked with "II" in Figure 5.8(a)), the lithium concentration in NMC reaches a value of 1.30 at the electrode surface and 1.24 close to the current collector (layer 26). This increase in lithium concentration is an indication that the starting point of degradation started exactly at this electrode position. It can be concluded that the cell failure was initiated due to the lithium concentration profile shown in area "II." For area "III" (Y_0(III)=7.5 mm–Y_1(III)=11.0 mm), the lithium concentration profile corresponds to the model described by Park *et al.* [56]: While the upper layer particles are exposed to higher current densities, the lower region of the electrode remains largely inactive during electrochemical cycling. The increased lithium concentration is due to cracking or formation of reaction products during the cycling process. The marked area "IV" is equivalent to area "I," whereby the increased lithium concentration can already be observed starting from the fourth layer. It is worth mentioning that lithium concentration gradients were observed in both directions. On the one hand, an increased lithium concentration spread from the electrode surface in the direction to the current conductor (area "II") was detected, on the other hand, an increased lithium concentration profile was detected in deeper regions of the composite layer (area "IV" and "I").

5.5 3D-Battery-Concept and its impact on high-rate capability

The 3D-Battery-Concept for ultra-thick film nickel-rich NMC cathode materials could be recently established. NMC 622 electrodes with a thickness of up to 250 μm and a mass loading of 52 mg/cm^2 were realized for this purpose. Structuring using ultrafast laser ablation enables a high aspect ratio and defect-free pattern down to the current collector (Figure 5.9(a)). The active mass loss could be reduced to less than 5%. The periodicity of the line pattern is typically in the range of 200 μm–600 μm [22]. NMC 622 cathodes were assembled in half-cells versus metallic lithium using the coin cell design. The impact of electrode thickness and laser structuring on the rate capability was investigated. Figure 5.9(b) shows that for laser-structured NMC 622 electrodes with a thickness of up to 210 μm, the discharge capacity could be maintained even for high-power operation (C-rate of 2C). For cells with unstructured NMC 622 electrodes of the same thickness, the capacity drops dramatically even for low power operation. An increase in NMC

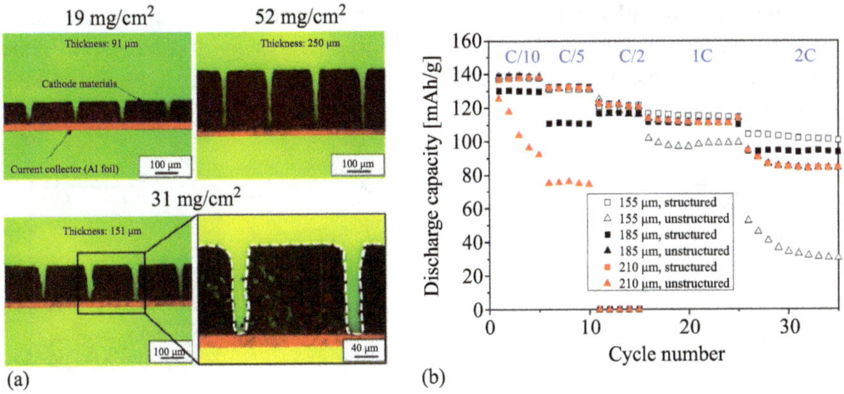

Figure 5.9 Laser-structured Li(Ni$_{0.6}$Mn$_{0.2}$Co$_{0.2}$)O$_2$ (NMC 622) cathodes and their impact on electrochemical performance: (a) cross-sectional view of NMC 622 cathodes with different film thicknesses (line distance 200 μm) [37]; (b) specific discharge capacity as a function of cycle number for lithium-ion cells with structured and unstructured NMC 111 thick-film electrodes for different discharge C-rates and electrode film thickness (155 μm, 185 μm, and 210 μm). Reproduced from [22] (open access).

film thickness from state-of-the-art 50 μm up to 210 μm corresponds to an increase in energy density of about 26% according to [6]. Thus, it could be proven for the first time for this type of battery system that high-energy and high-power operation are possible at the same time.

In recent research, the 3D-Battery-Concept was successfully transferred to next generation electrode materials such as silicon/graphite anodes and NMC 811 cathode materials by introducing line and grid patterns, respectively [28,30,36]. The results of 3D batteries based on thick-film NMC 811 electrodes with a thickness of 100 μm show a huge impact with regard to high-rate capability for structured electrodes as shown in Figure 5.10(a). It is quite obvious that in the presented case the high-rate capability is most efficient for a C-rate of 2C. The respective Nyquist plots (Figure 5.10(b)) which were taken prior to the galvanostatic measurements reveal that the charge transfer resistance is smaller for the 3D battery in comparison to the conventional 2D approach. In addition, a higher diffusion coefficient was measured for the 3D approach proving the enhancement of lithium-ion diffusion kinetics resulting in the increased high-rate capability (Figure 5.10(a)). For more details refer to [28].

There are several groups worldwide pushing the 3D-Battery-Concept toward application level. Currently, most of the research is focused on hole, line, and grid patterns [6,36,58–62]. Numerical studies of electrochemical processes prove that in a first approximation, the type of pattern does not matter [63]. The experimental evaluation of lithium concentration profiles in electrodes via LIBS is a suitable approach to support finding appropriate structure design rules (see next sections). However, with regard to wetting of electrode with liquid electrolyte and volume changes during lithiation/delithiation of the active materials, it is expected that free-standing structures and capillary features will be very beneficial with regard to battery lifetime. Another aspect to be taken into account for the selected pattern type is the laser process upscaling which might be an issue for selected structure designs.

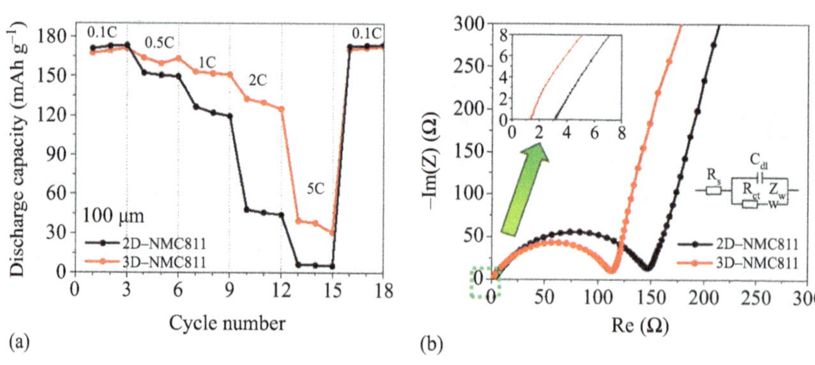

Figure 5.10 Electrochemical performance of cells with laser-structured NMC 811 electrodes: (a) rate capability performance and (b) respective Nyquist plots for unstructured and structured NMC electrodes with thicknesses of 100 μm. Reproduced from [28] (open access).

5.6 LIBS analysis applied to laser-structured model NMC thick-film electrodes

In order to investigate the impact of the 3D-Battery-Concept on lithium distribution by LIBS, a model electrode design was selected as described in detail by Smyrek *et al.* [45]. Free-standing electrode architectures consisting of micro-pillars with geometric dimensions of 600 μm × 600 μm × 100 μm were generated using ultrafast laser ablation, see Figure 5.11.

For this type of electrode architecture, due to the laser ablation, in addition to a uniform intrinsic porosity, additional cavity areas (electrolyte reservoirs) were created, which continuously replace the liquid electrolyte consumed during cell cycling. In addition, the porosity value (34% ± 3%) of the free-standing structures corresponds to that of standard NMC 111 thick-film electrodes. In order to be able to evaluate the lithium concentration distribution in free-standing structures and to correlate this together with the electrochemical data with regard to enhanced cell

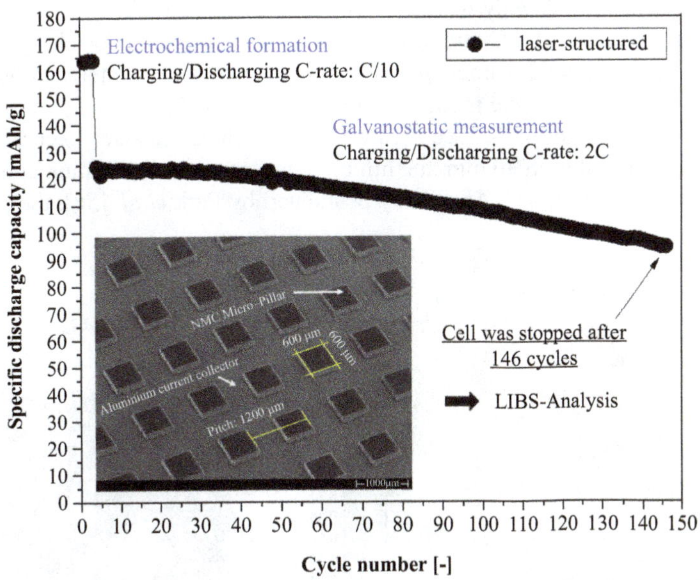

Figure 5.11 Electrochemical characterization of half-cells with laser-structured NMC 111 thick-film model electrode: specific discharge capacity as a function of the number of cycles. The formation was performed with a charge and discharge rate of C/10. From the 4th cycle, the charge and discharge rate was fixed to 2C. After a total number of 146 cycles, cell operation was stopped, the cell disassembled, and the electrode washed out in solvent DMC for subsequent LIBS investigation. The image insert represents an SEM image (top view) of the laser-structured NMC thick film at a tilt angle of 30°. For more details, refer to [45]. Reproduced from [8] (open access).

performance, the laser-modified NMC cathodes were assembled into half-cells. Galvanostatic measurements were performed by exposing the cells to C-rates of 2C (Figure 5.11). During the formation step, three cycles were performed with charging and discharging C-rate of C/10. The subsequent C-rate increase was undertaken to study lithium intercalation and its distribution in free-standing architectures under high-current conditions. After a total of 146 cycles, the electrochemical characterization was discontinued so that the electrode could be studied post mortem with LIBS after disassembling the cell. During the execution of the LIBS analysis, the NMC micro-pillars generated by means of fs laser ablation were removed very homogeneously down to the current collector.

For laser-structured electrodes, the lithium concentration distribution strongly depends on the charging and discharging current. Figure 5.12(a) shows the measured lithium concentration distribution in cross-sectional view of a single micro-pillar of the laser-structured NMC thick-film electrode (see SEM in Figure 5.11) after charging and discharging rate each at C/10 during formation (formation step of Figure 5.11). The shown LIBS lithium mapping was achieved after the respective cell has been completely discharged to a voltage of 3.0 V.

During formation at a C-rate of C/10, the long charging and discharging times of each 10 hours ensure a homogeneous intercalation and deintercalation of the lithium-ions as shown in the model scheme in Figure 5.12(b).

At elevated C-rates of 2C (see Figure 5.11), which corresponds to charging and discharging times of each 30 min, the lithium concentration distribution completely changes as reported in [45]. As already postulated by Park *et al.* [56], the particles

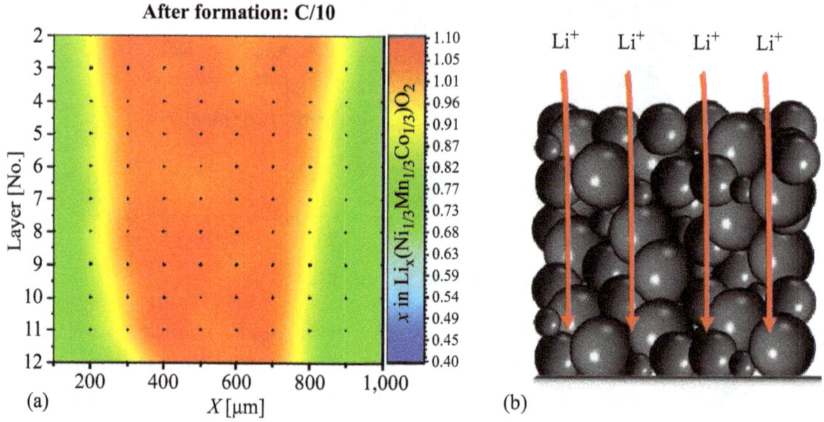

Figure 5.12 Lithium concentration profile in laser-generated NMC electrode (micro-pillar in cross-sectional view) after formation: (a) measured (postmortem) lithium concentration profile after performing the formation step (C-rate of C/10 during charging and discharging) and (b) model scheme of lithium-ion penetration into the micro-pillar. With modifications reproduced from [8] (open access).

which are in direct contact with the free electrolyte guarantee a reduction in cell polarization. It is assumed that the intercalation/deintercalation of lithium-ions in deeper regions of the electrode—for example, close to the current collector—mainly takes place via the laser-generated sidewalls in the liquid electrolyte. The tortuosity, that is, the ratio of the effective lithium-ion diffusion pathway to the theoretically minimum length of pathway (i.e., the electrode film thickness), decreases, leading to an increased effective diffusion coefficient for structured electrodes in comparison to unstructured ones.

It is worth mentioning that the maximum quantitative lithium concentration in the laser-generated micro-pillars of the 3D model electrode reaches maximum values of about 1.05, which is only a marginal deviation from the fully-lithiated state of NMC 111. Degradation would be expected for significantly higher lithium concentration values of about 1.3 as illustrated in Figure 5.8. In summary, it can be stated that no degradation processes could be detected for the laser-structured electrodes. However, due to the sensitivity of the LIBS measurement method, it could be clearly shown that new diffusion pathways are formed along the interface between liquid electrolyte and laser-structured side walls, which are activated at elevated C-rates. The change of lithium-ion diffusion pathways and respective lithium concentration profile in micro-pillars as a function of C-rate is schematically shown in Figure 5.13. The model scheme is related to LIBS studies presented in [45]. The center of the micro-pillars shows a drop in lithium concentration due to electrochemical cycling at increased C-rates. In current development, this is being compensated by geometric adjustments of the generated structure size.

Figure 5.14(b) shows the quantitative lithium concentration profile of a laser generated free-standing micro-pillar (Figure 5.14(a)) made of anode material [35] achieved post mortem after applying galvanostatic cycling at elevated C-rates. It can be clearly seen that the laser-generated sidewalls act as new lithium-diffusion pathways. Due to the anisotropy characteristics of flake-like graphite which result in a preferred lithium-diffusion parallel to the current collector plane, the impact of

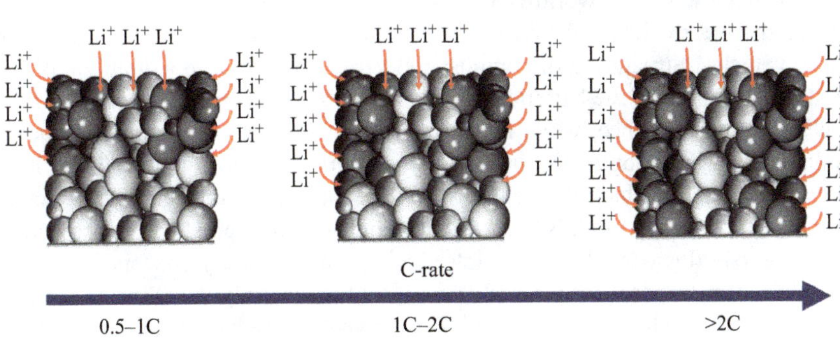

Figure 5.13 Model scheme for lithium-diffusion pathways in liquid electrolyte as function of C-rate. The model is referred to model electrode in Figure 5.11.

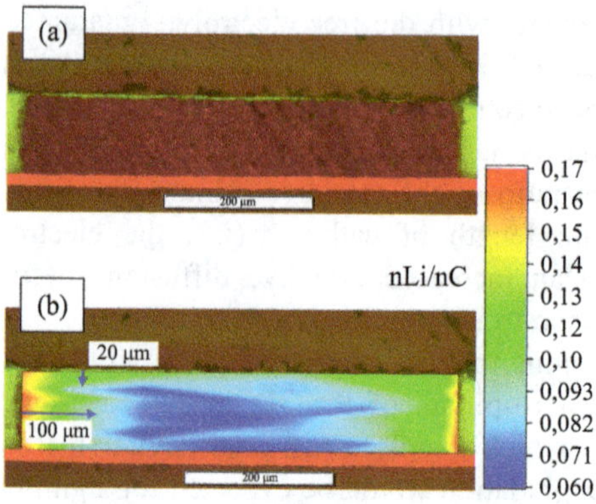

Figure 5.14 Cross-sectional view (a) and corresponding lithium mapping (b) of laser-structured flake-like graphite anode (micro-pillar). Modified from [35] (open access).

laser structuring is even more pronounced in comparison to NMC 111 cathode material [45]. The drop in lithium concentration in the center of the free-standing structure is more than a factor of two and the penetration depth (half of the maximum concentration) indicated by the arrows in Figure 5.14(b) is about 100 µm and 80 µm for anode [35] and cathode [45], respectively. This is in good agreement with the experimental results where the most benefit regarding high-rate capability is achieved for structure pitch distances in the order of 200 µm (see Figure 5.9).

5.7 Brief summary and perspective for upscaling electrode structuring in battery production

Laser materials processing and its upscaling in battery production are currently attracting huge attention worldwide as there is an increasing demand to further improve the electrochemical performance of next-generation batteries for use in electric vehicles. Structuring of composite electrodes can lead to a doubling of the operational lifetime of the batteries. Furthermore, the structuring of anodes boosts the charging performance, while the structuring of cathodes is required to enable simultaneously high-power and high-energy battery operation. For realizing high energy batteries, the thick-film concept will be flanked with high-energy active materials such as nickel-rich NMC cathodes, high-voltage spinel LMNO cathodes, and silicon-based anodes. However, introducing the 3D battery concept in parallel to the thick-film concept significantly improves the electrolyte wetting of the battery materials, which finally will lead to a remarkable reduction of scrap rate in battery production. In related research, it could be also shown that a positive impact regarding a reduced lithium plating and therefore a benefit with regard to battery safety can be expected as well [64]. Anyway, there is a strong need to be able to define suitable 3D electrode designs regarding application scenarios, respective degradation processes, and safety aspects. Therefore, the development of next generation anodes and cathodes architectures is flanked by post mortem LIBS capable to generate a rapid evaluation of the entire composite electrodes. It was shown that LIBS is able to visualize 3D lithium concentration and distribution providing a map of the state-of-the-charge of entire 2D and 3D electrodes. By combining those chemical data with data from electrochemical analyses an evaluation of thick-film electrode architectures becomes possible. 2D and 3D lithium concentration profiles were presented for unstructured and laser-structured electrodes and corresponding electrode design criteria can be derived.

Besides an optimization of electrode architectures regarding application scenario, the upscaling of the laser process for integration in battery manufacturing is quite a challenging and rather important task in current research and development. A first approach during upscaling would be simply increasing the laser power and using ultrashort laser pulses. For thin films and medium size footprints, this approach will provide a rather good pattern and surface quality. However, the processing speed has further to be increased as well as the film thickness and electrode footprint for high energy applications. It is worth mentioning, that the process efficiency gets a maximum with regard to the ablation rate per photon if the selected laser energy density is set almost one order of magnitude higher than the removal threshold, which is still a quite small value [6]. Therefore, in terms of ablation efficiency, a further increase in laser power is only meaningful if a multi-beamlet approach is chosen as schematically shown in Figure 5.15. For the next generation of ultra-thick film electrodes and large footprint batteries, the ultrafast laser structuring process needs to be further optimized regarding its processing speed. A first successful approach was demonstrated by industrial guided projects [34]. A suitable OEM ultrashort pulsed laser radiation source with an average laser power up to 500 W was developed while maintaining the laser pulse duration in the ultrafast regime.

The use of ultrafast lasers instead of classical nanosecond lasers is a main precondition to realize laser ablation without any thermal impact on the sensitive

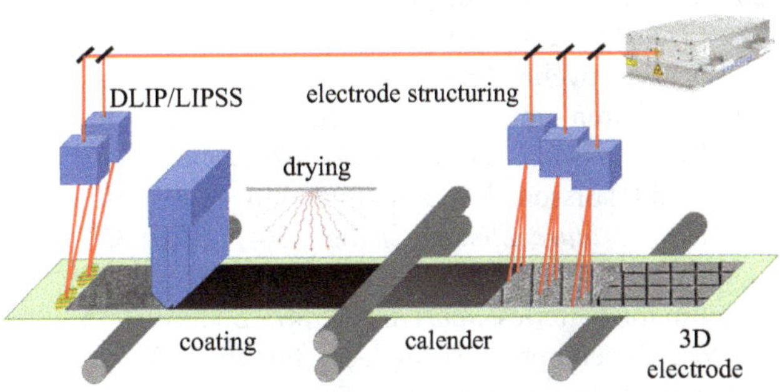

Figure 5.15 Schematic view of a 3D electrode manufacturing process consisting of coating, drying, calendering, and a multi-beamlet high-power laser approach for high-speed structuring of electrodes and current collectors.

active electrode materials. The high average laser power enables the use of a multi-beam laser approach by using diffractive optical elements and a multi-scanner approach (Figure 5.15). As shown in Figure 5.15, the same laser source can be applied for sub-micron structuring of the current collector by either direct ablation forming so-called laser-induced periodical surface structures (LIPSS) or by direct laser interference processing (DLIP) [65–67]. Laser structuring of current collector foils prior to coating process will gain an enhanced film adhesion which is especially important for silicon-based anodes with huge volume expansion during electrochemical cycling finally leading to life-extending battery operation. At KIT roll-to-roll infrastructures for laser processing of commercial and sophisticated electrodes with high precision and processing speed are enabling pilot line level. In the first technical approach, the processing speed could be increased by a factor of five. In current national and international projects pouch cells and cylindrical cells with structured electrodes and capacities of 20 Ah and beyond are being built.

Acknowledgments

We are grateful for the help of our colleagues, A. Reif, A. Meyer, Y. Zheng, P. Zhu, H. Besser, and M. Kapitz for their scientific and technical support and assistance in laser processing, material characterization, and battery analyses. This research was funded by the Federal Ministry of Education and Research (BMBF), project NextGen-3DBat, Project no. 03XP01798F.

References

[1] Nishi, Y., Lithium ion secondary batteries; past 10 years and the future, *Journal of Power Sources*, 100(1–2) (2001) 101–106

[2] Dühnen, S., Betz, J., Kolek, M., Schmuch, R., Winter, M., and Placke, T., Toward green battery cells: perspective on materials and technologies, *Small Methods*, 4(7) (2020) 2000039

[3] Partners, H., Weltweite Preisentwicklung für Lithium-Ionen-Akkus in ausgewählten Jahren von 2010 bis 2019 und eine Prognose bis 2025 (in Euro/kWh), Statista. Statista GmbH, 2020.

[4] Nykvist, B. and Nilsson, M., Rapidly falling costs of battery packs for electric vehicles, *Nature Climate Change*, 5(4) (2015) 329–332.

[5] Wood, D.L. III, Li, J., and Daniel, C., Prospects for reducing the processing cost of lithium ion batteries, *Journal of Power Sources*, 275 (2015) 234–242.

[6] Pfleging, W., Recent progress in laser texturing of battery materials: a review of tuning electrochemical performances, related material development, and prospects for large-scale manufacturing, *International Journal of Extreme Manufacturing*, 3 (2021) 012002 (20pp).

[7] Long, J.W., Dunn, B., Rolison, D.R., and White, H.S., Three-dimensional battery architectures, *Chemical Reviews*, 104(10) (2004) 4463–4492.

[8] Smyrek, P., Laserinduzierte Plasmaspektroskopie an strukturierten Lithium-Nickel-Mangan-Kobalt-Oxid Dickschichtelektroden. 2022: Karlsruher Institut für Technologie (KIT)

[9] Ferrari, S., Loveridge, M., Beattie, S.D., Jahn, M., Dashwood, R.J., and Bhagat, R., Latest advances in the manufacturing of 3D rechargeable lithium microbatteries, *Journal of Power Sources*, 286 (2015) 25–46.

[10] Baggetto, L., Niessen, R.A.H., Roozeboom, F., and Notten, P.H.L., High energy density all-solid-state batteries: a challenging concept towards 3D integration, *Advanced Functional Materials*, 18(7) (2008) 1057–1066.

[11] Xie, J., Oudenhoven, J.F.M., Li, D.J., Chen, C.G., Eichel, R.A., and Notten, P.H.L., High power and high capacity 3D-structured TiO_2 electrodes for lithium-ion microbatteries, *Journal of the Electrochemical Society*, 163(10) (2016) A2385–A2389.

[12] Kohler, R., Pfleging, W., and Proell, J., *Electrode Material for Lithium-Ion Batteries and Method for Its Production*, Germany: K.I.f. Technology (eds.), 2011. p. 12.

[13] Heungsoo, K., Proell, J., Kohler, R., Pfleging, W., and Pique, A., Laser-printed and processed $LiCoO_2$ cathode thick films for Li-ion microbatteries, *Journal of Laser Micro/Nanoengineering*, 7(3) (2012) 320–325.

[14] Kohler, R., Proell, J., Bruns, M., Ulrich, S., Seifert, H., and Pfleging, W., Conical surface structures on model thin-film electrodes and tape-cast electrode materials for lithium-ion batteries, *Applied Physics A*, 112(1) (2013) 77–85.

[15] Pfleging, W., Kohler, R., and Pröll, J., Laser generated microstructures in tape cast electrodes for rapid electrolyte wetting: new technical approach for cost efficient battery manufacturing, *Proceedings of the SPIE*, 8968 (2014) 89680B-1–89680B-8.

[16] Pröll, J., Kim, H., Mangang, M., Seifert, H.J., Piqué, A., and Pfleging, W., Fs-laser microstructuring of laser-printed LiMn2O4 electrodes for manufacturing of 3D microbatteries, *Proceedings of the SPIE*, 8968 (2014) 896805-1–896805-6.

[17] Mangang, M., Pröll, J., Tarde, C., Seifert, H.J., and Pfleging, W., Ultrafast laser microstructuring of LiFePO4 cathode material, *Proceedings of the SPIE*, 8968 (2014) 89680M-1–89680M-9.

[18] Pfleging, W. and Pröll, J., A new approach for rapid electrolyte wetting in tape cast electrodes for lithium-ion batteries, *Journal of Materials Chemistry A*, 2(36) (2014) 14918–14926.

[19] Xiong, W., Xia, Q.Y., and Xia, H., Three-dimensional self-supported metal oxides as cathodes for microbatteries, *Functional Materials Letters*, 7(5) (2014) 1430003-1–1430003-10.

[20] Mangang, M., Ultrakurzpulslaserstrukturierung von $LiFePO_4$- und $LiMn_2O_4$-Dickschichtelektroden für Lithium-Ionen-Zellen, in Fakultät für Maschinenbau, Institut für Angewandte Materialien – Angewandte Werkstoffphysik (IAM-AWP). Karlsruhe: Karlsruher Institut für Technologie (KIT), 2019.

[21] Mangang, M., Seifert, H.J., and Pfleging, W., Influence of laser pulse duration on the electrochemical performance of laser structured $LiFePO_4$ composite electrodes, *Journal of Power Sources*, 304 (2016) 24–32.

[22] Pfleging, W., A review of laser electrode processing for development and manufacturing of lithium-ion batteries, *Nanophotonics*, 7(3) (2018) 549–573.

[23] Pröll, J., Lasergestützte Modifikation von Lithiummanganoxid und Lithiumnickelmangankobaltoxid, in Fakultät für Maschinenbau, Institut für Angewandte Materialien – Angewandte Werkstoffphysik (IAM-AWP). Karlsruhe: Karlsruher Institut für Technologie (KIT), 2014.

[24] Pröll, J., Kim, H., Piqué, A., Seifert, H.J., and Pfleging, W., Laser-printing and femtosecond-laser structuring of $LiMn_2O_4$ composite cathodes for Li-ion microbatteries, *Journal of Power Sources*, 255 (2014) 116–124.

[25] Rakebrandt, J.H., Smyrek, P., Zheng, Y., Seifert, H.J., and Pfleging, W., Laser processing of thick Li(NiMnCo)O2 electrodes for lithium-ion batteries, *Proceedings of the SPIE*, 10092 (2017) 100920M1–100920M7.

[26] Song, Z.L., Zhu, P.H., Pfleging, W., and Sun, J.Y., Electrochemical performance of thick-film $Li(Ni_{0.6}Mn_{0.2}Co_{0.2})O_2$ cathode with hierarchic structures and laser ablation, *Nanomaterials*, 11(11) (2021) 2962-1–2962-16.

[27] Zhu, P.H., Seifert, H.J., and Pfleging, W., The ultrafast laser ablation of Li $(Ni_{0.6}Mn_{0.2}Co_{0.2})O_2$ electrodes with high mass loading, *Applied Sciences-Basel*, 9(19) (2019) 4067-1–4067-15.

[28] Tran, M.X., Smyrek, P., Park, J., Pfleging, W., and Lee, J.K., Ultrafast-laser micro-structuring of LiNi0.8Mn0.1Co0.1O2 cathode for high-rate capability of three-dimensional li-ion batteries, *Nanomaterials*, 12(21) (2022) 3897.

[29] Meyer, A., Ball, F., and Pfleging, W., The effect of silicon grade and electrode architecture on the performance of advanced anodes for next generation lithium-ion cells, *Nanomaterials*, 11(12) (2021) 3448-1–3448-26.

[30] Zheng, Y., Yin, D., Seifert, H.J., and Pfleging, W., Investigation of fast-charging and degradation processes in 3D silicon-graphite anodes, *Nanomaterials*, 12(1) (2022) 140.

[31] Zhu, P., Han, J., and Pfleging, W., Characterization and laser structuring of aqueous processed Li $(Ni_{0.6}Mn_{0.2}Co_{0.2})O_2$ thick-film cathodes for lithium-ion batteries, *Nanomaterials*, 11(7) (2021) 1840/

[32] Imashuku, S., Taguchi, H., Kawamata, T., *et al.*, Simpler method for acquiring quantitative state-of-charge distribution of lithium-ion battery cathode with high accuracy, *Journal of the Electrochemical Society*, 166(10) (2019) A1972–A1976.

[33] Park, J., Song, H., Jang, I., *et al.*, Three-dimensionalization via control of laser-structuring parameters for high energy and high power lithium-ion battery under various operating conditions, *Journal of Energy Chemistry*, 64 (2022) 93–102.

[34] Meyer, A., Sterzl, Y., Xiao, S., Rädel, U., and Pfleging, W., Ablation behaviour of electrode materials during high power and high repetition rate laser structuring, *Proceedings of the SPIE*, 11989 (2022) 119890G-1.

[35] Zheng, Y., Pfäffl, L., Seifert, H.J., and Pfleging, W., Lithium distribution in structured graphite anodes investigated by laser-induced breakdown spectroscopy, *Applied Sciences*, 9(20) (2019) 4218.

[36] Zheng, Y., Seifert, H., Shi, H., Zhang, Y., Kübel, C., and Pfleging, W., 3D silicon/graphite composite electrodes for high-energy lithium-ion batteries, *Electrochimica Acta*, 317 (2019) 502–508.

[37] Zhu, P., Seifert, H.J., and Pfleging, W., The ultrafast laser ablation of Li $(Ni_{0.6}Mn_{0.2}Co_{0.2})O_2$ electrodes with high mass loading, *Applied Sciences*, 9 (19) (2019) 4067.

[38] Song, Z., Zhu, P., Pfleging, W., and Sun, J., Electrochemical performance of thick-film Li $(Ni_{0.6}Mn_{0.2}Co_{0.2})O_2$ cathode with hierarchic structures and laser ablation, *Nanomaterials*, 11(11) (2021) 2962.

[39] Smyrek, P., Pröll, J., Seifert, H., and Pfleging, W., Laser-induced breakdown spectroscopy of laser-structured Li (NiMnCo) O_2 electrodes for lithium-ion batteries, *Journal of the Electrochemical Society*, 163(2) (2015) A19.

[40] Kramida, A., Ralchenko, Y., Reader, J., and Team, N.A., *NIST Atomic Spectra Database (version 5.8)*, National Institute of Standards and Technology, 2020.

[41] Li, Z., Wei, X., and Yang, Z., Pulsed laser 3D-micro/nanostructuring of materials for electrochemical energy storage and conversion, *Progress in Materials Science*, 133(2022) 101052.

[42] Armand, M., Axmann, P., Bresser, D., *et al.*, Lithium-ion batteries – current state of the art and anticipated developments, *Journal of Power Sources*, 479 (2020) 228708-1–228708-26.

[43] Bortolotti, M., Becker, B., Gil Bardají, M.E., *et al.*, European energy storage technology development roadmap towards 2030, in: Int. Energy Storage Policy Regul Work, 2017. https://ease-storage.eu/wp-content/uploads/2015/10/EASE-EERA-recommendations-Roadmap-LR.pdf, 108.

[44] Kabir, M.M. and Demirocak, D.E., Degradation mechanisms in Li-ion batteries: a state-of-the-art review, *International Journal of Energy Research*, 41(14) (2017) 1963–1986.

[45] Smyrek, P., Bergfeldt, T., Seifert, H.J., and Pfleging, W., Laser-induced breakdown spectroscopy for the quantitative measurement of lithium concentration profiles in structured and unstructured electrodes, *Journal of Materials Chemistry A*, 7(10) (2019) 5656–5665.

[46] Hou, H. and Zorba, V., 3D chemical imaging of Li-ion batteries using femtosecond laser plasma spectroscopy, in: *IEEE Conference on Lasers and Electro-Optics (CLEO) – Laser Science to Photonic Applications*, 2014, pp. 1–2.

[47] Zorba, V., Syzdek, J., Mao, X.L., Russo, R.E., and Kostecki, R., Ultrafast laser induced breakdown spectroscopy of electrode/electrolyte interfaces, *Applied Physics Letters*, 100(23) (2012) 234101-1–234101-5.

[48] Imashuku, S., Taguchi, H., Fujieda, S., Suzuki, S., and Wagatsuma, K., Three-dimensional lithium mapping of graphite anode using laser-induced breakdown spectroscopy, *Electrochimica Acta*, 293 (2019) 78–83.

[49] Imashuku, S., Taguchi, H., Kawamata, T., *et al.*, Quantitative lithium mapping of lithium-ion battery cathode using laser-induced breakdown spectroscopy, *Journal of Power Sources*, 399 (2018) 186–191.

[50] Lednev, V.N., Yakovlev, A.V., Labutin, T.A., Popov, A.M., and Zorov, N. B., Selection of an analytical line for determining lithium in aluminum alloys by laser induced breakdown spectrometry, *Journal of Analytical Chemistry*, 62(12) (2007) 1151–1155.

[51] Pfleging, W., Zheng, Y., Mangang, M., Bruns, M., and Smyrek, P., Laser processes and analytics for high power 3D battery materials, *Proceedings of the SPIE*, 9740 (2016) 974013 (9pp).

[52] Pfleging, W., Smyrek, P., Hund, J., Bergfeldt, T., and Pröll, J. Surface micro-structuring of intercalation cathode materials for lithium-ion batteries: a study of laser-assisted cone formation, in: *Laser-Based Micro-and Nanoprocessing IX*. International Society for Optics and Photonics, 2015.

[53] Park, J., Jeon, C., Kim, W., Bong, S.J., Jeong, S., and Kim, H.J., Challenges, laser processing and electrochemical characteristics on application of ultra-thick electrode for high-energy lithium-ion battery, *Journal of Power Sources*, 482 (2021) 228948-1–228948-9.

[54] Singh, M., Kaiser, J., and Hahn, H., Thick electrodes for high energy lithium ion batteries, *Journal of the Electrochemical Society*, 162(7) (2015) A1196–A1201.

[55] Lee, B.S., Wu, Z.H., Petrova, V., *et al.*, Analysis of rate-limiting factors in thick electrodes for electric vehicle applications, *Journal of the Electrochemical Society*, 165(3) (2018) A525–A533.

[56] Park, K.Y., Park, J.W., Seong, W.M., *et al.*, Understanding capacity fading mechanism of thick electrodes for lithium-ion rechargeable batteries, *Journal of Power Sources*, 468 (2020) 228369-1–228369-8.

[57] Liu, X.M., Fang, A., Haataja, M.P., and Arnold, C.B., Size dependence of transport non-uniformities on localized plating in lithium-ion batteries, *Journal of the Electrochemical Society*, 165(5) (2018) A1147.

[58] Matsumoto, F., Yamada, M., Tsuta, M., Nakamura, S., Ando, N., and Soma, N., Review of the structure and performance of through-holed anodes and cathodes prepared with a picosecond pulsed laser for lithium-ion batteries, *International Journal of Extreme Manufacturing*, 5 (2022) 012001.

[59] Kriegler, J., Hille, L., Stock, S., *et al.*, Enhanced performance and lifetime of lithium-ion batteries by laser structuring of graphite anodes, *Applied Energy*, 303 (2021) 117693.

[60] Chen, K.-H., Namkoong, M.J., Goel, V., *et al.*, Efficient fast-charging of lithium-ion batteries enabled by laser-patterned three-dimensional graphite anode architectures, *Journal of Power Sources*, 471 (2020) 228475.

[61] Dunlap, N., Sulas-Kern, D.B., Weddle, P.J., *et al.*, Laser ablation of Li-ion electrodes for fast charging: material properties, rate capability, Li plating, and wetting, *Journal of Power Sources*, 537 (2022) 231464.

[62] Park, J., Hyeon, S., Jeong, S., and Kim, H.-J., Performance enhancement of Li-ion battery by laser structuring of thick electrode with low porosity, *Journal of Industrial and Engineering Chemistry*, 70 (2019) 178–185.

[63] Schweighofer, L., Eschelmüller, B., Fröhlich, K., Pfleging, W., and Pichler, F., Modelling and optimisation of laser-structured battery electrodes, *Nanomaterials*, 12(9) (2022) 1574.

[64] Habedank, J.B., Kriegler, J., and Zaeh, M.F., Enhanced fast charging and reduced lithium-plating by laser-structured anodes for lithium-ion batteries, *Journal of the Electrochemical Society*, 166(16) (2019) A3940.

[65] Zheng, Y., An, Z., Smyrek, P., *et al.,* Laser interference patterning and laser-induced periodic surface structure formation on metallic substrates, in: 2016 *IEEE International Conference on Manipulation, Manufacturing and Measurement on the Nanoscale (3M-NANO)*. IEEE, 2016.

[66] Zheng, Y., An, Z., Smyrek, P., *et al.,* Direct laser interference patterning and ultrafast laser-induced micro/nano structuring of current collectors for lithium-ion batteries, in: *SPIELaser-Based Micro- and Nanoprocessing X*, 2016.

[67] Zhang, N., Zheng, Y., Trifonova, A., and Pfleging, W., Laser structured Cu foil for high-performance lithium-ion battery anodes, *Journal of Applied Electrochemistry*, 47(7) (2017) 829–837.

Chapter 6

Engineering advanced Lithium-ion batteries with additive manufacturing

Corie L. Cobb[1] *and Michelle E.R. Katz*[1]

Additive manufacturing (AM) enables the fabrication of complex shapes and formfactors that are inefficient or impossible to produce with traditional subtractive machining tools. AM emerged in the 1980s to enable the rapid creation of functional prototypes (also known as rapid prototyping). The first commercial implementation of AM was a stereolithography (SLA) system developed by 3D Systems in 1987, wherein a laser solidified thin layers of a photoactive polymer solution [1]. In the early 1990s, fused deposition modeling (FDM), selective laser sintering, and other AM modalities began to emerge and have continued to grow in the decades since [2]. Within the last ten years, AM has gained traction as an approach to fabricate Lithium-ion batteries (LIBs) because it enables (1) novel three-dimensional (3D) electrodes that optimize energy and power performance and (2) customizable battery shapes for integrated and mechanically robust batteries for portable device applications. As energy storage demands grow, so does the need for LIBs to come in a multitude of sizes, shapes, and materials that meet the needs of a given application. In this chapter, we review the main AM approaches that have been used to produce LIBs with a focus on FDM, direct-ink write (DIW), inkjet printing (IJP), aerosol jet printing (AJP), electrostatic spray deposition (ESD), stereolithography (SLA), and newer field-assisted (FA) methods.

6.1 Introduction

Since first commercialized in 1991, traditional Lithium-ion batteries (LIBs) have consisted of two planar electrodes (an anode and a cathode) insulated by a polymer separator film. As demand for LIBs has grown over the last decade, improvements in performance have primarily relied on advancements in material chemistry and optimization of the anode and cathode electrode formulation. The thickness, porosity, and composition of the electrodes are carefully adjusted to meet the performance needs of the end application. In addition, the separator is often enhanced

[1]Department of Mechanical Engineering, University of Washington, USA

through different polymer or polymer-composite compositions to yield thinner microporous films that increase battery power [3].

Although Lithium-ion (Li-ion) is the chemistry of choice for high-performance energy storage devices, future advancements in applications such as the Internet of Things (IoT), electric vehicles (EVs), and grid storage are severely hindered by capacity, rate, cycle life, and safety limitations [4]. As the wearable electronics market grows and IoT design concepts expand, compact and efficient LIBs are needed to meet growing power and energy requirements. Moreover, most applications require batteries to be engineered for continuous operation for long periods of time. It is important to note that these needs have the potential to be addressed through advancements in capacitors, fuel cells, and other energy storage technologies [5,6]. However, LIBs offer the most potential in the near-term based on their high gravimetric and volumetric energy densities and thus are the focus of this chapter.

There have been extensive investigations into new electrode and electrolyte materials for LIBs, including exploration of anode-free and solid-state LIBs [7]. However, it is equally important to consider manufacturing approaches that can push the limits of LIBs materials. The need to improve battery performance beyond material changes has motivated a shift to a more rapid exploration of LIBs with additive manufacturing (AM) [8]. This chapter focuses on the application of AM to advance LIB technology. Current LIB manufacturing approaches will not achieve the cost, fast charge, lifetime, and duration requirements needed for both small- and large-scale applications. Achieving these targets requires dramatic innovations in manufacturing efficiency, battery design, and material chemistry.

Figure 6.1 (upper image) outlines today's commercial LIB manufacturing workflow. Conventional processing creates planar LIB electrodes that are

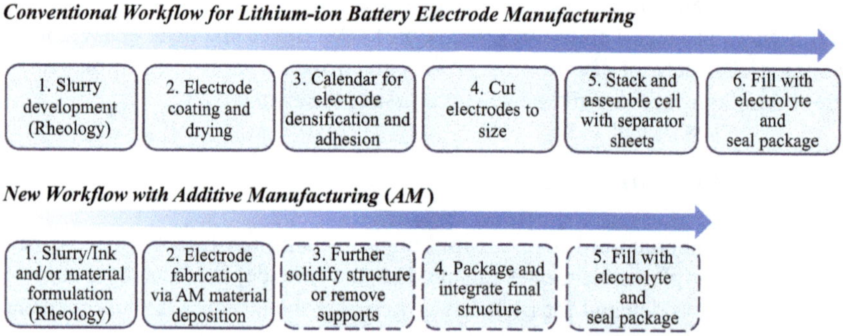

Figure 6.1 (Top image) Generalized diagram of a conventional LIB manufacturing workflow used in production today. (Bottom image) Proposed workflow changes with AM where the dashed outlined boxes (steps 3–5) represent steps that require further development beyond what we discuss in this chapter. Steps 3–5 will also differ with each of the AM processes covered in Table 6.1.

stochastic mixtures of binder, carbon, and active material. Slurries are made and coated onto current collectors via slot die coating. Heat is applied to thermally dry the coating and produce electrodes. The electrodes are cut to size, and separator sheets are inserted between the cathode and anode electrodes to insulate and prevent shorting. Depending on the end application, the LIBs are packaged into standardized coin, cylindrical, prismatic, or pouch cell formfactors. The cells are then flooded with liquid electrolyte and sealed. This conventional fabrication process forms planar electrode stacks. In addition, slot die coating yields very little control over electrode morphology with the main control parameters being speed and slurry coating thickness. With this process, we can create thin electrodes optimized for power density or thick electrodes optimized for energy density, but not both concurrently.

AM has the potential to revolutionize battery manufacturing, allowing for complex customized shapes, advanced device architectures, and finer micron-scale control over battery material systems [8]. Conventional LIB cell production cannot be easily deployed to create intricate shapes and non-planar LIB layouts. A cell manufacturer would need to have multiple coating lines or invest time and money into continuously modifying existing production lines for a dedicated formfactor, all adding up to a very expensive process. The freeform capability of AM [9] has the potential to create a more flexible and shorter workflow as shown in the bottom image of Figure 6.1. Future AM workflows for LIBs will still involve slurry or material formulation steps followed by material deposition. However, the subsequent steps from material solidification all the way through to final package sealing will be different between each AM process. These areas require further development beyond the content discussed in this chapter. The AM design space is vast but offers new opportunities to shorten and improve today's LIB manufacturing workflows for customized and high-performance batteries.

In this chapter, we look at emerging AM innovations in the context of customizing and controlling LIB architecture. In Section 6.2, we discuss the importance of controlling electrode architecture and creating custom-integrated batteries. In Section 6.3, we then cover the main AM approaches that have been used by researchers over the last decade and provide a topical review based on a selection of research studies. Finally, we conclude with our perspective on future opportunities for AM with LIBs in Section 6.4.

6.2 Battery architectures and formfactors

Optimizing energy and power in conventional, planar LIBs requires a delicate balance between electrode thickness and composition. As thickness increases, planar electrodes contain more active material but exhibit reduced ion transport. This results in lower performance with respect to rate capability, cycle life, and fast charge and discharge performance [10]. Electrodes thicker than 100 μm are challenging to fabricate with conventional manufacturing due to cracking and thermal shock issues that occur upon drying [11,12]. In addition, commercialized batteries are packaged in rigid, bulky prismatic, cylindrical, and coin cell formats that often dominate a LIB in both physical size and weight, limiting their ability to be seamlessly integrated with portable and wearable devices. This gives battery engineers a limited design space to tune for emerging LIB applications.

6.2.1 3D Batteries and electrode architectures

Three-dimensional (3D) batteries are one approach to break the trade-offs experienced in LIB design [13]. In 3D LIBs, the electrode structure is engineered to shorten Li-ion transport distances by controlling porosity and spacing between the anode and cathode on a micron-scale. As prior research has shown, this increases energy and power simultaneously over smaller footprint areas while enabling better rate capability performance relative to planar electrodes [14,15]. Controlled electrode architecture on the order of 10–200 μm offers the potential to advance battery performance in thick electrodes (>100 μm) by providing enhanced pathways that optimize the time- and length scales associated with ion transport. Figure 6.2 shows two general classes of 3D electrode designs fabricated with AM: (1) 3D structured electrodes (SEs) and (2) 3D interdigitated electrodes (IDEs). 3D SEs reduce tortuosity and enhance high-rate performance in thick electrodes by inserting micron-scale pore channels into the anode and/or cathode for additional electrolyte infiltration. 3D IDEs have a more radical engineering of electrode architecture wherein the anode and cathode are interwoven, allowing for significantly shorter ion transport distances between two electrodes.

A series of computational models have been conducted over the years to understand the behavior and limits of different 3D LIB electrode architectures [16–20].

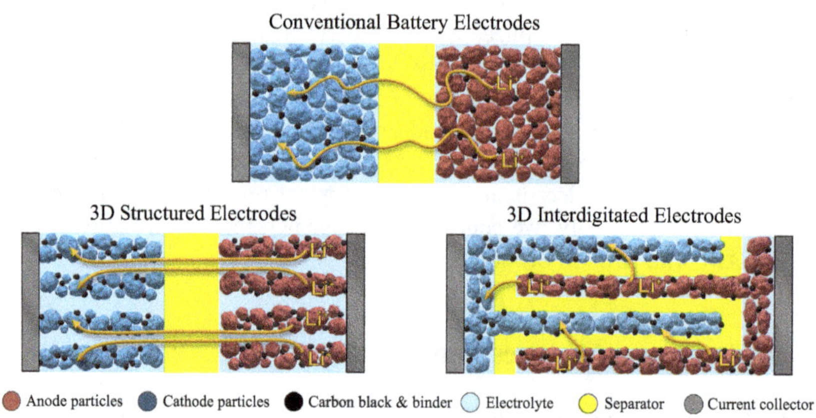

Figure 6.2 Illustration of Li-ion (Li^+) transport in conventional electrodes, 3D SEs, and 3D interdigitated electrodes (IDEs). Orange lines represent hypothetical Li^+ transport pathways during a discharge cycle, when ions shuttle from the anode to the cathode. Relative to conventional electrodes, 3D SEs reduce the overall tortuosity of Li^+ transport pathways, while 3D IDEs drastically shorten Li^+ transport pathways.

These models have validated the ability of cathode 3D LIBs made with $LiCoO_2$ (LCO), $LiMn_2O_4$ (LMO), $LiNi_{0.5}Mn_{0.3}Co_{0.2}O_2$ (NMC-532) and its composition variants to improve both energy and power density in thick electrodes. Prior work by Cobb and Blanco (shown in Figure 6.3(a) and (b)) conducted a modeling analysis on line-patterned SEs using a porous electrode model to calculate specific capacity and volumetric energy capacity for SEs made from LCO in a half-cell format [17]. Their analysis varied the distribution of a fixed amount of cathode mass over a range of electrode and electrolyte region widths (Figure 6.3(a)). Both the thickness and aspect ratio of the SEs were explored, and their results found that SEs 150–300 μm in thickness with features in the range of 25–100 μm yielded the greatest performance improvements. This work was further extended to $LiNi_{0.33}Mn_{0.33}Co_{0.33}O_2$ (NMC-111) paired with a graphite anode and focused on 1 Ah pouch cells [21]. The analysis demonstrated that gravimetric and volumetric capacity improvements > 10% are observed, relative to conventional thin electrode pouch cells, when thick 3D SE cathodes are mated with graphite electrodes and all packaging weight and volume are considered.

In addition, 3D IDEs were recently modeled with $Li_4Ti_5O_{12}$ (LTO) and $LiFePO_4$ (LFP) and Graphite and NMC-532 by Hung *et al.* [20] as shown in Figure 6.3(c). The authors compared four different IDEs to planar electrodes in two different LIB material systems. Their results demonstrated IDEs outperform planar electrodes based on specific capacity. Moreover, their work showed that the relative performance gains between different 3D IDEs and their current density distribution (Figure 6.3(c)) are driven by both electrode architecture and material selection.

3D SEs and IDEs open a new design space for LIBs with the potential to create solutions for future energy storage needs. However, most 3D electrode architectures are incompatible with today's conventional LIB manufacturing process. New manufacturing approaches like AM and fundamental processing science innovations are required to realize 3D LIBs over large areas with the precision needed for high-performing batteries while being compatible with current and emergent material systems.

6.2.2 Customized and integrated batteries

Traditional LIB packaging, protective casings, and components are rigid and stiff by design and are available in only a handful of shapes and sizes. This puts significant design constraints on products and devices, limiting their overall size, shape, and flexibility. As reported by Cobb and Ho, batteries are often 30% or more of a device's volume in the portable device market [8]. In addition, as highlighted in a recent roadmap article on printed electronics, flexible electronic devices require new materials for electrical conductivity, ionic conductivity, and mechanical integrity in addition to advancements in manufacturing approaches [6].

With AM, customized objects ranging from mechanical fasteners to functional material components can be fabricated and embedded in printed parts. When AM is applied to LIBs, the possibility of customized battery shapes and direct embedding

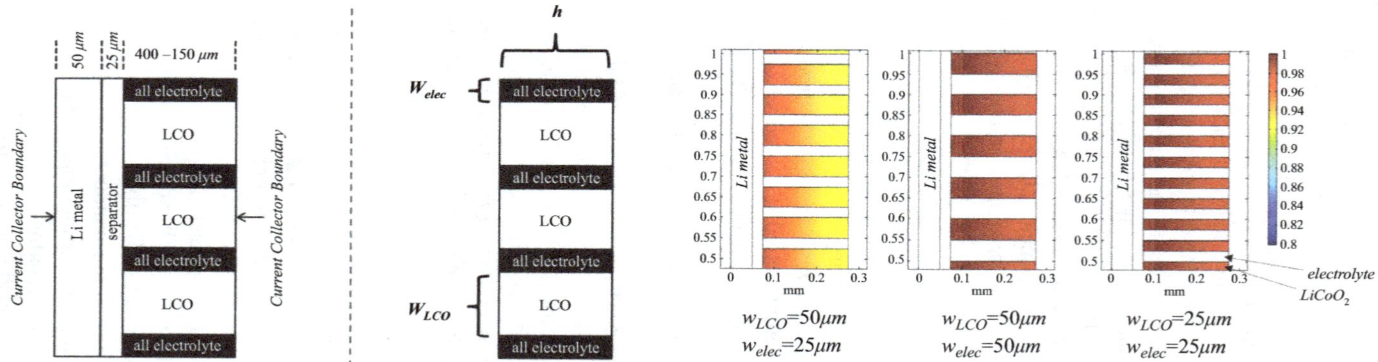

(a) (b)

Figure 6.3 *Electrochemical modeling of 3D SEs and 3D IDEs. (a) Plots show a parameterized schematic of a 3D SE LCO cathode electrode. (b) Shows Li-ion utilization plots normalized by the maximum concentration in the material at the end of a 1C discharge. Smaller feature widths yielded improved electrochemical performance. (c) Modeling images show current density plots for two different 3D IDEs. Design 4 demonstrated a high current density profile, which resulted in a highly non-uniform and unbalanced current density distribution. Transport pathways were longer and more tortuous in Design 4 than in Design 1, leading to Design 4 having a lower specific capacity performance than Design 1. (Images (a) and (b) are reprinted from Journal of Power Sources, from Ref. [17], Copyright 2014, with permission from Elsevier. The image in (c) is Copyright 2021 from Ref. [20], distributed under the terms of the Creative Commons Attribution 4.0 License (CC BY.)*

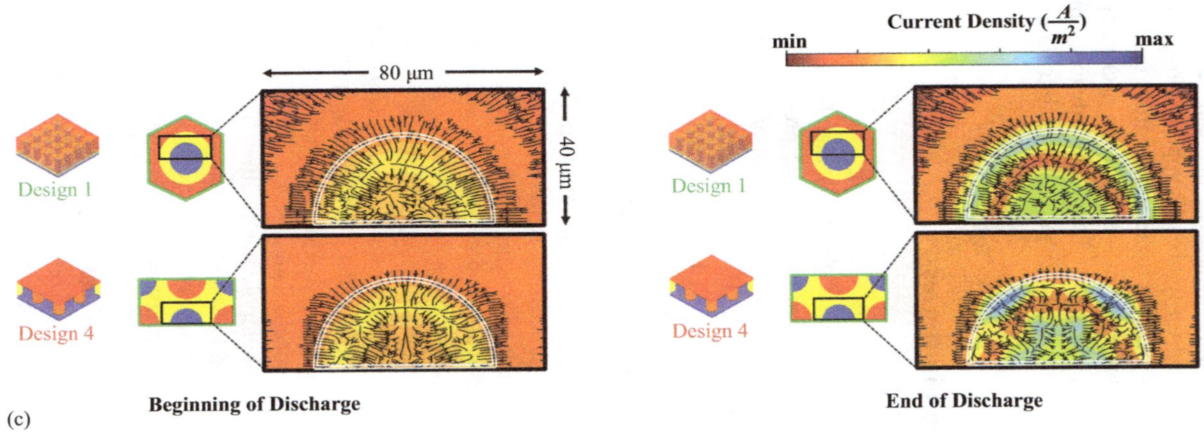

Figure 6.3 (*Continued*)

of energy storage materials during manufacturing gives engineers the opportunity re-envision how LIB-powered devices are designed and fabricated. In addition, as newer portable electronic devices require more seamless and wearable formfactors, moving toward conformal and flexible energy storage materials that can be customized with electronic components will be critical for future LIB applications [22]. While batteries need to have sufficient energy and power density for truly untethered device operation, considerations must also be given to customized packaging and potential mechanical deformation during operation. Thus, new advances in battery formfactors, packaging, and manufacturing are required to enable future potable electronic and IoT devices.

6.3 AM processes for Lithium-ion batteries

6.3.1 AM overview

In this section, we highlight the main AM processes for LIBs that have been reduced to practice in multiple research publications to date. AM processes for LIBs can broadly be divided into six categories: (1) fused deposition modeling (FDM), (2) direct-ink write (DIW), (3) inkjet printing (IJP), (4) aerosol jet printing (AJP) or electrostatic spray deposition (ESD), and (5) Field-Assisted Patterning (FA). FDM, DIW, IJP, SLA, and AJP/ESD are more established commercialized technologies, while FA is a new area that we believe has the potential to advance the AM field.

Figure 6.4 shows a schematic of each manufacturing process, while Table 6.1 describes each process and its associated post-processing steps. FDM deposits a solid polymer filament from a heated extruder onto a moving platform. The heated extruder melts the filament to enable fast deposition and adhesion to subsequent material layers. Upon deposition, the filament cools and solidifies. DIW uses high viscosity liquid-based inks that are extruded through a pressurized syringe and each layer is eventually cured or dried. IJP deposits micron-scale liquid droplets onto a substrate, and this process requires optimizing ink formulations to avoid separation and sedimentation. AJP and ESD are both spray-based AM processes that aerosolized droplets through different atomization mechanisms. SLA involves tuning laser or light sources to raster through a vat of photopolymer solution. This allows thin layers of photopolymer to crosslink and solidify as a part is being built. Lastly, FA patterning is a newer approach that uses electric or magnetic fields to align particles in a low-viscosity solution, giving AM processes another dimension of material patterning control during deposition.

AM process selection requires consideration of feature sizes, device footprint areas, and material compatibility. Table 6.2 summarizes the demonstrated capabilities of each AM process for LIBs from a selection of publications discussed in Sections 6.3.3–6.3.8. We have broken this comparison down into seven categories: (1) *LIB Components*, (2) *Active Materials*, (3) *Architectures*, (4) *Resolution*, (5) *Speed*, (6) *Footprint,* and (7) *Equipment*. *LIB Components* refers to the part of the LIB fabricated by the AM process, while *Active Materials* highlight the

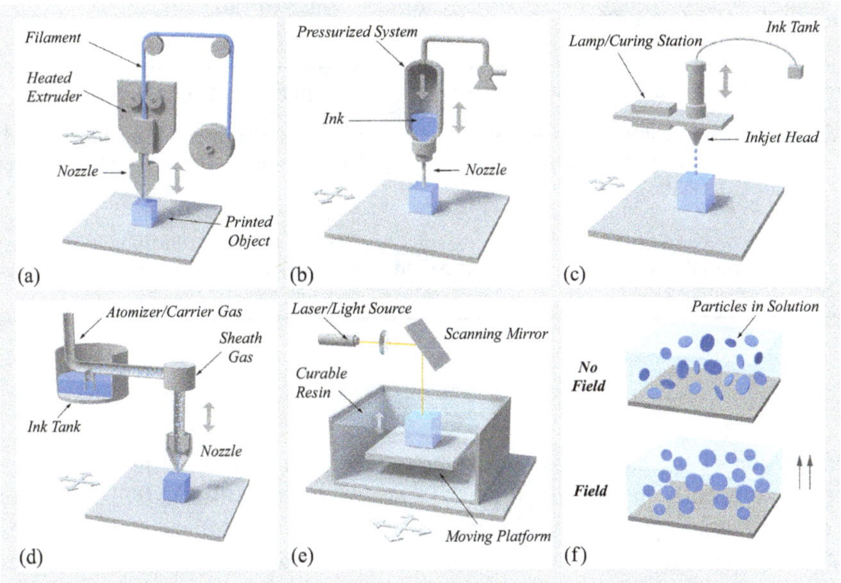

Figure 6.4 The six AM processes for LIBs reviewed in this chapter are: (a) FDM, (b) DIW, (c) inkjet printing (IJP), (d) spray-based AM which includes AJP and ESD, (e) stereolithography (SLA), and (f) field-assisted (FA) patterning

Table 6.1 Description of material behavior and post-processing operations for the AM processes shown in Figure 6.4 and discussed in Sections 6.3.3–6.3.8

Process	Material behavior during process	Post-processing operations
FDM	Solid polymer-composite filament is extruded through a heated extruder and nozzle.	If support material is used, it must be removed via mechanical or chemical means.
DIW	High viscosity, shear-thinning material is extruded by a pressurized system through a nozzle.	Material must be solidified via freeze drying, oven drying, curing, sintering, or a combination.
IJP	Low-viscosity colloidal suspension forms liquid droplets via oscillation, thermal, or piezoelectric actuation mechanisms.	Solvent in colloidal suspension must be removed via evaporation, thermal drying, or curing.
SLA	Vat of photosensitive polymer solution solidifies via crosslinking upon controlled exposure to a laser or light source.	Excess solution must be removed from solidified structure. Additional curing or treatment steps may be required to further solidify the material.
AJP	Low-viscosity colloidal suspension forms aerosol spray via pneumatic atomization.	Solvent in colloidal suspension must be removed.

(Continues)

Table 6.1 (Continued)

Process	Material behavior during process	Post-processing operations
ESD	Low-viscosity liquid mixture or dry powder mixture forms spray via electrostatic atomization.	Solvent must be removed from liquid based systems. If dry powder mixtures are used, they must be hot-pressed to solidify the structure.
FA	Low-viscosity liquid mixture is exposed to an electric or magnetic field.	Mixture must be solidified quickly to remove solvent and maintain patterned structure.

composition of the electrode components. *Architecture* is divided into three categories as originally shown in Figure 6.2: planar electrodes, SEs, and IDEs. We use * to denote flexible or shape conformal LIBs. *Resolution* is defined as the smallest reported dimension that researchers were able to control and electrochemically test in each publication for their selected AM method. The minimum value reported in each paper is used. For FDM and DIW, resolution is determined by the width of the extruded material after drying or curing. For IJP and AJP, the resolution is determined by the thickness of the deposited layer. For ESD, resolution is determined by layer thickness or layer width, depending on what value is controlled and reported by the authors. For SLA, the resolution is determined by the diameter of the cured photoactive polymer upon laser/light exposure, which may be controlled by the laser diameter or by the dimensions of any patterning stencils used. For FA, we define resolution using the average diameter of the particles that are aligned. *Speed* refers to how quickly the stage or printhead moves, and the average values from each paper are used. *Footprint* refers to the area over which the final structure resides. *Equipment* is divided into two categories: commercial or custom. Commercial means the equipment is purchased and used as an "off-the-shelf" setup; custom means the AM tool was built or customized by the authors of the publication.

Figure 6.5 visually compares the footprint, resolution, and deposition speed of each AM process from the papers summarized in Table 6.2. Figure 6.5(a) and Table 6.2 show that IJP has the highest resolution and demonstrated the ability to produce ultra-thin, flexible electrode layers with a similar footprint to other AM methods. For LIBs, the application of IJP has mostly been limited to planar structures, but IJP has been explored for 3D SEs with other material chemistries [54]. AJP has a lower process resolution and deposition speed than IJP but has been used to produce a wider array of architectures over larger footprint areas. ESD has produced the largest footprint of all AM methods discussed in this chapter but has limited resolution compared to AJP. Conversely, DIW has produced the greatest variety of LIB components and architectures at small to moderate footprint areas. DIW has similar deposition speeds to AJP, but higher resolution than FDM. FDM

Table 6.2 Comparison of AM processes from selected publications discussed in Sections 6.3.3–6.3.8

Process	LIB Component[a]	Active Material[b]	Architecture[c]	Resolution[d] (μm)	Speed[e] (mm/s)	Footprint[f] (cm^2)	Equipment[g]	References
FDM	Anode	LTO, Graphite	Planar, SE	50–1000	35	0.78–60	Commercial	[23–28]
	Cathode	LFP, LMO	Planar, SE					
	Separator		Planar, Structured					
DIW	Anode	LTO	Planar*, IDE	5–200	0.25–12.5	0.01–25	Commercial or Custom	[29–35]
	Cathode	LFP, LCO	Planar*, SE, IDE					
	Separator		Planar, Structured, IDE					
	Solid Electrolyte		Planar, Structured					
IJP	Anode	Graphene	Planar*	0.24–5	50	0.50–2.5	Commercial	[36–40]
	Cathode	LFP, LCO, NMC	Planar					
	Solid Electrolyte		Planar					
SLA	Anode	LTO, Graphite	SE	28–500	0.001–0.043	0.29–36	Commercial or Custom	[41–45]
	Cathode	LFP, LCO, NMC	SE					
	Separator		Structured					
	Solid Electrolyte		Structured*					
ESD	Anode	Graphite	Planar, IDE*	50–500	21	1.3–925	Custom	[46,47]
	Cathode	LCO, NMC	Planar, IDE					
AJP	Anode	LTO	IDE	1.76–22	0.58–2.8	0.4–56	Commercial	[48–51]
	Cathode	LFP	Planar, SE, IDE					
FA	Anode	Graphite	SE	7–10	200	1.5–1.75	Custom	[52,53]

Notes

[a]The part of the battery stack that was fabricated.

[b]Electrode composition.

[c]Architecture fabricated is planar, SE, or interdigitated electrode (IDE) correlating to Figure 6.2.

[d]Smallest reported dimension of control during deposition. Minimum reported values are used.

[e]How quickly the stage or printhead moves. Average reported values are used.

[f]Area over which the final structure resides.

[g]Commercial or Custom. Commercial: A purchased “off-the-shelf” setup. Custom: A new system built or modified by researchers.

*The architecture is flexible and conformal.

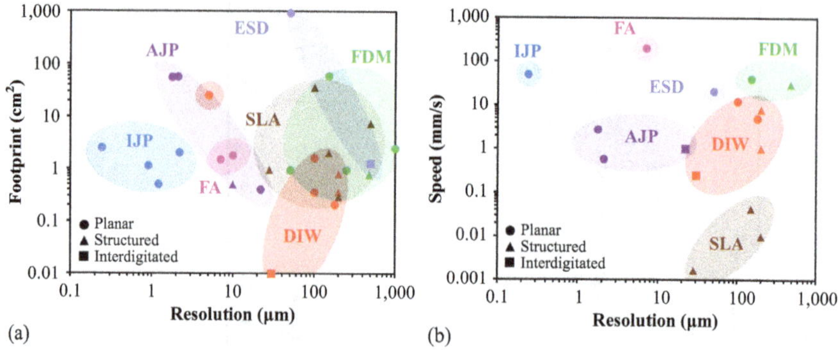

Figure 6.5 Plots summarizing the AM methods are compared in Table 6.2. (a) Relationship between footprint area and resolution. (b) Relationship between speed and resolution. Circles indicate a planar architecture. Triangles indicate a 3D SE, or structured separator or structured solid electrolyte. Squares indicate a 3D IDE. Each circle, triangle, or square corresponds to a publication. Data used in the formation of the plots came from the following references: FDM [23–28], DIW [29–35], IJP [36–40], SLA [41–45], ESD [46,47], AJP [48–51], and FA [52,53].

demonstrates the fastest deposition speed among the methods in Table 6.2, but FDM is currently more limited in resolution than the other AM processes. SLA has a similar resolution to both DIW and FDM but has more stringent material and processing requirements that make fabricating LIB components challenging. Lastly, FA patterning is a newer technique that we have chosen to incorporate in our discussion as it has the potential to further advance the AM field. FA patterning currently has a similar demonstrated resolution to AJP and rapid patterning speeds, making it a promising approach on its own or in combination with the other AM methods to make 3D LIBs.

6.3.2 Material rheology and solidification

Before covering the details of each process in Sections 6.3.3–6.3.8, it is necessary to discuss basic rheology principles and their impact on AM processing. Materials must be compatible with processing and post-processing operations for successful fabrication with the methods covered in Table 6.1. Broadly, rheology is the study of the flow of matter in various states under an applied stress. The rheology of a material is characterized by its viscosity, which is a measure of resistance to deformation by an applied stress [55]. For our purposes, an AM material is a fluid-like system composed of various colloidal or particulate additives in a liquid or polymer base. This is often referred to as an ink, slurry, paste, gel, or colloidal suspension depending on the process and material composition [56,57]. The relationship between viscosity and shear rate impacts material behavior during processing and is governed by fluid dynamics principles. Three types of rheological behavior are relevant to AM for LIBs: (1) Newtonian, (2) shear-thinning, and (3) plastic [58]. As shown in Figure 6.6(a), Newtonian fluids have constant viscosity with increasing shear rate, giving them consistent flowability. We define flowability as the ability of material to move or spread during processing. Shear-thinning and plastic fluids have variable viscosities with changes in shear rate. These AM materials are more resistant to flow at low shear stresses, but their flowability begins to approach that of Newtonian fluids at high shear stresses.

Next, storage (G') and loss modulus (G") represent a material's elastic and plastic behavior, respectively. The relationship between G' and G" determines the solid-like or liquid-like behavior of an AM material as shown in Figure 6.6(b) [59]. When G' exceeds G", the AM material will have solid-like behavior that enables it to retain its shape during processing. When G" exceeds G', the AM material will have liquid-like behavior with good flowability. In shear-thinning materials, there is a yield point where G' and G" intersect, allowing for flowability during deposition and shape retention post-deposition, which we describe as gel-like behavior in Figure 6.6(b).

Conventional LIB electrode slurries are composed of active material particles, a conductive carbon additive, and a polymer binder such as polyvinylidene fluoride

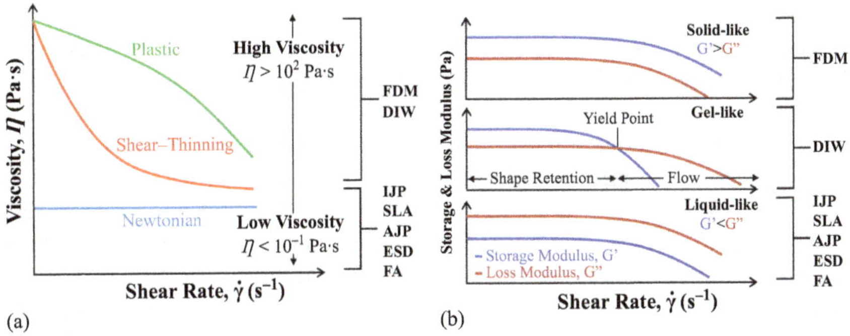

Figure 6.6 *Rheological behavior of material systems for various AM processes. (a) Relationship between viscosity and shear rate is categorized into plastic, shear-thinning, and Newtonian behavior (green, red, and blue lines, respectively). FDM and DIW require high-viscosity materials with plastic and shear-thinning behavior, respectively. IJP, SLA, AJ, ESD, and FA require low-viscosity material systems that may have either shear-thinning or Newtonian behavior. (b) Relationship between storage (G', blue) and loss (G", red) modulus. (Top) G' > G" supports solid-like behavior needed in FDM printing. (Middle) G' > G" at low shear rates until the yield point, at which G' < G". This creates gel-like behavior that enables shape retention needed for DIW. (Bottom) Liquid-like behavior where loss modulus is greater than storage modulus during deposition, needed for IJP, SLA, AJP, ESD, and FA.*

(PVDF) dispersed in a solvent [60]. *N*-Methyl-2-pyrrolidone (NMP) is the most common solvent used in commercial LIB manufacturing, but as the reader will see in Sections 6.3.3–6.3.8, NMP is one of many solvents used in the AM formulations developed by each research study. These electrode slurries have a gel-like consistency that allows the material to be deposited with slot die coating. Battery slurries possess a moderate to high viscosity with shear-thinning behavior in the 1–100 Pa-s range, although this varies depending on the electrode slurry composition. Slurries are then thermally dried to remove solvent and form a composite electrode.

AM fundamentally changes how we process battery materials, and each process has different requirements and limits. FDM and DIW both require high viscosity ($>10^2$ Pa-s) materials to enable extrusion-based processing. FDM uses a heated extruder to create plastic deformation and turn a solid polymer into a flowable melt for deposition. DIW uses pressure to overcome the yield point shown in Figure 6.6(b) and initiate flow in a shear-thinning material. IJP, AJP, ESD, SLA, and FA rely on low viscosity ($<10^{-1}$ Pa-s) fluid systems to facilitate the fabrication of intricate structures. IJP, AJP, and ESD require liquid-like material behavior to allow for the formulation of the droplet or atomized sprays. SLA requires liquid-like behavior to allow for light penetration and final part removal from the photopolymer vat. Lastly, FA patterning requires liquid-like behavior so that particles are capable of rotating to align with the applied field.

The nominal particle diameter (d_{50}) of an active material powder significantly affects the rheology of a conventional battery slurry and the final dried composition of an electrode. Additionally, the surface area of the conductive carbon additive influences viscosity and storage and loss moduli. On average, decreasing the d_{50} of active material particles while increasing the surface area of carbon additives will increase the viscosity of a battery slurry at low shear rates. From an electrode design standpoint, smaller active material particles are desired for power applications, and large particles are preferred for energy applications. Particle size and distribution can be tuned to balance these trade-offs, but this can be difficult to control during material synthesis. When moving down into d_{50} sizes of $\sim$1 μm or less, issues with agglomeration and forming a uniform interconnected network between the active material, binder, and conductive carbon additive while maintaining a reasonable dried electrode porosity becomes challenging; this must be carefully optimized when developing new LIB material compositions for AM. In Sections 6.3.3–6.3.8, we highlight these considerations and discuss a selected group of research papers to give the reader an understanding of the breadth of work in each AM area.

6.3.3 FDM

FDM is the most widely used AM process due to its ease of accessibility for prototyping and fabricating parts. Thermoplastic materials such as acrylonitrile butadiene styrene (ABS) and polylactic acid (PLA) are commonly used as build materials. Prior to fabricating a part, a computer-aided design (CAD) model of the part is created and imported into slicing software that generates the required code to control the stage and extruder movement in space. During the FDM process, a heated extrusion nozzle moves in Cartesian coordinate space over a build platform as shown in Figure 6.4(a). After one layer of melted material is deposited on the build platform, the nozzle or platform moves in the vertical direction to add the next layer of material. In some instances, a support material may be used to protect delicate features and overhangs on a part. Support material is removed during post-processing steps either through physical or chemical means. Overall, FDM part quality is tuned through print parameters, raster spacing, layer thickness, nozzle offset, and overall part orientation during fabrication [61]. Recently, researchers have explored the use of FDM to create LIB structures as standalone 3D LIBs or for integration with portable and wearable devices.

The filament material is critical to enabling a functional LIB. Early studies focused on creating composite filaments with a thermoplastic host matrix embedded with LIB anode or cathode materials. Foster *et al.* used a commercially available filament comprised of 8% graphene and 92% PLA to fabricate planar LIB anodes with a RepRap FDM printer [23]. The anode was paired with lithium metal, a Celgard separator, and a solution of 1 M $LiPF_6$ in 1:1 ethylene carbonate (EC): diethyl carbonate (DEC) (v/v) was used as the electrolyte to create a 2016 coin cell for testing. Discharge specific capacities up to 15.8 mAh/g were realized under rate capability testing. The FDM electrodes also functioned as supercapacitors due to the graphene, showing a dual purpose for filament. Maurel *et al.* also created LIB anodes via FDM using a graphite/PLA filament [26]. Their work examined how adding polyethylene glycol dimethyl ether (PEGDME500) as a plasticizer impacted both mechanical and electrochemical performance. Round disks approximately 250 μm thick were printed and tested in a coin cell with lithium metal as the reference electrode. A PLA host matrix with 40% plasticizer (PEGDME500) improved the mechanical properties of the filament while maintaining good electrochemical results and printability with a MicroDelta Rework 3D printer. The final printed anode consisted of 26/63/11 wt% PLA/graphite/plasticizer and demonstrated a specific capacity of 200 mAh/g at a C/20 rate.

Exploring integrated LIBs for IoT devices, Reyes *et al.* printed LTO-LMO cells [25]. Both coin cells and wearable electronic devices with integrated LIBs were fabricated and tested as shown in Figure 6.7(a)–(c). The current collectors, anode, cathode, separator, and casing were all FDM printed to form a LIB cell (Figure 6.7(a)). The anode, cathode, and separator were each 150 μm thick. Conventional PLA filament was used for the separator and the test cell was filled with 1 M $LiClO_4$ in ethyl methyl carbonate (EMC): propylene carbonate (PC) electrolyte. Figure 6.7(d) shows the printed coin cell averaged a specific areal capacity of 3.91 mAh/cm^3. A wearable battery bangle with an LED light was printed with a curved LTO anode and LMO cathode, allowing the integrated battery to fit around the user's wrist (see Figure 6.7(c)). A Celgard separator was used between the electrodes, leading to a device that powered an LED for about 60 s.

In another paper by Maurel *et al.*, the authors created new LFP/PLA and SiO_2/PLA filaments to form the cathode and separator, respectively, as a path toward a full FDM-printed battery [27]. The graphite filament previously reported in [26] was used as the mating anode to enable a full FDM-printed LIB cell. Poor ionic conductivity of

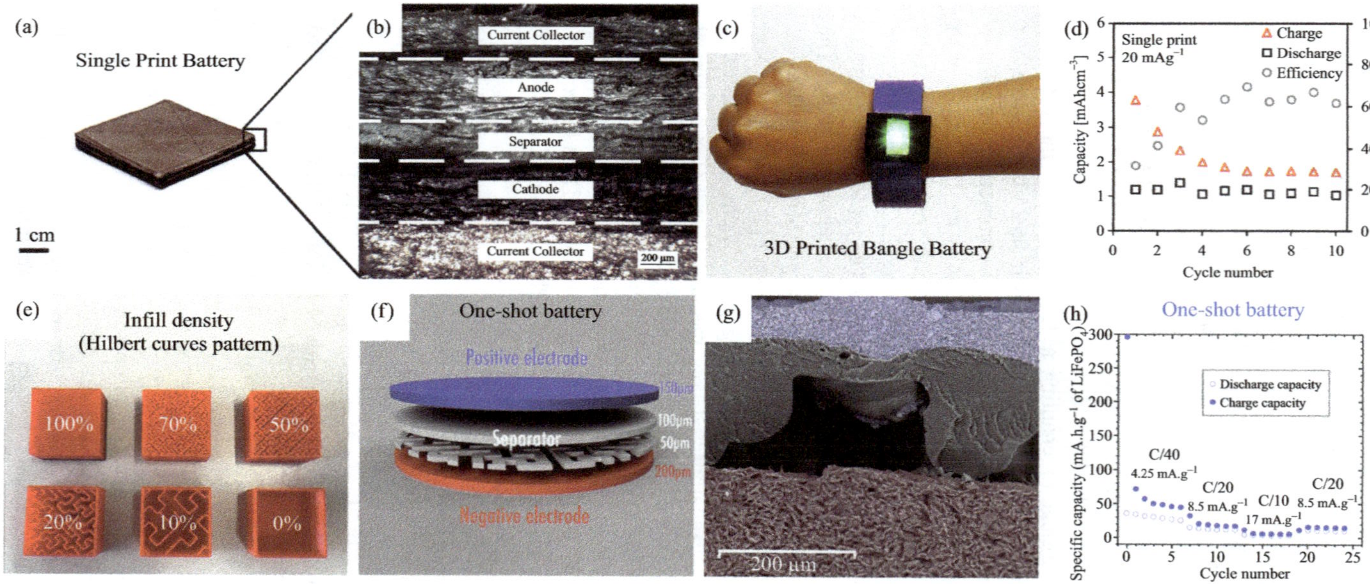

Figure 6.7 *This image displays FDM-printed LIBs made by Reyes* et al. *[25] and Maurel* et al. *[27]. (a) shows an FDM battery created in a single print session by swapping out the LIB component filaments at the completion of each layer. LIB was then vacuum filled with electrolyte for testing. (b) Cross-sectional image showing the layers of LIB made in (a). (c) A prototype curved bangle using the battery technology developed in (a) and (b). (d) Specific capacity and coulombic efficiency of the LIB in (a) where an applied current density of 20 mA/g for 10 cycles is used. (e) Various infill densities possible with FDM. (f) Illustration of the FDM "one-shot" LIB. (g) SEM image close-up view of the FDM-printed battery. (h) Capacity retention test results at different C-rates for the battery in (f) and (g). (Images in (a)–(d) are reprinted with permission from Ref. [25]. Copyright 2018 American Chemical Society. Images in (e)–(h) are reprinted with permission from Ref. [27]. Copyright 2019 and distributed under the terms of the Creative Commons Attribution 4.0 License (CC BY).)*

the filaments is one of the greatest challenges to overcome, and the authors focus on understanding the impact of carbon black in the cathode and ceramic additives in the separator filament. Different 3D patterns and infill density in the separator were explored to improve liquid electrolyte uptake in the cell while avoiding short circuits between the anode and cathode, shown in Figure 6.7(a) and (b). A printed in "one-shot battery" shown in Figure 6.7(f) and (g) was created by pausing the FDM printer and manually changing the filament between each layer. A specific capacity of 30 mAh/g of active material at C/40 was obtained as shown in Figure 6.7(h). While these results are low compared to conventional slurry cast electrodes, this shows a promising path toward enabling customizable and integrated batteries with an FDM printer. Beyond filament material optimization, creating a more streamlined, multi-printhead FDM machine may have additional benefits for improving the performance of fully printed FDM LIB cells.

Researchers have also explored FDM for 3D SEs. Ragones *et al.* focused on creating LTO and LFP filaments in different 3D spiral-shaped SE configurations [24]. LTO/PLA and LFP/PLA filaments consisted of 50–70% active material, 10% carbon, and 20–40% PLA. Upon electrochemical testing, the FDM SEs showed only ~ 50–60% of the theoretical capacity of the LFP and LTO materials when tested individually. While promising, this shows the need to further optimize the filament composition and active material utilization during charge/discharge. Gupta *et al.* explored 3D SE cathodes, composed of PLA/LFP/carbon nanotubes (CNTs) fabricated in a 3D grid-like pattern with a Creator Pro 3D printer. The authors focused on creating 3D SEs, with the goal of balancing electrode thickness, porosity, and CNT loading to achieve high specific and areal capacities. The optimal filament composition of PLA/LFP/CNT was 45/45/10 wt%. The FDM process deposited layers in 100 μm increments, resulting in 3D SEs with thicknesses of 100, 200, and 300 μm. The 300 μm 3D SE had a porosity of ~70% resulting in a specific and areal capacity of 151 mAh/g and 2.3 mAh/cm^2, respectively. The 100 μm 3D SE had a specific and areal capacity of 156 mAh/g and 1.7 mAh/cm^2, respectively, at 10% porosity.

FDM is the most accessible form of AM to the everyday user. The creation of LIB filaments opens opportunities for this userbase to print customized and integrated energy storage devices. The research work discussed herein is a step toward true freeform 3D printed batteries where all critical components—anode, cathode, current collectors, the separator, and a solid electrolyte—are deposited concurrently on the same print system. However, FDM for LIBs has shown limitations in achieving the small feature sizes and material loading needed to achieve high capacity batteries. Of the processes shown in Figure 6.5 and Table 6.2, FDM has the highest print resolution. In the next section, we discuss DIW as an alternative to FDM that can circumvent some of these challenges.

6.3.4 DIW

DIW, also known as single nozzle extrusion, is one of the most widely used AM methods for fabricating intricate electrode structures and customized battery formfactors. DIW leverages principles of operation similar to FDM, but instead of a heated extruder and filament, a pressurized system is used to extrude paste or ink through a smaller nozzle. This nozzle translates over a control stage to build a part up layer-by-layer based on an imported CAD model. DIW material is a wet, paste-like, or gel-like material, often referred to as an "ink," that requires thermal drying, sintering, or curing to remove solvents and additives to solidify the final structure. Of the AM processes discussed in this chapter, DIW currently has the highest level of research activity. We believe this is due in part to DIW's process compatibility with existing LIB slurry materials. In this section, have selected seven unique research studies to highlight with the goal of giving the reader a broad understanding of the range of LIB components and materials that can be fabricated with DIW.

In 2013, Sun *et al.* published the first DIW demonstration of 3D IDEs [29]. The anode and cathode were composed of LTO and LFP active materials, respectively. No separator was used in this design; only an air gap prevented contact between the anode and cathode. High viscosity inks with shear-thinning behavior in a 10^5–10^3 Pa-s range were created. These inks were composed of nano-sized active materials mixed with deionized water, ethylene glycol, glycerol, and cellulose-based viscosifiers. The LIBs were printed with a pressurized 30 μm diameter nozzle (Figure 6.8(a)) on a custom three-axis micro-positioning stage from Aerotech to create an 8-layer IDE (see Figure 6.8(b)) and a 16-layer IDE. On average, the electrode feature widths were 60 μm and spacing between features was 50 μm. To solidify the inks, the printed LIBs were heated up to 600°C to remove solvent and decompose the organic additives into conductive carbon. The 8-layer IDEs demonstrated specific capacities of 131 and 160 mAh/g at a 1C rate. Areal capacities up to 1.5–1.6 mAh/cm^2 were realized as shown in Figure 6.8(e) for the 8-layer structures. While the 16-layer structure showed areal capacities up to 2.0 mAh/cm^2, the thicker structures experienced more ohmic drops, which can lead to lower gravimetric and volumetric energy density relative to the 8-layer structure.

The study by Sun *et al.* [29] was foundational to DIW for LIBs. However, reliance on only an air gap between the electrodes to prevent short circuiting presents a safety concern if the anode and cathode physically contact. As a path toward solving this issue, Fu *et al.* [30] created an IDE from LTO and LFP, but their work focused on creating a solid polymer electrolyte material made from poly(vinylidene fluoride)-*co*-hexafluoropropylene (PVDF-HFP) and Al_2O_3 nanoparticles to serve as a separator-like material between the anode and cathode electrodes. The authors also explored adding graphene oxide (GO) to the electrode ink formulations to improve LIB conductivity over Sun *et al.*'s formulation. Electrodes were freeze dried to remove water solvent and solidify the 3D structures. A thermal annealing treatment process was used thereafter to reduce the GO. While a promising demonstration, the authors were not able to achieve the same electrochemical performance or compact formfactor demonstrated by Sun *et al.* [29], both of which are critical to maximizing battery energy and power performance. Fu *et al.*'s IDEs were an order of magnitude larger than Sun *et al.*, with electrode widths and spacing on a millimeter rather than micron-scale. Specific capacities were reported

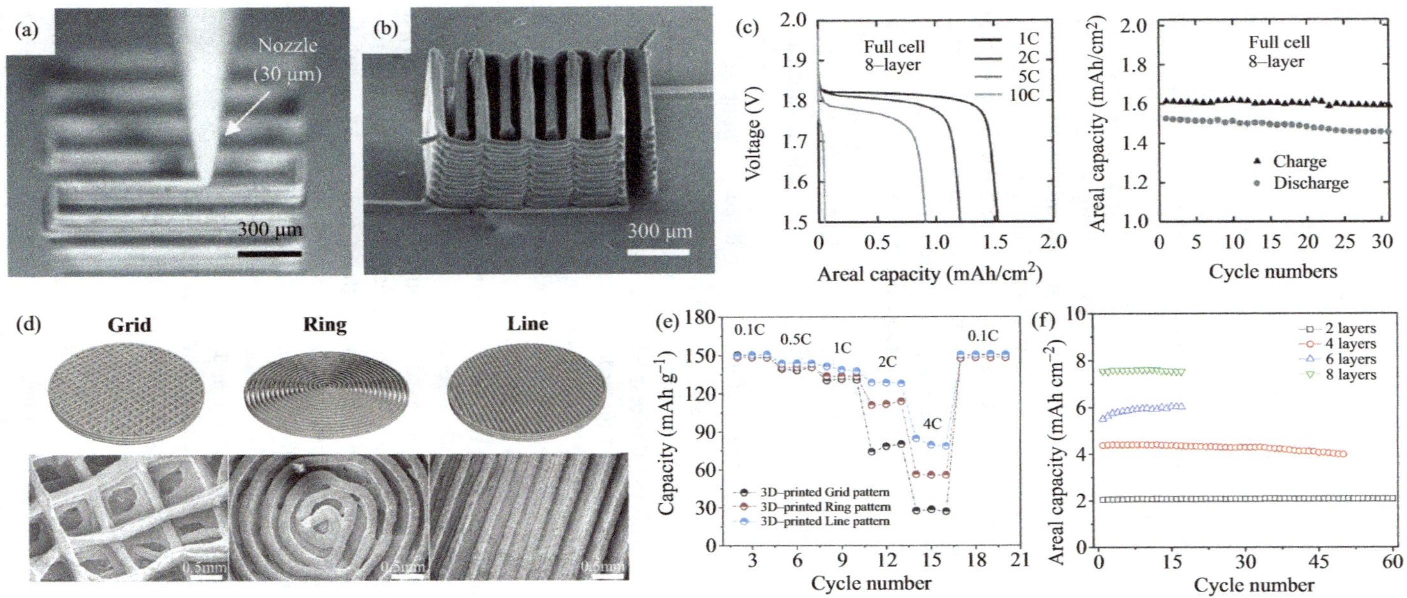

Figure 6.8 *DIW components of 3D IDEs (a–c) and 3D SEs (d–f). (a) Shows a 30 μm nozzle fabricating an IDE electrode while (b) shows the final printed 3D LIB IDE made with LTO-LFP. (c) Contains the electrochemical response of the IDE. (e) Shows three SE structures fabricated by DIW and (e) shows the rate capability test results while (f) shows the cycle life test results as a function of number of layers of electrode material printed. (Images (a)–(c) are reprinted with permission from Ref. [29]. Copyright 2013 Wiley-VCH Verlag GmbH & Co. KGaA, Weinheim. Images (d)–(e) are reprinted and adapted with permission from Ref. [35]. Copyright 2018 American Chemical Society.)*

near 100 mAh/g, and based on our own estimates, areal capacities were near 1.27 mAh/cm^2. The solid polymer electrolyte for the IDEs from a design standpoint improved battery safety but sacrificed some electrochemical performance due to the electrolyte's lower ionic conductivity and larger thickness relative to a conventional separator. While promising work, this study points toward the need for further investigation into robust polymer electrolytes and separator replacements for IDEs.

Wei *et al.* [33] extended the work done by Sun *et al.* and investigated thick electrode fabrication with DIW with a focus on fabricating all LIB components: anode, cathode, separator, and packaging. LTO and LFP electrode active materials were used, and four different inks were formulated to have high-viscosity, shear thinning behavior. Biphasic electrode inks were formulated by mixing LFP or LTO with conductive carbon into a solution of lithium bis(trifluoromethane) sulfonamide (LiTFSI) and PC. Ultraviolet (UV)-curable composite inks for the LIB packaging and separator materials were also created. Different LIB electrode thicknesses were explored up to 1 mm. Their work resulted in high areal energy density LIBs (4.45 mAh/cm^2 at an applied current density of 0.14 mA/cm^2) while also maintaining the power density that would be observed in a conventional thin electrode LIB. The authors created high-performance LIBs and showed a path toward customized LIB shapes for direct integration with other objects fabricated via AM.

Wang *et al.* used DIW to study the performance of three different 3D SE architectures made with LFP as shown in Figure 6.8(d) [35]. A series of grid, ring, and line-shaped SEs were examined. A number of deposited layers were printed and tested (see Figure 6.8(f)), with the thickest SE case consisting of eight layers for a total thickness of 1.5 mm, which is an order of magnitude larger than conventional slot die coated electrodes. The thickest design had an areal capacity of ~7.5 mAh/cm^2. The electrochemical performance is displayed in Figures 6.8(e) and (f). SEs with 2 layers, 4 layers, and 6 layers, had areal capacities of ~2.07, 4.3, and 6.0, respectively. The authors found that the line-patterned SE had performed better than the grid and ring designs (Figure 6.8(e)). Similarly, Lacey *et al.* explored grid SEs using a material system based on reduced holey GO (r-hGO), a highly porous 2D nanomaterial synthesized by the authors [32]. The authors formulated the r-hGO into an aqueous printable ink without the use of additives or binders. When printed into a grid SE using DIW, the r-hGO electrode by design had multiple length scales of porosity, from the nano- to millimeter-scale. These electrodes were cycled in 2032 coin cells with lithium metal and demonstrated a specific and areal capacity of 4,000 mAh/g and 13.3 mAh/cm^2 at a C/50 rate, respectively. Their work showed that creating multi-scale porosity by combining DIW SEs and nanoporous materials can advance the performance of next generation LIB materials.

Kohlmeyer *et al.* used a custom Aerotech DIW system to create flexible DIW LIBs by modifying the ratios of binder and other material components in their ink formulation [31]. The authors studied the impact of these composition changes on rheology, electrochemical performance, and mechanical properties. Their end goal was to fabricate free-standing flexible films of LFP, LTO, and LCO. Higher binder content (40–50%) in the dried electrode led to improved mechanical performance. While the high binder content electrodes showed high specific capacity performance, increased binder content fundamentally runs counter to increasing volumetric and gravimetric energy densities; more active material and less polymer binder material are needed in the dried electrodes to maximize overall LIB performance. The authors also found that at high discharge rates, rate performance dropped as binder loading was increased. Although the study shows promise toward DIW for flexible batteries, it highlights the trade-off space that battery engineers must balance between mechanical and electrochemical performance.

Lastly, McOwen *et al.* recently applied DIW with a commercial nScrypt 3D printer to create solid electrolyte structures based on ceramic $Li_7La_3Zr_2O_{12}$ (LLZO) for future solid-state lithium metal batteries [34]. LLZO inks were used to make grid, line, and column-shaped solid electrolyte structures. Two ink compositions with two different binder systems were explored as binders impact ink rheology and the printed structures. The first ink made with a polyvinyl butyral and benzyl butyl phthalate (PVB-BBP) binder system was easily wetted and conformed to the substrate surface, making it suitable for depositing thin layers of LLZO. The second ink used a proprietary texanol-based binder (made by manufacturer ESL) to create Bingham plastic behavior, enabling printed structures to maintain their shape during deposition. This allowed for thicker, structured LLZO prints. The printed features ranged from 12.5 to 125 μm in size and the height of LLZO structures was controlled by printing additional layers in the vertical direction. Electrochemical results indicated that the LLZO structures maintained high conductivity, similar to their bulk counterparts. This works opens up an opportunity to study 3D solid-state electrolyte structures as a potential solution to electrode–electrolyte interface issues and dendrite growth in solid-state lithium metal batteries [7].

DIW presents a means to customize 3D and flexible LIBs for IoT-scale and microelectronic applications with better resolution than FDM. The research work conducted with DIW to date has shown promising results with respect to material formulation and battery design to enable flexible LIBs and high-performance thick electrode IDEs and SEs. However, manufacturing over meter-scale areas with the micron-scale resolution benefits of DIW requires the development of new multi-nozzle arrays or other technology capable of rapid extrusion beyond the limits of a single DIW nozzle [62–64].

6.3.5 IJP

With the aid of computers, IJP can generate digital images with a variety of inks and materials. IJP is one of the most studied techniques for producing printed and flexible electronic components [6,65]. This AM method presents opportunities for finer scale control of LIB material films and integrated microbatteries composed of ultra-thin (< 10 μm) material layers. Inkjet printers deposit droplets of ink on substrates and employ low-viscosity inks. A thermal or UV lamp is sometimes mounted near the inkjet nozzles for rapid curing of the material to enable shape retention (Figure 6.4(c)). IJP can be classified into two modes of operation: continuous inkjet (CIJ) and drop-on-demand (DoD) [66]. CIJ relies on a continuous flow of ink passing through a nozzle. The ink is separated into droplets through oscillation via electrostatic charge prior to existing the nozzle. DoD, on the other hand, generates small droplets which are individually deposited via thermal or piezoelectric actuation mechanisms. DoD allows for multiple compact nozzles to be arranged together, enabling faster print speeds than what can be realized with a single CIJ nozzle.

IJP allows for the deposition of ultra-thin battery films that are not possible with FDM, DIW, or traditional slot die coating. IJP has been applied to printing thin layers of functional material for a range of energy storage applications [67], but the extension to LIB chemistries has been limited due to challenges with creating stable colloidal suspensions and synthesizing particles of sufficient size to avoid nozzle clogging. Digital control of ultra-thin layers allows a battery engineer to create flexible LIBs that can resist multiple modes of bending and stretching deformation, which is critical for advanced IoT devices.

An early demonstration of IJP for LIBs was conducted by Huang *et al.* in 2008 [36]. Huang *et al.* were able to successfully deposit ultra-thin LCO cathodes with a thickness of 1.2 μm by leveraging prior work on SnO_2 anodes [68]. Their primary goal was to create stable IJP inks with LIB materials. LCO powder was synthesized with sol-gel processing to yield particles with an average size of 93 nm. To produce a stable IJP ink, the LCO powder and conductive carbon were ultrasonically mixed into an aqueous solution containing carboxymethyl cellulose sodium (CMC) binder, monoethanolamine as a pH stabilizer, and a commercial surfactant. Their electrodes were printed with a Canon printer and showed that at an applied current density of 394 $\mu A/cm^2$, a specific capacity of 105 mAh/g could be attained. In addition, the cells maintained 95% of their capacity at 100 cycles when tested against lithium metal.

Ben-Barak *et al.* similarly demonstrated IJP with a Nordson PICO Pulse for LFP cathodes and obtained ~140 mAh/g over a C/10 to 1C rate [38]. The authors varied binder content and solid and dispersant agents to enable a stable ink suspension when using deionized water as the ink solvent. CMC was the binder while Triton X-100 was the dispersant. The authors performed impedance spectroscopy analysis to quantify the diffusion resistance of the LFP electrodes made with IJP, showing that this AM method does not alter charge and discharge mechanisms. Also focused on IJP of cathode materials, Kolchanov *et al.* used a Fujifilm Dimatix printer to explore the use of Li-manganese cathode materials doped with different concentrations of aluminum, sodium, and potassium and their influence on thin-film IJP electrodes and compared their results to conventional doctor blade electrodes [39]. The electrodes doped with potassium showed the best performance and were printed up to 15 layers in height with each layer averaging 0.46 μm in thickness (Figure 6.9).

Focusing on the anode, Kushwaha *et al.* created graphene anode films and used ethanol as a solvent with ethyl-cellulose to stabilize their ink and control the viscosity [40]. Ethyl-cellulose decomposes during thermal drying leaving behind a

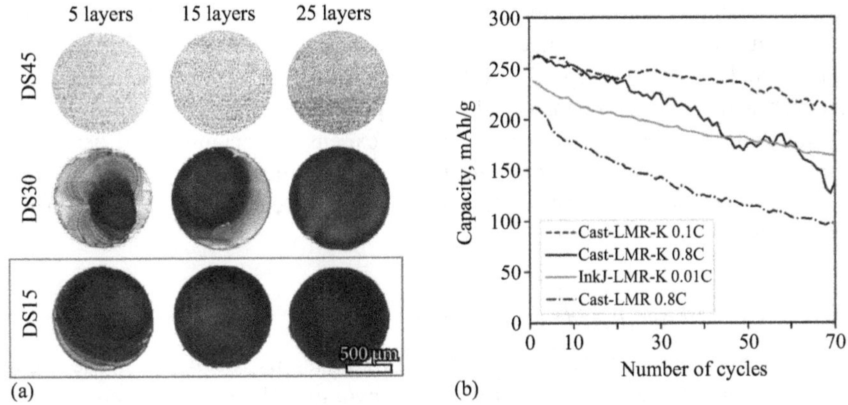

Figure 6.9 (a) Images of IJP LIB cathodes deposited with a 0.03 wt% ink composition. Images show the impact of number of deposited layers on the print quality with more layers yielding more uniform coverage over the substrate. The best results are achieved with drop spacing (DS) of 15 μm and are highlighted. (b) shows electrochemical results for capacity retention over 70 cycles where the "InkJ" label represents the IJP sample and "Cast" are conventional electrodes. (Reprinted with permission from Ref. [39]. Copyright 2019 WILEY-VCH Verlag GmbH & Co. KGaA, Weinheim.)

binder-free highly loaded composition of active particles. Their electrodes were printed with a Jetlab-4 inkjet printer and demonstrated that ~942 mAh/g could be reached at a C/10 rate, yielding a high-capacity IJP anode.

While promising work for the cathode and anode has been shown, creating an IJP-compatible solid-state electrolyte to enable full-cell fabrication is critical to streamlining the fabrication process for a fully printed IJP cell. As a path toward this, Delannoy *et al.* created an IJP-compatible silica-based ionogel ink [37]. This ionogel was printed with a Fujifilm Dimatix printer on top of tape-cast LFP cathodes and LTO anodes to create a solid-state LIB full cell for testing. The ionogel LIB full cell had an areal capacity of 300 μAh/cm^2 when tested at a C/10 rate, which corresponds to a specific capacity of 60 mAh/g for the LFP positive electrode. It is worth noting that is ~2.5 times lower than what one would expect from an equivalent cell with liquid electrolyte. Nonetheless, this is a promising path toward creating a full LIB IJP workflow for depositing anodes, cathodes, and electrolyte as printable solid-state electrolytes alleviate issues with handling and packaging liquid electrolyte for microelectronic applications.

IJP is critical for circumventing expensive thin-film microfabrication processes. Relative to conventional microfabrication and thin-film fabrication techniques [69–71], IJP is a low-cost method with the ability to fabricate small-scale patterns quickly. It is important to note that the electrochemical results of most IJP batteries created to date fall short of meeting the energy density needs of LIBs for microelectronic and IoT applications, and multiple passes of an IJP nozzle are required to build up thicker electrodes which can slow down production time. However, depending on the materials used, deposition can take place in an ambient environment and does not require the use of expensive vacuum-based deposition equipment. This presents a promising path toward printed microbatteries if processing challenges around stable LIB inks with good electrochemical performance can be solved.

6.3.6 Spray-based AM: AJP and ESD

AJP and ESD are other means to deposit LIB materials for 3D electrode architectures or microbatteries for direct device integration. Herein, we group AJP and ESD as two similar methods of spray-based AM that employ the atomization of inks into droplets for printing. However, their mechanisms of operation are slightly different. In AJP, ink is pneumatically atomized with a carrier gas to create droplets as shown in Figure 6.4(d). These droplets are focused through a nozzle into a stream and deposited on a substrate. An inert gas (typically nitrogen) is used to keep the flow of droplets away from the nozzle surface. Unlike AJP, ESD relies on electrostatic atomization, and ink droplets are deposited from the tip of a nozzle due to an applied electric field between the nozzle and a grounded substrate.

Similar to IJP, AJP has been used to create flexible, planar LIB electrodes. Lee *et al.* used AJP to create planar LTO anodes and LFP cathodes over a large 25 by 25 cm^2 area [72]. The focus of the study was on the optimization of ink and spray parameters for AJP LIBs. For the ink, active material and conductivity enhancers were dispersed into a mixture of DI water and isopropyl alcohol (IPA) with CMC binder to create a suspension. A DI:IPA volume ratio of 60:40 was used due to its stability and mitigation of ink sedimentation. The authors deposited multiple AJP layers to build up electrodes ranging from 10 to 80 μm thick while maintaining a heated current collector substrate to enable rapid drying of each layer after deposition. At C/10-2C charge/discharge rates, the AJP and regular cast electrodes had similar capacities, but at higher 20C rates, the AJP electrode had $\sim$ 25 % higher capacity than the cast reference electrode. Similarly, Ludwig *et al.* used a variant of ESD to create a solvent-free spray process to fabricate planar LCO electrodes over large areas [46]. A spray nozzle was used to charge dry battery particle mixtures. These charged particles were then deposited onto a grounded current collector and hot pressed to bind LCO active material particles and carbon black through the melting of PVDF binder. Overall, the ESD electrodes demonstrated similar electrochemical performance as conventional electrodes make via casting, with the ESD electrodes performing slighting better while maintaining flexibility. By removing solvent and drying steps, the authors created a spray-based AM process that is potentially more economical for large-scale manufacturing.

Examining 3D batteries, Saleh *et al.* used AJP to fabricate a lattice-like SE electrode with silver nanoparticles with an Optomec AJ-300 printer [48]. The lattice architecture by design infuses controlled porosity into the electrode enabling areal capacities that are up to four times higher than the bulk equivalent form. Deiner *et al.* [49] used AJP to create LFP SE electrodes with aligned pore channels for enhanced ion transport, although the structures had less feature definition than the ones fabricated by Saleh *et al.* Using an Optomec AJ-200 printer, 80 passes of aerosolized LFP ink were applied to a 75 by 75 mm^2 area, yielding a 170 μm thick electrode that had a specific capacity of 145 mAh/g in a half-cell format at C/10. For low discharge current densities, the areal capacity of the samples was lower than those made by DIW with similar materials [33], but the performance improves at higher rates due to the pore channels. Rodriguez *et al.* [51] also used an Optomec AJ-200 printer to create low tortuosity 220 μm thick LFP electrodes similar to Deiner *et al.* Their AJP process took 125 passes to create a thick LFP electrode. The AJP electrode demonstrated 5–14% improved rate capability from C/5-2C rates than the equivalent cast form with the same loading.

Lee *et al.* used a custom ESD apparatus to fabricate thin LCO-Graphite 3D IDEs [47], similar to Sun *et al.* [29]. Conventional slurry components of NMP, PVDF, and active material were used but the overall ink was more dilute than a conventional slurry due to the higher amount of NMP used to enable the spray process. An IDE mask was held over the current collector and ESD was sprayed on top of the mask, leaving behind an LCO-Graphite IDE once the mask was removed. Feature widths ~500 μm were created yielding a specific capacity of 79.1 mAh/g for the full flexible cell. Although the capacity is lower than that expected for LCO graphite, it is a promising alternative to expensive microfabrication techniques. However, AJP 3D LIBs require more process optimization to create structures that are as controlled and defined as those created with DIW. When optimizing for areal capacity, building a tall electrode structure is critical to increasing the total mAh capacity of a battery over small footprint areas. Due to the thin layers enabled by AJP and ESD, creating tall structures is a time-intensive process relative to DIW.

AJP is a competing technology to IJP given its ability to deposit a range of low-viscosity inks in thin layers. Unlike IJP, AJP is compatible with printing on complex non-planar surfaces due to the larger distance and area covered by atomized droplet streams. Printing on surfaces and substrates that are not planar requires very elaborate control schemes and controls development for FDM, DIW, and IJP. However, AJP removes these requirements, allowing standard three-axis motion stages to be used to print on curved surfaces due to the ability of atomized spray droplets to form thin conformal layers over large areas. As a path toward this, Yu et al. fabricated a LIB and showed how AJP could be used to print on non-planar substrates without a multi-axis printer, as shown in Figure 6.10(b–d) [50]. The authors used a CH Technologies nebulizer for their aerosol printing. Their results showed batteries with a specific capacity up to 148 mAh/g. While the battery was made with AJP, the packaging was made with FDM which shows how two different AM methods can be combined to create a fully package LIB. The ability to use AJP on non-planar surfaces removes the need to create more complex software and workflows with DIW and FDM.

Spray-based AM is promising for fabricating 3D and conformal energy storage devices. However, relative to slot die coating, there are diminishing returns for

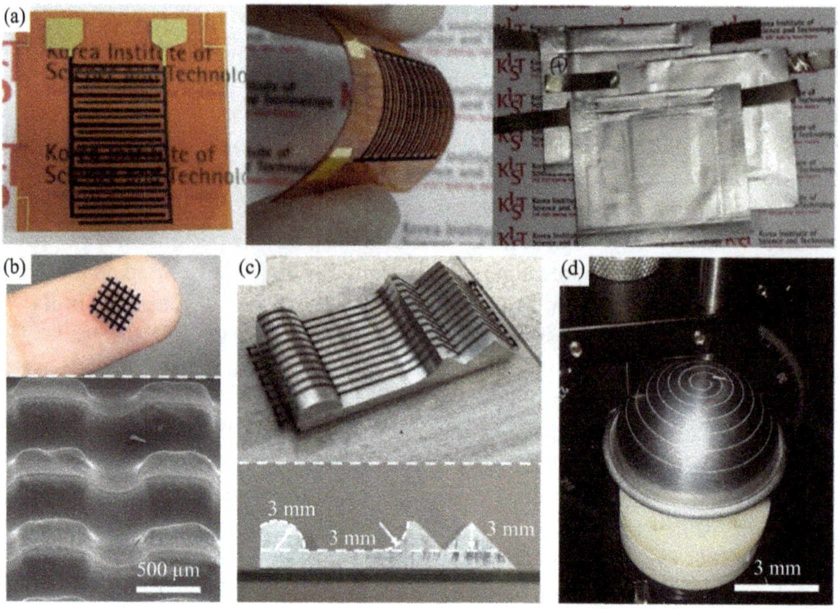

Figure 6.10 Images of research studies conducted by Lee et al. *and Yu* et al. *(a) Photographs of a flexible LIB cell made with IDEs with LCO for the cathode and graphite for the anode. Non-planar capabilities AJP are shown with three examples in (b)–(d). In (b), a 3D SE with an array of 500 μm wide squares is displayed. (c) Shows the AJP process on an aluminum target with complex surface topologies. (d) Shows a hemispherical substrate with a spiral line deposition. (Image (a) is Reprinted with permission from Ref. [47], 2019 WILEY-VCH Verlag GmbH & Co. KGaA, Weinheim. Images (b)–(d) are reprinted from Ref. [50], Copyright 2020, with permission from Elsevier.)*

creating conventional planar electrodes. Spray methods require additional layer-by-layer fabrication to make 40–100 μm thick electrodes that slot die coating can deposit in one pass. Moreover, spray-based AM lacks the resolution that can be achieved with IJP as shown in Figure 6.5.

6.3.7 SLA

Here, we define SLA as any process that involves vat-based photopolymerization to create 3D LIB electrodes and solid electrolyte architectures, including micro-stereolithography and high-resolution two-photon polymerization systems. Like FDM and DIW, a CAD model of the patterned object is imported into software that digitally slices the model into layers. These layers are then translated into machine code. A UV light source or laser is then used to crosslink and solidify layers of material in a vat of photopolymer solution. Each cured layer hardens and adheres to the prior layer until the full 3D object is fabricated. The final object is removed from the vat and often baked or fired in an oven as a post-processing step to further solidify the structure. Sintering may be used to shrink the architecture and convert the cured material into conductive carbon. Depending on the structure and materials used, subsequent post-processing steps may be required.

Developing suitable photopolymer resins with LIB active materials is a challenging aspect of creating 3D LIBs with SLA. An SLA resin must have low viscosity and good flowability, as shown in Figure 6.6, so the final fabricated object can be easily removed from the vat without damage and the vat can be refilled during processing. The resin must also have enough optical clarity to allow sufficient UV light penetration into the vat. These necessary properties are contradicted by the incorporation of battery active material particles. Battery active materials are opaque and dark in color, with micron-scale diameters. These properties all lead to challenges with light scattering and penetration during the curing process. Additionally, particles can settle at the bottom of the vat if viscosity is too low. This leads to inhomogeneous material compositions that are detrimental to battery performance. These trade-offs between material requirements must be carefully balanced to enable SLA manufacturing of LIBs. Until recently, these stringent material requirements have limited the ability of researchers to print LIBs with SLA.

Initial work with SLA centered on electrolyte materials. Chen *et al.* created a 3D gel polymer electrolyte (GPE) structure using polyethylene glycol diacrylate (PEGDA) and a photoinitiator mixed with a 1M $LiClO_4$ solution [41]. The authors built an SLA system capable of micron-scale patterning, known as micro-stereolithography, to fabricate a 3D structured GPE which was then fully cured with a UV lamp. Upon curing, the PEGDA forms a network of crosslinked poly (ethylene oxide) (PEO). The LTO anode and LFP cathode materials were mixed with the GPE resin, then manually coated onto either side of the 3D structured polymer electrolyte, infiltrating into the cured SLA electrolyte structure to form a 3D IDE LIB. Once tested, the 3D LIBs demonstrated low cyclability, lasting 2–5 cycles, with an areal discharge capacity of 1.4 $\mu Ah/cm^2$. Zekoll *et al.* examined ceramic–polymer–composite electrolytes with SLA and developed a 3D structured GPE with a lattice-like architecture using $Li_{1.4}Al_{0.4}Ge_{0.6}(PO_4)_3$ (LAGP) [73]. Relative to Chen *et al.*, the solid-state cells showed better capacity, cyclability, and mechanical durability due to both material composition and intricate lattice architectures. Taking the work done by Chen even further, He *et al.* used SLA to fabricate 3D spiral-shaped solid polymer electrolyte designs using PEGDA mixed with plasticizers and lithium bis(trifluoromethanesulfonyl)imide (LiTFSI) (Figure 6.11 (a–c)). The LFP cathode was manually infiltrated via a doctor blade into the electrolyte layer to form a 3D cathode and separator that were cycled in a half-cell with lithium metal [42]. These cells demonstrated improved cyclability and specific capacities, reaching 115 mAh/g with 85% capacity retention after 25 cycles at 25 °C and a discharge rate of 0.05C, but with a low areal capacity of 0.575 mAh/cm^2. Figure 6.11(c) shows the same cell cycled at 50 °C and a C/10 rate with a gravimetric and areal capacity of 166 mAh/g and 0.83 mAh/cm^2, respectively.

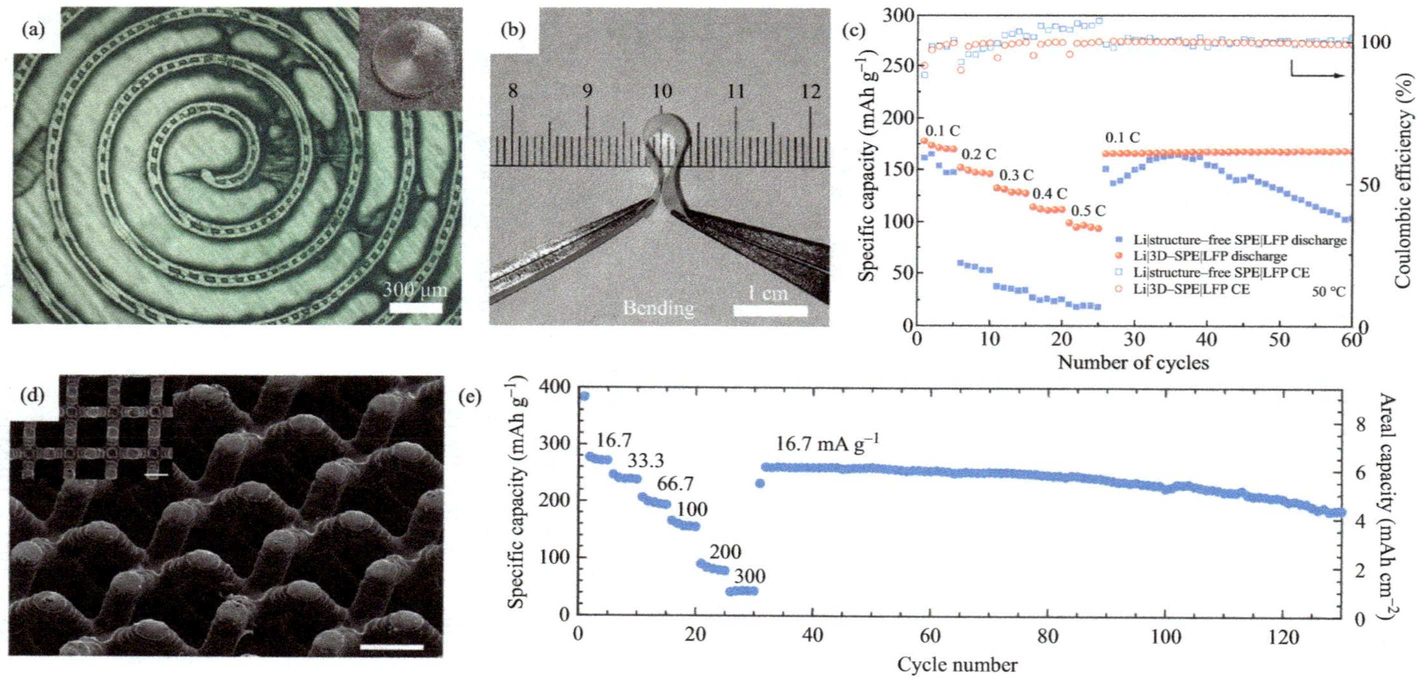

Figure 6.11 *SLA electrolyte and electrodes made by He et al. and Nartia et al., respectively. (a) Top view of the 3D structured polymer electrolyte. (b) Flexibility of the final solid polymer electrolyte material is demonstrated with a bend test. (c) Rate capability performance is shown for a Li metal/3D electrolyte/LFP and Li metal/planar electrolyte/LFP cell stack when held at 50°C. (d) 3D micro-architected SE made from carbon where scale bars are 100 μm for the main image and inset image. (e) Discharge capacities at different currents are indicated by the number above each blue line with one long cycle conducted at a current of 16.7 mA/g at the end. (Images in (a)–(c) are reprinted (adapted) with permission from Ref. [42]. Copyright 2020 American Chemical Society. Images in (d)–(e) are reprinted with permission from Ref. [43]. Copyright 2020 Wiley-VCH GmbH.)*

Examining recent advances in electrode fabrication with SLA, progress has been made with both carbon-based anodes and cathode oxide materials. Narita *et al.* used SLA to create carbon-based 3D LIB anodes (Figure 6.11(d) and (e)) with acrylate resins to increase the mechanical integrity of electrodes while also yielding a high areal capacity of 3.2 mAh/cm^2. Cycle life performance of 80% capacity retention after 117 cycles and 70% capacity retention after 130 cycles was also observed [43]. The authors used a commercial digital light processing (DLP) SLA printer from Autodesk. Yee *et al.* extended this concept and worked with LCO [44]. LCO 3D lattices were fabricated by creating lithium and cobalt nitrate aqueous photoresins. This resin was then processed in the SLA system to create 3D lithium and cobalt nitrate containing hydrogels, which were processed at high temperatures up to 1,000°C to form LCO lattices and shrink lattice dimensions. The LCO lattices demonstrated capacity retention ~ 76% after 100 cycles with a specific capacity of 121 mAh/g at C/40 and 115 mAh/g at C/20. Most recently, Martinez *et al.* published the first 3D SE cathode made NMC-111 material using SLA where a standard SLA process was combined with a precursor material approach. The SLA resin was prepared with metal precursor salts in addition to a photoinitiator and polymer host material. The electrodes created are lattice-like in geometry, similar to prior work by Narita *et al.* and Yee *et al.* High-temperature post-processing was used to remove the polymer matrix and photoinitiator and synthesize the NMC-111 material.

It is important to note that 3D SEs with lattice geometries are more amenable to SLA than DIW or FDM. However, 3D lattice geometries tend to have lower rate capability relative to other 3D LIB architectures [14]. In addition, lattice-based electrodes contain a substantially large amount of volume occupied by electrolyte compared to other 3D LIB designs, which reduces volumetric energy density. SLA electrode designs require further topology optimization or new high-capacity materials to meet or exceed the volumetric energy density of other 3D LIBs made with DIW. In addition, most electrodes made with SLA depend on a 600–1,000°C sintering step to synthesize the active material.

While SLA technology presents challenges with materials development, SLA shows promise for pushing the boundaries of microbattery development for the IoT and future wearable devices due to the ability to achieve high resolution features [74]. However, investigating ways to streamline the front-to-end process to reduce the number of manufacturing steps while finding ways to efficiently integrate and package these batteries requires further investigation by the research community. DIW, FDM, and spray-based AM are more adaptable to direct deposition of a LIB onto a target device substrate for quick and efficient packaging. SLA manufacturing requires pick-and-place operations and discrete processing steps based on what can be inferred by current research literature. This has the potential to increase manufacturing cost and time longer-term relative to the other technologies discussed and requires further research advancements to mitigate these issues.

6.3.8 FA patterning

Prior sections of this chapter reviewed conventional AM methods that have been adapted to LIB materials. We conclude our discussion of AM by highlighting an emerging approach to engineer LIB architectures using electric or magnetic fields (EMFs). FA patterning broadly refers to the use of magnetic or electric fields to rapidly pattern and engineer material microstructure "on-the-fly" during material deposition [75]. FA patterning can be used in tandem with FDM, DIW, IJP, AJP/ESD, and SLA, or in isolation to achieve finer control over the material structure. During FA patterning, an EMF is applied to a low-viscosity solution of particles. After the field is applied, particles in the solution are driven and aligned into unique 3D patterns. Microstructure can be fine-tuned in multiple directions by assembling particles to create porous networks that enhance electrical and ionic conductivity. While this approach has limited demonstrations for LIBs, we believe FA patterning offers great promise for the rapid manufacturing of 3D LIBs with engineered microstructures at micron and sub-micron scales.

Magnetic FA patterning has been used to create 3D SEs with anode materials. Anode material systems must first be altered to have para- or ferromagnetic behavior. A rotating magnetic field is applied perpendicular to a magnetically responsive slurry cast onto a current collector. The magnetic field pulls and aligns particles perpendicular to the current collector. Figure 6.12 shows some of the earliest work conducted in this area by Billaud *et al.* [52]. The authors demonstrated that FA patterning can be used to enhance the electrochemical performance of a 200 μm thick graphite electrode by lowering the effective tortuosity as shown in Figure 6.12(a), leading to a specific capacity that is about three times greater than the planar equivalent electrode at discharge rates between C/2 to 2C. They accomplished this by diluting the viscosity of the graphite slurry and functionalizing it with superparamagnetic iron oxide nanoparticles (SPIONs) to create paramagnetic properties. Upon exposure to a magnetic field, graphite flakes were vertically aligned, as seen in Figure 6.12(b) and (c).

Zhang *et al.* [53] used an approach similar to Billaud *et al.* but instead used graphite flakes coated with a ferrofluid to generate a magnetic field response. The authors achieved a specific capacity that was approximately four times higher than the equivalent planar electrode. Sander *et al.* [76] also used magnetic fields to pattern LCO cathode electrodes into similar patterns as Billaud *et al.* and Zhang *et al.*, but used the magnetically generated pattern as a sacrificial template for creating electrode porosity. Magnetic microrods and droplets were mixed with an LCO suspension. After the material formulation was patterned, the material was sintered at a temperature of 950–1,035°C, burning off the sacrificial microrods and droplets and leaving behind an LCO electrode. In its current form, magnetic FA patterning creates an additional step in the fabrication of a 3D LIBs as magnetic material additives must be included and later removed to enable a functional battery, adding process time and complexity.

During electric FA patterning, two electrode plates are oriented parallel to each other, and one of the plates must be grounded. A liquid solution of particles is suspended between the two plates, and an electric field is then driven by a high voltage amplifier to align particles in the solution. This process requires rapid curing of the solution with UV light or heat prior to removing the electric field. Due to the specific dielectric properties and particle morphologies required, research on

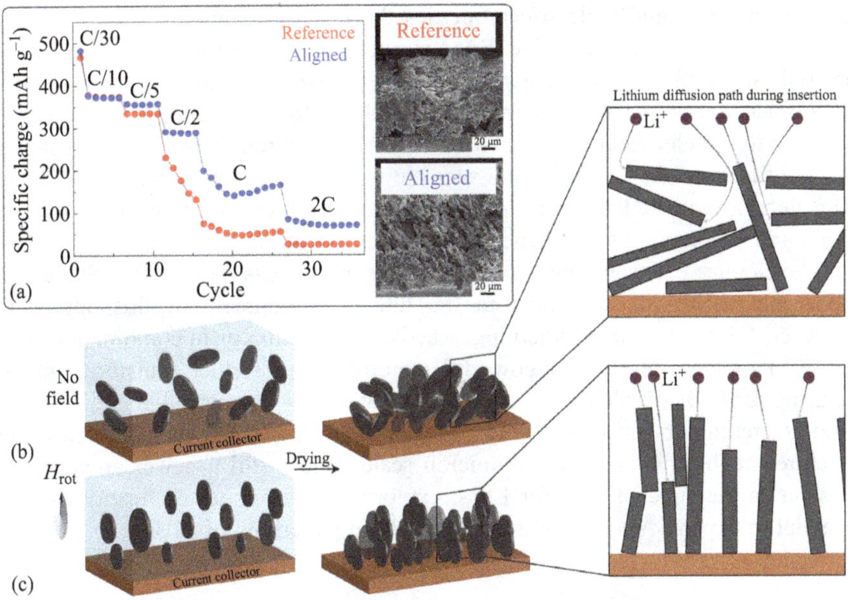

Figure 6.12 Magnetic FA patterning of graphite particles for a LIB anode is shown. Electrochemical performance for a planar reference electrode and an aligned FA electrode are displayed in (a). The FA electrode outperforms the conventional reference electrode. A standard slurry of graphite particles is illustrated in (b) while (c) shows the alignment of graphite particles relative to the current collector if a rotating magnetic field is applied. (Reprinted by permission from Springer Nature Customer Service Centre GmbH: Springer Nature, Nature Energy, Ref. [52], Copyright 2016.)

electric FA techniques for LIB anodes and cathodes is limited. Li *et al.* applied an electric field to wet-state LMO prints to align LMO particles prior to drying to create additional microscale architecture in the electrode [77]. The authors used conventional NMP-based LMO inks in their work. However, $Li_{1.3}Al_{0.3}Ti_{1.7}(PO_4)_3$ (LATP) ceramic–polymer composites have shown more promise with electric FA patterning as recently shown by Liu *et al.* [78]. Liu *et al.* dispersed LATP particles in a PEGDA precursor solution and added polydimethylsiloxane (PDMS) as a thermocuring agent. Two ITO-coated glass electrode plates were placed on a hot plate and connected to a function generator and high-voltage amplifier to generate an electric field (200–1,000 V/mm) for patterning. Their work aligned LATP particles into vertical channels and demonstrated that a 3D patterned composite electrolyte increased conductivity by almost an order of magnitude relative to the bulk counterpart.

As noted recently by Johnson *et al.* [79], FA techniques that leverage EMFs are only applicable to a narrow range of ink compositions and particle types, due to specific magnetic and dielectric properties the materials must possess. Utilizing acoustic waves is an alternative FA approach. Recent work in the composite materials space [80,81] has demonstrated how acoustic assembly of rods or platelets formed connected networks for structural reinforcement, enhanced thermal conductivity, or electrical conductivity. In addition, theoretical work [82,83] shows how the fundamental theory behind acoustic waves can generate complex 2D and 3D patterns. Acoustic FA patterning is an alternative to consider for future LIBs if EMFs limit the range of LIB materials that can be patterned.

Developing FA techniques that are compatible across most LIB cathode and anode materials will be critical to the future development of this approach. However, FA printing and patterning, whether in isolation or in combination with AM technologies, presents a powerful opportunity to further transform manufacturing for LIBs. Particle packing and orientation can be controlled and adjusted during material deposition, enabling a range of anisotropic and controlled material structures at the micron- and sub-micron scale. While still a newer approach to consider in the field of AM for LIBs, exploring these processes further has the potential to expand AM capabilities for the rapid patterning of LIBs.

6.4 Conclusion and future outlook

In this chapter, we reviewed the most common AM processes for engineering LIBs beyond homogenized rectangular shapes. The development of AM to produce engineered electrode architectures and flexible, embedded energy storage devices opens a new design space to maximize the performance of existing material chemistries. Modeling research has shown that 3D LIBs (>100 μm in thickness) can provide a 10–20% or greater improvement in energy and power over a conventional battery if electrode architecture is controlled down to 25–50 μm [17,20,21]. FDM and DIW present advantages for customized LIBs for wearable and portable devices, with DIW having a finer process resolution than FDM. SLA, IJP, and spray-based AM all present unique opportunities to engineer electrode architecture but have limited compatibility with LIB materials due to the smaller particle size, low viscosity, or curing mechanisms required. Due to the limits of extrusion and traditional slurry coating operations, the ability to quickly tune microstructure and battery architecture "on-the-fly" below 1 μm is limited. FA techniques, while still early-stage for LIBs, present interesting opportunities to circumvent these challenges and engineer microstructure at smaller length scales than what can be achieved with the conventional AM process.

To further advance the field of AM for LIBs, more materials development must occur in tandem with more streamlined manufacturing workflows. While each AM process offers distinct advantages and disadvantages, near-term solutions may involve creating a hybridized process that leverages combinations of each process to advance customizable electrodes and battery formfactors for small-scale applications. Larger-scale EV and grid-based energy storage applications require substantial investment and engineering development in processing scale and speed to be viable beyond microbattery scales. Moreover, finding ways to integrate and adapt LIBs to curved surfaces with integrated and efficient packaging remains a challenge that will limit the end utility of some of the AM LIBs covered in this chapter.

Glossary

ABS	acrylonitrile butadiene styrene
AJP	aerosol jet printing
AM	additive manufacturing
BBP	benzyl butyl phthalate
CAD	computer-aided design
CIJ	continuous inkjet
CMC	carboxymethyl cellulose sodium
CNT	carbon nanotubes
DEC	diethyl carbonate
DI	deionized
DIW	direct-ink write
DLP	digital light processing
EC	ethylene carbonate
EMC	ethyl methyl carbonate
EMF	electromagnetic field
ESD	electrostatic spray deposition
EV	electric vehicle
FA	field-assisted
FDM	fused deposition modeling
GO	graphene oxide
GPE	gel polymer electrolyte
IDE	interdigitated electrodes
IJP	inkjet printing
IoT	Internet of Things
IPA	isopropyl alcohol
LAGP	$Li_{1.4}Al_{0.4}Ge_{0.6}(PO_4)_3$
LATP	$Li_{1.3}Al_{0.3}Ti_{1.7}(PO_4)_3$
LCO	$LiCoO_2$
LED	light emitting diode
LFP	$LiFePO_4$
LIB	Lithium-ion battery
LiTFSI	lithium bis(trifluoromethane) sulfonamide
LLZO	$Li_7La_3Zr_2O_{12}$

LMO	$LiMn_2O_4$
LTO	$Li_4Ti_5O_{12}$
NMC	$LiNi_xMn_yCo_zO_2$
NMP	*N*-methyl-2-pyrrolidone
PC	propylene carbonate
PDMS	polydimethylsiloxane
PEGDA	polyethylene glycol diacrylate
PEGDME	polyethylene glycol dimethyl ether
PEO	poly(ethylene oxide)
PLA	polylactic acid
PVB	polyvinyl butyral
PVDF	polyvinylidene fluoride
PVDF-HFP	poly(vinylidene fluoride)-*co*-hexafluoropropylene
SE	structured electrode
SEM	scanning electron microscopy
SLA	stereolithography
SPION	superparamagnetic iron oxide nanoparticles
UV	ultraviolet

Acknowledgment

This work was funded in part by a Defense Advanced Research Projects Agency (DARPA) Young Faculty Award under grant number D19AP00038. The views, opinions, and findings expressed in this work are those of the authors and should not be interpreted as representing the official views or policies of the Department of Defense or the U.S. Government and no official endorsement should be inferred. This is approved for public release and distribution is unlimited.

References

[1] Wohlers, T., Caffrey, T., and Campbell, R.I. Wohlers Report 2016: 3D Printing and Additive Manufacturing State of the Industry Annual Worldwide Progress Report, Fort Collins, CO: Wohlers Associates, 2016.

[2] Beaman, J.J., Bourell, D.L., Seepersad, C.C., and Kovar, D. 'Additive manufacturing review: early past to current practice', *Journal of Manufacturing Science and Engineering*, 2020, 142, (11), p. 110812.

[3] Arora, P. and Zhang, Z. (John). 'Battery separators', *Chem. Rev.*, 2004, 104, (10), pp. 4419–4462.

[4] Tarascon, J.-M. and Armand, M. 'Issues and challenges facing rechargeable lithium batteries', *Nature*, 2001, 414, (6861), pp. 359–367.

[5] Chu, S. and Majumdar, A. 'Opportunities and challenges for a sustainable energy future', *Nature*, 2012, 488, (7411), pp. 294–303.

[6] Bonnassieux, Y., Brabec, C.J., Cao, Y., *et al.* 'The 2021 flexible and printed electronics roadmap', *Flex. Print. Electron.*, 2021, 6, (2), p. 023001.

[7] Hatzell, K.B., Chen, X.C., Cobb, C.L., *et al.* 'Challenges in lithium metal anodes for solid-state batteries', *ACS Energy Lett.*, 2020, 5, (3), pp. 922–934.

[8] Cobb, C.L. and Ho, C.C. 'Additive manufacturing: rethinking battery design', *Electrochem. Soc. Interface*, 2016, 25, (1), p. 75.

[9] Gao, W., Zhang, Y., Ramanujan, D., *et al.* 'The status, challenges, and future of additive manufacturing in engineering', *Comput.-Aided Design*, 2015, 69, pp. 65–89.

[10] Zheng, H., Li, J., Song, X., Liu, G., and Battaglia, V.S. 'A comprehensive understanding of electrode thickness effects on the electrochemical performances of Li-ion battery cathodes', *Electrochim. Acta*, 2012, 71, pp. 258–265.

[11] Rollag, K., Juarez-Robles, D., Du, Z., Wood, D.L.I., and Mukherjee, P.P. 'Drying temperature and capillarity-driven crack formation in aqueous processing of Li-ion battery electrodes', *ACS Appl. Energy Mater.*, 2019, 2, (6), pp. 4464–4476.

[12] Sahore, R., Wood, D.L.I., Kukay, A., Grady, K.M., Li, J., and Belharouak, I. 'Towards understanding of cracking during drying of thick aqueous-processed $LiNi_{0.8}Mn_{0.1}Co_{0.1}O_2$ cathodes', *ACS Sustainable Chem. Eng.*, 2020, 8, (8), pp. 3162–3169.

[13] Long, J.W., Dunn, B., Rolison, D.R., and White, H.S. 'Three-dimensional battery architectures', *Chem. Rev.*, 2004, 104, (10), pp. 4463–4492.

[14] Hung, C.-H., Huynh, P., Teo, K., and Cobb, C.L. 'Are three-dimensional batteries beneficial? Analyzing historical data to elucidate performance advantages', *ACS Energy Lett.*, 2022, 8, pp. 296–305.

[15] Ashby, D.S., Choi, C.S., Edwards, M.A., Talin, A.A., White, H.S., and Dunn, B.S. 'High-performance solid-state lithium-ion battery with mixed 2D and 3D electrodes', *ACS Appl. Energy Mater.*, 2020, 3, (9), pp. 8402–8409.

[16] Zadin, V., Kasemägi, H., Aabloo, A., and Brandell, D. 'Modelling electrode material utilization in the trench model 3D-microbattery by finite element analysis', *J. Power Sources*, 2010, 195, (18), pp. 6218–6224.

[17] Cobb, C.L. and Blanco, M. 'Modeling mass and density distribution effects on the performance of co-extruded electrodes for high energy density lithium-ion batteries', *J. Power Sources*, 2014, 249, pp. 357–366.

[18] Allu, S., Kalnaus, S., Simunovic, S., Nanda, J., Turner, J.A., and Pannala, S. 'A three-dimensional meso-macroscopic model for Li-ion intercalation batteries', *J. Power Sources*, 2016, 325, pp. 42–50.

[19] Trembacki, B., Duoss, E., Oxberry, G., Stadermann, M., and Murthy, J. 'Mesoscale electrochemical performance simulation of 3D interpenetrating lithium-ion battery electrodes', *J. Electrochem. Soc.*, 2019, 166, (6), p. A923.

[20] Hung, C.-H., Allu, S., and Cobb, C.L. 'Modeling current density non-uniformities to understand high-rate limitations in 3D interdigitated lithium-ion batteries', *J. Electrochem. Soc.*, 2021, 168, (10), p. 100512.

[21] Cobb, C.L. and Solberg, S.E. 'Communication—Analysis of thick co-extruded cathodes for higher-energy-and-power lithium-ion batteries', *J. Electrochem. Soc.*, 2017, 164, (7), p. A1339.

[22] Lv, Z., Li, W., Yang, L., Loh, X.J., and Chen, X. 'Custom-made electrochemical energy storage devices', *ACS Energy Lett.*, 2019, 4, (2), pp. 606–614.

[23] Foster, C.W., Down, M.P., Zhang, Y., *et al.* '3D Printed graphene based energy storage devices', *Sci. Rep*, 2017, 7, (1), p. 42233.

[24] Ragones, H., Menkin, S., Kamir, Y., *et al.* 'Towards smart free form-factor 3D printable batteries', *Sustain. Energy Fuels*, 2018, 2, (7), pp. 1542–1549.

[25] Reyes, C., Somogyi, R., Niu, S., *et al.* 'Three-dimensional printing of a complete lithium ion battery with fused filament fabrication', *ACS Appl. Energy Mater.*, 2018, 1, (10), pp. 5268–5279.

[26] Maurel, A., Courty, M., Fleutot, B., *et al.* 'Highly loaded graphite–polylactic acid composite-based filaments for lithium-ion battery three-dimensional printing', *Chem. Mater.*, 2018, 30, (21), pp. 7484–7493.

[27] Maurel, A., Grugeon, S., Fleutot, B., *et al.* 'Three-dimensional printing of a LiFePO4/graphite battery cell via fused deposition modeling', *Sci. Rep.*, 2019, 9, (1), p. 18031.

[28] Gupta, V., Alam, F., Verma, P., Kannan, A.M., and Kumar, S. 'Additive manufacturing enabled, microarchitected, hierarchically porous polylactic-acid/lithium iron phosphate/carbon nanotube nanocomposite electrodes for high performance Li-ion batteries', *J. Power Sources*, 2021, 494, p. 229625.

[29] Sun, K., Wei, T.-S., Ahn, B.Y., Seo, J.Y., Dillon, S.J., and Lewis, J.A. '3D Printing of interdigitated Li-ion microbattery architectures', *Adv. Mater.*, 2013, 25, (33), pp. 4539–4543.

[30] Fu, K., Wang, Y., Yan, C., *et al.* 'Graphene oxide-based electrode inks for 3D-printed lithium-ion batteries', *Adv. Mater.*, 2016, 28, (13), pp. 2587–2594.

[31] Kohlmeyer, R.R., Blake, A.J., Hardin, J.O., *et al.* 'Composite batteries: a simple yet universal approach to 3D printable lithium-ion battery electrodes', *J. Mater. Chem. A*, 2016, 4, (43), pp. 16856–16864.

[32] Lacey, S.D., Kirsch, D.J., Li, Y., *et al.* 'Extrusion-based 3D printing of hierarchically porous advanced battery electrodes', *Adv. Mater.*, 2018, 30, (12), p. 1705651.

[33] Wei, T.-S., Ahn, B.Y., Grotto, J., and Lewis, J.A. '3D Printing of customized Li-ion batteries with thick electrodes', *Adv. Mater.*, 2018, 30, (16), p. 1703027.

[34] McOwen, D.W., Xu, S., Gong, Y., *et al.* '3D-Printing electrolytes for solid-state batteries', *Adv. Mater.*, 2018, 30, (18), p. 1707132.

[35] Wang, J., Sun, Q., Gao, X., *et al.* 'Toward high areal energy and power density electrode for Li-ion batteries via optimized 3D printing approach', *ACS Appl. Mater. Interfaces*, 2018, 10, (46), pp. 39794–39801.

[36] Huang, J., Yang, J., Li, W., Cai, W., and Jiang, Z. 'Electrochemical properties of $LiCoO_2$ thin film electrode prepared by ink-jet printing technique', *Thin Solid Films*, 2008, 516, (10), pp. 3314–3319.

[37] Delannoy, P.-E., Riou, B., Lestriez, B., Guyomard, D., Brousse, T., and Le Bideau, J. 'Toward fast and cost-effective ink-jet printing of solid electrolyte for lithium microbatteries', *J. Power Sources*, 2015, 274, pp. 1085–1090.

[38] Ben-Barak, I., Kamir, Y., Menkin, S., *et al.* 'Drop-on-demand 3D printing of lithium iron phosphate cathodes', *J. Electrochem. Soc.*, 2019, 166, (3), pp. A5059–A5064.

[39] Kolchanov, D.S., Mitrofanov, I., Kim, A., *et al.* 'Inkjet printing of Li-rich cathode material for thin-film lithium-ion microbatteries', *Energy Technol.*, 2020, 8, (3), p. 1901086.

[40] Kushwaha, A., Jangid, M.K., Bhatt, B.B., Mukhopadhyay, A., and Gupta, D. 'Inkjet-printed environmentally friendly graphene film for application as a high-performance anode in Li-ion batteries', *ACS Appl. Energy Mater.*, 2021, 4, (8), pp. 7911–7921.

[41] Chen, Q., Xu, R., He, Z., Zhao, K., and Pan, L. 'Printing 3D gel polymer electrolyte in lithium-ion microbattery using stereolithography', *J. Electrochem. Soc.*, 2017, 164, (9), pp. A1852–A1857.

[42] He, Y., Chen, S., Nie, L., Sun, Z., Wu, X., and Liu, W. 'Stereolithography three-dimensional printing solid polymer electrolytes for all-solid-state lithium metal batteries', *Nano Lett.*, 2020, 20, (10), pp. 7136–7143.

[43] Narita, K., Citrin, M.A., Yang, H., Xia, X., and Greer, J.R. '3D architected carbon electrodes for energy storage', *Adv. Energy Mater.*, 2021, 11, (5), p. 2002637.

[44] Yee, D.W., Citrin, M.A., Taylor, Z.W., Saccone, M.A., Tovmasyan, V.L., and Greer, J.R. 'Hydrogel-based additive manufacturing of lithium cobalt oxide', *Adv. Mater. Technol.*, 2021, 6, (2), p. 2000791.

[45] Martinez, A.C., Maurel, A., Aranzola, A.P., *et al.* 'Additive manufacturing of $LiNi_{1/3}Mn_{1/3}Co_{1/3}O_2$ battery electrode material via vat photopolymerization precursor approach', *Sci. Rep.*, 2022, 12, (1), p. 19010.

[46] Ludwig, B., Zheng, Z., Shou, W., Wang, Y., and Pan, H. 'Solvent-free manufacturing of electrodes for lithium-ion batteries', *Sci. Rep.*, 2016, 6, (1), p. 23150.

[47] Lee, S.C., Jeong, J., Park, H.G., Min, B.-C., Chan Jun, S., and Chung, K.Y. 'Binder-assisted electrostatic spray deposition of $LiCoO_2$ and graphite films on coplanar interdigitated electrodes for flexible/wearable lithium-ion batteries', *J. Power Sources*, 2020, 472, p. 228573.

[48] Saleh, M.S., Li, J., Park, J., and Panat, R. '3D printed hierarchically-porous microlattice electrode materials for exceptionally high specific capacity and areal capacity lithium ion batteries', *Add. Manufact.*, 2018, 23, pp. 70–78.

[49] Deiner, L.J., Jenkins, T., Powell, A., Howell, T., and Rottmayer, M. 'High capacity rate capable aerosol jet printed Li-ion battery cathode', *Adv. Eng. Mater.*, 2019, 21, (5), p. 1801281.

[50] Yu, X., Liu, Y., Pham, H., *et al.* 'Customizable nonplanar printing of lithium-ion batteries', *Adv. Mater. Technol.*, 2019, 4, (11), p. 1900645.

[51] Rodriguez, R., Deiner, L.J., Tsao, B.H., and Fellner, J.P. 'Aerosol jet-printed LFP cathodes with bimodal pore distribution improve the rate capability of LIB cells', *ACS Appl. Energy Mater.*, 2021, 4, (9), pp. 9507–9512.

[52] Billaud, J., Bouville, F., Magrini, T., Villevieille, C., and Studart, A.R. 'Magnetically aligned graphite electrodes for high-rate performance Li-ion batteries', *Nat Energy*, 2016, 1, (8), pp. 1–6.

[53] Zhang, L., Zeng, M., Wu, D., and Yan, X. 'Magnetic field regulating the graphite electrode for excellent lithium-ion batteries performance', *ACS Sustain. Chem. Eng.*, 2019, 7, (6), pp. 6152–6160.

[54] Ho, C.C., Murata, K., Steingart, D.A., Evans, J.W., and Wright, P.K. 'A super ink jet printed zinc–silver 3D microbattery', *J. Micromech. Microeng.*, 2009, 19, (9), p. 094013.

[55] Armenta, J.L.R. and Salazar-Cruz, B.A. *Polymer Rheology*, London: Intechopen, 2018.

[56] Lewis, J.A. 'Colloidal processing of ceramics', *J. Am. Ceram. Soc.*, 2000, 83, (10), pp. 2341–2359.

[57] Lewis, J.A. 'Direct ink writing of 3D functional materials', *Adv. Funct. Mater.*, 2006, 16, (17), pp. 2193–2204.

[58] Glasser, A., Cloutet, É., Hadziioannou, G., and Kellay, H. 'Tuning the rheology of conducting polymer inks for various deposition processes', *Chem. Mater.*, 2019, 31, (17), pp. 6936–6944.

[59] M'Barki, A., Bocquet, L., and Stevenson, A. 'Linking rheology and printability for dense and strong ceramics by direct ink writing', *Sci. Rep*, 2017, 7, (1), p. 6017.

[60] Li, J., Daniel, C., and Wood, D. 'Materials processing for lithium-ion batteries', *J. Power Sources*, 2011, 196, (5), pp. 2452–2460.

[61] Thrimurthulu, K., Pandey, P.M., and Venkata Reddy, N. 'Optimum part deposition orientation in fused deposition modeling', *Int. J. Mach. Tools Manuf.*, 2004, 44, (6), pp. 585–594.

[62] Richter, L.P., Fischer, G., Sylla, L., *et al.* 'Progress in fine-line metallization by co-extrusion printing on cast monosilicon PERC solar cells', *Solar Energy Mater. Solar Cells*, 2015, 142, pp. 18–23.

[63] Izumi, A., Sanada, M., Furuichi, K., *et al.* 'Development of high capacity lithium-ion battery applying three-dimensionally patterned electrode', *Electrochim. Acta*, 2012, 79, pp. 218–222.

[64] Skylar-Scott, M.A., Mueller, J., Visser, C.W., and Lewis, J.A. 'Voxelated soft matter via multimaterial multinozzle 3D printing', *Nature*, 2019, 575, (7782), pp. 330–335.

[65] Gao, M., Li, L., and Song, Y. 'Inkjet printing wearable electronic devices', *J. Mater. Chem. C*, 2017, 5, (12), pp. 2971–2993.

[66] Derby, B. 'Inkjet printing of functional and structural materials: fluid property requirements, feature stability, and resolution', *Annu. Rev. Mater. Res.*, 2010, 40, (1), pp. 395–414.

[67] Deiner, L.J. and Reitz, T.L. 'Inkjet and aerosol jet printing of electrochemical devices for energy conversion and storage', *Adv. Eng. Mater.*, 2017, 19, (7), p. 1600878.

[68] Zhao, Y., Zhou, Q., Liu, L., Xu, J., Yan, M., and Jiang, Z. 'A novel and facile route of ink-jet printing to thin film SnO_2 anode for rechargeable lithium ion batteries', *Electrochim. Acta*, 2006, 51, (13), pp. 2639–2645.

[69] Dudney, N.J. and Jang, Y.-I. 'Analysis of thin-film lithium batteries with cathodes of 50 nm to 4 μm thick $LiCoO_2$', *J. Power Sources*, 2003, 119–121, pp. 300–304.

[70] Dudney, N.J. 'Thin film micro-batteries', *Electrochem. Soc. Interface*, 2008, 17, (3), p. 44.

[71] Min, H.-S., Park, B.Y., Taherabadi, L., *et al.* 'Fabrication and properties of a carbon/polypyrrole three-dimensional microbattery', *J. Power Sources*, 2008, 178, (2), pp. 795–800.

[72] Lee, S.H., Huang, C., Johnston, C., and Grant, P.S. 'Spray printing and optimization of anodes and cathodes for high performance Li-Ion batteries', *Electrochim. Acta*, 2018, 292, pp. 546–557.

[73] Zekoll, S., Marriner-Edwards, C., Ola Hekselman, A.K., *et al.* 'Hybrid electrolytes with 3D bicontinuous ordered ceramic and polymer microchannels for all-solid-state batteries', *Energy Environ. Sci.*, 2018, 11, (1), pp. 185–201.

[74] Maurel, A., Martinez, A.C., Grugeon, S., *et al.* 'Toward high resolution 3D printing of shape-conformable batteries via vat photopolymerization: review and perspective', *IEEE Access*, 2021, 9, pp. 140654–140666.

[75] Begley, M.R., Gianola, D.S., and Ray, T.R. 'Bridging functional nanocomposites to robust macroscale devices', *Science*, 2019, 364, (6447), p. eaav4299.

[76] Sander, J.S., Erb, R.M., Li, L., Gurijala, A., and Chiang, Y.-M. 'High-performance battery electrodes via magnetic templating', *Nat. Energy*, 2016, 1, (8), pp. 1–7.

[77] Li, J., Liang, X., Liou, F., and Park, J. 'Macro-/micro-controlled 3D lithium-ion batteries via additive manufacturing and electric field processing', *Sci. Rep.*, 2018, 8, (1), p. 1846.

[78] Liu, X., Peng, S., Gao, S., *et al.* 'Electric-field-directed parallel alignment architecting 3D lithium-ion pathways within solid composite electrolyte', *ACS Appl. Mater. Interfaces*, 2018, 10, (18), pp. 15691–15696.

[79] Johnson, K., Melchert, D., Gianola, D.S., Begley, M., and Ray, T.R. 'Recent progress in acoustic field-assisted 3D-printing of functional composite materials', *MRS Adv.*, 2021, 6, (25), pp. 636–643.

[80] Melchert, D.S., Collino, R.R., Ray, T.R., *et al.* 'Flexible conductive composites with programmed electrical anisotropy using acoustophoresis', *Adv. Mater. Technol.*, 2019, 4, (12), p. 1900586.

[81] Melchert, D.S., Tahmasebipour, A., Liu, X., *et al.* 'Anisotropic thermally conductive composites enabled by acoustophoresis and stereolithography', *Adv. Funct. Mater.*, 2022, 32, (31), p. 2201687.
[82] Melchert, D.S., Johnson, K., Giera, B., *et al.* 'Modeling meso- and micro-structure in materials patterned with acoustic focusing', *Mater. Design*, 2021, 202, p. 109512.
[83] Drinkwater, B.W. 'Dynamic-field devices for the ultrasonic manipulation of microparticles', *Lab Chip*, 2016, 16, (13), pp. 2360–2375.

Chapter 7
Dry processing for lithium-ion battery electrodes

Kubra Uzun[1], Ming Wang[2] and Yang-Tse Cheng[1]

7.1 Introduction

The conventional way of making lithium-ion battery (LIB) electrodes relies on the slurry-based manufacturing process, for which the binder is dissolved in a solvent and mixed with the conductive agent and active material particles to form the final slurry composition [1]. Polyvinylidene fluoride (PVDF) is the most widely utilized binder material in LIB electrode manufacturing, especially for positive electrodes. *N*-Methyl-2-pyrrolidone (NMP) is the preferred solvent for dissolution of the PVDF binder, facilitating the slurry properties. However, a well-known downside of NMP is its toxicity and energy consumption because of its expensive processing requirements for solvent evaporation and recovery [2,3]. Lately, adopting aqueous processing and using green solvents have been suggested as effective solutions for slurry-based manufacturing to tackle issues resulting from toxic and costly solvents [4]. For the negative electrodes, water has started to be used as the solvent, which has the potential to save as much as 10.5% on the pack production cost [1,3]. For the positive electrodes, on the other hand, the adoption of water as a solvent would require alternative binders, since PVDF is insoluble in water. Yet, a higher operating voltage window for the positive electrode limits the number of binders as viable replacements. In addition, water-based systems may affect the electrochemical performance of both positive and negative electrodes of LIBs, such as crack formation [5], transition metal dissolution [6], and current collector corrosion [3]. Similar to aqueous systems, employing nontoxic organic (green) solvents offers advantages in terms of safety and processing. Although switching to green solvents for the slurry-based fabrication process is an environmentally friendly approach, it is still not cost-efficient due to inevitable processing requirements such as solvent drying.

Using a dry process for electrode manufacturing would avoid toxic solvent usage completely. Recently, significant advances have been made in the dry process for electrode manufacturing of LIBs. Thin film (less than several micrometers

[1]Department of Chemical and Materials Engineering, University of Kentucky, USA
[2]Materials and Manufacturing Systems Research Laboratory, General Motors Research & Development Center, USA

in thickness) deposition methods, for example, pulsed laser deposition, sputtering deposition, and atomic layer deposition, and thick film (greater than several micrometers but less than a millimeter in thickness) coating methods, for example, electrostatic spray deposition, fibrillation, dry pressing, melting extrusion, and cold plasma have been investigated as dry electrode manufacturing processes [7,8].

Automakers, including Tesla and Volkswagen Group, have implemented dry processes for electrode manufacturing in their electric vehicle (EV) batteries. Moreover, equipment for dry processes has also come to market [9], LiCAP Technologies, Inc. (LiCAP), a developer of affordable and environmentally friendly dry electrode manufacturing solutions for lithium-ion batteries and solid-state batteries, has partnered with Siemens and BW Papersystems (BWP) to provide dry electrode manufacturing machinery [9].

7.2 Electrostatic spray deposition (ESD) process

Emerging in the 1940s as an electrostatic powder coating technology, the ESD method is extensively applied in the pharmaceutical, food, paint, and wood coating industry. In the ESD process, dry powders are charged using a spray gun and drawn to the grounded substrate surface [10,11]. This process is usually followed by heat treatment to enable the film formation of the deposited powder layers. The ESD process typically consists of three main steps, including charge, mobilization, and accumulation of the coating powders. After mixed powders are charged by the spray gun, particles begin to move toward the grounded substrate due to aerodynamic, gravitational, and electrostatic forces. The electrical force allows the particles to surpass gravitational and aerodynamic forces, moving the charged particles to the grounded substrate surface. The adhesion of the particles to the substrate is primarily due to Coulomb attraction between the particles and charges built up on the substrate. In general, electrostatic charge can be generated by two different mechanisms, corona and tribo charging. Corona charging is driven by high voltage, while tribo charging is created by friction within the gun. For corona charging, the voltage in the tip of the spraying gun creates an electrostatic field between the gun and the substrate [11]. For tribo charging, no free ions, and an electrical field exist between the spray gun and the substance. Corona charging is more widely adopted due to its versatility. Several parameters affecting the efficiency of the ESD process based on corona charging include electrical potential, gun to substrate distance, coating time, carrier gas (e.g., air) pressure, and powder feeding pressure. A general schematic of the ESD process is depicted in Figure 7.1.

The transfer efficiency is the ratio of the mass of powder adhered to the substrate to the mass of powder sprayed by the gun. Increasing electric potential can produce a strong electric field around the particles. This facilitates the transfer of charged particles near the coating substrate [12]. Thus, stronger adhesion to the substrate and increased transfer efficiency are provided. However, if overvoltage is applied, then a well-known breakdown process often referred to as back ionization

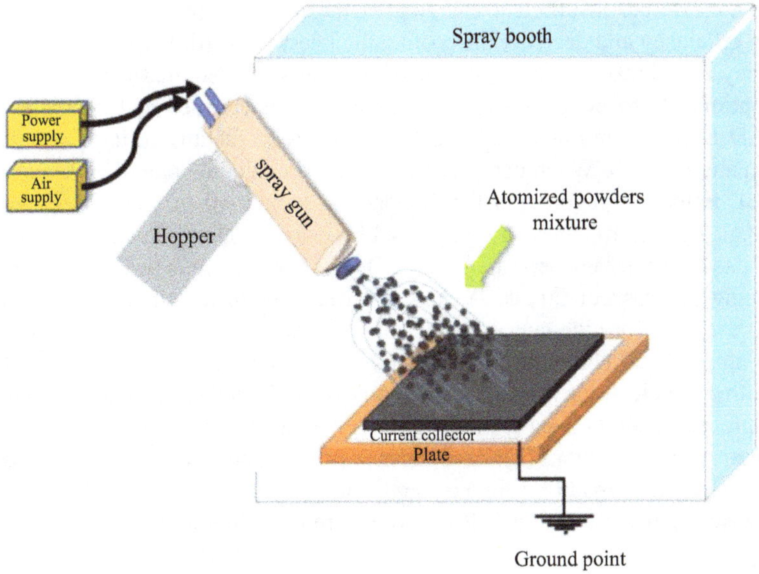

Figure 7.1 Schematic of ESD process, reprinted with permission from Ref. [12]. Copyright 2017 Elsevier.

can occur, leading to decreased accumulation rates and thereby lowered coating thicknesses [13]. Similarly, low voltage levels can cause weak powder adhesion or poor surface coverage on the substrate and deteriorated overall structure, because particles will be blown away by aerodynamics due to insufficient charges. Paturi *et al.* showed that an increasing coating thickness was achieved when electrostatically coated carbide cutting tools were deposited using increasing charging voltage from 50 to 70 kV; however, a decrease in thickness was observed when voltage was increased from 70 to 90 kV.

Usually, the distance between the gun deflector and the substrate slightly affects the thickness of the coating. Paturi *et al.* suggested that this small effect could be due to the higher magnitude of electrostatic force compared to the gravitational force [12].

The third process parameter, coating time, controls the coating thickness. The relationship between the coating time and thickness of the coating can be explained as follows: at first, the coating thickness increases with increased coating time; then as the coating duration is further extended, slowdown in the growth of the thickness and eventually decrease in the thickness happens, which can be ascribed to extended back ionization effect [14,15].

Powder feeding pressure and air pressure are also important for the transfer efficiency of the ESD. It is a common agreement from the literature that too low feeding pressure or too high auxiliary pressure should be avoided to prevent a reduction in the amount of powder coming out from the gun and weakened aerodynamic force. Insufficient charging of the particles can adversely affect the coating efficiency and decrease the coating thickness [14–16]. For improved transfer efficiency and thicker coatings, optimum pressures are required coupled with the other optimized process parameters that are usually determined by trial and error.

ESD is considered as an appealing dry electrode manufacturing process as it offers many benefits. When compared to fluidized or liquid spray deposition methods, ESD introduces higher transfer efficiencies of about 80–90 % [17]. Fast growth rates (e.g., 1 μm/min) can be accomplished by using ESD, producing both thin and thick films on a micrometer scale [18]. ESD does not require expensive and energy-consuming process conditions; it can be carried out in an ambient environment without the need for high temperatures and high vacuum. In addition to these benefits, using ESD can eliminate solvent recovery and reduce processing time [11,19]. Hence, this dry electrode manufacturing process can be integrated into existing roll-to-roll processes and utilized to produce structured electrodes with complex physical structures and compositions that have superior performance and safety characteristics than the electrodes made by the conventional wet slurry process. Table 7.1 summarizes studies using ESD for LIB electrodes from the literature.

7.3 Fibrillation (Maxwell type)

This dry process is based on the fibrillation of the polymeric binder, which is, in most cases, polytetrafluorethylene (PTFE). The method was mainly developed by Maxwell Technologies [7]. This method comprises three main production steps for LIB electrode production: (1) the electrode materials including active material, conductive agent, and binder are mixed, fibrillating the binder through shearing forces, (2) the mixed powders are rolled to obtain a free-standing electrode film, and (3) obtained film is laminated onto the current collector with adhesive primer [29,30]. Figure 7.2 shows an example of the process.

For batteries, high energy density can be achieved by reducing the number of inactive electrode components, such as binder and conductive agent, and increasing the active material content. Since using fibrillation as a dry electrode manufacturing process can potentially reduce the binder content, it can help increase the energy density. Indeed, it was shown by Hippauf *et al.* [32] that the composite electrode binder amount could be decreased to 0.1 wt%. In addition, similar to the ESD method, this method can be implemented in the existing commercial roll-to-roll production lines for large-scale mass production of electrodes [31]. Moreover, the fibrillation approach can create uniform binder distribution [33].

The binder is the most critical material in this technique as it relies on the fibrillation of polymers. PTFE is widely used as it can form 3D networks through fibrillation in LIB electrodes, providing robust mechanical properties and high temperature resistance [34]. However, the high chemical inertness makes PTFE non-sticky as the fluorine atoms repel any other kinds of molecules. To laminate the free-standing electrode film on the current collector, an additional adhesive primer that contains polymeric binders (e.g., PVDF) and conductive additives is

Table 7.1 Summary of studies using ESD for LIB electrodes

Electrode materials	Cell type	ESD parameters	Comments	References
$LiCoO_2$ (LCO), carbon black (CB), poly(methyl methacrylate) (PMMA)	Half cell	Tribo charging	The initial capacity was about 140 mAh/g	Abe *et al.* [20]
LCO, CB, PMMA	Half cell	Tribo charging	They investigated the effect of the nanoparticle coating on LCO. The initial Capacity was about 140 mAh/g	Abe *et al.* [21]
LCO or $LiNi_{0.33}Co_{0.33}Mn_{0.33}O_2$ (NMC111), CB, PVDF	Half cell	Corona charging	**NMC111:** Cycled between 2.8 V and 4.3 V at 0.5°C. The initial discharge capacity of 138 mAh/g. The capacity retention of 87% after 50 cycles. **LCO:** Cycled between 2.5 V and 4.2 V at 0.5°C. The initial discharge capacity of 114 mAh/g. The capacity retention of 70% after 50 cycles	Ludwig *et al.* [22]
LCO, CB, PVDF	Full cell	Corona charging	Cycled between 2.5 V and 4.2 V at 0.5°C for 150 cycles. The initial discharge capacity of 127.8 mAh/g. The capacity retention of 77% after 100 cycles	Ludwig *et al.* [23]
NMC111, CB, PVDF	Half cell	Corona charging	Cycled between 3.0 V and 4.3 V at 0.5°C for 300 cycles. The initial discharge capacity of 155 mAh/g. The capacity retention of 80% after 300 cycles	Al-Shroofy *et al.* [24]
NMC111, CB, PVDF	Full cell	Corona charging	Thin dry electrodes (e.g., 55 μm) had an initial capacity of 150 mAh/g at 0.5°C. Capacity retention of $>$80% after 500 cycles	Liu *et al.* [25]

(Continues)

Table 7.1 (Continued)

Electrode materials	Cell type	ESD parameters	Comments	References
NMC111, CB, PVDF (Alfa Aesar or HSV-900)	Half cell	Corona charging	Investigated the effect of the different polymeric binders on the ESD process. Cycled between 3.0 V and 4.3 V at 0.5°C for 50 cycles **Alfa Aesar:** 160 mAh/g at 0.2°C, 93% after capacity retention 50 cycles at 0.5°C. **HSV-900:** 160 mAh/g at 0.2°C, 91% capacity retention after 50 cycles	Wang *et al.* [26]
Artificial graphite, CB, PVDF/FEP-copolymer/THVcopolymer	Half cell	Gunless ESD	Cycled between 0.02 V and 1.5 V at 0.5°C. The initial discharge capacity of 370 mAh/g. 345 mAh/g after 50 cycles	Schälicke *et al.* [27]
$LiNi_{0.5}Mn_{0.3}Co_{0.2}O_2$ (NMC532), CB, PVDF	Half cell	Corona charging	Cycled between 3.0 V and 4.4 V at 1°C for 300 cycles. The initial discharge capacity of 151.2 mAh/g. The capacity retention of 69% after 300 cycles	Zhen *et al.* [28]
$LiMn_2O_4$ (LMO)/ $LiNi_{0.8}Co_{0.1}Mn_{0.1}O_2$ (NMC811) blend, CB, PVDF	Half cell	Corona charging	Investigated the effects of the premixing before the ESD processes on the performance of the dry electrodes. Electrodes showed comparable results to slurry made electrodes at low C-rates	Wang *et al.*

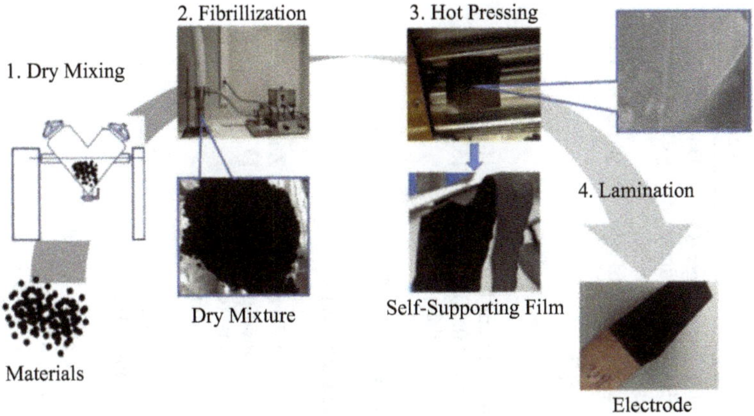

Figure 7.2 Schematic of dry pressing technique adopted from Ref. [31]. Open access.

required. Solvents such as NMP are still needed during the primer coating process, which undermines a fully solvent-free process and increases the manufacturing cost in addressing the regulation of volatile organic compound (VOC) [35]. Another problem with using the fibrillation method is that depending on the equipment used to perform the binder fibrillation, structural differences in samples occur, which in turn can affect the battery performance [36]. For example, a sample ground manually or in a roll mill produces flakes or blocks, whereas a sample ground in a jet mill produces flocculants. Furthermore, the PTFE binder is incompatible with negative electrode materials. In a study, researchers have addressed the instability issue of PTFE against graphite electrodes by using a blend of PVDF and PTFE binders which produced electrodes comparable in performance to that made by the slurry method [37]. However, the limited choice of binder makes the widespread adoption of this technique still difficult. Further research efforts may help widen the choices of binders for the dry processes based on polymer fibrillation. Table 7.2 summarizes studies using fibrillation for LIB electrodes.

7.4 Dry pressing and melt extrusion

Dry pressing and melt extrusion techniques are also being developed for LIB electrode manufacturing. In this section, dry pressing and melt extrusion techniques are reviewed separately.

Dry pressing of the powders can be performed both at hot and cold temperatures. In this technique, high pressure is applied to a mixture of the electrode materials to create a self-supporting electrode film. Figure 7.3 shows a schematic of this process. It is an effective technique, especially for dense polymer binder-containing electrode manufacturing. However, electrodes with extremely low polymer concentrations are difficult to produce using this technique. Moreover, the

Table 7.2 Summary of studies using fibrillation for LIB electrodes

Active material	Cell type	Binder	Comments	References
NMC111/graphite	Full cell	PTFE	Cycled between 2.7 V and 4.2 V at 0.5°C charge/1C discharge for more than 2,000 cycles. The capacity retention of 90% after 2,000 cycles	Duong *et al.* [30]
$LiNi_{0.6}Mn_{0.2}Co_{0.2}O_2$ (NMC622)/graphite	Full cell	PTFE and nitrile butadiene rubber (XNBR)	Cycled between 2.5 V and 4.3 V at a current of 1.0 mA/cm^2. **PTFE:** fast drop is observed at cycle 250 **XNBR:** the energy density had 80% retention after 400 cycles	Wu *et al.* [38]
NMC/Li_6PS_5Cl composite solid electrode	All-solid-state battery	PTFE	The electrodes were prepared with different binder contents (e.g., 0.1, 0.3, 0.7, and 1.0 wt%). The electrode with a binder content of 0.3 wt% showed 93.6% capacity retention after 100 cycles	Hippauf *et al.* [32]
$LiNi_{0.5}Mn_{1.5}O_4$ (LNMO)	Both half and full cell	PTFE	LNMO electrodes with different loadings (3, 4, 6, and 9.5 mAh/cm^2) were fabricated. 3.0 mAh/cm^2 LNMO/graphite electrodes achieved 80% capacity retention after 300 cycles	Yao *et al.* [33]
Graphite	Both half cell and full cell	PTFE/PVDF blend	Full cells with commercial $LiFePO_4$ (LFP) were cycled between 2.5 and 3.65 V.Prelithiated graphite anodes with PVDF and PTFE had capacity retention of about 80% after 110 cycles under a current of 0.7 mA/cm^2	Zhang *et al.* [37]
Graphite or soft carbon or hard carbon	Half cell	PTFE	**Graphite:** Cycled at 0.1°C. Fast capacity fading was observed. The capacity retention of only 30% after 35 cycles. **Hard carbon:** Cycled between 0.01 V and 1.5 V at 0.3°C. Capacity retention of 80% after 100 cycles. **Soft Carbon:** Cycled between 0.01 V and 1.5 V at 0.3°C. Capacity retention of roughly 80% after 100 cycles	Zhang *et al.* [31]

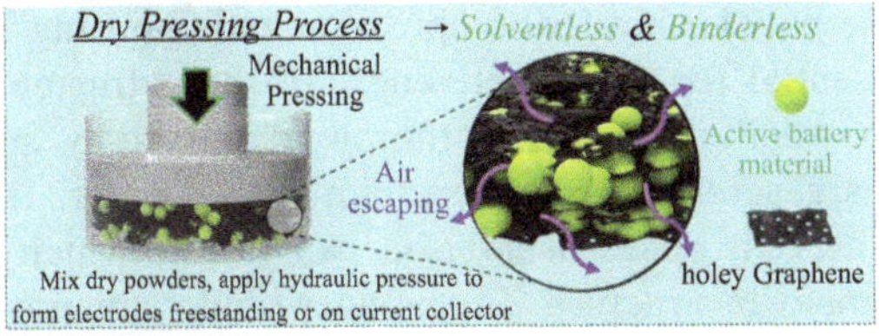

Figure 7.3 Schematic of dry pressing technique by Kirsch et al. [39]. Copyright 2019 American Chemical Society.

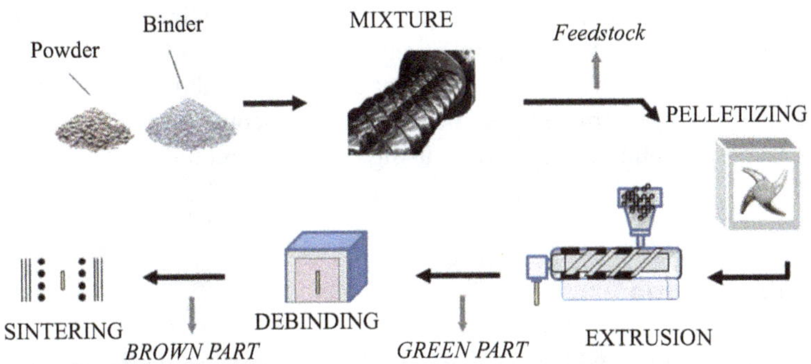

Figure 7.4 Schematic of melt extrusion technique by Sotomayor et al. [41]. Copyright 2019 Journal of Power Sources.

dry pressing technique is complicated for several reasons: (1) control of the homogeneity is difficult, (2) high pressure can introduce structural changes, and (3) materials need to be compressible and deformable. As such, holey graphene was usually preferred in this technique due to its ability to be compression molded into dense, porous, durable architectures of complicated shapes [40]. Even though this technique can be scaled up via existing industrial manufacturing equipment, these limitations hinder its widespread utilization, especially for roll-to-roll electrode manufacturing.

7.4.1 Melt extrusion

This technique consists of (1) creating a homogenous mixture of raw materials consisting of a blended thermoplastic polymeric binder, active material, and ceramic or metallic powder, (2) extrusion of the mixed powders, (3) removal of the binder by thermal treatment (thermal debinding), and (4) sintering the material to obtain the final electrode. As a completely dry process, this technique has been employed since 2019 (8). Thermoplastic polymeric materials, such as polyethylene (PE), polypropylene (PP), and PMMA, are typically used in this process. An example of the melt extrusion process is illustrated in Figure 7.4.

Optimizing the powder-binder composition and choosing the right powder and binder are essential for the effectiveness of the extrusion process. In this process, blended powder and binder are thoroughly mixed by specialized mixing equipment such as a screw extruder or rotor blade mixer. If mixture homogeneity is not achieved, phase separation can occur, which can disrupt the integrity of the composite structure. The powder-binder composition should also be optimized for the desired rheological characteristics that influence the extrusion process. In addition to the powder and binder control, precise control of temperature, shear force, and extrusion duration are also necessary. Temperature differences and a rise in internal pressure may result in cracks during thermal debinding [42,43].

Melt extrusion is a scalable process for producing electrodes with high loading. Nonetheless, its implementation in real-world electrode manufacturing is constrained by the high polymer content, the laborious manufacturing method, and the high temperatures needed for the debinding and sintering procedures. Table 7.3 summarizes studies using dry press and melt extrusion for LIB electrodes.

7.5 Cold plasma

Cold plasma deposition, which has been employed extensively in the production of semiconductors, magnetic media, specialty glasses, and metal coatings, is being considered as a viable technology for LIB electrode production. Cold plasma generation occurs when the gas is partially ionized at atmospheric pressure or low-level vacuum [47].

The cold plasma process was recently used by Liang *et al.* to make binder-free thick (around 200 microns) LFP electrodes [48]. In this study, both half and full cells were used to evaluate the performance of the LFP electrodes. Half cells with 60-micron and 230-micron thick LFP electrodes showed capacity retention of 91.3% and 84.4%, respectively. Full cells with a graphite negative electrode and LFP positive electrodes had high cycling performance, maintaining 81.6% of their capacity over 500 cycles. This study also showed that the cold plasma technique could produce electrodes without using polymeric binders and toxic solvents. The authors indicated that cost reduction could be about 40% if this technique is used for electrode manufacturing. Furthermore, the mechanical properties of the materials remain unaffected because of the low deposition temperature.

Despite its novelty, this technique has the potential for useful applications.

Recently, extensive research has focused on rechargeable thin film micro-batteries for applications such as medical devices, wireless sensors, flexible electronics, portable electronics, and smart cards [49]. Thin film electrodes can be manufactured using chemical and physical deposition techniques. Physical vapor deposition and magnetron sputtering are examples of physical deposition techniques, while chemical deposition methods include chemical vapor deposition, atomic layer deposition (ALD), etc.

Table 7.3 Summary of studies using dry press and melt extrusion techniques for LIB electrodes

Active material	Cell type	Method	Comments	References
Silicon oxide	Half cell	Dry pressing	Four different silicon oxide electrodes were manufactured	Yang *et al.* [44]
$LiFePO_4$ (LFP) and holey graphene	Full cell	Dry pressing	Different pressures (at 500, 200, and 20 MPa) were used to press binderless materials. The LFP showed no structural changes and comparable rate capability to the traditionally produced LFP	Kirsch *et al.* [39]
NCM532 and holey graphene	Half cell	Dry pressing	High mass loading (up to 26 mg/cm^2 in this work) NCM cathodes were assembled. They exhibited excellent retention of specific capacity over 200 cycles	Walker *et al.* [40]
$Li_4Ti_5O_{12}$ (LTO)/LFP	Half cell and full cell	Melt extrusion	Half LTO and LFP cells had a capacity retention of 60% after 10 cycles. Full cell with LTO and LFP had capacity retention of 76% after 50 cycles at C/12	Sotomayor *et al.* [41]
LTO/LFP or NMC111	Half cell and full cell	Hot melt extrusion	Hydrogenated nitrile butadiene rubber (HNBR) served as the binder, and polypropylene carbonate (PPC) was utilized as a polymer processing aid. Electrodes showed a comparable electrochemical performance to the conventional slurry coating	Khakani *et al.* [45]
Lithium nickel cobalt aluminum oxide (NCA) and graphite	Half cell	Melt extrusion	PPC and HNBR were used as binders. Electrodes showed a comparable electrochemical performance to the conventional slurry coating	Astafyeva *et al.* [46]

7.6 Pulsed laser deposition

Pulsed laser deposition is a physical vapor deposition technique that is commonly used for thin film deposition. In this technique, a high-power ultraviolet pulsed laser beam is used as an external energy source to vaporize materials and deposit a thin film on the substrate. The deposition of the films can be performed using either oxygen or other background gases, or ultra-high vacuum conditions. During the PLD process, four distinct phenomena occur (1) laser beam adsorption on the target; (2) the formation of a plasma plume; (3) interaction of the ablated species with the substrate; and (4) nucleation and growth of a thin film on the surface of the substrate [49]. The main components of PLD are shown in Figure 7.5.

PLD parameters include a target-to-substrate distance, substrate temperature, laser frequency, and gas environment and pressure. These parameters are interrelated, and they influence the stoichiometry and the structure of the films. It was demonstrated that as target-to-substrate distance increases, the number of deposited materials decreases. Moreover, it was found that the crystal structures of the films are also affected by the target-to-substrate distance. When the particles moving to the substrate are subject to low energies caused by large distances, the energy supply becomes inadequate [50]. Whereas high energies provided by short distances enable the relocation of the particles to minimize the binding energy, thereby forming a crystal structure [50]. Regarding the impact of the substrate temperature, elevated temperatures give rise to high surface mobilization of adatoms which allows the formation of crystalline films [51].

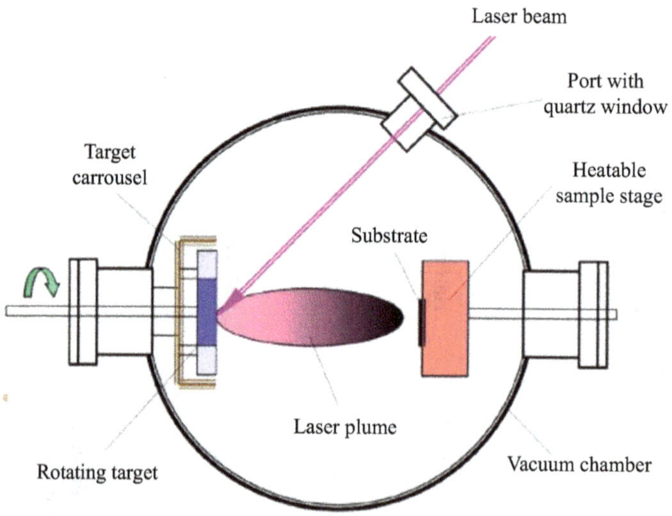

Figure 7.5 Schematic of PLD process. Source: Courtesy of http://groups.ist.utl.pt/rschwarz/rschwarzgroup_files/PLD_files/PLD.htm.

If the laser pulse frequency is too high, there is not sufficient time for relocation before new particles arrive. This can eventually lead to an inhomogeneous distribution of the material. Lastly, preferred gas also affects the kinetic energies of the applied substances, thus affecting the morphology [51].

Following the discovery of lasers in the 1960s, PLD has been investigated as a thin film deposition technique. This technique made several noteworthy contributions to various applications such as electronic materials, optical devices, sensors and actuators, biomaterials, and organic polymers. PLD can allow the construction of thin film microbatteries using conventional liquid electrolytes and separators as well as all-solid-state thin film microbatteries that use solid electrolytes in place of liquid electrolytes. Commonly grown positive electrode materials via PLD are layered-type LCO, olivine-type LFP, and spinel-structured LMO. Some of the common negative electrode materials include silicon, and LTO [52].

PLD has several attractive features including its versatility and stoichiometric growth of thin films. First, it allows the deposition of a broad range of materials such as semiconductors, nitrides, oxides, carbides, polymers, and metals. Second, stoichiometric growth of the film can be attained as a result of the strong heating of the target [53,54]. Apart from these features, another noteworthy characteristic of PLD is its ability to provide strong adhesion through the high energy of ablated species. On the other hand, difficulty in producing thin films with uniform thicknesses, particularly over a large substrate surface, is a significant weakness of PLD that restricts its wider implementation. This technique is more effective for small substrates [51]. Another drawback attributed to PLD is energy-intensive processing conditions as it requires high vacuum and high temperatures [54].

7.7 Magnetron sputtering

Magnetron sputtering is another extensively studied method to fabricate thin negative and positive electrodes. During the basic sputtering process, the surface of the target, e.g., the cathode is bombarded by energetic ions produced from the inert gas to eject the atoms. These atoms will then condense on the surface of the substrate as a thin film [55,56]. Magnetron sputtering systems construct a strong magnetic field near the target [56]. Sputtering parameters mainly include working gas types, gas flow rate, sputtering supplied power, gas pressure, substrate bias potential, target-substrate distance, and target orientation. There are two types of sputtering systems involving radio frequency (RF) magnetron sputtering and direct current (DC) magnetron sputtering. RF magnetron sputtering uses radio frequency power, while DC magnetron sputtering uses direct current power to generate plasma. DC magnetron sputtering is not suitable for insulating target materials. Alternatively, RF magnetron sputtering can be used for the deposition of thin films from non-conductive targets [55]. A schematic of RF magnetron sputtering is shown in Figure 7.6.

First reported by Bates *et al.* in 1993, RF magnetron sputtering was used for the fabrication of binder-free thin film positive electrode materials for LIB

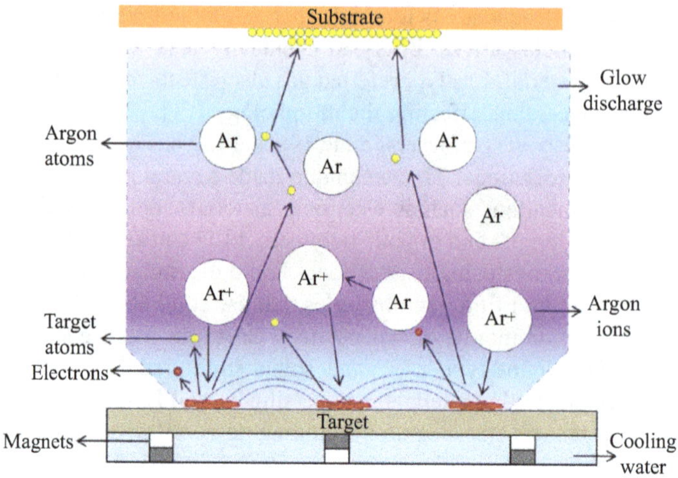

Figure 7.6 Schematic of RF magnetron sputtering [56]. Open access.

applications. Especially in the last decade, the application of magnetron sputtering for battery materials experienced explosive growth. It has been used for different battery cell constituents such as positive electrodes (e.g., LMO, LCO, and LFP), negative electrodes (e.g., LTO, $LiNiVO_4$, silicon (Si), antimony (Sb), and tin (Sn), and electrolytes (e.g., lithium phosphorous oxynitride glass electrolyte (LIPON)), $Li_7La_3Zr_2O_{12}$ (LLZ), $Li_{1+x}Al_xTi_{2-x}(PO_4)_3$ (LATP)). The magnetron sputtering process is adopted for these battery components for various applications involving thin film binder-free electrode fabrication, all-solid-state thin film lithium battery manufacturing, and artificial solid-electrolyte interphase film formation.

Magnetron sputtering benefits from the advantages of fast deposition rates ranging from 1 nm/s to 10 nm/s, conformality for scaling up, and capability to deposit a wide range of materials such as metal and metal oxide nanoparticles, nanoclusters, composites, insulators, alloys, and organic compounds depending on the type of magnetron used (e.g., DC or RF), and a high degree of adhesion between the film and the substrate. Nevertheless, one of the major problems associated with this technique is poor coating uniformity resulting from the non-homogeneous ion current distribution across the target surface. Furthermore, the high price of target materials and large target material requirements are the other existing limitations [57,58].

7.8 Atomic layer deposition (ALD)

ALD is a vapor phase thin film deposition technique used for a variety of materials. The origins of this technique date to the 1960s, being first introduced by Prof. Kol'tsov from the Russian Academy of Sciences. The working principle of this method is based on the successive alternating pulses of precursors generating half-reactions with the substrate [59]. One sequence of the deposition of films by ALD is made up of four stages: (1) pumping of the first precursor into the chamber to allow the formation of a monolayer, (2) purging of the reaction chamber to remove the reaction by-products, (3) pumping of the second precursor to into the chamber to allow the formation of a wanted layer, and (4) further purge of the reaction chamber [59–61]. A schematic of typical ALD cycle consisting of two half-cycles is shown in Figure 7.7.

The important process parameters for ALD include the heating temperature of the precursor, deposition temperature window, precursor pulse time, and purge time for precursor and oxidant [63,64]. The film thickness can be adjusted depending on the number of sequential cycles used for layer formation. Temperature can affect the crystal structure of the materials; the crystallinity of the deposited materials changes depending on the heating temperature used. Adequate precursor exposure time is required in order to accomplish accurate deposition by ALD; when the precursor exposure time is short, gas phase precursor mixing can occur and the coating quality can result in poor conformality [65].

ALD is adopted in different application areas such as catalysts, nanotechnology, and electronic and optoelectronic materials [66]. Owing to the growing interest in recent times, ALD has emerged as a promising candidate for LIB studies. It is widely used for microbattery applications as it is possible to fabricate nanostructured positive and negative electrode materials using ALD. Furthermore, it can also be used for surface modification and 3D-structured all-solid-state microbatteries.

ALD technique offers several advantages: (1) multilayer depositions can be more efficient than other methods due to its self-limiting growth mechanism (e.g., by the sequential nature of the process); (2) high quality and homogenous films are

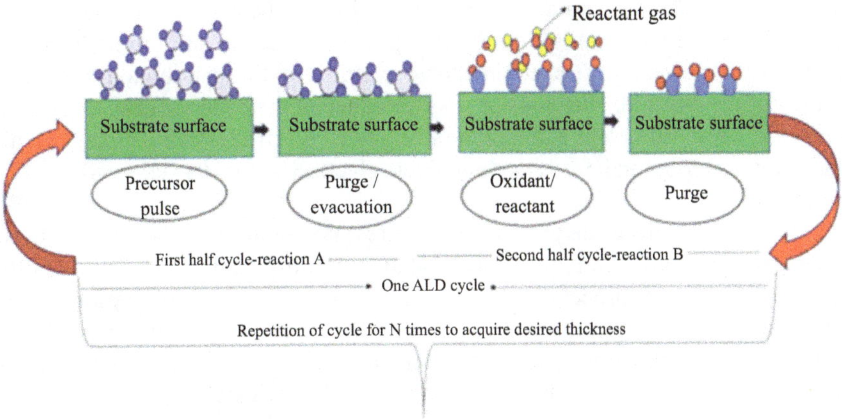

Figure 7.7 Schematic illustration of a typical ALD cycle consisting of two half cycles, reprinted with permission from Ref. [62]. Copyright 2020 Elsevier.

produced; (3) it does not require high temperatures (e.g., 100°C); (4) provides repeatable and controllable structures [61, 65].

Despite the many benefits that ALD offers, it suffers from several major drawbacks: low deposition rates, the energy-consuming nature of the process, expensive processing requirements due to the use of costly precursors, and scalability challenges due to the inability of batch processes [58,65].

7.9 The advantages and disadvantages of dry processes for the LIB manufacturing

Integration of dry electrode manufacturing processes to current roll-to-roll LIB manufacturing systems will enable the manufacturing of both thin and thick electrodes. Thin film electrodes are particularly desirable for compact applications where higher power densities are needed. Various thin film deposition techniques can be used to generate thin film electrodes with various microstructures or crystalline structures. Thin film electrodes enable more precise control of composition, microstructures, and interfaces. Unique microstructures and interfaces can improve the kinetics of lithium storage and charge transfer reactions due to the shortened diffusion paths and reduction of interfacial impedance. Additionally, thin film electrodes can be used for all-solid-state thin film batteries that do not employ flammable liquid electrolytes, increasing safety.

Thick electrodes, on the other hand, can increase energy density by minimizing the proportion of inactive components. Furthermore, cost reduction and higher areal capacities can be achieved by increasing the thickness of electrodes. Thus, manufacturing thick electrodes is particularly desirable for the EV industry. The ionic and electronic transport paths for thick electrodes can be engineered with dry electrode manufacturing. The binder-conductive agent distributions affecting the active surfaces of the electrode can be tuned. This can be achieved by the mixing step of dry processing as different mixing processes can lead to different microstructures. Therefore, with a comprehensive study of the distribution and morphology of the electrode materials, the performance of the battery can be enhanced. Moreover, compositionally and structurally graded electrodes compatible with all solid-state batteries and free-standing thick electrodes can be achieved by dry processing.

Although dry electrode processing has great potential, it has not yet achieved complete industrialization or demonstrated durability and safety in mobility applications. The dry process in the energy sector is still in its infancy and needs to be improved. It is required to investigate the conformity to the current wet slurry method production line. For practical applications, experimental parameters, and suitable materials should be further investigated. Especially, thickness control, homogeneity, and adhesion properties of the electrodes should be controlled by careful adjustments of the process parameters. Overall, with meticulous studies, a dry process can be adopted for LIB technology to enable high performance LIBs.

References

[1] Duffner F, Mauler L, Wentker M, Leker J, and Winter M. Large-scale automotive battery cell manufacturing: analyzing strategic and operational effects on manufacturing costs. *International Journal of Production Economics*. 2021;232:107982.

[2] Li J, Fleetwood J, Hawley WB, and Kays W. From materials to cell: state-of-the-art and prospective technologies for lithium-ion battery electrode processing. *Chemical Reviews*. 2022;122(1):903–56.

[3] Wood DL, Quass JD, Li J, Ahmed S, Ventola D, and Daniel C. Technical and economic analysis of solvent-based lithium-ion electrode drying with water and NMP. *Drying Technology*. 2017;36(2):234–44.

[4] Salini PS, Gopinadh SV, Kalpakasseri A, John B, and Thelakkattu Devassy M. Toward greener and sustainable Li-ion cells: an overview of aqueous-based binder systems. *ACS Sustainable Chemistry & Engineering*. 2020;8 (10):4003–25.

[5] Rollag K, Juarez-Robles D, Du Z, Wood DL, and Mukherjee PP. Drying temperature and capillarity-driven crack formation in aqueous processing of Li-ion battery electrodes. *ACS Applied Energy Materials*. 2019;2(6):4464–76.

[6] Julien CM and Mauger A. NCA, NCM811, and the route to Ni-Richer lithium-ion batteries. *Energies*. 2020;13(23):6363.

[7] Lu Y, Zhao C-Z, Yuan H, Hu J-K, Huang J-Q, and Zhang Q. Dry electrode technology, the rising star in solid-state battery industrialization. *Matter*. 2022;5(3):876–98.

[8] Verdier N, Foran G, Lepage D, Prebe A, Ayme-Perrot D, and Dolle M. Challenges in solvent-free methods for manufacturing electrodes and electrolytes for lithium-based batteries. *Polymers (Basel)*. 2021;13(3):323.

[9] BWP, LiCAP, and Siemens Partner on Mass Production of Dry Electrode Manufacturing Equipment Green Car Congress 2022. https://www.greencarcongress.com/2022/12/20221207-licap.html.

[10] Prasad LK, McGinity JW, and Williams RO, 3rd. Electrostatic powder coating: Principles and pharmaceutical applications. *International Journal of Pharmaceutics*. 2016;505(1–2):289–302.

[11] Yang Q, Ma Y, Zhu J, Chow K, and Shi K. An update on electrostatic powder coating for pharmaceuticals. *Particuology*. 2017;31:1–7.

[12] Paturi UMR, Reddy NS, Cheruku S, Narala SKR, Cho KK, and Reddy MM. Estimation of coating thickness in electrostatic spray deposition by machine learning and response surface methodology. *Surface and Coatings Technology*. 2021;422:127559.

[13] Sampuran-Singh, O'Neill BC, and Bright AW. A parametric study of electrostatic powder coating. *Journal of Electrostatics*. 1978;4(4):325–34.

[14] Barletta M, Gisario A, and Tagliaferri V. Electrostatic spray deposition (ESD) of polymeric powders on thermoplastic (PA66) substrate. *Surface and Coatings Technology*. 2006;201(1–2):296–308.

[15] Meng X, Zhu J, and Zhang H. Influences of different powders on the characteristics of particle charging and deposition in powder coating processes. *Journal of Electrostatics*. 2009;67(4):663–71.

[16] Barletta M, Gisario A, Rubino G, and Tagliaferri V. Electrostatic spray deposition (ESD) of 'self organizing' TiO2-epoxy powder paints: experimental analysis and numerical modeling. *Surface and Coatings Technology*. 2006;201(6):3212–28.

[17] Luo Y, Zhu J, Ma Y, and Zhang H. Dry coating, a novel coating technology for solid pharmaceutical dosage forms. *International Journal of Pharmaceutical*. 2008;358(1–2):16–22.

[18] Zhu C, Fu Y, and Yu Y. Designed nanoarchitectures by electrostatic spray deposition for energy storage. *Advances Materials*. 2019;31(1):e1803408.

[19] Dhuppe S, Mitkare SS, and Sakarkar DM. Recent techniques of pharmaceutical solventless coating: a review. *Journal of Pharmaceutical Science and Research*. 2012;3:1976.

[20] Abe H, Kondo A, Naito M, Wakimoto T, and Yamaguchi M (eds.). Electrostatic dry coating of cathode materials for Li ion battery. In *Additional Conferences (Device Packaging HiTEC HiTEN & CICMT)*, 2013.

[21] Abe H, Naito M, Kondo A, and Yamaguchi M. Electrostatic spray deposition for fabrication of Li-ion batteries. *Transactions of JWRI*. 2015;44:9–12.

[22] Ludwig B, Zheng Z, Shou W, Wang Y, and Pan H. Solvent-free manufacturing of electrodes for lithium-ion batteries. *Science Report*. 2016;6:23150.

[23] Ludwig B, Liu J, Chen IM, *et al.* Understanding interfacial-energy-driven dry powder mixing for solvent-free additive manufacturing of Li-ion battery electrodes. *Advanced Materials Interfaces*. 2017;4(21):1700570.

[24] Al-Shroofy M, Zhang Q, Xu J, Chen T, Kaur AP, and Cheng Y-T. Solvent-free dry powder coating process for low-cost manufacturing of LiNi1/3Mn1/3Co1/3O2 cathodes in lithium-ion batteries. *Journal of Power Sources*. 2017;352:187–93.

[25] Liu J, Ludwig B, Liu Y, *et al.* Scalable dry printing manufacturing to enable long-life and high energy lithium-ion batteries. *Advanced Materials Technologies*. 2017;2(10):1700106.

[26] Wang M, Hu J, Wang Y, and Cheng Y-T. The influence of polyvinylidene fluoride (PVDF) binder properties on LiNi0.33Co0.33Mn0.33O2 (NMC) electrodes made by a dry-powder-coating process. *Journal of The Electrochemical Society*. 2019;166(10):A2151–A7.

[27] Schälicke G, Landwehr I, Dinter A, Pettinger K-H, Haselrieder W, and Kwade A. Solvent-free manufacturing of electrodes for lithium-ion batteries via electrostatic coating. *Energy Technology*. 2019;8(2):1900309.

[28] Zhen E, Jiang J, Lv C, *et al.* Effects of binder content on low-cost solvent-free electrodes made by dry-spraying manufacturing for lithium-ion batteries. *Journal of Power Sources*. 2021;515:230644.

[29] Grant PS, Greenwood D, Pardikar K, *et al.* Roadmap on Li-ion battery manufacturing research. *Journal of Physics: Energy*. 2022;4(4):042006.

[30] Duong HM, Shin J, and Yudi Y (eds.). Dry electrode coating technology. In *48th Power Sources Conference*; 2018.

[31] Zhang Y, Huld F, Lu S, Jektvik C, Lou F, and Yu Z. Revisiting polytetrafluorethylene binder for solvent-free lithium-ion battery anode fabrication. *Batteries*. 2022;8(6):57.

[32] Hippauf F, Schumm B, Doerfler S, *et al.* Overcoming binder limitations of sheet-type solid-state cathodes using a solvent-free dry-film approach. *Energy Storage Materials*. 2019;21:390–8.

[33] Yao W, Chouchane M, Li W, *et al.* A 5V-class cobalt-free battery cathode with high loading enabled by dry coating. *Energy & Environmental Science*. 2022;16(9):1620–1630.

[34] Gardiner J. Fluoropolymers: origin, production, and industrial and commercial applications. *Australian Journal of Chemistry*. 2014;68(1):13–22.

[35] Bader M, Wrbitzky R, Blaszkewicz M, and van Thriel C. Human experimental exposure study on the uptake and urinary elimination of N-methyl-2-pyrrolidone (NMP) during simulated workplace conditions. *Archives of Toxicology*. 2007;81(5):335–46.

[36] Li Y, Wu Y, Wang Z, *et al.* Progress in solvent-free dry-film technology for batteries and supercapacitors. *Materials Today*. 2022;55:92–109.

[37] Zhang Y, Lu S, Lou F, and Yu Z. Leveraging synergies by combining polytetrafluorethylene with polyvinylidene fluoride for solvent-free graphite anode fabrication. *Energy Technology*. 2022;10(11):2200732.

[38] Wu Q, Zheng JP, Hendrickson M, and Plichta EJ. Dry process for fabricating low cost and high performance electrode for energy storage devices. *MRS Advances*. 2019;4(15):857–63.

[39] Kirsch DJ, Lacey SD, Kuang Y, *et al.* Scalable dry processing of binder-free lithium-ion battery electrodes enabled by holey graphene. *ACS Applied Energy Materials*. 2019;2(5):2990–7.

[40] Walker BA, Plaza-Rivera CO, Sun S-S, Lu W, Connell JW, and Lin Y. Dry-pressed lithium nickel cobalt manganese oxide (NCM) cathodes enabled by holey graphene host. *Electrochimica Acta*. 2020;362:137129.

[41] Sotomayor ME, Torre-Gamarra Cdl, Levenfeld B, *et al.* Ultra-thick battery electrodes for high gravimetric and volumetric energy density Li-ion batteries. *Journal of Power Sources*. 2019;437: 26923.

[42] Gonzalez-Gutierrez J, Cano S, Schuschnigg S, *et al.* Additive manufacturing of metallic and ceramic components by the material extrusion of highly-filled polymers: a review and future perspectives. *Materials*. 2018;11(5): 840.

[43] Patil H, Tiwari RV, and Repka MA. Hot-melt extrusion: from theory to application in pharmaceutical formulation. *AAPS PharmSciTech*. 2016;17 (1):20–42.

[44] Yang J, Takeda Y, Imanishi N, *et al.* SiO_x-based anodes for secondary lithium batteries. *Solid State Ionics*. 2002;152–153:125–9.

[45] El Khakani S, Verdier N, Lepage D, *et al.* Melt-processed electrode for lithium ion battery. *Journal of Power Sources*. 2020;454 227884.

[46] Astafyeva K, Dousset C, Bureau Y, Stalmach SL, and Dufour B. High energy Li-ion electrodes prepared via a solventless melt process. *Batteries & Supercaps*. 2020;3(4):341–3.

[47] Domonkos M, Tichá P, Trejbal J, and Demo P. Applications of cold atmospheric pressure plasma technology in medicine, agriculture and food industry. *Applied Sciences*. 2021;11(11):4809.

[48] Liang Z, Li T, Chi H, *et al.* Solvent-free manufacturing of lithium-ion battery electrodes via cold plasma. *Energy & Environmental Materials*. 2022. https://doi.org/10.1002/eem2.12503.

[49] Xia H, Tang SB, and Lu L. Thin film microbatteries prepared by pulsed laser deposition. *Journal of the Korean Physical Society*. 2007;51(3):1055–1062.

[50] Plonczak P, Bieberle-Hütter A, Søgaard M, *et al.* Tailoring of LaxSr1-xCoyFe1-yO3-δ nanostructure by pulsed laser deposition. *Advanced Functional Materials*. 2011;21(14):2764–75.

[51] Haider AJ, Alawsi T, Haider MJ, Taha BA, and Marhoon HA. A comprehensive review on pulsed laser deposition technique to effective nanostructure production: trends and challenges. *Optical and Quantum Electronics*. 2022; 54(8):Article: 534.

[52] Fenech M, Sharma N. Pulsed Laser Deposition-based Thin Film Microbatteries. *Chem Asian J*. 2020;15(12):1829–47.

[53] Indrizzi L, Ohannessian N, Pergolesi D, Lippert T, and Gilardi E. Pulsed laser deposition as a tool for the development of all solid-state microbatteries. *Helvetica Chimica Acta*. 2021;104(2):e2000203.

[54] Uzakbaiuly B, Mukanova A, Zhang Y, and Bakenov Z. Physical vapor deposition of cathode materials for all solid-state li ion batteries: a review. *Frontiers in Energy Research*. 2021;9:625123.

[55] Wang F, Zhao H, Liang J, *et al.* Magnetron sputtering enabled synthesis of nanostructured materials for electrochemical energy storage. *Journal of Materials Chemistry A*. 2020;8(39):20260–85.

[56] Maurya D, Sardarinejad A, and Alameh K. Recent developments in R.F. magnetron sputtered thin films for ph sensing applications—an overview. *Coatings*. 2014;4(4):756–71.

[57] Ma Y, Li L, Qian J, *et al.* Materials and structure engineering by magnetron sputtering for advanced lithium batteries. *Energy Storage Materials*. 2021; 39:203–24.

[58] Oke JA and Jen T-C. Atomic layer deposition and other thin film deposition techniques: from principles to film properties. *Journal of Materials Research and Technology*. 2022;21:2481–514.

[59] Yang G, Abraham C, Ma Y, *et al.* Advances in materials design for all-solid-state batteries: from bulk to thin films. *Applied Sciences*. 2020;10(14):4727.

[60] Johnson RW, Hultqvist A, and Bent SF. A brief review of atomic layer deposition: from fundamentals to applications. *Materials Today*. 2014;17(5):236–46.

[61] Leskela M and Ritala M. Atomic layer deposition chemistry: recent developments and future challenges. *Angewandte Chemie International Edition in English*. 2003;42(45):5548–54.

[62] Madhuri KV. Thermal protection coatings of metal oxide powders. In: Al-Douri Y (eds.). *Metal Oxide Powder Technologies*. New York, NY: Elsevier; 2020. p. 209–31.

[63] Fei C, Liu H, Wang X, and Fan X. The influence of process parameters and pulse ratio of precursors on the characteristics of La1-x Al x O3 films deposited by atomic layer deposition. *Nanoscale Research Letters*. 2015; 10:180.

[64] Marin E, Lanzutti A, Andreatta F, Lekka M, Guzman L, and Fedrizzi L. Atomic layer deposition: state-of-the-art and research/industrial perspectives. *Corrosion Reviews*. 2011;29(5–6):101.

[65] Oviroh PO, Akbarzadeh R, Pan D, Coetzee RAM, and Jen TC. New development of atomic layer deposition: processes, methods and applications. *Science and Technology of Advanced Materials*. 2019;20(1):465–96.

[66] Leskelä M and Ritala M. Atomic layer deposition (ALD): from precursors to thin film structures. *Thin Solid Films*. 2002;149(1):138–46.

Chapter 8

Electrode drying and calendering

W. Blake Hawley[1], *William Kays*[2] *and Jianlin Li*[3]

8.1 Introduction

Once the lithium-ion battery (LIB) electrode slurry is applied to the current collector, the solvent from the coating must be removed in the drying step. Occasionally, the qualifier "primary" is used to differentiate this step from downstream secondary drying, which is used to reduce the moisture content of electrodes. Electrode drying most commonly occurs in long convection ovens that are placed in-line with the coating apparatus. In other words, the current collector starts as a bare foil, then is coated, and fed into the drying oven all in one step without any need for switching the foil onto a new roll. Knowing the speed of the substrate, the dryer can be sized appropriately to secure that there will be sufficient time for the solvent to be removed. The synergy between coating and drying is beneficial for process efficiency but requires a high degree of tuning and quality assurance to ensure that the electrodes are being dried well.

After the electrode is dried, it is passed to a machine that reduces the thickness of the electrode. The machine, commonly called a calender and sometimes called a rolling press, consists of two rollers that are rotated in opposite directions (i.e., one clockwise and the other counterclockwise). A gap is set between the rollers that can be correlated to the desired thickness of the electrode. Since energy density is essential for LIB electrodes in many applications, calendering is a useful process for reducing excessive void space in the electrodes that contribute nothing to energy density. However, the balance between thickness/porosity and electron/ion transport, as well as the effect on electrode mechanical properties, must be considered carefully when determining the optimal calendering conditions.

Further downstream in manufacturing, typically right before the electrodes are built into a cell, a secondary drying step is often employed. This secondary drying step is necessary to keep the moisture content of the electrodes low since moisture can harm the performance and safety characteristics of the cell. However, the

[1]Sparkz, Inc., USA
[2]R.W. Baron Process Equipment, USA
[3]Electrification and Energy Infrastructures Division, Oak Ridge National Laboratory, USA

secondary drying step time should be minimized to avoid process bottlenecks and excessive manufacturing energy costs.

In this chapter, these three processes will be discussed. The effect of each process on the electrode microstructure and resulting electrode performance is considered, as are the characterization methods of electrodes after each stage.

8.2 Drying

As mentioned, the electrode drying process occurs in-line with and immediately following the coating process. Electrode dryers can range in length from 20 to 60 m. At coating speeds of 10–30 $m \cdot min^{-1}$, the residence time of a coating in the dryer is on the order of 1–3 min [1].

The physical process of electrode drying proceeds similarly to the drying of other coatings in similar industries, such as latex coatings and paper (see Figure 8.1) [2,3]. In the first step, generally called "film shrinkage," solvent is removed from the top surface of the electrode and the thickness of the coating decreases. As the film thickness decreases, the particles begin to contact each other with gaps of void space between them. Next, in a step called "pore emptying," the solvent in the pores between particles is evaporated. These solvent molecules diffuse upward through the thickness of the electrode to the surface where they are vaporized. Finally, in "coalescence," the final solvent molecules are dried from the bottom of the coating and the microstructure of the electrode is formed, as the active materials, conductive additives, and binders coalesce [4,5].

For an electrode coating and drying system, because the dryer length is fixed, the characteristics of the electrode and the settings of the dryer are the factors influencing the speed of drying and the quality of the resulting product. Regarding the quality of the product, there are two kinds of defects that can arise from electrode drying: those

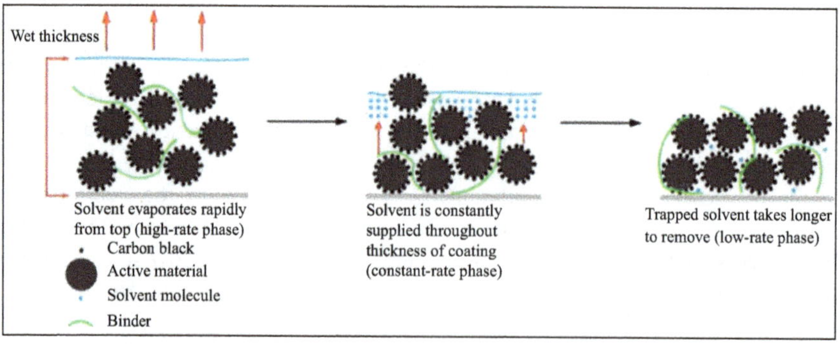

Figure 8.1 Cartoon representation of the stages of electrode drying. Reprinted with permission from [1]. Copyright 2023 American Chemical Society.

Figure 8.2 Image of the surface of a severely mud cracked electrode. Reprinted with permission from [1]. Copyright 2023 American Chemical Society.

that are plainly visible on the electrode surface and those that are less obvious but occur at the microstructural level. The most common visible defect is electrode cracking, while defects that may not be visible include binder and conductive additive migration. Cracking occurs during the film shrinkage stage of drying. As a result of the receding thickness during film shrinkage, stresses are imparted into the coating. Since the coating is constrained by a rigid current collector, if the stresses generated exceed the mechanical properties of the coating, then the stresses will be relieved as cracks in the surface of the coating [6]; an example of a badly cracked electrode is shown in Figure 8.2.

The less apparent defect in electrode coatings is binder migration. During drying, as solvent molecules diffuse to the surface of the coating for evaporation, they can drag segments of polymer with them to the surface [5,7–12]. A binder concentration gradient in which the concentration is highest at the surface of the coating and the lowest at the coating/current collector interface is undesirable for two reasons. First, given that among the primary roles of the binder is to provide adhesion between the coating and the current collector, fewer connections are made between the binder and the current collector when this gradient exists. Second, with more non-conductive binder located at the surface, the conductivity of the electrode is reduced, and the ionic pathways become more tortuous. This migration phenomenon is also observed with conductive additive particles, which also impacts the electronic conductivity of the electrode [13,14].

In Section 8.2.1, we will discuss how electrode properties and dryer parameters affect the final electrode microstructure and performance. In Section 8.2.2, we will present some basics of dryer and solvent recovery design.

8.2.1 Factors affecting dry electrode properties

One of the themes of this text is that the roll-to-roll manufacturing process is interconnected, with each step having influence on the downstream processes and the properties of the resulting electrode. The decisions made in the slurry design and preparation process will influence the drying process. For example, recall from Chapter 3 that the selected mixing order and instrumentation will influence the achievable solid loading of a slurry. A higher solid loading means that less solvent will need to be removed from the coating, making the drying process faster and less energy intensive. The properties of the electrode materials selected for the slurry, such as particle size, particle size distribution, and surface tension will also influence the drying process, particularly the formation and propagation of cracks [15]. The slurry viscosity will also impact the prevalence of binder and conductive additive migration since motion of the materials will be impeded at higher viscosities [16].

The slurry component that has the greatest influence on the drying process is the solvent. The state-of-the-art process mixes cathode slurries in *N*-methyl-2-pyrrolidone (NMP) and graphite anode slurries in deionized (DI) water. There is growing interest in processing cathodes in water, which is discussed in Chapter 10. The boiling point, vapor pressure, and heat of vaporization of these two solvents, the most relevant properties related to drying, are shown in Table 8.1.

Comparing the boiling point and vapor pressure of the two solvents would suggest that DI water would dry much more quickly than NMP. On the other hand, the heat of vaporization is nearly four times higher for water than NMP, meaning less energy is needed per unit mass to vaporize NMP without raising the temperature. Empirically, NMP has been shown to require a greater amount of energy to dry than water [13,17]. The outcome of this is that, for a constant drying temperature, the water-based electrode will dry more quickly, though the web temperature will not get as hot as quickly as for the NMP-based electrode [1,17]. Additionally, regarding the previous point about solid loading, water-based slurries may be capable of higher solid loadings than NMP-based slurries [17], depending on the water-soluble binder system selected.

The mass loading of the electrode will also dictate the coating/drying speed and the residence time in the dryer. Electrodes with a higher mass loading are thicker, and thus have more solvent to dry than an electrode with a lower mass loading, assuming the solid loading of the slurries for these two electrodes is equal. Therefore, for a constant dryer temperature, the thick electrode will need more time in the dryer than the thin electrode. If these electrodes are made on the same coating and drying line, then the dryer length is fixed, and the coating speed will need to be reduced for the thick electrode. Thick electrodes also are more susceptible to cracking, as more drying-induced stress is present in thick electrodes [15]. Cracking of thick electrodes is one of the processing challenges that must be overcome to enable more energy-dense cells [7]. This problem is particularly

Table 8.1 Relevant parameters of NMP and DI water for electrode drying. Values according to [17] unless otherwise indicated

Property	NMP	DI water
Boiling point (°C)	204.3	100.0
Vapor pressure (Pa)	133	7373
Heat of vaporization (kJ·kg^{-1})	510	2,260
Surface tension (mN·m^{-1}) [18]	72.8	40.8

troublesome for water-based electrodes since the surface tension of water is significantly higher than that of NMP (see Figure 8.3) [15,18]. A final consideration for thick electrodes (~ 10.5 mg·cm^{-2} total mass loading or greater) is that they are more prone to adhesion failures [19], making careful consideration of the drying process parameters even more important. Adhesion characterization of electrodes is discussed in Section 8.3.2.

If we combine the discussion regarding solvent selection and electrode thickness, we can say that for current state-of-the-art production, anodes will dry faster than cathodes for three reasons. First, water dries faster than NMP; second, water-based slurries likely have a higher solid loading than NMP-based slurries; and third, anodes are thinner than cathodes, since graphite has a higher capacity (320–350 mAh·g^{-1}) than all commercially used cathode materials (130–190 mAh·g^{-1}) [20]. This means that a blanket design of a coating and drying system for both electrodes is unlikely to produce the best electrodes, not to mention that coating and drying units for NMP-based electrodes will require a solvent recovery system (see Section 8.2.2).

In addition to the slurry and electrode properties dictating drying speed and effectiveness, the parameters of drying are a major determinant of drying process efficiency and success. The most significant of these process parameters is the temperature. If the goals of the drying process are to create a uniform, defect-free film quickly and at as low a cost as possible, there is a balance to strike when selecting a drying temperature. A high drying temperature will increase the drying rate [21], which will improve process speed. However, at a certain point, increasing the drying temperature will require additional energy and cost input that does not

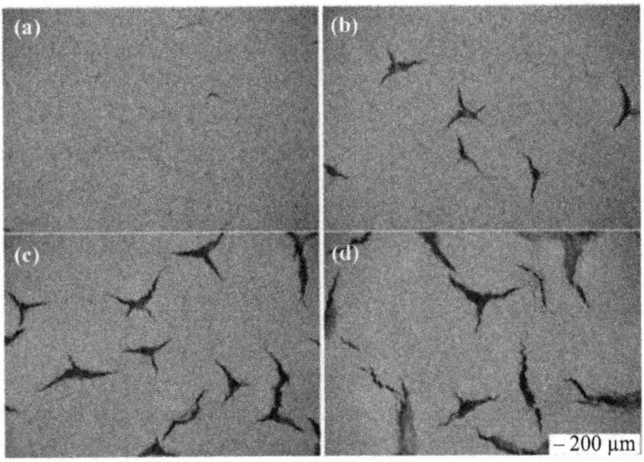

Figure 8.3 Cracking of dried aqueous-processed NMC532 cathodes as a function of total mass loading: (a) 15 mg·cm^{-2}, (b) 17.5 mg·cm^{-2}, (c) 20 mg·cm^{-2}, and (d) 25 mg·cm^{-2}. Reprinted with permission from Elsevier [18].

lead to faster drying and thus is not justified. A balance must be found between keeping energy costs low and maintaining expedient production.

Complicating temperature selection further is that binder migration has been routinely found to be exacerbated at high drying rates [5,7,8,19,22,23]. Exceptionally high drying rates may also lead to worsened electrode cracking [1]. Electrodes allowed to dry at room temperature have shown much more homogeneous binder distributions than those dried at high temperatures (~ 150°C) and thus better electrochemical properties [24]. Of course, drying electrodes at low temperatures greatly impedes the throughput of electrode manufacturing. As with all aspects of electrode processing, there exists a series of complex trade-offs in selecting a drying temperature. In state-of-the-art production, this trade-off has been optimized by implementing multi-zone dryers, each tuned to a different temperature, which will be discussed in the subsequent section. One factor that simplifies the drying process somewhat is the generally high thermal conductivity of the electrode components and solvent [3,5]. This minimizes the presence of thermal gradients down the thickness of the electrode during drying.

The dependence of binder migration on drying temperature has been characterized empirically in a few different ways, mostly through direct observation (i.e., Raman spectroscopy, scanning electron microscopy/energy dispersive X-ray spectroscopy, SEM/EDX) of different cross-sectional slices of an electrode [8], through indirect correlation of properties (i.e., adhesion strength) to drying conditions [19,23], or through modeling [5]. One SEM/EDX investigation of an electrode cross-section is shown in Figure 8.4(a), in which the fluorine content of each layer is examined. Since the graphite anodes in this study were processed in NMP, fluorine is the distinct indicator of the polyvinylidene fluoride (PVDF) binder. An interesting observation regarding electrode adhesion is that, for thin electrodes (mass loading < 10.5 mg·cm^{-2}), higher drying temperatures may have no impact or even a positive impact on adhesion strength; however, at moderate-to-high thickness (≥ 10.5 mg·cm^{-2}), higher drying temperatures adversely affect adhesion (see Figure 8.4(b)) [19].

Empirical studies that study the drying process in situ tend to be qualitative while those that can study dried electrodes post-drying tend to be quantitative [7]; developing quantitative approaches that can be employed in situ would complement the findings found thus far. Early drying simulations were able to provide insights and complement experimental findings, though sometimes had to rely upon simplifying assumptions about material properties or the diffusion of solvent and binder phases [7]. More recent drying simulations have produced three-dimensional electrode mesostructures [25].

In addition to temperature, it is worth mentioning that the drying environment can influence drying speed and effectiveness. Drying in humid air can inhibit coating cracking [15], likely due to less stress being imparted to the coating. Conversely, a given volume of humid air dries less solvent than an equivalent volume of dry air from a coating due to the air already being partially saturated with water which leads to slower drying. Inert gas drying could speed up drying, as heavier gases (such as argon) could supplant the solvent during drying [4]. Such

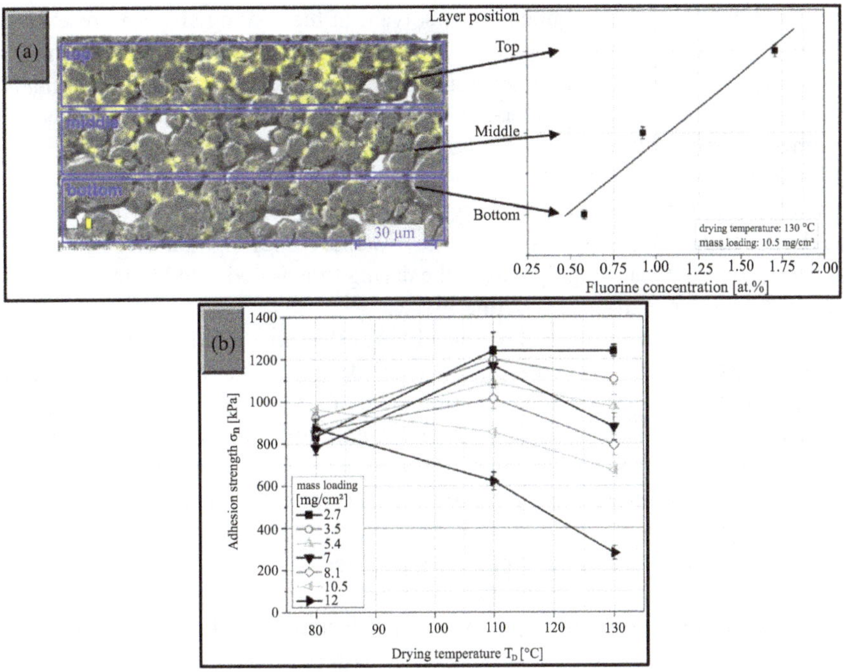

Figure 8.4 (a) SEM/EDX of three layers of a graphite anode with PVDF binder, with the bottom layer referring to the layer at the interface between the current collector and the electrode. (b) The influence of drying temperature and mass loading on adhesion strength for graphite anodes. Reprinted with permission from Elsevier [19].

vacuum drying is used in secondary drying to reduce the moisture content of electrodes to a safe level for cell assembly, as will be discussed in Section 8.4.

8.2.2 Design of electrode dryers

Due to the complex trade-offs between process speed, cost, and product quality, drying electrodes at a single temperature is not optimal. Instead, multi-zone dryers are the best choice for convection drying of electrodes [1,7,9–12]. These multi-zone dryers are designed such that the first zone is at the lowest temperature, with each zone progressively becoming warmer until the final zone, which is the hottest. The first zone is typically set to a temperature around 75–90°C and the final zone is often set to approximately 125–140°C. Just like the temperatures of the zones, the length of each zone may be customized, though it is common to see dryers designed with zones of equal length.

To understand why the multi-zone dryer is a good tool to optimize the drying process, and why the temperatures progressively get warmer in the zones, consider the underlying physical mechanism of the drying process. The film shrinkage stage occurs rapidly, due to the abundance of solvent in the coating and the presence of solvent at the surface of the coating. Pore emptying is slower than film shrinkage, though the rate of drying through pore emptying is relatively constant. Finally, removing the final solvent molecules during coalescence is a slow process that becomes progressively slower until the coating is dry. Sometimes the three drying stages are instead designated by the drying rate of each stage, that is, the first phase is the high-rate (or increasing rate) drying stage, the second phase is the constant-rate drying stage, and the third phase is the low-rate (or falling rate) drying stage. It has been estimated that roughly half of the drying time is dedicated to removing the final 10% of solvent from the coating [3].

The film shrinkage stage is responsible for the most binder migration in the electrode [5]. So, if binder migration is made worse in the initial stages of drying and at high drying stages, the first zone of a dryer should have the lowest temperature. Once film shrinkage has ended and pore emptying has begun, the risk of binder migration decreases considerably. Therefore, the temperature should be increased to provide a boost to the drying rate, which has fallen from the film shrinkage stage. Finally, to extract the final solvent molecules from the bottom of the coating, the temperature should be increased further to expedite the process.

In addition to the multi-zone dryer, the placement and airflow velocity of the nozzles is important. Airflow velocity contour maps have been used to optimize the placement of dryer nozzles within a system [1]. The placement of the nozzles determines the amount of heat transfer to the coating and must work in concert with the selected air temperature. Nozzles placed too far from the coating and the substrate will lead to a coating exiting the dryer that is still wet, while nozzles placed too closely will impart too much heat and can lead to electrode cracking [1].

Embedded in the drying process of NMP-based coatings is a solvent recovery system which is necessary due to the toxicity of NMP [26,27]. An NMP recovery system consists of three components in sequence: (1) a condenser, (2) a scrubber, and (3) a carbon adsorption system [1]. As the NMP-saturated air is passed through each unit operation, the NMP concentration in the air decreases. Most of the NMP is removed in the condenser, where it is recovered as a liquid and sent to a tank. Later, the collected liquid will be distilled to obtain purer NMP; depending on the size of the NMP distillation needs and the resulting economics, NMP purification can be done onsite or outsourced [1]. After the air has passed through the condenser, it is sent to a scrubber, which further reduces the concentration of NMP in the air. Even after scrubbing, it is common for the air to contain too much NMP, requiring a final carbon adsorption system to finish the job. This system contains active carbon media designed to adsorb more NMP, which may be either replaced with fresh carbon media when saturated or installed with a regeneration system that desorbs the NMP from the media [1].

Dryers constitute a much larger footprint of an electrode production floor than do mixers, coaters, and calendering rolls, and account for one of the largest footprints in the entire battery manufacturing process [1]. In addition to the size of the operation, the cost of drying and solvent recovery is the highest of all electrode processing steps. For an $LiNi_xMn_yCo_{1-x-y}O_2$ (NMC) electric vehicle (EV) pouch cell battery, one case study found that coating, drying, and solvent recovery is slightly more expensive than but comparable in cost to electrolyte filling and sealing, tab welding, packaging, and formation, and is only less expensive than lamination and stacking (see Figure 8.5) [1,28]. Another study found that 38% of energy consumption in 24 kWh battery pack manufacturing comes from electrode drying [29]. While the exact cost depends significantly on specifics of the material, electrode, cell, and pack, it has routinely been demonstrated that electrode drying and solvent recovery is a far more energy-intensive, expensive process relative to the other stages of electrode processing. To mitigate against excessive energy consumption during electrode drying, preventing heat loss from the dryer is essential by monitoring and replacing deteriorating insulation beneath external cladding of ducts [1].

The capital investment required for the NMP recovery system in electrode manufacturing will cost in excess of $3,000,000 at the GWh scale of production; though less significant, energy cost savings are also present for water-based processing [17]. This, when combined with the advantages of better environmental

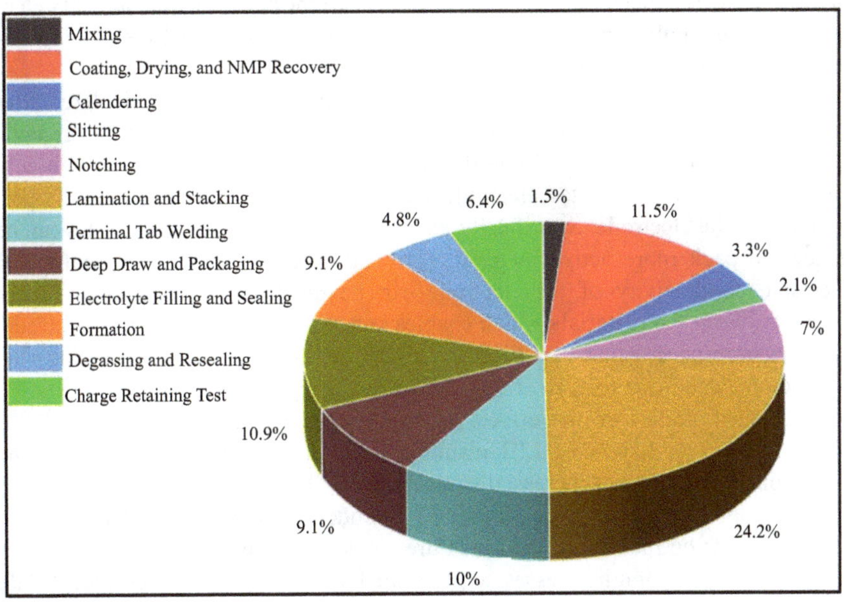

Figure 8.5 Cost of each LIB cell manufacturing step represented as a percentage of the entire process cost. These figures come from a case study of an EV pouch cell with NMC cathode and assuming 4,700 h·y^{-1} of production time. Data from [28]. Reprinted with permission from [1]. Copyright 2023 American Chemical Society.

guardianship and worker safety, has motivated the transition of graphite anode processing from NMP to water and is presently motivating the effort to transition cathode processing to water as well; further discussion on the latter topic is presented in Chapter 10.

8.3 Calendering

After drying, the relevant properties of the electrode sheet include the mass loading, thickness, porosity, density, ionic conductivity, electronic conductivity, and mechanical properties (i.e., adhesion/cohesion). Of these properties, the only property that cannot be modified is the mass loading. The term "mass loading" refers to the mass of electrode components (i.e., active material, conductive additive, binder, and any other solid additives that may have been included in the slurry) per unit area. Technically, the mass loading can be changed because active material particles can be embedded into the current collector foil under high compressive loads, which has been used to fabricate electrodes intended for use in analytical electrochemistry [30–33]. That said, it is generally a fair assumption that the solid mass loading of an electrode will not change from the moment the slurry is coated, since the only reason the mass will change is due to the evaporation of solvent. Besides, embedding particles into the current collector foil for LIB electrodes requires high compressive loads, which are not advisable to apply for reasons that will be discussed.

Besides the mass loading, the other electrode properties are not finalized after the drying stage. The electrode after the drying stage is referred to as an "as-coated" or "uncalendered" electrode. While the as-coated electrode may be used in a cell, the properties of an as-coated electrode are not conducive to maximizing the performance of the electrode. Specifically, the thickness of as-coated electrodes can be reduced to pack more active materials into a smaller volume. Additionally, the electronic conductivity of the electrode can be improved by promoting better contact between the materials which enables better rate capability of the electrode. These performance gains can be realized by calendering, or roll-pressing, the electrode between two rolls rotating in opposite directions (see Figure 8.6).

Published studies on the effects of calendering originate in the early 2000s [34,35]. However, like most LIB manufacturing research topics, interest in the subject of calendering has increased in recent years. This is reflected by the number of publications concerning the subject of calendering increasing recently. Early studies concerning the influence of pressure on electrodes post-drying used simpler instruments, like hydraulic presses, to press small area samples of electrodes. While this practice is still employed occasionally at lab scale, smaller calendering machines with heated rolls and adjustable gap are available that can process small sheets of electrodes and more closely simulate the LIB manufacturing process.

In the subsequent two sections, we will present the properties of calendered electrodes in further detail and explain how calendered electrodes are characterized, respectively.

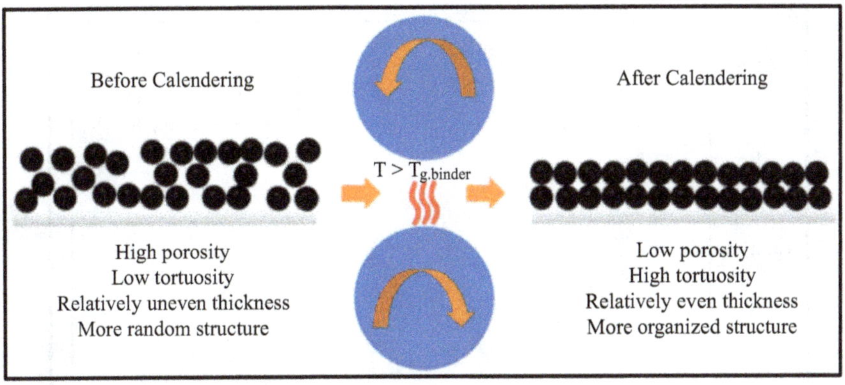

Figure 8.6 Cartoon representation of the electrode calendering process; conductive additives and binders omitted from the electrode microstructure for simplicity. Reprinted with [1]. Copyright 2023 American Chemical Society.

8.3.1 Calendered electrode properties

The properties of calendered electrodes are generally correlated with one another. For instance, because of the previously mentioned conservation of mass loading in the electrode, the thickness, porosity, and density of the electrode are all related. As the electrode is calendered, the thickness decreases as a result of filling in void space with electrode materials, thereby leading to a reduction in porosity and an increase in electrode density. Similarly, the ionic conductivity may be positively correlated to the porosity of the electrode and the electronic conductivity may be positively correlated to the density of the electrode. In the case of ionic conductivity, the electrolyte fills the pores in the electrode to facilitate transfer of ions to and from active materials. With more porosity in an electrode, there is a more expansive network of channels for the electrolyte to reach the active materials, and thus the transport of ions is easier. Conversely, the electronic conductivity is mostly a function of the conductive additive networks in an electrode. If the materials are in closer proximity to one another, then the conductive additive networks can more easily shuttle electrons from the current collector to the active materials, and vice-versa.

Based on the preceding discussion, a recurring theme of LIB design principles is clear once again: There exist trade-offs in every stage of electrode processing. An as-coated electrode will have a randomly distributed porosity of around 40–55%. The as-coated porosity is influenced primarily by slurry properties, such as the percentage of active material and the solid loading of the slurry. Generally, slurries with less solvent and more active material have lower as-coated porosities [36]. This lower as-coated porosity also correlates with a lower minimum achievable porosity post-calendering (see Figure 8.7) [36].

The as-coated electrode represents the case where porosity, thickness, and ionic conductivity are maximized, and density, electronic conductivity, and

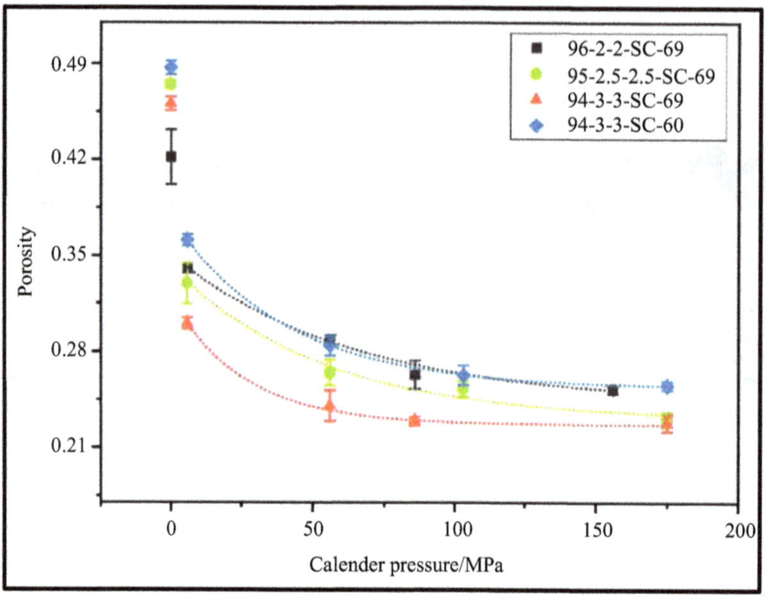

Figure 8.7 Porosity of four NMC111 cathodes as a function of calendering. The first number in the code represents the mass fraction of NMC111 in the electrode, the second number in the code represents the mass fraction of carbon black in the electrode, and the third number in the code represents the mass fraction of PVDF in the electrode. The number after "SC" refers to the solid loading of the slurry. For example, the black trend "96-2-2-SC-69" is an electrode with 96 wt% NMC111, 2 wt% carbon black, 2 wt% PVDF, and was processed in a slurry with 69 wt% solid loading. Reprinted from [36]. Article is open access under Creative Commons license (https://creativecommons.org/licenses/by-nc-nd/4.0/).

volumetric energy density are minimized. As the electrode is calendered, the porosity, thickness, and ionic conductivity are diminished and the density, electronic conductivity, and volumetric energy density are improved. Finding the "sweet spot" in which a desirable balance of properties is achieved is an optimization project worthwhile to undergo for any electrode design. Depending on the needs of the application, different degrees of calendering may be appropriate. For example, if an application requires high volumetric energy density and fast charging capability, such as for an EV, the electrodes may need to be calendered to higher densities than if the application was not as constrained by volume and did not require as fast of charging.

The electrode thickness is different at three distinct points in the calendering process: (1) before calendering, when it is highest, (2) during calendering, when it lowest, and (3) after calendering, when it is between the two previous values. The thickness increases after exiting the calender because of the springback effect of the electrodes. The amount of springback is dependent on the electrode formulation; for instance, electrodes with more binder have less springback. Reducing the amount of springback is beneficial since springback is an additional stress on the electrode that can cause deformation and because more springback constrains the minimum achievable density and thickness of the electrode [37].

The mechanical properties of the electrode also depend on the degree of calendering in the electrode. Among the most relevant mechanical properties of the electrode is the adhesion, which quantifies the strength of the bond between the current collector foil and the coating. Under light-to-moderate calendering conditions, the electrode adhesion decreases approximately linearly from the adhesion of the as-coated electrode. This is because the modulus of the current collector foil is substantially higher than that of the coating [1,38]. During calendering, the coating stretches to a much greater extent than the foil does which interrupts connections made between the foil and the binder in the coating. This unequal stretching occurs until high degrees of calendering are achieved. Eventually, the coating is pressed into the foil to such an extent that the binder re-bonds with the foil to create mechanical interlocks, which causes the adhesion between the foil and the electrode to increase from the minimum it reaches at moderate calendering [1,38]. Interestingly, this "adhesion recovery" phenomenon is more pronounced at higher mass loadings ($\geq$ 16.1 mg·cm^{-2}) [38].

At this point, we have discussed calendered electrodes without distinguishing between different electrode materials. Studies have shown that the largest determinant of what causes the optimal amount of calendering are the physical properties of the electrode constituents, with the roll temperature and roll gap also mattering but to a lesser extent [39]. Of course, with the majority of the electrode percentage assignable to the active material, there is a noticeable difference between cathodes and anodes in their response to calendering. There are likely differences between different cathode materials, such as $LiCoO_2$ (LCO) [34], $LiFePO_4$ (LFP) [40,41], $LiMn_2O_4$ [42], NMC [36,37,43–48], and emerging Co-free materials like $LiNi_{0.9}Mn_{0.05}Al_{0.05}O_2$ (NMA) [49], to name a few. The same could be said for anode materials, such as graphite [34,35,48,50], silicon, graphite/silicon composites, and $Li_4Ti_5O_{12}$.

Several studies have shown that graphite particles are softer than cathode active material particles [45,48]. The effect of this is that graphite anodes should be calendered more lightly than cathodes to preserve good electrochemical performance. Cyclic voltammetry (CV) curves conducted on graphite anodes and LCO cathodes under the same high pressure showed that graphite performance suffered compared to the as-coated anode, while the LCO performance was improved compared to the as-coated cathode [34]. In fact, the electronic conductivity of the anode material itself could be a larger determinant of the electronic conductivity than the degree of calendering in part due to the softness of the graphite particles [35]. The binder must also be considered when discussing the properties of an electrode as they pertain to calendering. For example, in graphite anodes, higher electrode densities are achievable when PVDF is used as opposed to water-soluble binders carboxymethyl cellulose, styrene butadiene rubber, and polyacrylic acid [37].

In addition to the material properties, a secondary factor in the calendering behavior is the temperature of the rolls. The adhesion of the electrodes can be benefited by heated rolls due to the dependence of the elasticity of the binder on temperature [36,38,45]. Studies have shown that for PVDF-containing electrodes roll temperatures are best around 75–90°C with little benefit at temperatures higher than 90°C [36,45]. Additionally, warmer rolls can reduce the minimal achievable porosity in the electrode [38].

To this point, we have considered the primary benefits of calendering as an improvement in electrode density and in improvement of electronic conductivity, as well as the variation in electrode adhesion with calendering. However, there are other effects of calendering that must be considered, namely that the surface of the electrode become smoothed, the mechanical properties of the electrode become more homogenous, and the pore size distribution is altered [50]. Rough electrode surfaces are more likely to protrude through the separator when being built into a cell, causing an electrical short circuit (i.e., unimpeded flow of electrons between the electrodes). This short circuit generates heat quickly, which can ignite the flammable liquid electrolyte and cause a fire and/or explosion that produces harmful gases. In summary, it is important to have smooth, clean electrode surfaces so as to not damage the separator; light calendering is an effective treatment to prevent this.

The ideal electrode would have a homogeneous distribution of active material, conductive additive, and binder from the beginning of the electrode roll to the end of the roll, with equal mechanical and electrical properties from point-to-point. While point-to-point variation in an electrode roll is inevitable, it is possible to reduce some variation in mechanical properties, thickness, and density by calendering the electrode. Finally, the larger pores in the electrode tend to be closed during calendering and the surface porosity is reduced [45,46,50]. While calendering has many of the aforementioned benefits, it can impact the wettability of the electrolyte at high compaction [51,52]. This is because with reduced porosity some active materials may be less accessible to the liquid electrolyte. A final consideration of calendering is the impact it has on moisture uptake of the electrode. After calendering, the surface area of particles in the electrode increases, which consequently increases the amount of moisture that can be adsorbed by the electrode [46].

The preceding discussion begs the following question: What amount of calendering is best? While there are a myriad of factors to consider, it is possible that thinner, lower-loading electrodes achieve better performance when calendered to lower porosities (20–30%) while higher-loading electrodes are better at slightly higher porosity (30–35%) [47]. Graphite anodes should be calendered to higher porosities than cathodes since graphite is more sensitive to high pressure than cathode materials. The rate capability and cyclability of electrodes has been shown to be better when calendered than when uncalendered, across a variety of active materials, likely due to better balance between the electronic and ionic conductivity [40–42,44,46,49].

8.3.2 Characterizing dried and calendered electrodes

Assuming that the mass of the electrode is conserved post-calendering (i.e., no particles are embedded into the current collector foil), the porosity of the electrode may be calculated according to (8.1) [53,54]:

$$\epsilon = 1 - \frac{m_{areal}}{L}\left(\frac{x_{AM}}{\rho_{AM}} + \frac{x_{CA}}{\rho_{CA}} + \frac{x_B}{\rho_B}\right) \tag{8.1}$$

In (8.1), ε is the porosity (expressed as a decimal), m_{areal} is the total mass loading in g·cm^{-2}, L is the thickness of the coating in cm (not including thickness of current collector foil), x is mass fraction, and ρ is density in g·cm^{-3}. The subscripts *AM*, *CA*, and *B* refer to the active material, conductive additive, and binder components, respectively. Additional mass fraction divided by density terms may be included if the electrode contains other components besides the active material, conductive additive, and binder. Because the m_{areal}, x, and ρ terms do not change, and the current collector may be assumed to be incompressible, a direct correlation between porosity and thickness may be established.

Of these two parameters, electrode thickness is much simpler to quantify than porosity. At the lab-scale, micrometer screw gauges may be used to measure the thickness of electrodes. By lab-scale, we generally mean sheets of electrode that are on the order of ≤ 1 m long and ~few 100 mm wide. To determine the mass loading of these sheets, a circular punch tool of certain diameter (typically in the 12–18 mm range) may be used to collect a few samples from the electrode. Then, the mass of these samples can be measured, and the mass loading may be calculated, assuming that the current collector mass is known or able to be determined. The mass fraction of each component is determined prior to mixing the slurry and the density of each component is generally known or measured prior to mixing. With these parameters, the porosity may be calculated.

To relate the density of the electrode to the porosity, we first define that porosity is the volume fraction of voids in the electrode. In a perfectly dense electrode, the porosity is zero and the volume of this electrode is calculated by dividing the mass of the electrode by the theoretical density ($\rho_{theoretical}$). The theoretical density of the electrode can be estimated by calculating a weighted average based on the mass fraction of the electrode components; though this method of calculating theoretical density is technically erroneous, it is a simple approximation to make that yields values that are approximately correct. The volume of a real electrode, then, is the volume of this perfectly dense electrode divided by the volume fraction of electrode components (i.e., one minus the porosity). Finally, the density of the real electrode ($\rho_{electrode}$) will be the mass of the electrode divided by its volume. Condensing all of these relations to one term, we arrive at (8.2a). The mass of electrode terms on each side of the equation offset and the equation is simplified to express how the density of the real electrode is a function of the other components in (8.2b):

$$\left(\frac{m_{electrode}}{\rho_{theoretical}}\right)\left(\frac{1}{(1-\epsilon)}\right) = \frac{m_{electrode}}{\rho_{electrode}} \tag{8.2a}$$

$$\rho_{electrode} = \rho_{theoretical}(1-\epsilon) \tag{8.2b}$$

Measuring the thickness and mass of electrode rolls larger than those made at the lab scale with a micrometer tool and balance becomes tedious, time-consuming, and infeasible. In-line thickness measuring tools for pilot- and manufacturing-scale electrodes, such as beta transmission lines and X-ray radiography, are tools that can also be used to calculate mass loading.

The mechanical properties of an electrode can be characterized in a few different ways. A straightforward approach is to bend and/or scratch finished electrode rolls and determine if the electrodes delaminate or crack, which are indicative of adhesion or cohesion failures, respectively. These tests are not particularly quantitative as they are appraised according to "pass/fail" criteria but are still effective and good checks to perform. A quantitative approach to determining the mechanical properties of electrodes is peel testing, which has been used for decades on LIB electrodes to appraise different binder systems, electrode formulations, and manufacturing processes [53,55–58]. In peel testing, a sample of electrode is fixed to a test platen using tape. For the sake of repeatability, it is advisable to apply a known force for a predetermined amount of time to the sample to ensure that results are not influenced by the sample being insufficiently stuck to the tape on the platen. Next, tape is attached to a hook and the top of the sample. The hook is displaced at a fixed rate and the coating is peeled from the substrate. The force required to remove the coating is measured and is divided by the width of the tape used to peel the coating to determine the adhesion in $N{\cdot}m^{-1}$. The angle at which the sample is peeled may vary, but 90° and 180° tests are common. Another testing system for adhesion measurements, called surface and interfacial cutting analysis system (SAICAS), has been implemented for LIB electrodes [59,60]. In SAICAS, a V-shaped blade pierces the coating at a specified depth and location and measures the force needed in horizontal and vertical directions to remove the coating [59]. Simple schematics of SAICAS and 180° peel testing are shown in Figure 8.8(a) and (b), respectively, for graphite anodes.

Traditional galvanostatic charge and discharge methods may be used to quantify the performance gain/loss in an electrode after being subjected to a particular calendering protocol. An interesting technique to isolate the effects of calendering on either the anode or the cathode is the construction of symmetric cells [40,49]. A symmetric cell may be constructed by building two cells, typically half cells, that feature the electrode of interest and a counter electrode. A few cycles are run for each cell such that the capacity can be determined as a function of cell potential. Then, each cell is charged or discharged to the desired state of charge such that the electrode has the intended amount of lithium. The cells are disassembled in a glovebox and the electrodes of interest, now lithiated to a

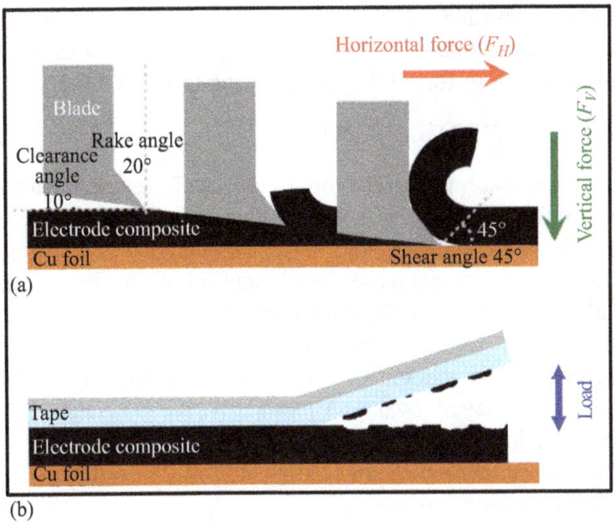

Figure 8.8 Cartoon schematics of (a) SAICAS and (b) 180° peel testing for graphite anodes. Reprinted with permission from [59]. Copyright 2023 American Chemical Society.

predetermined amount, are removed. Generally, the electrodes are washed in a solvent to remove any solid electrolyte interphase (SEI) or side products formed in the few cycles performed previously. Finally, the two electrodes of interest are built into a cell and may be tested. Symmetric cells studied with electrochemical impedance spectroscopy (EIS) have shown that the contact resistance at the electrode/current collector interface is reduced in calendered cathodes compared to as-coated cathodes [40]. Symmetric cell testing has also shown that the cathodic surface area of NMA is increased with calendering due to the pulverization of active particles; it was further shown that this deformation of particles did not negatively impact the cycle life in full cells [49].

8.4 Secondary drying

If we were to track the moisture content of an electrode from the moment it is coated until the moment it is built into a cell, the value would be the highest at the beginning of the process, the lowest at the moment the electrode exits the dryer, then would steadily rise in the time between exiting the dryer and being built into the cell. Assuming a constant temperature and humidity, the rate of moisture uptake for a dry electrode is significantly higher in the first hour of exposure to humid air than it is at times beyond one hour [61]. The rate of moisture uptake is a function of the electrode properties and the humidity and temperature of the surrounding air [62], meaning it could be minimized if all subsequent processing took place in a dry room. Regardless of processing environment, it is essential to keep the moisture content of the electrodes (and separators and electrolyte, for that matter) low. Water reacts with the fluorine-containing lithium salt (e.g., lithium hexafluorophosphate, $LiPF_6$) in the electrolyte to form HF gas according to Eq. 8.3 [63]:

$$LiPF_6 + H_2O \rightarrow 2HF + LiF + POF_3 \tag{8.3}$$

There are performance and safety consequences of these side reactions. First of all, the reaction of the $LiPF_6$ salt with water reduces the concentration of the salt in the electrolyte, therefore reducing the ionic conductivity of the electrolyte. HF poisoning of the cathode can increase the dissolution of transition metals from the active materials, particularly those containing manganese, which also leads to increased capacity loss. HF also interferes with the development of the SEI by forming a disproportionate amount of LiF compared to Li_2CO_3, which harms the conductivity of the SEI layer [64,65]. Additionally, the generation of HF gas will increase the pressure inside the cell, making it more susceptible to explosion. If the cell were to experience thermal runaway and catch fire, then the release of this HF gas would be highly toxic to humans.

As a consequence of this sensitivity to moisture, all cell components (electrodes, separator, and electrolyte) should be dry. Separators are most commonly comprised of hydrophobic polymers, like polyethylene and polypropylene, though separators sometimes contain surface coatings that can impact their hygroscopicity. Despite this, it is still best practice to store separator in a dry environment, as it is with electrodes. Because state-of-the-art electrolytes contain the moisture-sensitive fluorine-containing lithium salts, it is particularly important that they be kept dry. Most electrolytes are produced with ≤ 20 ppm moisture content [62,63]. While dry separators and electrolytes are key to cell performance and safety as well, further discussion in this section will be limited to the electrodes.

To reduce the moisture content and thus avoid these undesirable side reactions with water, a drying stage for the electrodes is often implemented immediately prior to cell assembly. This drying stage is referred to by a few names, most commonly secondary drying, or post-drying; in this text, we use the term secondary drying. Secondary drying generally occurs in a vacuum oven or on large belt dryers using radiative drying [62]. Radiative drying techniques are discussed in further detail in Chapter 9.

As mentioned, the rate of moisture sorption and desorption is dependent on factors such as electrode properties, temperature, and humidity. For starters, physical and chemical differences exist between electrode active materials. In general, LFP particles are smaller and have a higher specific surface area than other active materials, like NMC and NCA. Smaller particles are preferred for low-conductivity materials, such as LFP, to overcome the conductivity limitations [1,66]. The higher surface area does tend to lead to increased moisture uptake, since the adsorption of water of active materials tends to follow Brunauer–Emmett–Teller (BET) theory [67]. Interestingly, in one study, although LFP cathodes had a moisture content roughly 2.5× greater than that of NMC cathodes, the specific surface area of the LFP material was 3.6× greater than that of NMC. It was hypothesized that the heat of adsorption, another factor in BET adsorption, may explain this difference [67]. Another outcome of BET adsorption is that there exists an equilibrium moisture content that is achieved; essentially, an electrode can only be so dry, regardless of the temperature and humidity of the environment and the time of drying [67].

Another strategy to boosting the electronic conductivity of the LFP material is to coat the material with carbon. This also influences the hygroscopic behavior of LFP when compared with other active materials that are more electrically conductive and do not require a carbon coating. Carbonaceous anodes have been shown to have active surface sites that more readily adsorb water [68], which could be the case for carbon-coated LFP cathode materials as well. Despite the hygroscopicity of graphite anode, they tend to dry relatively quickly due to weakly-bound water molecules [61].

Because most of the moisture uptake for an electrode is highest in the first hour of exposure, the humidity of the air is a larger predictor of electrode moisture content than is the time of exposure for extended periods of time. Conversely, for a moisture-free environment, the temperature is a significant predictor of the drying rate and, thus, the amount of time needed to dry an electrode to low moisture content. A similar moisture content has been achieved in LFP and NMC cathodes stored in a glovebox (<1 ppm moisture) at room temperature for five weeks as cathodes stored in a vacuum environment for 15 h at 120°C [67] (see Figure 8.9(a) and (b)). In a manufacturing context, when time is of the essence, using higher secondary drying temperatures to reduce the time needed to dry the electrodes is important. However, binders may degrade at excessively high temperatures, which is why it is advisable to limit secondary drying temperatures to around 120°C for PVDF-containing electrodes [67].

Besides maintaining temperatures below a critical temperature that may induce degradation of the binder, PVDF-containing electrodes see little contribution to the moisture content from the conductive additive or binder. However, for water-

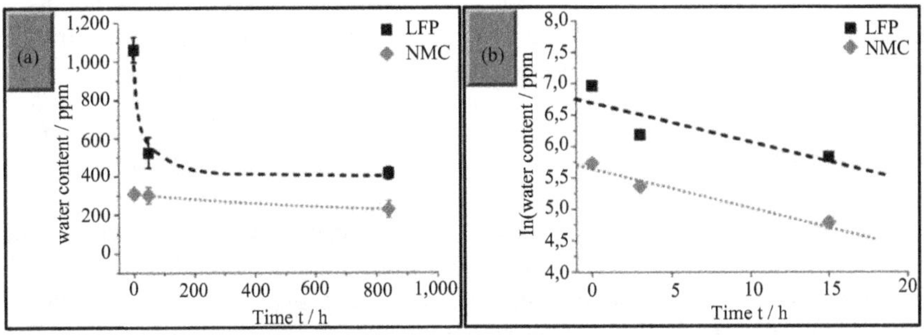

Figure 8.9 Moisture content of LFP (squares, black trend) and NMC (diamonds, gray trend) cathodes under (a) room temperature, vacuum conditions, and (b) 120°C, vacuum conditions. Reprinted from [67]. Article is open access under Creative Commons license (https://creativecommons.org/licenses/by-nc-nd/4.0/).

processed electrodes, the hygroscopicity of the binder tends to be the largest determinant of electrode moisture content [69]. Secondary drying of water-processed electrodes is discussed further in Chapter 10.

Above all else, the influence of electrode moisture content on the electrochemical performance of a cell is of utmost importance. High moisture contents ($\geq$ 10,000 ppm, or 1%) are detrimental to cycle life [67]. EIS data has shown that cells with high amounts of water impurities have increased SEI and charge transfer resistance originating from water molecules negatively impacting the formation of the SEI layer [61]. However, there is evidence that a small moisture content can actually be beneficial to the cycle life of a cell [62,67]. The amount of moisture that may lead to a cyclability benefit for cells is a function of the cell materials selected, including the electrodes and the electrolyte, since they all behave differently in the presence of water.

At the laboratory level, moisture content can be measured in a few ways, such as halogen moisture analyzers or coulometric Karl Fischer titrations. A halogen moisture analyzer is comprised of a balance and a halogen heating unit; the more sensitive the balance, the more precise the measurement can be. Halogen moisture analyzers measure the mass of a sample at room temperature, then heat the sample to a specified temperature for a predetermined amount of time and collect the mass of the sample at user-defined intervals. The temperature selected should be high enough to evaporate moisture from the sample and the sample should be spread thinly over a sample pan to ensure even drying. It is important to deconvolute the influence of moisture from the sample pan by analyzing the mass loss of a few "blank" samples consisting of only the sample pan. In Karl Fischer titration, the sample is dissolved into a titration cell that consists of a reagent and a solvent. When electrical current is flowed through the cell, the reagent that reacts with water to form a non-conductive compound. Based on the stoichiometry of the reactions, the amount of current (i.e., moles of electrons) can be correlated to the amount of water in the sample. Typically, Karl Fischer titration is better than moisture analyzers at measuring the moisture content of samples containing low amounts of moisture making it necessary for characterizing the moisture of electrolyte. Conversely, moisture analyzers are simpler to set up and less expensive.

8.5 Summary

The final steps in state-of-the-art LIB electrode processing, drying, calendering, and secondary drying were presented in this chapter. The discussion in this chapter highlights the interdependency of the electrode processing steps; electrode mixing and coating parameters clearly influence electrode drying and calendering, while drying and calendering more directly dictate the electrode microstructure. The electrode microstructure, meaning the arrangement of active materials, conductive additives, binders, and void space, determines the electrode's electronic and ionic conductivity, as well as the mechanical properties of the electrode, which impact both the electrochemical performance and the ease of downstream cell assembly processes. Deconvoluting this complex web of cause-and-effect is a non-trivial task, although empirical and simulation work in industry and academia have provided several clues that have led to the developments used in the state-of-the-art electrode manufacturing process used today.

The coating and drying processes work in tandem as a continuous roll-to-roll operation, making balancing the speed of these processes of utmost importance. A poorly calibrated coating and drying line will lead to electrodes that are still wet, cracked, or that perform sub-optimally due to binder and/or conductive additive migration. Obtaining a good, dried electrode is also a function of the slurry design and mixing process, with parameters like solid loading, viscosity, and surface tension all affecting the final microstructure. The cell design, which is dictated by the application for which the cell is being produced, will impact the thickness and mass loading of the electrodes. Thicker electrodes are more prone to cracking during drying, which is a further consideration.

The drying process is understood to occur in three phases, with the rate of drying decreasing as the process proceeds. To balance the quality of the electrode and the need for high throughput, multi-zone dryers with progressively increasing temperature have been implemented. Solvent recovery systems are used to recover and recycle the toxic solvent NMP, though such systems are not necessary for water-based coatings. In addition to safety, environmental stewardship, and significant capital equipment cost savings, water-based coatings generally dry faster with less energy consumption, though they have higher surface tension and face technical barriers with cathode active materials, as will be discussed in Chapter 10.

After drying, a calender is used to press the electrodes into thinner sheets, which promotes electronic conductivity and energy density. However, the electrodes must be calendered carefully, as some amount of porosity is needed to enable good ionic conductivity and the mechanical properties of electrodes decline (though, with enough force and mass loading, can improve slightly from a minimum) when calendered. The amount of calendering is dictated by the choice of active material; for instance, graphite is mechanically weaker than most cathode active materials, so it cannot be calendered with the same force without damaging the particles and hindering electrochemical performance.

The final treatment for an electrode prior to sealing the cell is secondary drying. Secondary drying ensures that the moisture content of the electrodes is low enough to not react detrimentally with the fluorine-containing lithium salt in the electrolyte. This reaction produces HF, which can cause the cell to swell, leach metals from the cathode, and is dangerous if released from the cell. Conversely, some moisture in the electrodes may be beneficial, as extremely dry electrodes may crack and small amounts of water can produce helpful SEI products.

References

[1] Li, J., J. Fleetwood, W.B. Hawley, and W. Kays, From materials to cell: state-of-the-art and prospective technologies for lithium-ion battery electrode processing. *Chemical Reviews*, 2022. 122: p. 903–956.

[2] Jaiser, S., M. Muller, M. Baunach, W. Bauer, P. Scharfer, and W. Schabel, Investigation of film solidification and binder migration during drying of Li-ion battery anodes. *Journal of Power Sources*, 2016. 318: p. 210–219.

[3] Susarla, N., S. Ahmed, and D.W. Dees, Modeling and analysis of solvent removal during Li-ion battery electrode drying. *Journal of Power Sources*, 2018. 378: p. 660–670.

[4] Zhang, Y.S., N.E. Courtier, Z. Zhang, *et al.*, A review of lithium-ion battery electrode drying: mechanisms and metrology. *Advanced Energy Materials*, 2021. 12: p. 2102233.

[5] Font, F., B. Protas, G. Richardson, and J.M. Foster, Binder migration during drying of lithium-ion battery electrodes: Modelling and comparison to experiment. *Journal of Power Sources*, 2018. 393: p. 177–185.

[6] Xu, P., A.S. Mujumdar, and B. Yu, Drying-induced cracks in thin film fabricated from colloidal dispersions. *Drying Technology*, 2009. 27: p. 636–652.

[7] Hawley, W.B. and J. Li, Electrode manufacturing for lithium-ion batteries – analysis of current and next generation processing. *Journal of Energy Storage*, 2019. 25. p: 100862.

[8] Hagiwara, H., W.J. Suszynski, and L.F. Francis, A Raman spectroscopic method to find binder distribution in electrodes during drying. *Journal of Coatings Technology and Research*, 2014. 11(1): p. 11–17.

[9] Jaiser, S., A. Friske, M. Baunach, P. Scharfer, and W. Schabel, Development of a three-stage drying profile based on characteristic drying stages for lithium-ion battery anodes. *Drying Technology*, 2017. 35(10): p. 1266–1275.

[10] Jaiser, S., L. Funk, M. Baunach, P. Scharfer, W. Schabel, Experimental investigation into battery electrode surfaces: the distribution of liquid at the surface and the emptying of pores during drying. *Journal of Colloid and Interface Science*, 2017. 494: p. 22–31.

[11] Jaiser, S., J. Kumberg, J. Klaver, *et al.*, Microstructure formation of lithium-ion battery electrodes during drying – an ex-situ study using cryogenic broad ion beam slope-cutting and scanning electron microscopy (Cryo-BIB-SEM). *Journal of Power Sources*, 2017. 345: p. 97–107.

[12] Jaiser, S., N.S. Salach, M. Baunach, P. Scharfer, and W. Schabel, Impact of drying conditions and wet film properties on adhesion and film solidification of lithium-ion battery anodes. *Drying Technology*, 2017. 35(15): p. 1807–1817.

[13] Li, C.-C. and Y.-W. Wang, Binder distributions in water-based and organic-based $LiCoO_2$ electrode sheets and their effects on cell performance. *Journal of The Electrochemical Society*, 2011. 158(12): p. A1361–A1370.

[14] Pfaffmann, L., S. Jaiser, M. Muller, *et al.*, New method for binder and carbon black detection at nanometer scale in carbon electrodes for lithium ion batteries. *Journal of Power Sources*, 2017. 363: p. 460–469.

[15] Scheepers, F., A. Stahler, M. Stahler, M. Carmo, W. Lehnert, D. Stolten, Steering and in situ monitoring of drying phenomena during film fabrication. *Journal of Coatings Technology and Research*, 2019. 16: p. 1213–1221.

[16] Buss, F., C.C. Roberts, K.S. Crawford, K. Peters, and L.F. Francis, Effect of soluble polymer binder on particle distribution in a drying particulate coating. *Journal of Colloid and Interface Science*, 2011. 359: p. 112–120.

[17] Wood, D.L., J.D. Quass, J. Li, S. Ahmed, D. Ventola, and C. Daniel, Technical and economic analysis of solvent-based lithium-ion electrode drying with water and NMP. *Drying Technology*, 2018. 36(2): p. 234–244.

[18] Du, Z., K.M. Rollag, J. Li, *et al.*, Enabling aqueous processing for crack-free thick electrodes. *Journal of Power Sources*, 2017. 354: p. 200–206.

[19] Westphal, B.G. and A. Kwade, Critical electrode properties and drying conditions causing component segregation in graphitic anodes for lithium-ion batteries. *Journal of Energy Storage*, 2018. 18: p. 509–517.

[20] Armand, M., P. Axmann, D. Bresser, *et al.*, Lithium-ion batteries – current state of the art and anticipated developments. *Journal of Power Sources*, 2020. 479. 228708.

[21] Stein IV, M., A. Mistry, and P.P. Mukherjee, Mechanistic understanding of the role of evaporation in electrode processing. *Journal of The Electrochemical Society*, 2017. 164(7): p. A1616–A1627.

[22] Westphal, B.G., H. Bockholt, T. Gunther, W. Haselrieder, and A. Kwade, Influence of convective drying parameters on electrode performance and physical electrode properties. *ECS Transactions*, 2015. 64(22): p. 57–68.

[23] Baunach, M., S. Jaiser, S. Schmelzle, H. Nirschl, P. Scharfer, and W. Schabel, Delamination behavior of lithium-ion battery anodes: influence of drying temperature during electrode processing. *Drying Technology*, 2016. 34(4): p. 462–473.

[24] Liu, Z., V. Battaglia, and P.P. Mukherjee, Mesoscale elucidation of the influence of mixing sequence in electrode processing. *Langmuir*, 2014. 30: p. 15102–15113.

[25] Lombardo, T., A.C. Ngandjong, A. Belhcen, and A.A. Franco, Carbon-binder migration: a three-dimensional drying model for lithium-ion battery electrodes. Energy Storage *Materials*, 2021. 43: p. 337–347.

[26] Sitarek, K. and J. Stetkiewicz, Assessment of reproductive toxicity and gonadotoxic potential of *N*-Methyl-2-pyrrolidone in male rats. *International Journal of Occupational Medicine and Environmental Health*, 2008. 21(1): p. 73–80.

[27] Sitarek, K., J. Stetkiewicz, and W. Wasowicz, Evaluation of reproductive disorders in female rats exposed to N-Methyl-2-Pyrrolidone. *Birth Defects Research (Part B)*, 2012. 95: p. 195–201.

[28] Chaluvegowda, M., Cell manufacturing – a paradigm shift to meet the demand forecasts. *The Battery Show Conference*, 2019.

[29] Yuan, C., Y. Deng, T. Li, and F. Yang, Manufacturing energy analysis of lithium ion battery pack for electric vehicles. *CIRP Annals*, 2017. 66: p. 53–56.

[30] Totir, D.A., B.D. Cahan, and D.A. Scherson, Electrochemical characterization of lithiated transition metal oxide cathode particles in the absence of carbon, binders and other additives. *Electrochimica Acta*, 1999. 45: p. 161–166.

[31] Clemencon, A., A.T. Appapillai, S. Kumar, and Y. Shao-Horn, Atomic force microscopy studies of surface and dimensional changes in Li_xCoO_2 crystals during litihium de-intercalation. *Electrochimica Acta*, 2007. 52: p. 4572–4580.

[32] Chung, M.D., J.H. Seo, X.C. Zhang, and A.M. Sastry, Implementing realistic geometry and measured diffusion coefficients into single particle electrode modeling based on experiments with single $LiMn_2O_4$ spinel particles. *Journal of the Electrochemical Society*, 2011. 158(4): p. A371–A378.

[33] Hawley, W.B., Z. Du, A.J. Kukay, N.J. Dudney, A.S. Westover, and J. Li, Deconvoluting sources of failure in lithium metal batteries containing NMC and PEO-based electrolytes. *Electrochimica Acta*, 2022. 404, 139579.

[34] Gnanaraj, J.S., Y.S. Cohen, M.D. Levi, and D. Aurbach, The effect of pressure on the electroanalytical response of graphite anodes and $LiCoO_2$ cathodes for Li-ion batteries. *Journal of Electroanalytical Chemistry*, 2001. 516: p. 89–102.

[35] Wang, C.-W., Y.-B. Yi, A.M. Sastry, J. Shim, and K. A. Striebel, Particle compression and conductivity in Li-ion anodes with graphite additives. *Journal of the Electrochemical Society*, 2004. 151: p. A1489–A1498.

[36] Primo, E.N., M. Chouchane, M. Touzin, P. Vazquez, and A.A. Franco, Understanding the calendering processability of $Li(Ni_{0.33}Mn_{0.33}Co_{0.33})O_2$-based cathodes. *Journal of Power Sources*, 2021. 488: 229361.

[37] Diener, A., S. Ivanov, W. Haselrieder, and A. Kwade, Evaluation of deformation behavior and fast elastic recovery of lithium-ion battery cathodes via direct roll gap detection during calendering. *Energy Technology*, 2022. 10: 2101033.

[38] Meyer, C., M. Weyhe, W. Haselrieder, and A. Kwade, Heated calendering of cathodes for lithium-ion batteries with varied carbon black and binder contents. *Energy Technology*, 2020. 8: 1900175.

[39] Faraji Niri, M., G. Apachitei, M. Lain, M. Copley, and J. Marco, The impact of calendering process variables on the impedance and capacity fade of lithium-ion cells: an explainable machine learning approach. *Energy Technology*, 2022. 10: 2200893.

[40] van Bommel, A. and R. Divigalpitiya, Effect of calendering $LiFePO_4$ electrodes. *Journal of the Electrochemical Society*, 2012. 159: p. A1791–A1795.

[41] Oladimeji, C.F., P.L. Moss, and M.H. Weatherspoon, Analyses of the calendering process for performance optimization of Li-ion battery cathode. *Advances in Chemistry*, 2016. 2016.

[42] Schilcher, C., C. Meyer, and A. Kwade, Structural and electrochemical properties of calendered lithium manganese oxide cathodes. *Energy Technology*, 2016. 4: p. 1604–1610.

[43] Zheng, H., G. Liu, X. Song, P. Ridgway, S. Xun, and V.S. Battaglia, Cathode performance as a function of inactive material and void fractions. *Journal of the Electrochemical Society*, 2010. 157: p. A1060–A1066.

[44] Kang, H., C. Lim, T. Li, Y. Fu, *et al.*, Geometric and electrochemical characteristics of $LiNi_{1/3}Mn_{1/3}Co_{1/3}O_2$ electrode with different calendering conditions. *Electrochimica Acta*, 2017. 232: p. 431–438.

[45] Schreiner, D., M. Oguntke, T. Gunther, and G. Reinhart, Modelling of the calendering process of NMC-622 cathodes in battery production analyzing machine/material-process-structure correlations. *Energy Technology*, 2019. 7: 1900840.

[46] Huttner, F., A. Diener, T. Heckmann, *et al.*, Increased moisture uptake of NCM622 cathodes after calendering due to particle breakage. *Journal of the Electrochemical Society*, 2021. 168.

[47] Appiah, W.A., J. Park, S. Song, S. Byun, M.-H. Ryou, and Y.M. Lee, Design optimization of $LiNi_{0.6}Co_{0.2}Mn_{0.2}O_2$/graphite lithium-ion cells based on simulation and experimental data. *Journal of Power Sources*, 2016. 319: p. 147–158.

[48] Meyer, C., H. Bockholt, W. Haselrieder, and A. Kwade, Characterization of the calendering process for compaction of electrodes for lithium-ion batteries. *Journal of Materials Processing Technology*, 2017. 249: p. 172–178.

[49] Sim, R., S. Lee, W. Li, and A. Manthiram, Influence of calendering on the electrochemical performance of $LiNi_{0.9}Mn_{0.05}Al_{0.05}O_2$ cathodes in lithium-ion cells. *ACS Applied Materials & Interfaces*, 2021. 13: p. 42898–42908.

[50] Haselrieder, W., S. Ivanov, D.K. Christen, H. Bockholt, and A. Kwade, Impact of the calendering process on the interfacial structure and the related electrochemical performance of secondary lithium-ion batteries. *ECS Transactions*, 2013. 50: p. 59–70.

[51] Sheng, Y., C.R. Fell, Y.K. Son, B.M. Metz, J. Jiang, and B.C. Church, Effect of calendering on electrode wettability in lithium-ion batteries. *Frontiers in Energy Research*, 2014. 2. p: 1–8.

[52] Davoodabadi, A., J. Li, H. Zhou, D.L. Wood III, T.J. Singler, and C. Jin, Effect of calendering and temperature on electrolyte wetting in lithium-ion battery electrodes. *Journal of Energy Storage*, 2019. 26: 101034.

[53] Hawley, W.B., H.M. Meyer III, and J. Li, Enabling aqueous processing for $LiNi_{0.80}Co_{0.15}Al_{0.05}O_2$ (NCA)-based lithium-ion battery cathodes using polyacrylic acid. *Electrochimica Acta*, 2021. 380. p: 1–54.

[54] Parikh, D., T. Christensen, and J. Li, Correlating the influence of porosity, tortuosity, and mass loading on the energy density of $LiNi_{0.6}Mn_{0.2}Co_{0.2}O_2$ cathodes under extreme fast charging (XFC). *Journal of Power Sources*, 2020. 474: 228601.

[55] Chen, L., X. Xie, J. Xie, K. Wang, and J. Yang, Binder effect on cycling performance of silicon/carbon composite anodes for lithium-ion batteries. *Journal of Applied Electrochemistry*, 2006. 36: p. 1099–1104.

[56] Haselrieder, W., B. Westphal, H. Bockholt, A. Diener, S. Hoft, and A. Kwade, Measuring the coating adhesion strength of electrodes for lithium-ion batteries. *International Journal of Adhesion and Adhesives*, 2015. 60: p. 1–8.

[57] Park, H.-K., B.-S. Kong, and E.-S. Oh, Effect of high adhesive polyvinyl alcohol binder on the anodes of lithium ion batteries. *Electrochemistry Communications*, 2011. 13: p. 1051–1053.

[58] Du, Z., C.J. Janke, J. Li, C. Daniel, and D.L. Wood III, Electron beam curing of composite positive electrode for Li-ion battery. *Journal of the Electrochemical Society*, 2016. 163(13): p. A2776–A2780.

[59] Son, B., M.-H. Ryou, J. Choi, *et al.*, Measurement and analysis of adhesion property of lithium-ion battery electrodes with SAICAS. *ACS Applied Materials & Interfaces*, 2014. 6: p. 526–531.

[60] Kim, K., S. Byun, J. Choi, S. Hong, M.-H. Ryou, and Y.M. Lee, Elucidating the polymeric binder distribution within lithium-ion battery electrodes using SAICAS. *ChemPhysChem*, 2018. 19: p. 1627–1634.

[61] Stich, M., N. Pandey, and A. Bund, Drying and moisture resorption behaviour of various electrode materials and separators for lithium-ion batteries. *Journal of Power Sources*, 2017. 364: p. 84–91.

[62] Huttner, F., W. Haselrieder, and A. Kwade, The influence of different post-drying procedures on remaining water content and physical and electrochemical properties of lithium-ion batteries. *Energy Technology*, 2020. 8: 1900245.

[63] Yang, H., G.V. Zhuang, and P.N. Ross Jr., Thermal stability of $LiPF_6$ salt and Li-ion battery electrolytes containing $LiPF_6$. *Journal of Power Sources*, 2006. 161: p. 573–579.

[64] Edstrom, K., M. Herstedt, and D.P. Abraham, A new look at the solid electrolyte interphase on graphite anodes in Li-ion batteries. *Journal of Power Sources*, 2006. 154: p. 380–384.

[65] An, S.J., J. Li, C. Daniel, D. Mohanty, S. Nagpure, and D.L. Wood III, The state of understanding of the lithium-ion battery graphite solid electrolyte interphase (SEI) and its relationship to formation cycling. *Carbon*, 2016. 105: p. 52–76.

[66] Kraytsberg, A. and Y. Ein-Eli, Conveying advanced Li-ion battery materials into practice: the impact of electrode slurry preparation skills. *Advanced Energy Materials*, 2016. 6, 1600655.

[67] Langklotz, U., M. Schneider, and A. Michaelis, Water uptake of tape-cast cathodes for lithium-ion batteries. *Journal of Ceramic Science and Technology*, 2013. 4: p. 69–76.

[68] Fu, L.J., H.P. Zhang, Y.P. Wu, H.Q. Wu, and R. Holze, Surface active sites: an important factor affecting the sensitivity of carbon anode material towards humidity. *Electrochemical and Solid-State Letters*, 2005. 8: p. A456–A458.

[69] Li, J., C. Daniel, S.J. An, and D. Wood, Evaluation of residual moisture in lithium-ion battery electrodes and its effect on electrode performance. *MRS Advances*, 2016. 1(15): p. 1029–1035.

Chapter 9
Future electrode processing

W. Blake Hawley[1] and Jianlin Li[2]

9.1 Introduction

The current state-of-the-art lithium-ion battery (LIB) electrode manufacturing process has been explained in detail in the preceding chapters. Through these chapters, the state-of-the-art process has been demonstrated to be highly scalable and relatively efficient. However, nearly all steps have aspects that could use improvement, be it with regards to cost, safety, process speed, efficiency, product quality, or a combination thereof.

A recent study of LIB manufacturing revealed that there are five determinants of a particular plant's technical economies of scale: (1) cathode material, (2) electrode thickness, (3) roll-to-roll speed, (4) end-of-line scrap rate, and (5) number of operating days [1]. Improvement of the cathode material would involve increasing the material properties and performance including the specific capacity, the average potential, the cycle life, the high-rate charge and discharge performance, and/or reducing the cost of the material, either in its material cost or production cost. These improvements can be achieved in a number of ways, with these four being among the most relevant: (a) modifying the chemistry of an existing material, (b) modifying the properties of an existing material (e.g., concentration gradient materials, surface coatings), (c) novel material and material synthesis methods, or (d) improving the accessible electrode surface area through advanced processing. Improving the cathode active material through these means will require innovation and optimization in material synthesis, which lies outside the scope of this text. The other four determinants of a plant's economies of scale, on the other hand, are influenced significantly by the electrode manufacturing process. Clearly, with this many key determinants of economies of scale impacted by electrode manufacturing, advancements in electrode processing will drive optimal plant sizing. It is important to also consider that future electrode manufacturing innovations should, ideally, (6) reduce capital expenses (CapEx) and (7) operation expenses (OpEx), as well as (8) maintain or improve worker safety [2].

[1]Sparkz, Inc., USA
[2]Electrification and Energy Infrastructures Division, Oak Ridge National Laboratory, USA

In this chapter, six innovations in LIB electrode manufacturing are considered with respect to these eight criteria. The technical challenges associated with these innovations are also discussed. Ideally, an electrode manufacturing innovation would improve as many of these criteria as possible with minimal detriment to the remaining criteria. The six innovations considered are continuous mixing (Section 9.2), ultraviolet (UV) and electron beam (EB) curing (Section 9.3), multilayer coating (Section 9.4), aligned pore structure processing (Section 9.5), and radiative drying (Section 9.6). Other chapters in this book are dedicated to other LIB electrode manufacturing innovations, including laser structuring [3–9] (Chapter 5), 3D printing [10–12] (Chapter 6), dry processing [13–16] (Chapter 7), and aqueous and alternative solvent processing [17–26] (Chapter 10).

The discussion in this chapter is based on data available from published academic work and patents. Additionally, a broad theme of this chapter is that it is imperative that the testing of electrodes made from these technologies be performed simulating as realistic of conditions as possible. The best assessment of an electrode's performance is in a full cell with realistic areal capacities, porosities, amount of added electrolyte, areal negative-to-positive (N:P) capacity ratio, and cell form factors. While all of these conditions may vary depending on application, areal capacities $\geq 2.0\ \text{mAh}\cdot\text{cm}^{-2}$ and calendered to porosities $\leq 40\%$ are common [2]. The fill factor, which is the ratio between the electrolyte volume and the pore volume in the electrode, is optimized for most industrially-relevant cells at a value around 2.0 [27] while the areal N:P ratio is typically best between 1.0 and 1.2 [28–30]. It is worthy to note that the electrolyte amount varies on cell capacity and chemistry. It is also commonly normalized to capacity (i.e., 3–3.5 $\text{g}\cdot\text{Ah}^{-1}$). Finally, although coin cells are generally the preferred cell format for initial testing of materials due to their low cost and ease of assembly, to advance the technology readiness level (TRL) of a manufacturing process innovation requires testing in more practical cell formats, such as pouch, prismatic, or cylindrical cells. Additionally, testing the product of a new electrode processing innovation is only complete when it is compared with a "baseline" electrode, i.e., one that is prepared with similar specifications using the state-of-the-art process and tested under identical circumstances. Throughout this chapter, many studies will be discussed, though those that adhere most closely to these rigorous testing standards will be most consequential in assessing the prospects of a particular technology.

9.2 Continuous mixing

Industrially, slurry mixing is a batch process, in which a mixing unit disperses slurry components according to a specified order of addition. As components are added, the mixing intensity can change, according to the "sweet spot" of energy input that is sufficient to disperse agglomerates and create a homogeneous distribution of each component. However, the mixing intensity must also not be too high to avoid damaging particles or binder strands or incurring unnecessary energy costs, as the same quality of mixing could be achieved at lower energy input. After several hours of mixing, when the batch is ready, it is fed to the coater or stored until the coater is ready to receive the slurry. The particulars of batch mixing are discussed in detail in Chapter 3.

While batch mixing has been demonstrated to work at scale, it faces many of the issues of other industrial batch processes. The first is the trade-off between sizing of batch mixers and the loss of consistency within the batch. As mixers increase in size, it becomes increasingly difficult to input the power needed to impart necessary work on the slurry to disperse it. On the other hand, while smaller mixers can better mix a batch of slurry at a reasonable power input, they require a greater footprint and additional workers to clean and operate these mixers [2]. Regardless of the size of the mixer, it is impossible to guarantee that all slurry components experience the same amount of force in a batch mixer, meaning there could be inconsistency within the product. Some parts of the slurry may be dispersed well, while others may be over- or under-dispersed, even though the entire slurry was subject to the same mixing conditions. Essentially, the biggest opportunities for improvement in batch mixing are in time, footprint, energy demand, and product consistency of the mixing process.

Continuous mixing offers a solution to the problems associated with batch mixing. In continuous mixing, electrode ingredients are dosed into a twin-screw extruder. The twin-screw extruder concept has been used in other fields, such as the food and pharmaceutical industries, for decades [31]. The binder solution may be dosed in separately after the active materials and conductive additives have been dry-mixed. As with batch mixing, the accuracy of the raw material feeding system is of utmost importance and will impact the slurry properties and quality. The extruder is also outfitted with an atmospheric degassing station to remove air bubbles from the slurry during processing.

The twin-screw extruder is designed with screws made up of different types of elements along the shaft. One kind of element imparts relatively little shear while another imparts a high amount of shear. The low-shear screws are designed like traditional screws with a helix pattern, while the high-shear screws have flat edges and are called kneading discs [2]. The screws rotate in the same direction, simultaneously pushing the slurry down the extruder barrel and mixing the slurry components. Because the screws are composed of different elements, the choice of screw elements becomes a process variable of interest, much like the sizing of the impeller and design of the scrapers in a batch planetary mixer. The process may be optimized by designing screws comprised of the right elements strategically placed down the length of the extruder barrel.

How does continuous mixing compare to batch mixing across the four areas of improvement identified earlier for batch mixing? Those areas were: (1) process time, (2) equipment footprint, (3) energy demand, and (4) product consistency. The exact time and space savings of the continuous mixing process compared to the batch mixing depend on many factors, such as the throughput of the facility and the design of the mixers themselves. However, studies and continuous mixing line producers suggest that the mixing time can be reduced by a factor of 4–6 and the footprint can be reduced by a factor of 2–4 for a continuous mixing line producing the same amount of slurry per hour as a batch mixing line [32,33]. The energy cost savings could be as high as 60% when switching from batch mixing to continuous mixing. These savings come from a mixer that can produce the same volume of slurry in a shorter amount of time and because of a significantly lower mixer footprint resulting in less need for dry room space [33]. Additionally, twin-screw extruders have been reported to require significantly less solvent than a batch mixing system to achieve desirable slurry properties and electrochemical performance in the finished electrode [32]. Using less solvent in the continuous mixing process would lead to a smaller dryer footprint, quicker drying process, and lower drying energy than the batch mixing process.

Finally, all slurry mixed in a twin-screw extruder experiences the same amount of shear at the same time in the mixing process, which is not necessarily the case for batch mixing [2]. If there is a disturbance in the continuous slurry mixing process, the raw material feeding system can be suspended and troubleshooting or maintenance can begin. Since the barrel of the continuous mixing extruder contains a significantly lower volume of slurry than a large planetary mixer in a batch process does, the slurry waste is minimal. Meanwhile, if a problem occurs in a batch mixer, the volume of slurry in the mixer will likely all be wasted. Because the volume of the batch mixer is usually larger than the continuous mixer, cleaning the batch mixer will also take a longer amount of time. Therefore, not only is the product likely to be more consistent in a continuous mixer but it is also less detrimental in terms of slurry waste if there is an issue.

Though the advantages could be significant, implementation of continuous mixing faces two consequential hurdles. The first is that the continuous mixing process is constantly providing slurry to the coating/drying line, which will require the mixing process to be balanced with these processes [2]. In other words, the velocity of slurry being provided to the coater must be perfectly tuned to the velocity of the current collector and the shear experienced by the slurry during coating to achieve a coating with appropriate thickness. Already in the state-of-the-art electrode production process, this coating speed must be calibrated to match the dryer length and energy, such that the dry electrode is not too wet or too dry. Adding the mixing process to this delicate balancing act is very challenging. While the continuously mixed slurry could be re-routed to a storage tank to overcome this requirement, these storage tanks would require additional floorspace and would lead to variation in slurry consistency, which would mostly defeat the advantages of continuous mixing. Another consequence of this balancing requirement is that if a disturbance is experienced in the mixing process, the coating and drying lines associated with that mixer must also be shut down. In the batch mixing process, one mixer being offline may slightly slow down production, but will not lead to a complete stop as with continuous mixing since other batch mixers can feed slurry to the coating and drying line.

The other issue with the scale-up of continuous mixing is that the hardware of continuous mixing is a serious departure from that of batch mixing. The collective industrial know-how of planetary mixers and the batch mixing process has taken decades to compile and is still in a state of continuous improvement, as are all industrial processes. To essentially "start over" with a continuous mixer would entail a dramatic learning curve and significant engineering costs. This highlights the "drop-in solution" conflict that is a major theme of this chapter. A manufacturing innovation that requires less modification to the existing process will always be favorable, as it will require less time to scale up at the R&D and pilot scale levels and to integrate into existing manufacturing line designs. Still, even the most ready-made, drop-in solutions require years to be brought up from the R&D level into a manufacturing line. Continuous mixing with a twin-screw extruder is not a total departure from the current state-of-the-art process but is different enough to necessitate serious engineering effort to realize at scale.

9.3 Ultraviolet and electron beam curing

As discussed in Chapter 3, what is considered to be the "best" composition of an electrode slurry is often governed by the rheological properties of the slurry. The composition refers to the volume ratio between the slurry ingredients, as well as the overall solid loading of the slurry. All else being equal, a slurry with high solid loading would be preferable to a slurry with low solid loading since the drying time and energy would be reduced. Despite being present in low quantities (2–5 wt% of the final electrode composition), the binder has significant influence on the rheological properties of the slurry. The property of the binder that has the most influence on rheology is its molecular weight (MW). As the MW of the binder increases, so does its viscosity. Therefore, more solvent is necessary to disperse a high-MW binder than a low-MW binder. Polyvinylidene fluoride (PVDF) for LIBs is often present in roughly an 8–10 wt% solution in *N*-methyl-2-pyrrolidone (NMP). For water-soluble binders, such as sodium carboxymethyl cellulose (CMC), the solid loading is even lower, often around 2–5 wt%. These low-solid fraction binder solutions constrain the maximum achievable solid loading of a slurry since a significant amount of solvent is introduced just in the binder solution.

Therefore, a higher solid loading could be achieved if a low-MW binder were used. However, longer binder strands are useful in achieving long-range connectivity between particles and the current collector, promoting better cohesion and adhesion. By using curable oligomers as binder, a high solid loading slurry can be processed and an electrode with good mechanical properties can be achieved by crosslinking shorter oligomer strands into longer polymer strands. This crosslinking process can occur in <1 s in a roll-to-roll, continuous process, making it an excellent candidate for scale-up [2].

Two techniques of curing oligomers for electrode binders have been considered: (1) UV curing and (2) EB curing. Both techniques have the advantage of being able to cure water-soluble oligomers; the advantages of water-processed electrodes are elaborated in Chapter 10 but can be summarized as promoting worker safety and costing significantly less in CapEx and OpEx. UV curing has a lower cost than EB curing, although it requires a photoinitiator to be included in the slurry and has a low penetration depth. Because UV rays do not have great penetration depth, they cannot cure oligomers close to the current collector in thick electrodes. Conversely, EB curing does not require a photoinitiator and has a high penetration depth. Since thick electrodes are increasingly required for energy-dense applications, EB curing is the preferable route. It is worth mentioning that UV treatment could have benefits in LIB production, such as increasing the oxygen content of graphite anodes and therefore improving their electrolyte wettability [34,35]. Advanced LIB separators have also been fabricated at the research and development (R&D) level using UV treatment and curing techniques [36,37]. Solid-state batteries with UV-cured polymer electrolytes have also been demonstrated [38,39]. The interested reader is directed to the references cited here, as these topics lie outside the scope of the present discussion.

In addition to not requiring photoinitiators and having high penetration depths due to their high energy, EB curing boasts incredibly fast line speeds, up to 300 $m \cdot min^{-1}$ [2]. For comparison, conventional electrodes are coated at $\sim$30 $m \cdot min^{-1}$ with physical vaporization of solvent, a whole order of magnitude slower than the EB curing process [2]. A cartoon showing the curing mechanism of EB curing is shown in Figure 9.1.

Though a promising technology, there are technical challenges to implementing EB curing. The first is that a curable binder must be identified that can deliver performance similar to or better than PVDF. Acrylated polyurethane resins (APRs) have been studied and have been found to be stable between 2.0 and 4.6 V according to cyclic voltammetry results [40]. This wide window of electrochemical stability makes the oligomer compatible with the most common cathode active materials, such as $LiFePO_4$ (LFP), $LiNi_xMn_yCo_{1-x-y}O_2$, and $LiNi_xCo_yAl_{1-x-y}O_2$, and the current state-of-the-art electrolyte formulations.

For a given electron beam energy and dry electrode thickness (or mass loading), the amount of crosslinking of the APRs can be studied using Fourier-transform infrared spectroscopy (FT-IR) (see Figure 9.2). For this oligomer, the 810 cm^{-1} peak is of interest, representing the C=C bond of the acrylate group. The intensity of this peak will increase as the amount of crosslinking in the polymer

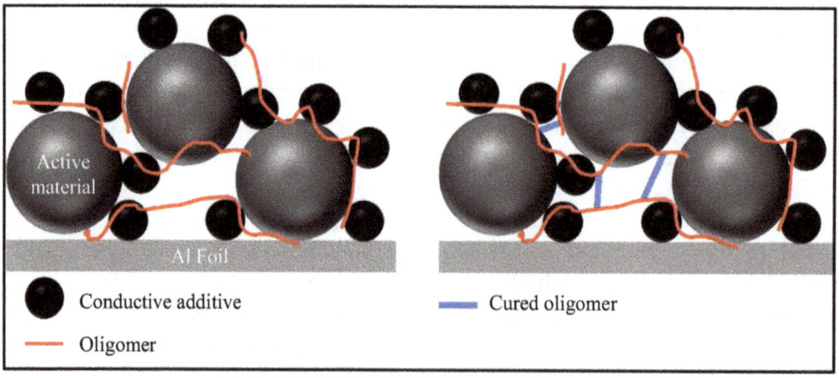

Figure 9.1 Cartoon image of EB cross-linked, cured oligomers in LIB electrodes

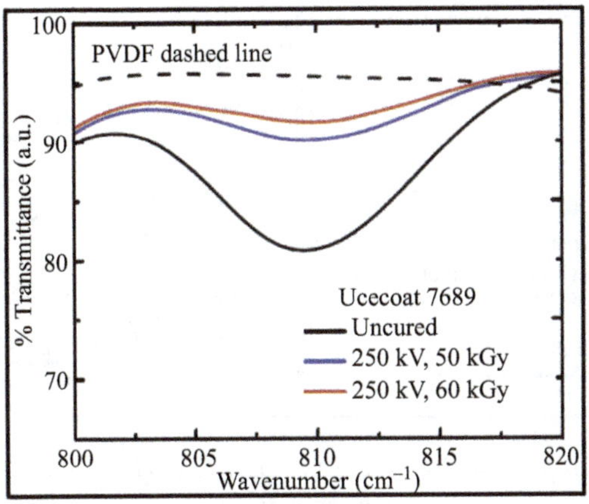

Figure 9.2 FT-IR spectra showing the degree of crosslinking as a function of radiation dose for a LIB cathode containing APRs. This cathode was coated at 150 m·min^{-1} using an EB pilot line. Reprinted from Ref. [41].

increases, which will scale with the intensity of the radiative dose. It has been found that, eventually, the intensity of the peak does not increase with radiative dose, indicating that the sample has been crosslinked as much as possible. At this point, increasing the dose beyond the point at which further crosslinking is achieved only leads to unnecessary additional process cost. For a $LiNi_{0.5}Mn_{0.3}Co_{0.2}O_2$ (NMC532) cathode with a mass loading of 12.5 mg·cm^{-2}, the radiative dose necessary to completely crosslink the sample was found to be 30 kGy at an electron beam energy of 200 keV [40].

Lab-scale testing of NMC532 cathodes with fully cured APRs shows nearly identical traits as an analogous cathode made using the state-of-the-art process. These cathodes, with a mass loading of 12.5 mg·cm^{-2}, have good mechanical properties and electrochemical performance, as measured by room-temperature rate capability (up to 3C discharge) and cycle life (100 cycles at +0.33C/−0.33C) testing between 2.5 and 4.2 V in full cells [40]. A later study from the same researchers doubled the loading of the NMC532 cathode (25 mg·cm^{-2}), which required an increase in the electron beam energy from 200 keV to 275 keV. The EB curing process was performed on a pilot-scale line operated at 150 m·min^{-1}. These electrodes achieved similar, although lower, rate capability when compared with the NMP-processed baseline [41] (see Figure 9.3). The cell with the EB-cured cathode polarized at a higher rate early in cycling than the baseline cell, as evident by the more significant capacity loss in the initial 100 cycles, though the capacity fade from cycles 101 to 1,000 was similar between the two cells (see Figure 9.3) [41].

A direction of future research for EB curing could be on the selection of the oligomer. APRs are certainly a good choice, given the electrochemical stability

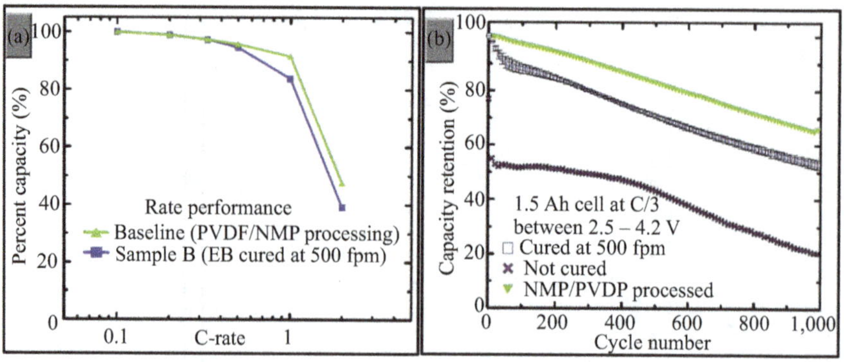

Figure 9.3 (a) Rate capability and (b) cycle life performance of a cell featuring an EB-cured LIB electrode (blue squares) compared to a cell featuring a cathode processed in NMP (green triangles). Testing occurred in 1.5 Ah pouch cells using NMC532 cathodes and graphite anodes. The data from a cell featuring a cathode with uncured APR binders is shown in the cycle life plot (purple Xs). Reprinted from Ref. [41].

mentioned earlier, and their high reactivity when subject to the EB line. The curing reaction with APRs is a radical polymerization that proceeds according to (9.1a) and (9.1b):

$$R^{\cdot} + R'CH = CH_2 \rightarrow R'C^{\cdot}H - CH_2 - R \tag{9.1a}$$

$$R'C^{\cdot}H - CH_2{-}R + R'CH = CH_2 \rightarrow R'C^{\cdot}H - CH_2 - R'CH - CH_2 - R \tag{9.1b}$$

This polymerization reaction can be disrupted by atmospheric oxygen as oxygen from the air reacts quickly with the carbon-centered radical to generate slow-reacting peroxyl radicals. The side reaction of the free radical with oxygen is shown in (9.2). Therefore, EB curing with APRs as binder must be conducted in an inert atmosphere, which is usually accomplished by purging the reaction chamber with high-purity nitrogen.

$$R^{\cdot} + O_2 \rightarrow R - O - O^{\cdot} \tag{9.2}$$

Because of APR's reactivity with oxygen, and because the electrochemical performance of the EB-cured cathodes in the studies discussed here could be improved, further work on EB-curable oligomers would be beneficial. One suggestion for future research could be the addition of thiols and amines to the chemistry of the oligomer since they are unaffected by atmospheric oxygen [41–44]. Another would be to consider a different reaction scheme entirely, such as EB curable cationic resins [41,45–47].

9.4 Multilayer coatings

In the state-of-the-art electrode manufacturing process, the entire thickness of the electrode is coated, dried, and calendered continuously to create a uniform electrode. However, if the electrode can be designed into multiple layers with specific characteristics to improve performance, the overall electrode performance could be improved. Depending on the specific design, the multilayers can be coated, dried, and calendered one by one, simultaneously, or a combination of both. First, we will walk through the production of a theoretical multilayer electrode and how its gradients impact performance; then, we will explain how multilayer coatings can be achieved in practice and discuss experimental results of tailored electrodes.

9.4.1 Slot-die coating the theoretical multilayer electrode

Consider a theoretical electrode cut into x layers, with x being constrained by the active material particle diameter and the minimum wet thickness that can be achieved by a slot-die coater, each with a unique composition. The layers are numbered such that Layer 1 is the interface between the current collector and Layer 2, and Layer x is the interface between Layer $x-1$ and the separator. This theoretical electrode is to be structured such that it is capable of greater energy density and rate capability than a single-layer electrode of identical total thickness and total active mass loading.

To achieve such high performance with this theoretical electrode, an example is Layer 1 should contain the greatest mass fraction of binder among all layers since the binder serves to provide adhesion to the current collector and cohesion between the electrode components. Layer 2 should contain the second greatest mass fraction of binder, so that binder strands in Layer 2 can connect with binder strands in Layer 1 and provide sufficient cohesion between the active material and conductive additive particles within the layer. The binder fraction should continually decrease until Layer x, where the least binder is used, only needing to connect with a relatively few number of binder strands in Layer $x-1$ and providing cohesion between the active material and conductive additive particles within the layer.

There are additional reasons to create such a binder gradient. For instance, to maintain sufficient adhesion strength, a certain content of binder is needed for the interface between the coating and current collector whereas the binder content could be reduced for other layers. This means the total binder content in a uniform electrode could be reduced with multilayer design. Additionally, most binders are electronically insulating and block ion pathways within the electrode. Having a high mass fraction of binder within Layer x would reduce the rate capability of the electrode, since the ionic resistance is great immediately after diffusing through the separator. This binder gradient would also be preferable during the drying process, as binder migration (see Chapter 8) could be alleviated within such thin layers of electrode. Since the binder content is the highest in Layer 1, the active material and conductive additive fraction will be the lowest in Layer 1 and the greatest in Layer x.

In addition to the binder gradient, a porosity (or density) gradient would also be possible. According to the Bruggeman relationship, the porosity of an electrode is inversely proportional to the tortuosity of the electrode. The tortuosity of an LIB electrode refers to the average distance travelled by a Li-ion from the electrode/separator interface to the current collector/electrode interface divided by the thickness of the electrode. Tortuosity, then, is a general measure of the "ease" with which an ion can diffuse through an electrode (or an individual layer). The proportionality constant within the Bruggeman relationship is a function of electrode microstructure, which is random for wet-processed electrodes, and is unimportant for the purpose of this exercise.

With this in mind, Layer x is theoretically the least tortuous layer since it contains the least ion-blocking binder assuming same porosity and pore structure within all layers. Layer $x-1$ is the next least tortuous, then Layer $x-2$, and so on until Layer 1 is reached, which is the most tortuous electrode. Therefore, to keep the tortuosity of Layer x low, it should be the most porous of the layers. Layer 1, then, should be the least porous and most tortuous. Although ion diffusion is the most difficult in Layer 1, it is also very energy-dense despite the lowest active material fraction of all layers since it will contain relatively little void space. Packing the active materials close together in Layer 1 is necessary, since there are no layers beneath Layer 1 that are blocked by it. It is also necessary to have dense electrode layers, otherwise the electrode will be far thicker and ultimately not competitive in energy density with calendered, dense, single-layer electrodes. In practice, this porosity gradient would be achieved by calendering Layer 1 with the greatest force and subsequent layers with diminishing force, leaving them more porous than the layer below which, until Layer x is calendered with the least force, or perhaps not at all.

The third gradient in the electrode to consider is an active material particle size gradient. Layer x, which is optimized for high-rate charge and discharge, would be best suited with small active material particles. The size of the particles decreases until Layer 1 is reached, which contains the largest particles. Small particles are preferred for high rate capability since the solid-state diffusion of the Li-ion through the particle is shorter than for large particles and the interface area between particles and electrolyte is larger for small particles to facilitate charge transfer [48]. However, the larger surface area of small particles could facilitate more side reactions with electrolyte and compromise cycle life [49].

In summary, the theoretical multilayer electrode is crafted with (see Figure 9.4):

1. A binder gradient in which Layer 1 has the *greatest* binder content, with binder content *decreasing* until Layer x, which has the *lowest* binder content.
2. A porosity gradient in which Layer 1 has the *lowest* porosity, with porosity *increasing* until Layer x, which has the *greatest* porosity.
3. An active material particle size gradient in which Layer 1 has the *largest* particles, with active material particle size *decreasing* until Layer x, which has the *smallest* particles.

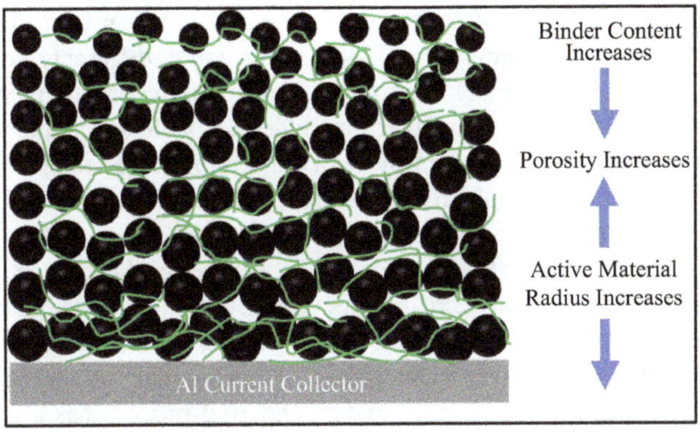

Figure 9.4 Cartoon illustration of the cross-section of a theoretical multilayer electrode showing the binder content, porosity, and active material particle size gradients. Active material particles are represented by black spheres, binder strands are represented by green freeform lines, and the Al current collector is represented by a gray rectangle. Conductive additive particles are omitted from the figure for simplicity.

Other gradients and/or materials are also possible. For instance, varying the conductive additive by including carbon nanofibers or nanotubes could bolster the electronic conductivity of the dense bottom layers. Using binders with varying MWs could replace or smooth out the binder content gradient. Different active material particle morphologies, such as single-crystal, concentration-gradient, or particle shapes could be used in conjunction with the size gradient. It has been suggested that reversing the active material particle size gradient described earlier, such that the small particles are closest to the current collector and the large particles are closest to the separator, could be a better approach. In this scenario, the sluggish solid-state diffusion through the large particles is balanced by placing these particles closer to the separator, reducing the diffusion distance of the Li-ions through the electrolyte [50].

9.4.2 Multilayer coatings in practice

The theoretical multilayer electrode is constrained by at least five practical considerations that must be mentioned. First, as mentioned in the previous section, the thickness of a layer is limited by the minimum wet thickness, as determined by the setup of the slot-die coater (see Chapter 4). Second, the theoretical multilayer electrode requires either x number of slurry mixers, coaters, dryers, and calendering rolls or a single slurry mixer, coater, dryer, and calendering roll to be used x times to produce one coating. In the state-of-the-art case, only one set of equipment is needed to be used exactly one time to produce an electrode roll of a given length, meaning the amount of equipment (in the former case) and time (in both cases) to produce a coating of the same length increases as a function of x. Third, as the gradients become more complicated, such as utilizing different particle sizes, shapes, and morphologies, material synthesis costs and/or procurement costs are likely to increase. Fourth, as more layers are added to an electrode, the possibilities of layer compositions to be studied by R&D and pilot-scale teams increase substantially, adding to the complexity of finding the best electrode morphology and delaying the product's arrival to market.

Finally, with each layer that is added to an electrode, the chance of the electrode needing to be scrapped increases, assuming the quality assurance tools are equally reliable in the single-layer and multilayer case (i.e., the scrap rate is the same for each individual layer independent of the number of electrode layers). Assume one mixing, coating, drying, and calendering process has a scrap rate of 10%, meaning that 10% of the time a coating is ruined during one of these four stages and 90% of the time a coating is successful after these four stages. This means a single-layer coating has a 90% chance of being built into a cell, a dual-layer coating has an 81% chance of being built into a cell, and a 10-layer coating has just a 35% chance of being built into a cell. These gaps in scrap rate as a function of layer number are significantly smaller if the scrap rate is lower; conversely, they are wider when the scrap rate is higher (see Table 9.1), meaning the feasibility of multilayer coating is somewhat a function of improved quality assurance tools and lower scrap rates. These probabilities are calculated according to the simple formula shown in (9.3).

$$(1 - \text{scraprate})^{\text{number of layers}} \tag{9.3}$$

Given these constraints, many empirical multilayer studies consider the dual-layer case [51–53]. Other empirical studies use a spray printing technique [54–56] in which two suspensions are mixed together at different rates, depending on the desired composition of an individual layer. As the suspensions are applied to the current collector, the flow rate of each suspension changes, therefore changing the overall composition of the layer of electrode being applied to the current collector. This technique is useful in evaluating the benefits of multilayer electrodes, albeit in small

Table 9.1 Probability of an electrode being scrapped as a function of its layer count and the scrap rate for a single layer

	Probability of electrode being built into a cell			
Number of layers	**20% Scrap rate**	**10% Scrap rate**	**5% Scrap rate**	**2% Scrap rate**
1	80%	90%	95%	98%
2	64%	81%	90%	96%
3	51%	73%	86%	94%
5	33%	59%	77%	90%
10	11%	35%	60%	82%

area samples, though the former dual-layer strategy has a higher TRL for use in practical applications.

In one study, a simultaneous dual-layer coating was achieved by using a custom-made slot-die with two exits [51]. The thickness of each layer was controlled using a shim to set the gap of each exit. This simultaneous multilayer coating strategy has the advantage over the consecutive coating approach in that only one coating and drying process are needed. It is possible a slot-die could be machined with more than two exits, allowing for additional layers to be coated. The simultaneous multilayer coating process increases the number of coatings made per unit time while reducing the number of coating and drying units needed for a single coating. However, there could be issues with interlayer mixing during the simultaneous multilayer coating process and the density (i.e., porosity) of each layer cannot be controlled since the entire coating will be calendered at once. In essence, the simultaneously coated multilayer electrode would be a cheaper and higher-throughput endeavor, though studies are needed to determine whether the simultaneous or consecutive multilayer coating approach leads to a better final product. There likely exists a trade-off between the cheaper, faster simultaneous multilayer coating process and the potentially higher-performing consecutive multilayer coating process, though understanding the extent of this trade-off is vital to determine the more commercially viable process.

In the simultaneous dual-layer coating study, the subject of investigation was how the content of styrene butadiene rubber (SBR) binder in each layer of an aqueous-processed graphite anode impacted the performance of the electrode. The amount of SBR varied from 0 to 7.5 wt%, with SBR mass being exchanged for graphite, while the amount of conductive additive and CMC binder remained constant across all layers [51]. The highest-performing anode contained no SBR in the top, separator-interfacing layer and 7.5 wt% SBR in the bottom, current collector-interfacing layer [51]. It is critical to note that the top layer did still contain a small amount of CMC binder, which provides vital dispersing function in the slurry and acts as a rigid binder. The highest-performing anode also exhibited the greatest amount of adhesive force, as measured by peel testing [51], which was likely the outcome of the high SBR content in the bottom layer. This outcome suggests that there is still some layer separation in the simultaneous multilayer coating process. As the amount of SBR in the bottom layer decreased and the amount in the top layer increased, the performance of the electrode suffered [51].

Studies on consecutive multilayer coatings have varied the electrode parameters outlined in Section 9.4.1, namely the binder content, porosity, and active material particle size of each layer. Generally, the cathode performs best when smaller particles are present in the top layer and larger particles are present in the bottom layer, as described in Section 9.4.1 [52,53]. In fact, the exact opposite configuration has been found to be worst for the cathode, likely because of larger particles in the top layer creating a tortuous layer through which Li-ions must travel to reach the smaller particles in the bottom layer [53]. The cathode also performs optimally when the bottom layer is calendered to a relatively high density (~30% porosity) and the top layer is only lightly calendered or remains as-coated (~48% porosity) [52].

Studies have also found that only optimizing the cathode leads to a modest performance gain in a full cell; to truly realize the potential of dual-layer electrodes, both the cathode and anode must be optimized. Interestingly, it was found that large active material particles in the top layer and small active material particles in the bottom layer of the anode achieved remarkably high discharge capacity at high current density (2C, 25 mg·cm^{-2} of $LiNi_{0.8}Mn_{0.1}Co_{0.1}O_2$, or NMC811, total in the cathode and 15 mg·cm^{-2} of graphite total in the anode) [53] (see Figure 9.5). The conclusion of the study was that the exact layering of the anode had less bearing on the result than whether the cathode was layered in the manner described [53]. More work is needed to determine the best combination of cathode and anode, which will certainly depend on the active material selected and the specific application of the cell. For instance, the cathode layering may shift slightly when the active material selected is more or less conductive than the example material used in the studies cited here (NMC811). Additionally, if the cell is intended as an energy-dense or power-dense cell will also influence the precise layering of the electrodes.

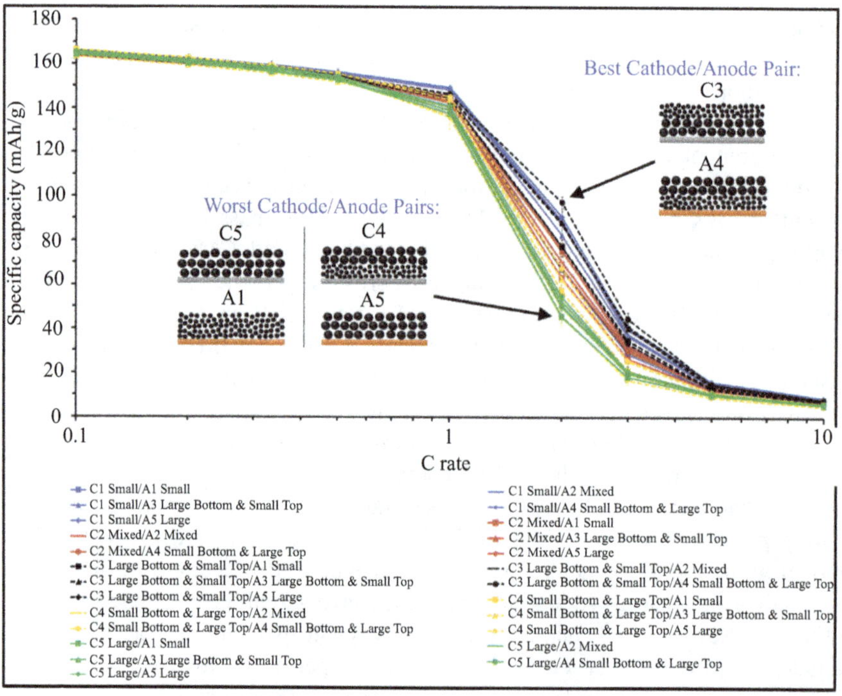

Figure 9.5 Rate capability of 25 cathode/anode pairs. Each electrode could feature one of five particle size distributions: (1) all small particles, (2) all large particles, (3) small particles in the top layer and large particles in the bottom layer, (4) large particles in the top layer and small particles in the bottom layer, or (5) all mixed particle sizes. Reprinted with permission from [53].

Despite the differences in experimental design, very similar findings were observed when the previously described spray deposition technique was used to fabricate multilayer electrodes [54–57]. This was consistently found to be due to a better spatial distribution of overpotential in the thick electrodes, which prolonged the life of the cells over the single-layer baseline by a significant margin.

9.5 Aligned pore structure processing

Just like multilayer electrodes are designed to reduce tortuosity by optimizing layer properties, more advanced manufacturing techniques have been designed to create straight channels down the thickness of the electrode. According to the definition of tortuosity in Section 9.4.1, an electrode with perfectly straight pores of sufficiently wide diameter would have a tortuosity of 1, which would be extraordinarily beneficial from a rate capability perspective, especially as the thickness of the electrode increases. Some methods of creating an aligned pore structure include laser structuring [3–9] and additive manufacturing [10–12], though these techniques are covered in detail in Chapters 5 and 6, respectively. Other possibilities include salt-templating [58–61] and wood-templating [61–63], which are discussed in the referenced papers. Two other techniques, freeze-tape casting and magnetic alignment, are discussed in the subsequent subsections.

9.5.1 Freeze-tape casting

Freeze-tape casting is a modified version of freeze casting, or ice templating, which has been used for decades to create ceramics with particular pore structures. In freeze casting, a slurry is deposited into a mold and then cooled to below the freezing point of the slurry. Once frozen, a vacuum is applied to the mold, which sublimes the frozen solvent from the material. The phase diagram of the solvent dictates the necessary temperature and pressure needed to convert the solvent from the liquid phase to the solid phase in the freezing step and from the solid phase directly to the vapor phase in the sublimation step. Therefore, the properties of the solvent will determine the ease (or difficulty) of freeze casting a particular material. The freezing and sublimation conditions for water are relatively easy to achieve compared to other organic solvents, making aqueous-based slurries among the most readily used with freeze casting. How different parameters, including freezing temperature, vacuum pressure, solvent selection, solid loading of the slurry, and the presence of additives, influence the resulting ceramic are covered elsewhere [64]. Freeze casting has been used in the past to make cathodes from $LiNi_{0.8}Co_{0.15}Al_{0.05}O_2$ [65], LFP [66], $LiCoO_2$ (LCO) [67], and anodes from graphite [68].

Freeze-tape casting works very similarly to freeze casting, except that instead of being applied to a static mold, the slurry is coated onto a moving substrate, as is done in tape casting (see Figure 9.6). Once applied to the current collector, the coating is rolled onto a freeze bed. Heat diffuses from the room temperature slurry through the substrate and onto the freeze bed, reducing the temperature of the slurry. The speed with which the slurry freezes depends on the thickness and

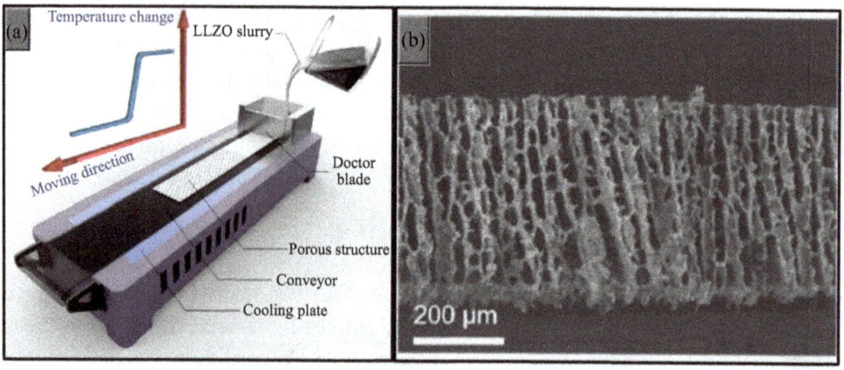

Figure 9.6 (a) Cartoon of freeze-tape casting for a $Li_7La_3Zr_2O_{12}$ (LLZO) solid electrolyte. The mechanism of freeze-tape casting is the same when used for a LIB electrode. Reprinted with permission from Ref. [69]. (b) Scanning electron microscope image of the vertically aligned pore structure formed from freeze-tape casting. Reprinted with permission from Ref. [70].

thermal properties of the substrate, the temperature of the freeze bed, the solvent properties, and the solid loading of the slurry. Ideally, as the front edge of the coating is frozen, the frozen solvent will grow in a semi-straight micelle from the substrate to the top of the coating. Once these micelles have grown, they push the solids in the slurry into a valley between the first and second row of ice. Once the whole coating has been passed onto the freeze bed, vacuum is drawn on the freeze bed and the solvent is removed via sublimation, just as with freeze casting.

Freeze-tape casting has been used since at least 2007 in a variety of applications [71,72]. The technique had not been adopted in LIBs for several years but was adopted relatively early in solid oxide fuel cells [73–79]. More recently, freeze-tape casting has been used to make components for LIBs and more advanced batteries, including $Li_4Ti_5O_{12}$ (LTO) anodes [80], graphite anodes [81], sulfur electrodes [82], and $Li_7La_3Zr_2O_{12}$ (LLZO) solid electrolytes [69,70,83,84].

While both freeze-tape cast anode studies used water as solvent for the slurry and the same water-soluble binder system (SBR and CMC), most of the other conditions are vastly different. For instance, the freezing temperatures vary significantly (−130 to −170°C for the LTO study, −30°C for the graphite study). The graphite study uses a bilayer approach, where the bottom layer is a 40 μm, conventionally coated (i.e., processed in NMP) layer and the top layer is a 200–300 μm, freeze-tape cast layer. This promoted better adhesion in the electrode and led to a 20% increase in charge capacity at 5C rate and a 17% increase in capacity in a full cell after 1,000 cycles at +5C/−0.33C current densities when compared with conventionally-processed graphite anodes [81]. The LTO study found that freeze-tape casting led to full utilization of LTO active materials, unlike in the conventionally-processed LTO anodes, even at low current densities [80].

In terms of commercial viability, freeze-tape casting is the preferred method to traditional freeze casting since it more closely mirrors the existing state-of-the-art roll-to-roll process and is effective at creating vertically aligned pores of near-unit tortuosity, as images from the previously cited studies show. In other words, freeze casting is a batch process, while freeze-tape casting could be conducted in a more continuous fashion. However, there are many technical and scalability hurdles that must be overcome for freeze-tape casting to achieve a high TRL. The first issue is that the porosity of freeze-tape cast electrodes is significantly higher than for tape cast electrodes (70–80% to approximately 50%, respectively). This high porosity is undesirable since the volumetric energy density is reduced significantly at high electrode porosity and because the mechanical properties of such a porous film are generally weaker [81]. While tape cast electrodes can be calendered to reduce their porosity, it is unclear to what extent the vertically aligned porosity would be damaged by the calendering process. The film could also be densified by sintering the electrode. In this scenario, the freeze-tape cast electrode would be coated onto a plastic film as opposed to the metal foil, then peeled off the substrate and sintered at high temperature in an atmosphere appropriate for the selected active material. This technique has proved successful in freeze-tape cast LLZO solid electrolytes [69].

The second hurdle for freeze-tape casting viability is the process speed. At present, studies on freeze-tape casting of LIB electrodes have occurred on small, benchtop units. The line speed of these units is slower than the state-of-the-art line speed [80] and, at any size unit, is constrained by how quickly the solvent can freeze and ice crystals can grow. While the freezing rate could be increased by using colder temperatures, the utility needed to reach these cold temperatures (such as $-150°C$) would add cost [80]. Additionally, the subsequent sublimation and (possible) sintering steps add lengthy steps to the process. Finally, significantly more work is required on freeze-tape casting slurry optimization. While there are generally understood trends on how solid content, solvent selection, freeze bed temperature, and line speed impact the morphology of the final product, more work is required to create a high-functioning, reproducible product.

9.5.2 Magnetic alignment

Magnetic alignment of LIB electrodes was first demonstrated in two publications in 2016 under two similar but different embodiments [85,86] (see Figure 9.7). In the first embodiment, the graphite particles for an anode were coated with superparamagnetic Fe_3O_4 nanoparticles (SPIONs) to make the graphite particles receptive to a relatively weak (100 mT) magnetic field during drying. When the field was applied, the particles became aligned vertically. As the solvent was dried, the gaps between the aligned particles became near-unit tortuosity channels [85]. In the second embodiment, LCO cathode active material particles were mixed with magnetic microrods or emulsion droplets. When the magnetic field was applied, the magnetically receptive additives and LCO particles became vertically aligned during drying, as with the first magnetic alignment embodiment. The cathodes must be sintered in the microrod case in order to pyrolyze the microrods and form the vertically aligned pore structure, though is optional in the emulsion droplet case [86].

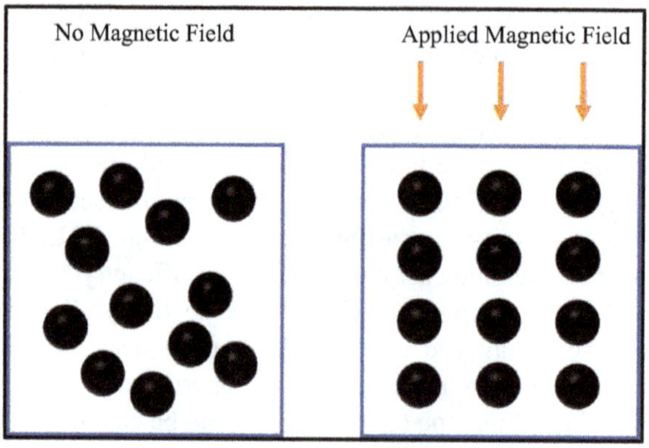

Figure 9.7 Cartoon representations of the alignment of magnetically receptive particles in solution due to magnetic field application

At high active material loading (9.1 mg·cm^{-2}), the SPION-coated graphite anode achieved a capacity of 100 mAh·g^{-1} after 50 cycles at +1C/−1C and room temperature cycling conditions in a half-cell. The capacity increased in the anode between the 10th and 50th cycles, possibly due to an insufficiently formed solid electrolyte interphase and/or a stabilizing ion transport pathway. Thus was a significant improvement over the baseline anode, which delivered just 65 mAh·g^{-1} after 50 cycles under identical conditions [85]. As for the second embodiment, 40% porous, 310 μm thick LCO cathodes can deliver an impressive 7–8 mAh·cm^{-2} at 1C current density, roughly double what the baseline cathodes achieved. These electrodes were tested in half-cells at room temperature. This performance was achieved with both the microrod and emulsion method [86].

Direct comparison of the two embodiments based on the results of the studies is challenging, since they were tested under very different conditions (active material, electrode thicknesses, electrolyte compositions, etc.) In the first embodiment, while only 0.3 wt% of SPIONs (relative to the graphite mass) are needed to functionalize the graphite, the SPION coating process has only been demonstrated at very small volumes of solution (100 mL). How the costs and time needed for this process will scale is an open question. In the second embodiment, sintering for the microrod approach is a nearly day-long process that requires very high temperatures (~1,000°C), which will lead to very high costs. For this reason, the microemulsion process appears to be the most scalable approach while still yielding high-performance electrodes. A later study from the same research group demonstrated high performance anodes made using mesocarbon microbeads made from the emulsion-based magnetic alignment process [87]. Of course, all embodiments of magnetic alignment will need to contend with the cost and logistical challenges of integrating a magnetic field into the roll-to-roll drying process to upgrade the process from a batch, lab-scale process to a continuous, industrial-scale process.

The next step in the development of magnetic alignment would be the evaluation of a full cell in which cathode and anode are both made via emulsion-based magnetic alignment. As shown in Section 9.4, the performance benefits of dual-layer electrodes were best elucidated when both electrodes in a full cell were dual-layer coated. It is likely that even greater strides in performance could be achieved at high current density ($\geq$ 1C) when both electrodes have low tortuosity.

9.6 Radiative drying

The importance of dry electrodes in LIB cells has been emphasized throughout this text and discussed in detail in Chapter 8. Dry electrodes are such an important prerequisite to a high-performance LIB that there are commonly two drying stages in electrode manufacturing; the first is immediately after coating to evaporate solvent from the slurry and the second is to remove any adsorbed water from the electrodes immediately prior to assembly in the cell. These stages are usually called the primary and secondary drying steps, respectively. Oftentimes the primary drying step is simply called the drying step since it is responsible for the majority of moisture and solvent removal, though since both steps will be discussed in this section, the primary and secondary terms will be used to distinguish between them. These steps often occur in a dehumidified room in order to further reduce moisture content in the electrodes.

The mechanism of heat transfer in both primary and secondary drying is convection. In primary drying, wet electrodes are fed in-line from the coater to a long convection oven, where warm, dry air absorbs the solvent from the coating and is carried out of the dryer as it becomes saturated with solvent. In secondary drying, the electrodes are placed in a vacuum oven. The warm air from the oven absorbs residual moisture in the electrode and is drawn out of the oven due to the applied vacuum. Both of these processes utilize air as a medium to transfer solvent and, as a result, are energy intensive. Primary drying occurs quickly, typically on the scale of 1–2 min, but requires high temperatures and leads to the possibility of binder and conductive additive migration, even if the system is designed well. Secondary drying, meanwhile, can take hours and high temperatures to reduce the moisture content to a desirable range.

These flaws in both stages of drying could be overcome with the institution of radiative drying methods. Radiative drying has been used in other industries, such as food and paper, for decades and has been used recently in some LIB electrode manufacturing [88]. The most commonly reported type of radiation used in electrode drying is infrared (IR) and microwave. There are two advantages of radiative drying when used instead of convection drying: (1) radiative drying is a more energy efficient process, and (2) the power of a radiation source can be tuned more easily and quickly than the temperature of a dryer. In the case of secondary drying, radiative drying can improve process speed significantly. A vacuum oven set to 100°C can reduce the moisture content of an electrode processed in NMP to ~90 ppm and an electrode processed in water to ~160 ppm in 2 h (the latter has a higher moisture content due to the hydrophilic nature of the binders [89]) [90]. Meanwhile, IR drying can achieve sufficiently dry electrodes in around 2 min [91]. The exact time required for secondary drying depends on how electrodes are arranged in the dryer, the initial moisture content of the electrodes, and the final moisture content desired from the electrodes, but the improvement in drying speed from IR drying to vacuum oven drying is present for otherwise identical conditions.

Recent studies have compared strictly IR drying with strictly convection drying, as well as a hybrid approach that uses both IR and forced convection in drying. One study showed that water-processed graphite anodes may be best dried strictly using IR drying [92]. This electrode had the highest film temperature during drying, which correlated favorably with process speed and adhesion properties [92]. The high adhesion of these electrodes is likely a consequence of limited binder migration in the electrode during drying. Another study investigated the optimal primary drying of a LCO cathode processed in NMP [93]. Primary drying may be divided into three stages: (1) high-rate drying, (2) constant-rate drying, and (3) falling-rate drying; these stages are discussed in detail in Chapter 8. It was found that IR drying was faster than convection drying in the first and second stages of primary drying, although convection drying was quicker in the third stage [93]. Considering that drying rate is dependent on the surface emissivity of the coating and the optical properties of the solvent [94], it is possible that a hybrid approach is more appropriate for drying some electrodes.

In summary, radiative drying provides an energy efficiency benefit in primary drying and a significant time saving in secondary drying. Since primary drying occurs in-line with coating, the implementation of radiative drying with manufacturing is more complicated than with secondary drying due to the larger number of process variables involved. At least in published literature, there is a lack of studies on IR drying to quantify the advantages in time and energy. Determining this benefit is important as a function of solvent, since water is commonly used for graphite anodes and NMP for cathodes. It is also unclear whether IR drying is the best kind of radiation to use compared to a lower-energy wave, such as microwaves [2]. With further studies on the topic of radiative drying, it could be an excellent candidate for use in LIB electrode manufacturing.

9.7 Summary

Innovations in LIB electrode manufacturing are difficult to incorporate into manufacturing for a myriad of reasons. First, the innovation must be scalable. Considering many plants are producing hundreds of MWh to tens of GWh of LIB cells per year, the electrode manufacturing equipment must be able to keep up with this level of production to satisfy market demand and make the plant efficient [1]. Technology that cannot be consistently replicated beyond the benchtop will never be suitable for production. Second, the innovation must provide a significant value for manufacturing. This value comes from providing marked improvement in several of the eight criteria discussed in the introduction, which were: (1) cathode material, (2) electrode thickness, (3) roll-to-roll speed, (4) end-of-line scrap rate, (5) number of operating days, (6) CapEx, (7) OpEx, and (8) worker safety. Because the state-of-the-art electrode manufacturing process has been demonstrated and repeated at scale frequently, there is rightful reservation to change it if the potential gains are marginal. Third, the amount of time and resources required to validate a new electrode manufacturing technology is immense. This effort can be reduced by inventing so-called "drop-in" processes that use existing production equipment and only require slight modifications. Whether a process is drop-in or revolutionary, it will require extensive characterization and validation at the R&D and pilot-scales.

With that said, as robust as the state-of-the-art process has been in producing LIB electrodes, there is an opportunity all along the process for improvement. Some innovations, including laser structuring (Chapter 5), additive manufacturing (Chapter 6), dry processing (Chapter 7), and aqueous processing for cathodes (Chapter 10) are presented elsewhere in this book. In this chapter, continuous slurry mixing, UV and EB curing, multilayer electrode coating, freeze-tape casting, magnetic alignment, and radiative drying are discussed.

Continuous slurry mixing offers a high-throughput alternative to batch mixing that can significantly reduce the floorspace dedicated to mixing in a manufacturing plant. However, the major drawback to continuous mixing is that maintenance issues or interruptions in mixing can halt the whole process, which is not necessarily the case for batch mixing. Fine-tuning of continuous mixing parameters is necessary for the process to be introduced seamlessly into production. If implemented perfectly, continuous mixing could offer benefits in roll-to-roll speed, end-of-line scrap rate, CapEx, and OpEx.

For LIB electrode production, EB curing is the superior technology to UV curing, particularly in the case of thick electrodes, since the penetration depth of EB is greater than UV. EB curing allows for slurries with extremely high solid loadings due to the implementation of short-chained oligomers as binder that are cured after exposure to the EB line. Initial trials of EB curing have shown that electrodes can be produced with higher thicknesses and at remarkably faster line speeds than the state-of-the-art process. On the other hand, EB curing requires inert atmospheres to avoid unwanted side reactions which are expensive and tricky to realize in practice. Literature data does also not show an improvement in electrochemical performance of electrodes made via EB curing when compared to the state-of-the-art baseline. Still, EB curing presents a major opportunity in increasing production speed and in boosting product energy density.

Multilayer electrode coating allows for the properties of individual layers to be optimized to yield an energy-dense or power-dense product. Although the concept is straightforward, dual-layer electrode coating requires two mixing, coating, drying, and calendering lines to make one electrode roll, while a single-layer electrode produced from state-of-the-art manufacturing only needs one line for an electrode roll. This requirement increases the floorspace needed per product and introduces more sources of variation and potential for error in manufacturing. In addition, lengthy R&D and pilot trials will be needed to optimize the multilayer electrode performance since there are limitless combinations of slurry and electrode properties to consider.

Freeze-tape casting and magnetic alignment are used to create aligned pore structures within electrodes. These pore structures enable thick electrodes with low Li-ion diffusion resistance through the electrode. This design allows for an energy dense electrode capable of fast charge and discharge, which are mutually exclusive in state-of-the-art manufacturing due to the random pore structure formed. However, both freeze-tape casting and magnetic alignment face significant scalability issues that must be addressed before they can be implemented at scale.

Radiative drying, whether used solely or in conjunction with forced convection drying, has been shown to reduce drying time and energy and assist in binder migration. Radiative drying can be used in either the primary electrode drying step that occurs directly after slurry coating or in the secondary drying step that occurs just before cell assembly. Radiative drying could be adaptable to different solvents or slurry compositions in-line as well, since tuning the power of an IR source is a less complicated procedure than adjusting temperatures in a multi-zone convection dryer. Although radiative drying has many advantages, there are few studies on the topic, indicating that there may not yet be a sufficient knowledge base on radiative drying to justify its use at scale.

References

[1] Mauler, L., F. Duffner, and J. Leker, Economies of scale in battery cell manufacturing: The impact of material and process innovations. *Applied Energy*, 2021. 286.

[2] Li, J., J. Fleetwood, W. B. Hawley, and W. Kays, From materials to cell: State-of-the-art and prospective technologies for lithium-ion battery electrode processing. *Chemical Reviews*, 2022. 122: p. 903–956.

[3] Smyrek, P., J. Pröll, H. Juergen Seifert, and W. Pfleging, Laser-induced breakdown spectroscopy of laser-structured $Li(NiMnCo)O_2$ electrodes for lithium-ion batteries. *Journal of the Electrochemical Society*, 2016. 163(2): p. A19–A26.

[4] Pfleging, W., A review of laser electrode processing for development and manufacturing of lithium-ion batteries. *Nanophotonics*, 2018. 7: p. 549–573.

[5] Habedank, J.B., L. Kraft, A. Rheinfeld, and C. Krezdorn, Increasing the discharge rate capability of lithium-ion cells with laser-structured graphite anodes: Modeling and simulation. *Journal of The Electrochemical Society*, 2018. 165: p. A1563–A1573.

[6] Habedank, J.B., F.J. Günter, N. Billot, *et al.*, Rapid electrolyte wetting of lithium-ion batteries containing laser structured electrodes: In situ visualization by neutron radiography. *The International Journal of Advanced Manufacturing Technology*, 2019. 102: p. 2769–2778.

[7] Zhu, P., H.J. Seifert, and W. Pfleging, The ultrafast laser ablation of Li $(Ni_{0.6}Mn_{0.2}Co_{0.2})O_2$ electrodes with high mass loading. *Applied Sciences*, 2019. **9**: p. 4067.

[8] Zhu, P., J. Han, and W. Pfleging, Characterization and laser structuring of aqueous processed $Li(Ni_{0.6}Mn_{0.2}Co_{0.2})O_2$ thick-film cathodes for lithium-ion batteries. *Nanomaterials*, 2021. 11: p. 1840.

[9] Song, Z., P. Zhu, W. Pfleging, and J. Sun, Electrochemical performance of thick-film $Li(Ni_{0.6}Mn_{0.2}Co_{0.2})O_2$ cathode with hierarchic structures and laser ablation. *Nanomaterials*, 2021. 11: p. 2962.

[10] Bae, C.-J., C.K. Erdonmez, J.W. Halloran, and Y.-M. Chiang, Design of battery electrodes with dual-scale porosity to minimize tortuosity and maximize performance. *Advanced Materials*, 2013. 25: p. 1254–1258.

[11] Cobb, C.L. and M. Blanco, Modeling mass and density distribution effects on the performance of co-extruded electrodes for high energy density lithium-ion batteries. *Journal of Power Sources*, 2014. 249: p. 357–366.

[12] Cobb, C.L. and S.E. Solberg, Communication – Analysis of thick co-extruded cathodes for high-energy-and-power lithium-ion batteries. *Journal of the Electrochemical Society*, 2017. 164(7): p. A1339–A1341.

[13] Ludwig, B., Z. Zheng, W. Shou, Y. Wang, and H. Pan, Solvent-free manufacturing of electrodes for lithium-ion batteries. *Scientific Reports*, 2016. p. 6.

[14] Ludwig, B., J. Liu, I.-M. Chen, *et al.*, Understanding interfacial-energy-driven dry powder mixing for solvent-free additive manufacturing of Li-ion battery electrodes. *Advanced Materials Interfaces*, 2017. 4.

[15] Al-Shroofy, M., Q. Zhang, J. Xu, T. Chen, A.P. Kaur, and Y.T. Cheng, Solvent-free dry powder coating process for low-cost manufacturing of $LiNi_{1/3}Mn_{1/3}Co_{1/3}O_2$ cathodes in lithium-ion batteries. *Journal of Power Sources*, 2017. 352: p. 187–193.

[16] Kirsch, D.J., S.D. Lacey, Y. Kuang, *et al.*, Scalable dry processing of binder-free lithium-ion battery electrodes enabled by holey graphene. *ACS Applied Energy Materials*, 2019. 2: p. 2990–2997.

[17] Li, J., B.L. Armstrong, J. Kiggans, C. Daniel, and D.L. Wood, III, Optimization of $LiFePO_4$ nanoparticle suspensions with polyethyleneimine for aqueous processing. *Langmuir*, 2012. 28: p. 3783–3790.

[18] Li, J., B.L. Armstrong, C. Daniel, J. Kiggans, and D.L. Wood, III, Optimization of multicomponent aqueous suspensions of lithium iron phosphate ($LiFePO_4$) nanoparticles and carbon black for lithium-ion battery cathodes. *Journal of Colloid and Interface Science*, 2013. 405(405): p. 118–124.

[19] Li, J., B.L. Armstrong, J. Kiggans, C. Daniel, and D.L. Wood, III, Lithium ion cell performance enhancement using aqueous $LiFePO_4$ cathode dispersions and polyethyleneimine dispersant. *Journal of the Electrochemical Society*, 2013. 160(2): p. A201–A206.

[20] Bresser, D., D. Buchholz, A. Moretti, A. Varzi, and S. Passerini, Alternative binders for sustainable electrochemical energy storage – The transition to aqueous electrode processing and bio-derived polymers. *Energy & Environmental Science*, 2018. 11: p. 3096–3127.

[21] Wood, D.L., J.D. Quass, J. Li, S. Ahmed, D. Ventola, and C. Daniel, Technical and economic analysis of solvent-based lithium-ion electrode drying with water and NMP. *Drying Technology*, 2018. 36(2): p. 234–244.

[22] Wood, M., J. Li, R.E. Ruther, *et al.*, Chemical stability and long-term cell performance of low-cobalt, Ni-rich cathodes prepared by aqueous processing for high-energy Li-ion batteries. Energy Storage *Materials*, 2020. 24: p. 188–198.

[23] Hawley, W.B., H.M. Meyer III, and J. Li, Enabling aqueous processing for $LiNi_{0.80}Co_{0.15}Al_{0.05}O_2$ (NCA)-based lithium-ion battery cathodes using polyacrylic acid. *Electrochimica Acta*, 2021. 380: p. 138203.

[24] Hawley, W.B., A. Parejiya, Y. Bai, H.M. Meyer, III, D.L. Wood, III, and J. Li, Lithium and transition metal dissolution due to aqueous processing in lithium-ion battery cathode active materials. *Journal of Power Sources*, 2020. 466: 228315.

[25] Kukay, A., R. Sahore, A. Parejiya, W.B. Hawley, J. Li, and D.L. Wood, III, Aqueous Ni-rich-cathode dispersions processed with phosphoric acid for lithium-ion batteries with ultra-thick electrodes. *Journal of Colloid and Interface Science*, 2021. 581: p. 635–643.

[26] Sahore, R., M. Wood, A. Kukay, *et al.*, Performance of different water-based binder formulations for Ni-rich cathodes evaluated in $LiNi_{0.8}Mn_{0.1}Co_{0.1}O_2$//graphite pouch cells. *Journal of The Electrochemical Society*, 2022. 169: 040567.

[27] An, S.J., J. Li, C. Daniel, *et al.*, Electrolyte volume effects on electrochemical performance and solid electrolyte interphase in Si-graphite/NMC lithium-ion pouch cells. *ACS Applied Materials & Interfaces*, 2017. 9(22): p. 18799–18808.

[28] Krueger, S., R. Kloepsch, J. Li, S. Nowak, S. Passerini, and M. Winter, How do the reactions at the anode/electrolyte interface determine the cathode performance in lithium-ion batteries? *Journal of The Electrochemical Society*, 2013. 160(4): p. A542–A548.

[29] Kim, C.-S., K.M. Jeong, K. Kim, and C.-W. Yi, Effects of capacity ratios between anode and cathode on electrochemical properties for lithium polymer batteries. *Electrochimica Acta*, 2015. 155: p. 431–436.

[30] Kleiner, K., P. Jakes, S. Scharner, V. Liebau, and H. Ehrenberg, Changes of the balancing between anode and cathode due to fatigue in commercial lithium-ion cells. *Journal of Power Sources*, 2016. 317: p. 25–34.

[31] Martin, C., Twin screw extruders as continuous mixers for thermal processing: A technical and historical perspective. *AAPS PharmSciTech*, 2016. 17: p. 3–19.

[32] Dreger, H., H. Bockholt, W. Haselrieder, and A. Kwade, Discontinuous and continuous processing of low-solvent battery slurries for lithium nickel cobalt manganese oxide electrodes. *Journal of Electronic Materials*, 2015. 44(11): p. 4434–4443.

[33] Spillman, A., *From coin cell to gigafactory – Scale-up of LIB slurry processing*. Buhler AG: Uzwil, Switzerland, 2017.

[34] An, S.J., J. Li, Y. Sheng, C. Daniel, and D.L. Wood, III, Long-term lithium-ion battery performance improvement via ultraviolet light treatment of the graphite anode. *Journal of the Electrochemical Society*, 2016. 163: p. A2866–A2875.

[35] An, S.J., J. Li, C. Daniel, and D.L. Wood, III, Effects of ultraviolet light treatment in ambient air on lithium-ion battery graphite and PVDF binder. *Journal of the Electrochemical Society*, 2019. 166: p. A1121–A1126.

[36] Na, W., A.S. Lee, J.H. Lee, *et al.*, Lithium dendrite suppression with UV-curable polysilsequioxane separator binders. *ACS Applied Materials & Interfaces*, 2016. 8: p. 12852–12858.

[37] Zhu, W., Z. Wang, W. Qiu, *et al.*, Improving the electrochemical performance of polypropylene separator through instantaneous photo-induced functionalization. *Journal of the Electrochemical Society*, 2018. 165(9): p. A1909–A1914.

[38] Du, Z., X.C. Chen, R. Sahore, X. Wu, J. Li, and N.J. Dudney, Effect of plasticizer content and ceramic addition on electrochemical properties of cross-linked polymer electrolyte. *Journal of the Electrochemical Society*, 2021. 168: 050549.

[39] Hawley, W.B., Z. Du, A.J. Kukay, N.J. Dudney, A.S. Westover, and J. Li, Deconvoluting sources of failure in lithium metal batteries containing NMC and PEO-based electrolytes. *Electrochimica Acta*, 2022. 404: 139579.

[40] Du, Z., C.J. Jankeb, J. Lia, C. Daniela, and D.L. Wood, III, Electron beam curing of composite positive electrode for Li-ion battery. *Journal of the Electrochemical Society*, 2016. 163(13): p. A2776–A2780.

[41] Du, Z., C.J. Janke, J. Li, and D.L. Wood, III, High-speed electron beam curing of thick electrode for high energy density Li-ion batteries. *Green Energy & Environment*, 2019. 4: p. 375–381.

[42] Lecamp, L., F. Houllier, B. Youssef, and C. Bunel, Photoinitiated cross-linking of a thiol-methacrylate system. *Polymer*, 2001. 42: p. 2727–2736.

[43] Cramer, N.B. and C.N. Bowman, Kinetics of thiol-ene and thiol-acrylate photopolymerizations with real-time Fourier transform infrared. *Journal of Polymer Science Part A: Polymer Chemistry*, 2001. 39: p. 3311–3319.

[44] O'Brien, A.K., N.B. Cramer, and C.N. Bowman, Oxygen inhibition in thiol-acrylate photopolymerizations. *Journal of Polymer Science Part A: Polymer Chemistry*, 2006. 44: p. 2007–2014.

[45] Lopata, V.J., C.B. Saunders, A. Singh, C.J. Janke, G.E. Wrenn, and S.J. Havens, Electron-beam-curable epoxy resins for the manufacture of high-performance composites. *Radiation Physics and Chemistry*, 1999. 56: p. 405–415.

[46] Saunders, C., V. Lopata, J. Barnard, and T. Stepanik, Electron beam curing – Taking good ideas to the manufacturing floor. *Radiation Physics and Chemistry*, 2000. 57: p. 441–445.

[47] Sangermano, M., N. Razza, and J.V. Crivello, Cationic UV-curing: Technology and applications. *Macromolecular Materials and Engineering*, 2014. 299: p. 775–793.

[48] Kraytsberg, A. and Y. Ein-Eli, Conveying advanced Li-ion battery materials into practice: The impact of electrode slurry preparation skills. *Advanced Energy Materials*, 2016. 6.

[49] Fang, X., Y. Lu, N. Ding, X.Y. Feng, C. Liu, and C.H. Chen, Electrochemical properties of nano- and micro-sized $LiNi_{0.5}Mn_{1.5}O_4$

synthesized via thermal decomposition of a ternary eutectic Li-Ni-Mn acetate. *Electrochimica Acta*, 2010. 55: p. 832–837.

[50] Chowdhury, R., Y. Zhao, Y. Xia, M. Ouyang, N. Brandon, and A. Banerjee, Revisiting the promise of bi-layer graded cathodes for improved Li-ion battery performance. *Sustainable Energy Fuels*, 2021. 5: p. 5193–5204.

[51] Diehm, R., J. Kumberg, C. Dörrer, *et al.*, In situ investigations of simultaneous two-layer slot die coating of component-graded anodes for improved high-energy Li-ion batteries. *Energy Technology*, 2020. 8: 1901251.

[52] Kalnaus, S., K. Livingston, W.B. Hawley, H. Wang, and J. Li, Design and processing for high performance Li-ion battery electrodes with double-layer structure. *Journal of Energy Storage*, 2021. 44.

[53] Wood, M., J. Li, Z. Du, *et al.*, Impact of secondary particle size and two-layer architectures on the high-rate performance of thick electrodes in lithium-ion battery pouch cells. *Journal of Power Sources*, 2021. 515: 230429.

[54] Cheng, C., R. Drummond, S.R. Duncan, and P.S. Grant, Micro-scaled graded electrodes for improved dynamic and cycling performance of Li-ion batteries. *Journal of Power Sources*, 2019. 413: p. 59–67.

[55] Cheng, C., R. Drummond, S.R. Duncan, and P.S. Grant, Combining composition graded positive and negative electrodes for higher performance Li-ion batteries. *Journal of Power Sources*, 2020. 448: 227376.

[56] Cheng, C., R. Drummond, S.R. Duncan, and P.S. Grant, Extending the energy-power balance of Li-ion batteries using graded electrodes with precise spatial control of local composition. *Journal of Power Sources*, 2022. 542:231758.

[57] Drummond, R., C. Cheng, P.S. Grant, and S.R. Duncan, Modelling the impedance response of graded $LiFePO_4$ cathodes for Li-ion batteries. *Journal of the Electrochemical Society*, 2022. 169.

[58] Deng, W., W. Shi, Q. Liu, J. Jiang, X. Li, and X. Feng, Constructing gradient porous structure in thick $Li_4Ti_5O_{12}$ electrode for high-energy and stable lithium-ion batteries. *ACS Sustainable Chemistry & Engineering*, 2020. 8: p. 17062–17068.

[59] Elango, R., A. Demortière, V. De Andrade, M. Morcrette, and V. Seznec, Thick binder-free electrodes for Li-ion battery fabricated using templating approach and spark plasma sintering reveals high areal capacity. *Advanced Energy Materials*, 2018. 8: 1703031.

[60] Zhao, H., Q. Yang, N. Yuca, *et al.*, A convenient and versatile method to control the electrode microstructure toward high-energy lithium-ion batteries. *Nano Letters*, 2016. 16: p. 4686–4690.

[61] Zhu, P., P.R. Slater, and E. Kendrick, Insights into architecture, design and manufacture of electrodes for lithium-ion batteries. *Materials & Design*, 2022. 223: 111208.

[62] Zhang, Y., W. Luo, C. Wang, *et al.*, High-capacity, low-tortuosity, and channel-guided lithium metal anode. *Proceedings of the National Academy of Sciences*, 2017. 114: p. 3584–3589.

[63] Lu, L.-L., Y.-Y. Lu, Z.-J. Xiao, *et al.*, Wood-inspired high-performance ultrathick bulk battery electrodes. *Advanced Materials*, 2018. 30.

[64] Li, W.L., K. Lu, and J.Y. Walz, Freeze casting of porous materials: Review of critical factors in microstructure evolution. *International Materials Reviews*, 2012. 57(1): p. 37–60.

[65] Delattre, B., R. Amin, J. Sander, J. De Coninck, A.P. Tomsia, and Y.-M. Chiang, Impact of pore tortuosity on electrode kinetics in lithium battery electrodes: Study in directionally freeze-cast $LiNi_{0.8}Co_{0.15}Al_{0.05}O_2$ (NCA). *Journal of the Electrochemical Society*, 2018. 165(2): p. A388–A395.

[66] Huang, C., M. Dontigny, K. Zaghib, and P.S. Grant, Low-tortuosity and graded lithium ion battery cathodes by ice templating. *Journal of Materials Chemistry A*, 2019. 7: p. 21421–21431.

[67] Huang, C. and P.S. Grant, Coral-like directional porosity lithium ion battery cathodes by ice templating. *Journal of Materials Chemistry A*, 2018. 6: p. 14689–14699.

[68] Amin, R., B. Delattre, A.P. Tomsia, and Y.-M. Chiang, Electrochemical characterization of high energy density graphite electrodes made by freeze-casting. *ACS Applied Energy Materials*, 2018. 1: p. 4976–4981.

[69] Shen, H., E. Yi, S. Heywood, *et al.*, Scalable freeze-tape-casting fabrication and pore structure analysis of 3D LLZO solid-state electrolytes. *ACS Applied Materials & Interfaces*, 2020. 12: p. 3494–3501.

[70] Yi, E., H. Shen, S. Heywood, *et al.*, All-solid-state batteries using rationally designed garnet electrolyte frameworks. *ACS Applied Energy Materials*, 2020. 3: p. 170–175.

[71] Sofie, S.W., Fabrication of functionally graded and aligned porosity in thin ceramic substrates with the novel freeze-tape-casting process. *Journal of the American Ceramic Society*, 2007. 90(7): p. 2024–2031.

[72] Ren, L., Y.-P. Zeng, and D. Jiang, Fabrication of gradient pore TiO_2 sheets by a novel freeze-tape-casting process. *Journal of the American Ceramic Society*, 2007. 90(9): p. 3001–3004.

[73] Cable, T.L., J.A. Setlock, S.C. Farmer, and A.J. Eckel, Regenerative performance of the NASA symmetrical solid oxide fuel cell design. *International Journal of Applied Ceramic Technology*, 2011. 8(1): p. 1–12.

[74] Gannon, P., S. Sofie, M. Deibert, R. Smith, and V. Gorokhovsky, Thin film YSZ coatings on functionally graded freeze cast NiO/YSZ SOFC anode supports. *Journal of Applied Electrochemistry*, 2009. 39: p. 497–502.

[75] Chen, Y., J. Bunch, T. Li, Z. Mao, and F. Chen, Novel functionally graded acicular electrode for solid oxide cells fabricated by the freeze-tape-casting process. *Journal of Power Sources*, 2012. 213: p. 93–99.

[76] Chen, Y., Q. Liu, Z. Yang, F. Chen, and M. Han, High performance low temperature solid oxide fuel cells with novel electrode architecture. *RSC Advances*, 2012. 2: p. 12118–12121.

[77] Chen, Y., Y. Lin, Y. Zhang, *et al.*, Low temperature solid oxide fuel cells with hierarchically porous cathode nano-network. *Nano Energy*, 2014. 8: p. 25–33.

[78] Chen, Y., Y. Zhang, J. Baker, *et al.*, Hierarchically oriented macroporous anode-supported solid oxide fuel cell with thin ceria electrolyte film. *ACS Applied Materials & Interfaces*, 2014. 6(7): p. 5130–5136.

[79] Chen, Y., Y. Zhang, Y. Lin, *et al.*, Direct-methane solid oxide fuel cells with hierarchically porous Ni-based anode deposited with nanocatalyst layer. *Nano Energy*, 2014. 10: p. 1–9.

[80] Ghadkolai, M.A., S. Creager, J. Nanda, and R.K. Bordia, Freeze tape cast thick Mo doped $Li_4Ti_5O_{12}$ electrodes for lithium-ion batteries. *Journal of the Electrochemical Society*, 2017. 164(12): p. A2603–A2610.

[81] Parikh, D. and J. Li, Bilayer hybrid graphite anodes via freeze tape casting for extreme fast charging applications. *Carbon*, 2022. 196: p. 525–531.

[82] Hwa, Y., E. Yi, H. Shen, *et al.*, Three-dimensionally aligned sulfur electrodes by directional freeze tape casting. *Nano Letters*, 2019. 19: p. 4731–4737.

[83] Shen, H., E. Yi, M. Amores, *et al.*, Oriented porous LLZO 3D structures obtained by freeze casting for battery applications. *Journal of Materials Chemistry A*, 2019. 7: p. 20861–20870.

[84] Shen, H., E. Yi, L. Cheng, *et al.*, Solid-state electrolyte considerations for electric vehicle batteries. *Sustainable Energy Fuels*, 2019. 3: p. 1647–1659.

[85] Billaud, J., F. Bouville, T. Magrini, C. Villevieille, and A.R. Studart, Magnetically aligned graphite electrodes for high-rate performance Li-ion batteries. *Nature Energy*, 2016. 1.

[86] Sander, J.S., R.M. Erb, L. Li, A. Gurijala, and Y.-M. Chiang, High-performance battery electrodes via magnetic templating. *Nature Energy*, 2016. 1: 151.

[87] Li, L., R.M. Erb, J. Wang, J. Wang, and Y.-M. Chiang, Fabrication of low-tortuosity ultrahigh-area-capacity battery electrodes through magnetic alignment of emulsion-based slurries. *Advanced Energy Materials*, 2019. 9: 1802472.

[88] Kwade, A., W. Haselrieder, R. Leithoff, A. Modlinger, F. Dietrich, and K. Droeder, Current status and challenges for automotive battery production technologies. *Nature Energy*, 2018. 3: p. 290–300.

[89] Heckmann, T., J.C. Eser, A. Altvater, N. Streller, P. Scharfer, and W. Schabel, Experimental investigation of the temperature, pressure, and binder system influence on vacuum post-drying processes and moisture management of Li-ion battery electrodes. *Energy Technology*, 2022 11: 2200859.

[90] Li, J., C. Daniel, S.J. An, and D. Wood, Evaluation of residual moisture in lithium-ion battery electrodes and its effect on electrode performance. *MRS Advances*, 2016. 1(15): p. 1029–1035.

[91] Daniel, C. and M. Wixom, Transformational electrode drying process. Final Report, 2012.

[92] Altvater, A., T. Heckmann, J.C. Eser, S. Spiegel, P. Scharfer, and W. Schabel, (N)IR-drying of lithium-ion battery electrodes – Influence of energy input on process speed and electrode adhesion. *Energy Technology*, 2022. 11: 2200785.

[93] Wu, X., X. Li, and X. Bei, Experiments on hot-air and infrared drying characteristics of $LiCoO_2$ cathode coating for lithium-ion battery. *Materials Science Forum*, 2020. 1003: p. 260–267.

[94] Sabau, A.S., C.I. Contescu, G.D. Jellison, *et al.*, Evaporation due to infrared heating and natural convection. *Heat and Mass Transfer*, 2020. 56: p. 2585–2593.

Chapter 10
Aqueous cathode processing

W. Blake Hawley[1] *and Jianlin Li*[2]

10.1 Introduction

The state-of-the-art lithium-ion battery (LIB) manufacturing process uses *N*-methyl-2-pyrrolidone (NMP) as solvent for the electrode slurry dispersing stage. NMP is a hazardous chemical, known particularly for its reproductive toxicity [1,2]. NMP has been the subject of recent government action in many countries and is likely to face increasingly restrictive measures curtailing its use. In December 2020, the United States Environmental Protection Agency (US EPA) concluded its most recent risk assessment for NMP, finding that it poses an unreasonable risk of injury to health in 26 of 37 use cases, including specifically for "industrial and commercial use… in lithium ion battery manufacturing" [3]. More direct action was taken in Europe in 2020 when the European Chemical Agency imposed a restriction on May 9, 2020, banning the use of NMP in concentrations greater than 0.3% without appropriate risk management measures [4].

Partially because of the toxicity of NMP, plant siting for LIB manufacturing plants is a relatively arduous and potentially costly exercise. Another outcome of the toxicity of NMP is the need to recycle it onsite within the production facility. The equipment needed for NMP recycle, including condensers, zeolite wheels, and scrubbers, can cost millions of dollars, depending on the size of the facility [5]. Operating the NMP recycling line is also expensive; alone, it is estimated to comprise about 3% of all manufacturing costs [6]. Finally, one must consider the cost of purchasing NMP and the binder commonly used with it, polyvinylidene fluoride (PVDF). While this solvent/binder pairing is highly effective, they are more expensive than many other solvent/binder pairings. Therefore, the costs associated with NMP-based electrode processing are multi-faceted, ranging from land acquisition, facility construction, facility operation, and material costs.

To circumvent the dangers and costs associated with NMP in electrode processing, alternative solvents have been proposed. By far the most common alternative solvent to NMP is water, which will be discussed in detail in this chapter.

[1]Sparkz, Inc., USA
[2]Electrification and Energy Infrastructures Division, Oak Ridge National Laboratory, USA

Other organic solvents will be covered in Section 10.8. Transitioning anode processing from NMP to water has been achieved at lab and industrial scales. Graphite anodes, which also use PVDF when processed in NMP, typically substitute PVDF for a combination of sodium carboxymethyl cellulose (Na-CMC) and styrene butadiene rubber (SBR). Na-CMC provides good adhesion as well as increases the viscosity of the slurry, while SBR gives the electrode film flexibility.

However, so-called "aqueous processing" for cathodes is a considerably more challenging prospect than it is for graphite anodes. The issues that face aqueous-processed cathodes, to varying degree depending on the material, are as follows:

1. Identifying an optimal water-soluble binder
2. Removing residual water after drying
3. Dispersing active material and conductive additive agglomerates
4. Achieving proper wetting of the slurry onto the current collector
5. Mitigating corrosion of the current collector
6. Mitigating leaching of Li and transition metal-ions during processing

Aqueous-processed cathodes have been researched since at least 2005 when $LiCoO_2$ (LCO) was processed with the addition of polyammonium acrylate (PAA-NH_4) as dispersant [7]. Since then, over 100 studies have been published about aqueous-processed cathodes, ranging from additional LCO studies [8–14], olivine materials (specifically $LiFePO_4$, or LFP) [15–50], $LiNi_xMn_yCo_{1-x-y}O_2$ (NMC) materials [5,51–76], $LiNi_xCo_yAl_{1-x-y}O_2$ (NCA) materials [77–85], $LiNi_xMn_{2-x}O_4$ (LNMO) materials [86–99], $LiMn_2O_4$ (LMO) [100,101], and Li-rich cathode materials of varying chemistries [102–113]. LCO was popularly studied in the mid-to-late aughts with little emphasis now, while aqueous processing of NMC111, NMC532, and NMC811 was first reported in 2013, 2017, and 2020, respectively. NCA has received attention in aqueous processing studies since 2017 while LFP was intensely studied for a decade until about 2018. These trends mostly reflect the trends seen in battery research at-large, with higher-capacity, less Co-reliant materials becoming favorable for energy density and cost reduction purposes, respectively. Aqueous-processed cathodes have also been used in conjunction with other next-generation manufacturing technologies, such as electron beam (EB) curing and multilayer coating, to name a few (see Chapter 9).

The discussion of the six issues with aqueous processing will be discussed in the following five subsections, with Issues (5) and (6) being coupled. While all six issues are interconnected and are difficult to decouple, this pairing is most strongly tied.

10.2 Water-soluble binders

In this chapter, aqueous processing is framed as originating from a desire to transition from NMP to water. While this largely reflects the modern motivation for aqueous processing, the researchers that initially studied the technique did so because they were interested in testing binders (most of which were bio-derived) that happened to be water-soluble. These studies identified many binders that were water-soluble that achieved similar, if not better, performance than PVDF.

Bresser *et al.* created three water-soluble binder classifications by the number of sustainability criteria they met [114]. The first class of binders do not contain fluorine (F) and are bio-polymers and derivatives of bio-polymers and include oligo- and polysaccharides and proteins. The second class of binders do not contain F but are synthetic and include polyacrylates, aliphatic polymers, and aromatic polymers. The third class, those that are water-soluble but are not naturally occurring and contain F, are the least sustainable and include binders such as the fluoro acrylic binder TRD202A from JSR Micro and polytetrafluoro-ethylene.

In addition to the sustainability criteria, water-soluble cathode binders must meet the characteristics necessary for any electrode binder, including: (1) low cost, (2) good adhesion to the Al current collector, (3) good cohesion between the active material and conductive additive particles, (4) electrochemical stability across a wide potential range, (5) chemical stability with the cathode material and the electrolyte, and (6) good rheological properties. These six characteristics must also be achieved while keeping the binder in a relatively low mass and volume fraction (on the order of 2–3 wt% or 5–7 vol%, respectively) of the cathode [6].

By far the most common binder used in aqueous-processed cathodes is Na-CMC, typically with a molecular weight (MW) $\geq$ 250,000 g·mol^{-1}. Na-CMC belongs to the first class of water-soluble binder (i.e., F-free and bio-derived polymers), making it desirable from a sustainability perspective. Additionally, the degree of substitution (DS) of Na-CMC is a relevant parameter; DS refers to the number of carboxymethyl groups that are substituted per monomer unit, meaning the DS can range from 0 to 3. It has been shown that a lower DS contributed to a higher capacity in half cells with an LFP cathode and is believed to be because of an improved dispersion quality [30]. Na-CMC has been studied as a solo binder in a cathode film; however, these films often lack in flexibility. Additionally, high-MW Na-CMC must be pre-dispersed in water in a low mass fraction ($\leq$5 wt%) due to its poor solubility. This can constrain the solid loading of a slurry when such a large amount of solvent must be included with the binder. Still, despite these drawbacks, Na-CMC as a solo binder has outperformed PVDF in cyclability tests in several studies (see Table 10.1). There are other studies that use Na-CMC solely as a binder, but the studies featured in Table 10.1 directly compare Na-CMC with PVDF. It should be noted that not all work on Na-CMC directly states that Na^+ salt is in the CMC chemistry; in these instances, the “Na” prefix is removed, though it is likely that Na-CMC is the binder being used.

Other binders that are bio-derived and F-free have been used as solo binders in aqueous-processed cathodes as well. In one instance, half cells with a Li-rich active material showed sodium alginate (SA) has better cyclability than CMC; after 100 cycles, the SA-containing cathode retained 94% of its capacity, while CMC retained only 84% of its capacity [102]. However, this is contrasted by results in another study in which CMC provides better initial discharge capacity and similar capacity retention to alginate binders [75]. Xanthan gum (XG) and guar gum have been demonstrated to perform similarly to CMC and significantly better than PVDF

Table 10.1 Summary of studies comparing PVDF and Na-CMC as solo binders in half cells. Some studies do not specifically mention that Na^+ salt is part of the CMC chemistry, so the "Na" prefix is omitted in these instances, though it may be reasonably assumed that these studies use Na-CMC.

Cathode recipe	Cell and testing conditions	Performance description	References
85 wt% $Li_{1.182}Ni_{0.182}Co_{0.091}Mn_{0.545}O_2$, 10 wt% carbon black (CB), 5 wt% binder (CMC or PVDF)	2.0–4.8 V, room temperature (RT), +0.2C/−0.2C, undisclosed areal capacity, undisclosed calendering status	The initial discharge capacity of both cells was roughly 245 $mAh{\cdot}g^{-1}$; after 100 cycles, the cell containing PVDF-based cathode retained ~61% of its capacity, while the cell containing CMC-based cathode retained ~84% of its capacity	[102]
90 wt% LFP, 5 wt% CB, 5 wt% binder (CMC or PVDF)	2.5–4.0 V, 25°C, +0.2C/−0.2C, 0.8 $mAh{\cdot}cm^{-2}$, undisclosed calendering status	The initial discharge capacity of the cell with the PVDF-containing cathode was 163 $mAh{\cdot}g^{-1}$ and retained 92.8% of its capacity after 100 cycles; the initial discharge capacity of the cell with the CMC-containing cathode was 155 $mAh{\cdot}g^{-1}$ and retained 97.6% of its capacity after 100 cycles	[24]
85 wt% $Li_{1.2}Mn_{0.56}Ni_{0.16}Co_{0.08}O_2$, 10 wt% CB, 5 wt% Na-CMC	2.5–4.8 V, 20°C, +1C/−1C, 0.4 $mAh{\cdot}cm^{-2}$, calendered but do not know porosity	The initial discharge capacity of the cell with the PVDF-containing cathode was 250 $mAh{\cdot}g^{-1}$ and retained ~60% of its capacity after 200 cycles; the initial discharge capacity of the cell with the Na-CMC-containing cathode was 205 $mAh{\cdot}g^{-1}$ and retained ~85% of its capacity after 200 cycles	[106]
88 wt% LFP, 7 wt% CB, 5 wt% binder (Na-CMC or PVDF)	2.8–4.2 V, 20 ± 2°C, +1C/−1C, 0.2 $mAh{\cdot}cm^{-2}$, uncalendered	The initial discharge capacity of the cell with PVDF-containing cathode was 156 $mAh{\cdot}g^{-1}$ and retained ~90% of its capacity after 50 cycles; the initial discharge capacity of the cell with Na-CMC-containing cathode was 140 $mAh{\cdot}g^{-1}$ and experienced negligible fade after 50 cycles	[38]

(Continues)

85 wt% $LiNi_{0.4}Mn_{1.6}O_4$, 10 wt% CB, 5 wt% binder (CMC or PVDF)	3.5–5.0 V, 20°C, +1C/–1C, 0.6 $mAh \cdot cm^{-2}$, undisclosed calendering status	The initial discharge capacity of the cell with PVDF-containing cathode was 120 $mAh \cdot g^{-1}$ and retained ~96% of its capacity after 100 cycles; the initial discharge capacity of the cell with CMC-containing cathode was 140 $mAh \cdot g^{-1}$ and retained 89% of its capacity after 100 cycles	[97]
85 wt% NMC111, 10 wt% CB, 5 wt% binder (CMC or PVDF)	2.5–4.6 V, RT, +0.5C/–0.5C, undisclosed areal capacity, calendered but do not know porosity	The initial discharge capacity of the cell with PVDF-containing cathode was 131 $mAh \cdot g^{-1}$ and retained ~85% of its capacity after 200 cycles; the initial discharge capacity of the cell with CMC-containing cathode was 159 $mAh \cdot g^{-1}$ and retained ~90% of its capacity after 200 cycles	[75]
80 wt% $Li_{1.2}Ni_{0.2}Mn_{0.6}O_2$, 10 wt% CB, 10 wt% CMC	2.0–4.8 V, RT, +0.1C/–0.1C, 0.7 $mAh \cdot cm^{-2}$, calendered but do not know porosity	The initial discharge capacity of the cell with PVDF-containing cathode was 220 $mAh \cdot g^{-1}$ and retained ~23% of its capacity after 80 cycles; the initial discharge capacity of the cell with CMC-containing cathode was 210 $mAh \cdot g^{-1}$ and retained ~119% of its capacity after 80 cycles	[113]

in some studies [24,101,111]. Finally, various chitosan materials (cyanoethylated chitosan, carboxymethyl chitosan, and chitosan derived from shrimp) have also outperformed PVDF [23,42,98]. While these tests have been performed for a wide variety of cathode materials, they almost exclusively occur in half cell formats with low areal capacities and high binder contents. Comparisons under more realistic conditions are necessary to judge whether F-free, bio-derived polymers are a scalable alternative to PVDF.

The shortcomings of Na-CMC may be overcome by adding a secondary binder that is more flexible and contributes less to viscosity. The second class of water soluble binders, those that are F-free but synthetic, includes many binders that have been used in conjunction with Na-CMC for cathodes. One of the most studied such binders is SBR. SBR may be dispersed in a much higher concentration in a water emulsion (about 50 wt%) and has a low viscosity, which complements Na-CMC well.

There has been hesitation to include SBR in non-LFP cathodes because the C=C bond in SBR's structure has been suggested to be prone to oxidation above 4.2 V vs. Li^+/Li [14]. Despite this, the Na-CMC/SBR hybrid binder has shown great cyclability with LCO [8,10–12], LNMO [90], and $LiNi_{0.83}Mn_{0.05}Co_{0.12}O_2$ [72] cycled to upper cutoff potentials of $\geq$ 4.2 V. In the latter case, the Na-CMC/SBR binder system was tested under industrially relevant conditions for 1,000 cycles. The ratio between Na-CMC and SBR was found to be best at around 3:7 [12] and most studies featuring both binders use a similar ratio. The degree of cross-linking in SBR is also important, as a less cross-linked SBR provides favorable adhesion and cohesion [8]. A less-often studied aromatic polymer, polyurethane (PU), was studied in an NMC111 cathode and a full cell format and found to perform best in a 1:2 ratio by mass with Na-CMC [67].

In addition to aromatic polymers like SBR and PU, polyacrylates have been widely-studied F-free, synthetic binders. These polyacrylates include polyacrylic acid (PAA) [16,63,70], polyacrylonitrile (PAN) [7,9,63,70,76,115,116], polybutyl acrylate (PBA) [34], polymethyl acrylate (PMA) [71], and polyvinyl acetate (PVA) [43], and have been used either as a standalone binder or in binary or ternary systems. Of these, PAA and PAN have been studied most readily. Two recent studies have compared these two binders in half cells featuring NMC811 cathodes. Both studies found that the resistance in cathodes featuring PAA were lower than those featuring PAN based on cyclic voltammetry and electrochemical impedance spectroscopy. As a result, the cycle life and rate capability of cathodes featuring PAA were significantly better than those featuring PAN [63,70].

The most widely-studied binder within the third class of binders, those that contain F and are synthetic, is TRD202A [51,54,64,77,78,83–85,96,104,107,108]. TRD202A is too low of a viscosity emulsion to be used alone, so it is often paired with Na-CMC in a ratio similar to that of SBR and Na-CMC. Several of these studies have shown the effectiveness of TRD202A, though when combined with Na-CMC, it was less effective in retaining cycle life when compared head-to-head with SA and Na-CMC as a solo binder in a cathode featuring a Li-rich active material [104]. Fluoro acrylic binders may be pursued less frequently in the future, due to their cost and the complications they cause with battery recycling (see Section 10.8), which will be necessary as LIBs become more ubiquitous. Therefore, for sustainability, cost, and performance reasons, binders from the first two categories are worthy of more study first.

Finally, Li-substituted variations of Na-CMC [56,117–120] and PAA [17,95, 110] have been studied in the literature. These binders provide enhanced Li-ion conductivity in the composite cathode and, in the case of LiPAA, have been studied as active material coatings [116] and next-generation binders for silicon-graphite anodes [121,122]. LiPAA is prepared by titrating a PAA solution with LiOH until the pH of the solution is ~7. However, there has yet to be a direct comparison between LiPAA and PAA for cathode materials.

In summary, there are many sustainable, water-soluble binders that show promise as being comparable to or perhaps better than PVDF. Often, combining these binders is necessary to obtain the optimal cathode film properties. However, it is inconclusive from the data provided by peer-reviewed studies which of these binder systems is the best. This is because there are relatively few direct comparisons between binders and few studies that explore these binders in realistic testing conditions, as discussed in Chapter 9.

10.3 Drying water-processed cathodes

Achieving a dry cathode is essential for successful LIB function, as discussed in Chapter 8. Removing residual moisture from aqueous-processed cathodes is more challenging than it is for NMP-processed cathodes since water-soluble binders retain more moisture than PVDF. Water-processed cathodes have roughly three times the moisture content of NMP-processed cathodes after the initial drying process [123]. To alleviate this, a secondary drying step is necessary, whether it be in a vacuum oven or using radiative drying techniques (see Section 9.6).

A secondary drying comparison study was conducted for aqueous-processed and NMP-processed NMC532 cathodes (see Figure 10.1). This study found that, after 2 h of drying in a vacuum oven at 25°C, the water-processed cathode had a moisture content greater than 600 ppm compared to just 200 ppm for the NMP-processed cathode. However, the moisture content decreases at a significantly higher rate for each additional degree of drying temperature for the water-based system than the NMP-based system. The NMP-based NMC532 cathode attains a residual moisture of 87 ppm after 2 h of drying at 100°C, while the water-processed cathode can reach this level of moisture at a temperature between 110 and 120°C [123]. This difference in slope is likely due to the differences in boiling point between NMP and water (204.3°C and 100°C, respectively) and vapor pressure (1.0 mm Hg and 55.3 mm Hg at 40°C, respectively) [5].

Another strategy to impede the negative influence of water on cell performance is to add water-scavenging particles to the electrode, such as Al_2O_3 or SiO_2. Half cells with either particle added to the NMC111-based cathode achieved similar rate capability to the state-of-the-art-processed baseline cathode (67% and

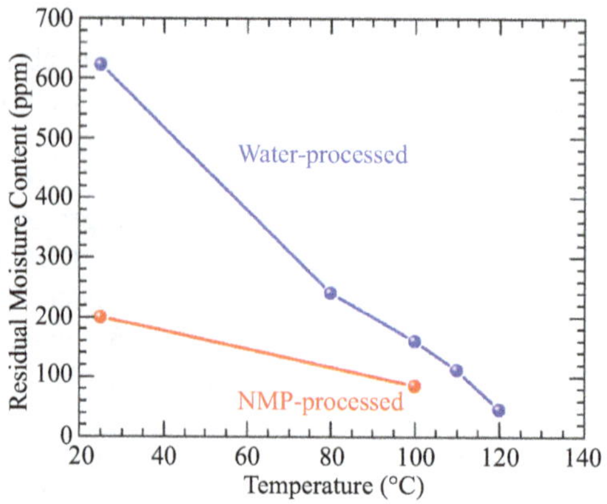

Figure 10.1 *Residual moisture content of a NMC532 cathode as a function of drying temperature in a vacuum oven after 2 h. The water-processed cathode is displayed in blue, and the NMP-processed cathode is displayed in red. Adapted and reprinted with permission from Ref. [123]. Copyright 2016 The Materials Research Society.*

73% of low-rate capacity at 5C, respectively), while the aqueous-processed cathode with no additives was significantly worse (42%) [69].

10.4 Achieving proper dispersion

As mentioned in previous chapters, most spherical active material primary particles are on the order of ~1 μm in diameter and secondary particles are on the order of 10–20 μm [124]. The secondary particles are held together by strong electrostatic forces that are difficult to break; thus, this level of agglomeration is accepted in LIB manufacturing. The mixing stage is designed to break any large agglomerates (>20 μm), which are held together by weak van der Waals attractions. However, in aqueous solution, the attractive electrostatic and hydrogen bonding forces between active material particles are greater than in NMP.

As discussed in Chapter 3, stability from agglomeration may be imparted through electrostatic, steric, or electrosteric means. Additives designed to disperse the particles (i.e., dispersants) are not needed in the state-of-the-art process but can be useful in aqueous processing. The efficacy of dispersants is often assessed by measuring the zeta potential using phase analysis light scattering (PALS).

To understand zeta potential and PALS, it is first important to understand what PALS is measuring [125] (see Figure 10.2). Positively charged ions in solution are attracted to the surface of particles in solution, which generally carry a negative surface charge. This thin layer of positively charged ions is called the Stern layer. Beyond the Stern layer, differently charged ions approach the Stern layer until an imaginary slipping plane is reached, at which there are few ions crowding the

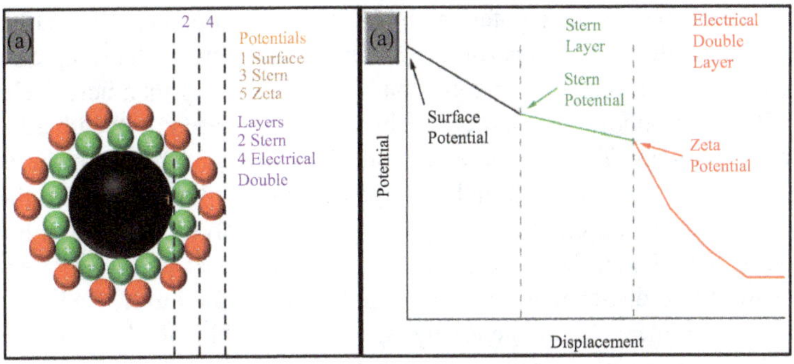

Figure 10.2 (a) Illustration of the charges surrounding a particle in solution, with the surface, Stern, and zeta potentials indicated by the orange numbers 1, 3, and 5, respectively. The Stern and electrical double layers are denominated by the purple numbers 2 and 4, respectively, constrained between the dashed gray lines. (b) Plot showing the decay of potential as the displacement from the particle surface increases, in arbitrary units. The surface, Stern, and zeta potentials are designated. The Stern and electrical double layers are designated by the dashed gray lines.

Table 10.2 The influence of zeta potential on dispersion stability [126]

Zeta potential (mV)	**Dispersion stability**
0 to ±5	Little to none
±5 to ±30	Developing stability
±30 to ±40	Moderate stability
±40 to ±60	Satisfactory stability
< -60 or > 60	Exceptional stability

negatively charged particle. This region of negatively charged ions surrounding the positively charged ions is called the electrical double layer (see Figure 10.2(a)). The potential energy decreases linearly from its value at the particle surface (called the surface potential) to its value at the edge of the Stern layer (called the Stern potential). The potential beyond the Stern layer continues to decrease linearly, though with a slower rate of change than within the Stern layer. The potential between the layer of solution at the edge of the negatively charged particle and the edge of the slipping plane is the zeta potential. The magnitude of the zeta potential is used to characterize the stability of a dispersion against agglomeration (see Table 10.2). At distances beyond the edge of the Stern layer, where the zeta potential is measured, the potential decreases rapidly before reaching a static value at sufficient distance within the electrical double layer (see Figure 10.2(b)).

PALS characterizes zeta potential by directing a laser beam towards a dispersion that is scattered by the particles within the dispersion. The samples used in PALS have a significantly lower solid loading ($<<1$ wt%) than actual electrode slurries. The electrophoretic mobility can be calculated based on the phase shift of the scattered photons. The zeta potential can then be calculated with the electrophoretic mobility, viscosity, and dielectric permittivity of the dispersion, the latter two of which can be assumed to be equivalent to the solvent due to the low concentration of particles in the dispersion. This, and the easier detection of scattered light, is why dilute dispersions are preferred for PALS. However, PALS and zeta potential measurements more generally are only capable of characterizing dispersants that impart electrostatic or electrosteric stability to particles in a slurry, since steric dispersants do not change the surface charge of the particles [125].

Since zeta potential changes as a function of positive ion concentration in the solution, it is best to measure zeta potential as a function of pH. Nitric acid (HNO_3) and ammonium hydroxide (NH_4OH) are used to decrease and increase the pH, respectively, of samples for PALS. The influence of pH on the active material and current collector is explored in further detail in Section 10.6.

In conjunction with finding water-soluble binders, dispersing cathode particles in water was the first problem identified with aqueous cathode processing. Early studies that featured LCO as active material used PAA-NH_4 to disperse the active material particles and had success at a low slurry concentration (0.01 wt%) [7,9]. PAA-NH_4 dissociates at a pH of 2–3, producing PAA^- species that adsorb onto the LCO particle surface. In fact, these species adsorb to the particle surface such that the zeta potential is negative, even at very acidic pH ($\sim$2), which indicates that the PAA^- causes the LCO surface charges to be more negative than without the dispersant.

Similarly, PAA may double as a binder and dispersant. PAA donates protons to solution, dissociating into PAA^- and H^+. Just as PAA^- from PAA-NH_4 provides electrostatic stability to the particles, PAA^- anions from PAA provide stability, with the added inclusion of steric stability due to the long polymer chains [63,77]. Other polyacrylates, like PAN, do not provide the same electrostatic stability to active material particles [7,63]. This provides further evidence as to why PAA may be the superior polyacrylate binder. Another binder that has been shown to provide some electrosteric stability is Na-CMC, though it provides exceptional stability over a more narrow pH range than PAA [77].

The cathode active materials that are most susceptible to agglomeration are those that are small and acicular-shaped [78], since they have a greater exposed surface area than micron-sized, spherical particles. Similarly, it is important to ensure that carbon black particles are sufficiently dispersed, since they are also small particles that are heavily prone to agglomerate. Anionic dispersants, such as sodium dodecyl sulfoxide, are ineffective in preventing carbon black agglomeration. However, there is some debate as to whether cationic dispersants, like hexadecyltrimethylammonium bromide (CTAB) and polyethylenimine (PEI), or non-ionic dispersants, like the isooctylphenylether of polyoxyethylene (commercial name Triton X-100), are the best for dispersing LFP and carbon black [36,37,40].

In terms of zeta potential, PEI is effective between pH of 2–5 for LFP, although at higher pH values the magnitude of the zeta potential is better for LFP particles without a dispersant [37]. For carbon black, Triton X-100 shows a similar particle size distribution as CTAB [40] and PEI was shown to improve the zeta potential up to pH = 9 [37]. As for electrochemical performance, both Triton X-100 and PEI are best around a concentration of 1–2 wt% [35–37,40].

10.5 Wettability to the current collector

Ensuring good wettability of the slurry to the current collector is essential to obtaining a crack-free finished product with good adhesion. More fundamental reading on the topic of wettability is available in Section 3.2.3 and the method of measuring contact angle is elaborated in Section 3.6.4. NMP-based slurries do not usually have many wettability issues with the current collector, however aqueous-based slurries can suffer from poor wettability. This may be owed to the significantly higher surface tension of water (72.8 $mN \cdot m^{-1}$ at 20°C) compared to NMP (40.8 $mN \cdot m^{-1}$ at 20°C) [59,127]. For comparison, the surface energy of an untreated Al current collector foil is about 48 $mN \cdot m^{-1}$ at room temperature [127]. The lower surface tension of NMP (relative to the surface energy of Al) makes for good wettability, while the much larger surface tension of water causes contact issues. Additionally, this large difference between liquid surface tension and solid surface energy is responsible for stronger capillary forces during drying, which leads to electrode cracking, particularly in thick electrodes [59].

Two solutions may be proposed by this conundrum: (1) increase the surface energy of the Al current collector or (2) decrease the surface tension of the aqueous slurry. The first solution has been achieved by implementing a corona plasma treatment of the Al current collector prior to coating [127]. Corona plasma treatment introduces polar groups to the surface of a substrate, thereby increasing the surface energy temporarily. A relatively low treatment energy density (0.4 $J \cdot cm^{-2}$) was able to increase the surface energy of the Al foil to the level of the water-based LFP slurry for the length of time needed to coat and dry the electrode. This also enabled superior performance in cyclability tests and at high current densities compared to untreated foils or foils treated at different energy densities [127]. This strategy has since been adopted in a few other studies with success [35,74].

Decreasing the surface tension of the slurry may be achieved by adopting a co-solvent that has a low surface tension, such as isopropyl alcohol (IPA, 23.0 $mN \cdot m^{-1}$ at 20°C) [59]. Pure water has a contact angle of 86° on Al foil, while a 4:1 mix by mass of water:IPA has a contact angle of 40° on Al foil. While a pouch cell featuring an NMC532 cathode (full cell, 4 $mAh \cdot cm^{-2}$) produced with this solvent still lagged behind a cell with a cathode produced in NMP in terms of rate capability (60% capacity retention at 2C discharge compared to 74%), the cycling performance between the two cells was comparable [59]. This was attributed to a crack-free film being produced due to the surface tension modification of the slurry.

10.6 Reactivity of cathode with water

To this point, most of the issues with aqueous cathode processing have been independent of the choice of active material. Alternatively, the reactivity of the cathode material with water is highly dependent on the chemistry of the active material. These reactions are strictly surface phenomena, as there have been no reported modifications to the crystal structure of active materials after water exposure [74,78,128].

The surface of an active material is partially dependent on its synthesis conditions. For example, Ni-rich cathode materials, like NCA and NMC811, are best synthesized using at least 5 mol% excess Li, whether the Li source is lithium hydroxide (LiOH), lithium carbonate (Li_2CO_3), or another precursor material. As a result, the unreacted precursors used in Ni-rich cathode synthesis remain on the surface of the active material particles. When exposed to water, these precursor materials are dissolved in solution and expose the cathode material to water [78]. A thermodynamically favorable process is cation exchange between the Li^+ ions in the cathode material and the protons (H^+) from the solution (see Figure 10.3). These leached Li^+ ions diminish the capacity of the cathode since the cathode is the source of reactive Li species in a full cell. Cation exchange can also lead to the formation of new surface species, such as NiOOH on NCA [129], that can impede Li^+ ion transport through the material.

In addition to reducing the capacity of the cathode, another negative consequence of cation exchange is the increase in pH of the slurry. As Li^+ ions are replaced by protons, the proton concentration in the solution decreases, causing the slurry to become more basic. Al is stable between pH of 4.0–8.5, according to the Pourbaix diagram, in aqueous media [130]. Many aqueous cathode slurries have a pH > 8.5, meaning some amount of corrosion is predictable. However, less reactive materials like LFP and LMO show less evidence of corrosion than more reactive, Ni-rich cathode materials. The pH of various cathode active materials when exposed to water is shown in Figure 10.4.

It is worth noting that exposure time [60,78] and mass fraction of active materials within an aqueous solution will have some influence on the measured pH; few of the studies cited report the elapsed time or mass fraction of active materials

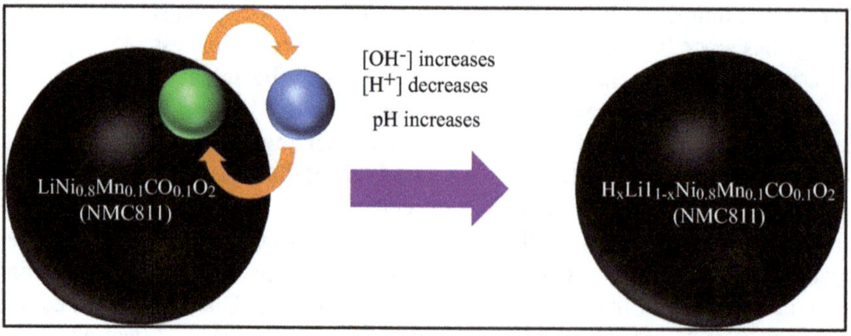

Figure 10.3 Simple schematic of cation exchange

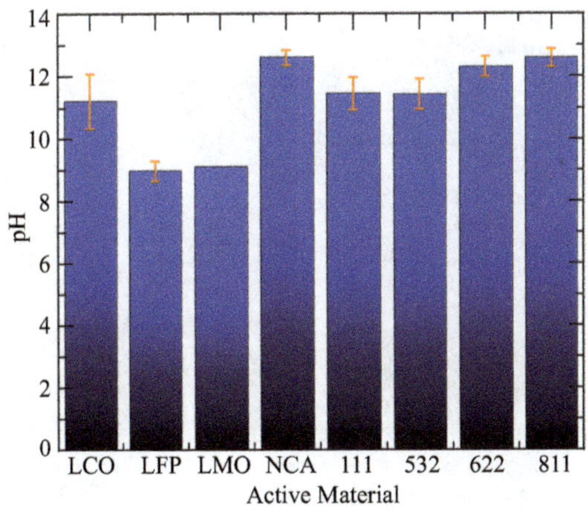

Figure 10.4 *The pH of aqueous solutions containing various cathode active materials. The average pH is calculated using values reported from the literature [7,9,10,29,31,34,41,53,60,64,77,78,128]. The error bars for the average pH are the standard deviation of the values reported. LMO does not have an error bar because, to the authors' best knowledge, there is only one study that measured the pH of LMO. The bars labelled, "111," "532," "622," and "811" represent NMC111, NMC532, NMC622, and NMC811, respectively.*

in the aqueous solution, which could cause some of the deviations. The pH of an aqueous suspension rises rapidly immediately upon contact with an active material due to the dissolution of these surface species. The gradual rise in pH that occurs after the first ~1 min of exposure is a function of the ongoing cation exchange that removes more protons from solution as time passes. On the other hand, slurry additives, like carbon black and non-acidic and non-basic binders, do not have much influence on slurry pH [74,77]. Any changes these additives cause may be attributed to the adsorption of these additives to the surface of the active material that partially block cation exchange of surface impurity dissolution, though this occurs at a small enough scale such that pH is hardly affected.

The corrosion reaction of the Al current collector proceeds according to (10.1):

$$Al + OH^- + 3H_2O \rightarrow Al(OH)_4^- + \frac{3}{2}H_2 \tag{10.1}$$

The H_2 gas bubbles that are produced are evident almost immediately upon the slurry contacting the Al current collector for Ni-rich cathode active materials (NCA, NMC811), though it can take over a minute for bubbles to appear for less reactive materials like NMC532 [131] (see Figure 10.5). H_2 gas evolution has been shown to be the primary culprit, as opposed to surface energy differences between

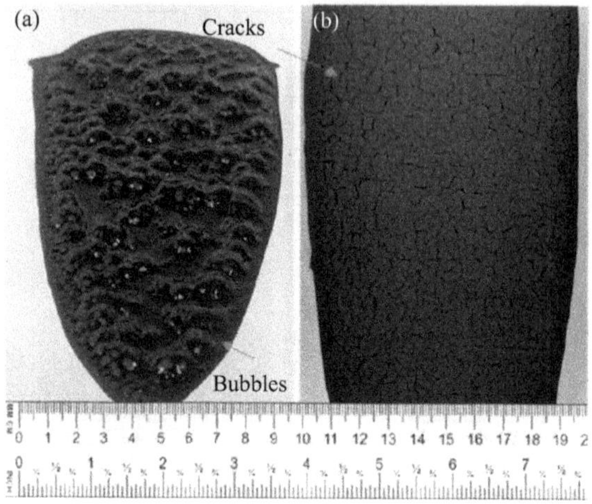

Figure 10.5 *(a) H_2 gas bubbles can appear within seconds of a basic cathode slurry contacting an Al current collector. (b) A dried cathode with many cracks due to the collapse of the bubbles formed. Reprinted with permission from Ref. [131]. Copyright 2020 American Chemical Society.*

slurry and substrate discussed in Section 10.5, for cracking in thick NMC811 cathodes. This was demonstrated by coating thick NMC811 slurries onto Cu substrates, which are stable in aqueous media at high pH. Most of the cracking had subsided by exchanging the Al current collector with Cu, with most of the remaining defects being cleared by adding IPA co-solvent to reduce the surface tension of the slurry [131].

In addition to Li leaching due to cation exchange, some transition metals can also leach from certain cathode active materials as well. In NMC materials, the dissolution of Ni, Mn, and Co has been shown to be relatively benign compared to Li leaching [60]. In fact, the dissolution of these transition metal ions has been found to be slightly more troublesome in NMP than in water, though this is certainly not the case for Li [78]. Similarly for LCO, LFP, and LMO, the Co, Fe, and Mn dissolution, respectively, is greater for the material when exposed to NMP than when exposed to water [78].

It seems that there is little time dependency on the leaching of transition metals from active materials when exposed to water, or at least that the magnitude of the leaching is so small that the time dependency cannot be established with certainty [60,78]. However, the amount of Li leached into solution does increase with time, though most is leached in the first few hours. About 60–70% of the Li present in solution after one week was leached in the first 2 h for NCA, NMC622, and NMC811, and nearly 90% of the Li present in solution after one week was leached in the first 2 h for NMC111 [60]. The amount of Li present in solution after 2 h is relatively low for NMC111 and NMC622 (228 $mg{\cdot}L^{-1}$ and 231 $mg{\cdot}L^{-1}$, respectively) and is significantly higher for NMC811 and NCA (825 $mg{\cdot}L^{-1}$ and 1,027 $mg{\cdot}L^{-1}$, respectively) [60].

The preceding discussion reveals that for a given active material in water, the amount of Li leached, and the pH measured after a particular amount of time will generally scale with the amount of Ni in the material. However, Ni content is not the sole predictor of an active material's reactivity in water. An interesting case study for this is the comparison between NMC811 and NCA, both of which feature the same amount of Ni. There is a difference in Co content, with Co comprising 15 mol% of the transition metal content of NCA and 10 mol% of the transition metal content of NMC811. However, as mentioned previously, Co dissolution in both materials is negligible [60,78]. Similarly, there is very little Mn dissolution from NMC811 in water [60]. On the other hand, Al dissolution from NCA is noticeably higher than the dissolution of Ni, Mn, and Co from NMC and NCA materials [60,78]. The dissolution of Li and Al-containing species (such as $LiAlO_2$) are believed to be the cause of this [77,129], which is supported by the finding that Li leaching is also greater in NCA than NMC811 [60].

To summarize, cathode active materials are quite different in their reactivity with water. The more reactive a material is, the more difficult is the prospect of processing that material in water. Many strategies to mitigate the reactivity of cathode materials with water have been proposed to varying levels of success. These strategies, including carbon-coated Al current collectors, pH modifying agents, and active material coatings formed both before and during slurry mixing have been proposed. These strategies are discussed further in the subsequent subsections.

10.6.1 Carbon-coated current collectors

To prevent corrosion of the Al current collector from contact with high pH cathode slurries, a thin carbon layer can be applied to the foil. Besides this purpose, thin carbon coatings on Al current collectors have been thought to be effective at improving the electronic conductivity of the electrode at the current collector/electrode interface. Literature and patents that discuss the process of fabricating carbon-coated current collectors usually describe a process that begins with the formation of a conductive glue, i.e. a carbon material such as carbon black and a binding polymer such as PVDF mixed in an organic solvent [57,132]. The glue is then applied to the surface of an Al foil using an extremely thin wet gap, then dried. Ideally, these coating layers would be made as thin as possible, as they add weight and volume to the cell without adding electrochemically active material. The carbon coating layers are ≤ 5 μm, with some layers reportedly as thin as 300 nm [132].

Carbon-coated Al current collectors were first studied for low-Ni active materials, such as NMC111 [57,68]. With this material, minimal corrosion is predicted for the current collector, although there is a roughly 15% increase in 1C discharge capacity when an aqueous-processed NMC111 cathode is coated on a 5-μm carbon-coated current collector as opposed to a bare Al foil [57]. A thinner 2 μm coating also provides a slight increase in the capacity compared to a bare Al foil. This increase in capacity is also noticeable at higher C-rates, up to at least 10C [57]. However, long-term cycling studies are lacking to demonstrate an effect of carbon coating thickness on the cyclability of an aqueous-processed cathode. Some recent studies have demonstrated excellent cyclability of LNMO cathodes that are coated on carbon-coated current collectors [92,93], though there is no comparison with a bare foil to determine the role the carbon coating plays in the cyclability. In one of these studies, the carbon coating was shown to provide the additional benefit of promoting cross-linking of the aqueous binder, which consisted of CMC and citric acid. These cathodes featuring cross-linked binders had better adhesion to the current collector and electrochemical performance compared to cathodes coated on bare Al foils [92].

10.6.2 pH modifying agents

Instead of protecting the surface of the Al current collector, another strategy is to reduce the pH of the slurry such that the current collector surface will not be corroded. Although most conductive additives and binders will not influence the pH of the slurry, some of them will, with the most prevalent example being PAA. As mentioned in Section 10.4, PAA dissociates into PAA^- and H^+ when exposed to water, with the additional protons driving the pH of the solution down. The addition of PAA has been shown to reduce the pH of water-processed slurries featuring NMC111 [51] and NCA [77,78]. However, depending on the amount of PAA added, the pH will only remain within the stability limits of the Al current collector for a few hours, emphasizing the need for quick mixing and coating [77]. Additionally, the performance of these PAA-modified cathodes is worse than the unmodified aqueous-processed baseline for NMC111 and the state of the art-processed baseline for NCA, respectively.

Another possible pH-modifying agent is acetic acid, which was also studied for NMC111. Acetic acid was able to boost the electrochemical performance compared to the unmodified and PAA-modified aqueous-processed cathodes, likely because PAA coats the active materials and increases resistance (discussed further in Section 10.6.3). Additionally, it was found that acetic acid delivered the best performance when the slurry was coated at a pH = 9 when compared to pH values in the range of 7–11 [51]. Additionally, the pH of the acetic acid-modified cathodes increases significantly when the pH rises above 9. This demonstrates the importance of not just being within a stable pH range but being at an optimal pH value.

While PAA and acetic acid have been used, the most common acidic additive used in the literature is phosphoric acid (PA) [52,53,64,65,71,91,104], for materials such as LNMO, NMC111, and NMC811. Like PAA, PA also coats the active material once mixed into the slurry, forming a Li_3PO_4 layer on the active material surface [65]. More discussion of PA as an active material coating will be had in Section 10.6.3.

Lowering the pH of the cathode slurry such that it is stable when coated onto the Al current collector is helpful in reducing corrosion; however, cathodes that are coated at pH $\leq$ 9 may have lower capacity than those coated at higher pH [51,61]. This is because the cation exchange mechanism is possibly exacerbated at lower pH [61,133]. A creative solution to allow for a high pH slurry and a pristine Al current collector is to use a carbon-coated foil and LiOH addition to the slurry. In this scheme, OH^- is donated to the solution, thus increasing the pH of the slurry while the Li inventory of the active material is retained. Indeed, the performance of a LiOH-containing, aqueous-processed NMC532 slurry (pH > 12.5) coated onto a carbon-coated Al current collector was comparable to a state-of-the-art-processed cathode and better than a NMC532 cathode reduced to a pH of 7.6 by PAA addition [61].

10.6.3 Active material coatings

As mentioned in Section 10.6.2, many pH modifying agents may also adsorb to the active material. For instance, when PAA dissociates into PAA^- and H^+ in water, and the PAA^- chains adsorb to the surface of the active material, as revealed by X-ray photoelectron spectroscopy [77]. In the case of PA, a Li_3PO_4 layer is formed on the surface of active materials [65], which has an additional advantage of being Li^+-ion conducting. In both cases, the thickness of the coating will depend on the amount of agent added to the slurry.

The question as to whether it is better to coat the active material or coat the current collector has been addressed in recent studies. For NMC532 tested under relevant conditions (2.5–4.2 V, 2.0 mAh·cm^{-2}, 30% porosity, full cells, +0.5C/−1C, room temperature), a cathode featuring Li_3PO_4-coated active material particles (due to PA addition) has better cyclability and rate capability than a cathode with a carbon-coated current collector [52]. This suggests that the reactions occurring at the NMC532 surface are more deleterious to cell performance than those occurring between the basic slurry and the current collector. Whether this is true for more Ni-rich cathodes, such as NMC811 and NCA, remains to be determined. Since these materials are more reactive with water than NMC532, it could be deduced that it is even more important to protect the surfaces of Ni-rich materials from reactions with water than it is to coat the current collector, though it is known that some modification is necessary for NCA to be processable at all [77,78].

10.7 Recycling end-of-life cathodes

The bulk of this chapter has been focused on the advantages that aqueous-processed cathodes can provide during the manufacturing stage of LIBs and on reviewing the research that has been performed to address the technical challenges of aqueous-processed cathodes. Additionally, many studies have sought to investigate whether these aqueous-processed cathodes can provide better, or at least comparable, electrochemical performance to state-of-the-art-processed cathodes. While emphasis on the production and use stages of the LIB are important, growing concerns about disposal of spent LIBs and accelerated mining of limited metals have led industry and academia to direct more attention to the third stage of the LIB, its end-of-life (EOL).

The most obvious course of action to take with LIBs at their EOL is to recycle them to avoid wasting metals and materials that could be reused. Typical recycling methods include pyrometallurgical and hydrometallurgical recycling, though direct recycling is the most cost-effective way to recycle LIBs. In direct recycling, the binder is recovered from the spent electrode by using a solvent to dissolve the binder and separate out the active material and conductive additives. In state-of-the-art-processed cathodes, PVDF will be best recovered by using NMP whose problems have been well-documented in this chapter.

Alternatively, when aqueous-processed cathodes reach their EOL, the water-soluble binders may be retrieved by using water. The active material, which will require relithiation once at its EOL, may be effectively relithiated using one of a few techniques that lie outside the scope of the present text. Spent aqueous-processed cathodes that have been directly recycled using water and re-coated to produce fresh cathodes have shown very promising performance [134]. If direct recycling of aqueous-processed cathodes can achieve similar performance to state-of-the-art-processed recycled cathodes, then an environmentally-friendly, closed-loop manufacturing system is possible with water-processed cathodes. Such a system will strongly mitigate the valid concerns of mining limited metal resources and the supply chain, cost fluctuation, and ethical concerns associated with procuring these metals.

10.8 Alternative solvent processing

Although water has certainly received the most attention as an alternative solvent to NMP, other green solvents have been proposed recently, including dimethyl sulfoxide (DMSO) [135], dihydrolevoglucosenone (Cyrene) [136], and γ-valerolactone (GVL) [137]. These solvents have an advantage over water in that they can all dissolve PVDF, meaning there is no need to seek an alternative binder to make the process scalable. In fact, these solvents were identified because of their low toxicity and similarity to NMP in their Hansen solubility parameters (HSPs). HSPs were developed by Charles Hansen in his Ph.D. thesis in 1967 and are used to determine whether a material will dissolve in a solvent [138]. Each material has three parameters, based on the intermolecular forces within the material: (1) the energy of the dispersive forces (δ_d), (2) the energy of the polar forces (δ_p), and (3) the energy of the hydrogen bonds (δ_h). The three HSPs may be plotted in three-dimensional space. To determine whether a material will dissolve in a solvent, each material is given a radius of interaction (R_0). The distance between the material and the solvent (R_a) in three-dimensional space can be calculated according to (10.2).

$$R_a = \sqrt{4(\delta_{d2} - \delta_{d1})^2 + (\delta_{p2} - \delta_{p1})^2 + (\delta_{h2} - \delta_{h1})^2} \tag{10.2}$$

Finally, the ratio between R_a and R_0 determines the relative energy difference (RED) between the material and the solvent. If RED < 1, then the solvent is predicted to dissolve the material. The HSPs and calculated R_a and RED of PVDF, NMP, DMSO, Cyrene, and GVL are shown in Table 10.3.

Table 10.3 The HSPs of PVDF, NMP, and three alternative solvents researched for cathode processing. The values for NMP are the average of the values from the two sources referenced

Polymer/solvent	δ_d (MPa)$^{0.5}$	δ_p (MPa)$^{0.5}$	δ_h (MPa)$^{0.5}$	References	R_a	RED
PVDF	17.2	12.5	9.2	[135]		
NMP	18.1	12.4	7.3	[135,137]	2.62	0.27
DMSO	18.4	16.4	10.2	[135]	4.69	0.49
Cyrene	18.1	12.4	7.3	[137]	3.79	0.39
GVL	18.5	14.1	6.6	[137]	4.01	0.42

Because all of these solvents have a RED < 1, they are considered appropriate for dissolving PVDF. However, one shortcoming of the HSPs is that they do not account for the influence of temperature on the solubility of a material in a solvent [138]. In fact, many studies have shown that PVDF is not soluble in DMSO, Cyrene, or GVL at room temperature, instead needing to be heated to 60°C, >80°C, and 70°C to dissolve PVDF, respectively [135–137,139]. Heating a slurry to these temperatures during the mixing, storage, and coating stages has been proposed previously as a means of improving the rheological properties of NMP-based cathode slurries [140], but it is an expensive endeavor that is not required. Another alternative is to use co-solvents that make PVDF soluble at lower temperatures, though this adds complexity to the solvent recycle stage and could add cost and hazards to the manufacturing process. Thus, these green solvents face a scalability challenge with their high-temperature constraints.

Despite these challenges, the electrochemical performance of cathodes processed in these alternative solvents is competitive with those processed in NMP. Table 10.4 shows a summary of the electrochemical performance of cathodes processed in DMSO, Cyrene, and GVL.

Table 10.4 Summary of studies using DMSO, Cyrene, and GVL as solvent and PVDF as binder in half cells

Solvent and cathode material	Cell and testing conditions	Performance description	References
DMSO, NMC111	3.0–4.3 V, undisclosed testing temperature, +0.5C/−0.5C, 1.4 mAh·cm^{-2}, calendered but do not know porosity	The cell featuring the DMSO-processed cathode had initial discharge capacity of 148 mAh·g^{-1} and faded to 144 mAh·g^{-1} after 50 cycles; the cell featuring the NMP-processed cathode had initial discharge capacity of 150 mAh·g^{-1} and faded to 147 mAh·g^{-1} after 50 cycles	[135]

(Continues)

Table 10.4 (Continued)

Solvent and cathode material	Cell and testing conditions	Performance description	References
Cyrene, NMC811	2.8–4.4 V, undisclosed testing temperature, +0.1C/−0.3C, 3.5 mAh·cm^{-2}, calendered but do not know porosity	The cell featuring the Cyrene-processed cathode had maximum discharge capacity of 182 mAh·g^{-1} and faded to 170 mAh·g^{-1} after 100 cycles; the cell featuring the NMP-processed cathode had maximum discharge capacity of 188 mAh·g^{-1} and faded to 176 mAh·g^{-1} after 100 cycles	[136]
GVL, NMC111	3.0–4.3 V, 25°C, +0.1C/−0.1C, 2.9 mAh·cm^{-2}, calendered to ~25% porosity	The cell featuring the GVL-processed cathode had initial discharge capacity of 142 mAh·g^{-1} and faded to 127 mAh·g^{-1} after 100 cycles; the cell featuring the NMP-processed cathode had initial discharge capacity of 144 mAh·g^{-1} and faded to 136 mAh·g^{-1} after 100 cycles	[137]

10.9 Summary

In the quest to reduce the manufacturing cost of LIBs, there have been many pathways proposed that are covered in this text. Some of these innovations make larger differences than others; conversely, many of the seemingly lower-impact innovations may be more realistic to attain, at least in the short term, than those that are more impactful. Perhaps one of the best examples of an effective combination between an *impactful* manufacturing innovation and a *realistic* manufacturing innovation is aqueous-processed cathodes. Aqueous-processed cathodes provide significant savings in material costs (binders and solvent), capital equipment costs (equipment required for solvent recovery system), and energy costs (no energy needed to operate recovery equipment, lower drying energy needed for water-based coating than NMP-based coating). Removing NMP from the process also permits a much safer work environment for workers, since NMP has been reconfirmed as a health risk by several international institutions in recent years.

The obstacles to aqueous-processed cathodes have been discussed in the preceding sections of this chapter. While there are many high-performing candidates as water-soluble binders to replace PVDF, there has yet to be a binder with the consensus approval of PVDF. Most likely, the binders used in aqueous-processed cathodes will feature Na-CMC and/or a synthetic polymer, such as a polyacrylate.

As the technology matures, a fully bioderived composite binder system with excellent performance is desirable from a sustainability and cost perspective. The use of water-processed cathodes also eliminates the need for all cathode processing to occur in a low-humidity environment, which will lead to further cost savings and simpler facility maintenance and operation. However, in these production scenarios, it is exceedingly important to ensure the cathodes are extremely dry since they may not have the benefit of being completely fabricated in a dry room and the water-soluble binders retain more water than PVDF does for a comparable NMP-processed cathode. This stresses the importance of high-quality secondary drying techniques that do not slow down or bottleneck the manufacturing process.

Additional concerns about aqueous-processed cathodes exist, including the proper dispersion of particles and the proper wettability of the electrode to the current collector. As discussed in this chapter, many common aqueous-processing binders can act as dispersants in small quantities. Wettability of the cathode to the current collector becomes more problematic as the thickness of the electrode increases, so for some thin cathodes, perhaps no particular attention to wettability is necessary. However, thicker cathodes may benefit from an agent to reduce the surface tension of the slurry or from a treatment of the current collector foil to increase its surface energy. Another possibility for thick aqueous-processed cathodes would be coating multiple layers of cathode in separate steps, as discussed in Section 9.4.

Nearly all cathode active materials are reactive with water, though, as discussed, there is significant deviance in the degree of reactivity. For a material like LFP, the reactivity is relatively low. However, Ni-rich materials tend to be highly reactive with water and require mitigation strategies to avoid surface reconstruction of the material and/or corrosion of the current collector. Much of the research on these strategies for high-capacity active materials, like NCA and NMC811, has been published within the last few years, and there is not yet a consensus on the best strategy to make aqueous-processed cathodes with these materials. Certainly, the best strategy will protect the surface of the material from Li-ion (and, in the case of NCA, Al-ion) leaching upon exposure to the solvent, which will keep the pH of the slurry stable and prevent the material from losing capacity and forming a high-impedance surface layer. Finding a strategy that is cost effective, simple to execute, and easy to replicate, however, will be challenging. Additionally in this chapter, alternative solvents to water were discussed that would provide cost savings and minimize health risks to workers. While these solvents are capable of dissolving PVDF, the solvents proposed thus far are only capable of doing so at high temperatures ($\geq 60°C$), which brings into question their feasibility, at least for now.

References

[1] Sitarek, K., J. Stetkiewicz, and W. Wasowicz, Evaluation of reproductive disorders in female rats exposed to N-Methyl-2-Pyrrolidone. *Birth Defects Research (Part B)*, 2012. 95: p. 195–201.

[2] Sitarek, K. and J. Stetkiewicz, Assessment of reproductive toxicity and gonadotoxic potential of *N*-methyl-2-pyrrolidone in male rats. *International Journal of Occupational Medicine and Environmental Health*, 2008. 21(1): p. 73–80.

[3] *N*-Methylpyrrolidone (NMP) Risk Evaluation and Risk Management Under TSCA Section 6. 2021.

[4] Commission Regulation (EU) 2018/588. 2018.

[5] Wood, D.L., J.D. Quass, J. Li, S. Ahmed, D. Ventola, and C. Daniel, Technical and economic analysis of solvent-based lithium-ion electrode drying with water and NMP. *Drying Technology*, 2018. 36(2): p. 234–244.

[6] Li, J., J. Fleetwood, W.B. Hawley, and W. Kays, From materials to cell: State-of-the-art and prospective technologies for lithium-ion battery electrode processing. *Chemical Reviews*, 2022. 122: p. 903–956.

[7] Li, C.-C., J.-H. Lee, C.-Y. Lo, and M.-S. Wu, Effects of PAA-NH_4 addition on the dispersion property of aqueous $LiCoO_2$ slurries and the cell performance of as-prepared $LiCoO_2$ cathodes. *Electrochemical and Solid-State Letters*, 2005. 8(10): p. A509–A512.

[8] Isozumi, H., T. Horiba, K. Kubota, *et al.*, Application of modified styrene-butadiene-rubber-based latex binder to high-voltage operating $LiCoO_2$ composite electrodes for lithium-ion batteries. *Journal of Power Sources*, 2020. 468: 228332.

[9] Li, C.-C., J.-T. Lee, and X.-W. Peng, Improvements of dispersion homogeneity and cell performance of aqueous-processed $LiCoO_2$ cathodes by using dispersant of PAA-NH_4. *Journal of The Electrochemical Society*, 2006. 153(5): p. A809–A815.

[10] Li, C.-C., J.-H. Lee, Y.-L. Yung, and C.-R. Yang, Effects of pH on the dispersion and cell performance of $LiCoO_2$ cathodes based on the aqueous process. *Journal of Materials Science*, 2007. 42: p. 5773–5777.

[11] Li, C.-C. and Y.-W. Wang, Binder distributions in water-based and organic-based $LiCoO_2$ electrode sheets and their effects on cell performance. *Journal of the Electrochemical Society*, 2011. 158(12): p. A1361–A1370.

[12] Li, C.-C. and Y.-W. Wang, Importance of binder compositions to the dispersion and electrochemical properties of water-based $LiCoO_2$ cathodes. *Journal of Power Sources*, 2013. 227: p. 204–210.

[13] Spreafico, M.A., P. Cojocaru, L. Magagnin, F. Triulzi, and M. Apostolo, PVDF latex as a binder for positive electrodes in lithium-ion batteries. *Industrial & Engineering Chemistry Research*, 2014. 53: p. 9094–9100.

[14] Yabuuchi, N., Y. Kinoshita, K. Misaki, T. Matsuyama, and S. Komaba, Electrochemical properties of $LiCoO_2$ electrodes with latex binders on high-voltage exposure. *Journal of The Electrochemical Society*, 2015. 162: p. A538–A544.

[15] Apraksin, R.V., S.N. Eliseeva, E.G. Tolstopjatova, A.M. Rumyantsev, V.V. Zhdanov, and V.V. Kondratiev, High-rate performance of $LiFe_{0.4}Mn_{0.6}PO_4$ cathode materials with poly(3,4-ethylenedioxythiopene):poly(styrene sulfonate)/carboxymethylcellulose. *Materials Letters*, 2016. 176: p. 248–252.

[16] Cai, Z.P., Y. Liang, W.S. Li, L.D. Xing, and Y.H. Liao, Preparation and performances of $LiFePO_4$ cathode in aqueous solvent with polyacrylic acid as a binder. *Journal of Power Sources*, 2009. 189(1): p. 547–551.

[17] Chong, J., S. Xun, H. Zheng, *et al.*, A comparative study of polyacrylic acid and poly(vinylidene difluoride) binders for spherical natural graphite/ $LiFePO_4$ electrodes and cells. *Journal of Power Sources*, 2011. 196: p. 7707–7714.

[18] Daigle, J.-C., F. Barray, C. Gagnon, *et al.*, Amphiphilic latex as a water-based binder for $LiFePO_4$ cathode. *Journal of Power Sources*, 2019. 415: p. 172–178.

[19] Das, P.R., L. Komsiyska, O. Osters, and G. Wittstock, Effect of solid loading on the processing and behavior of PEDOT:PSS binder based composite cathodes for lithium ion batteries. *Synthetic Metals*, 2016. 215: p. 86–94.

[20] Das, P.R., L. Komsiyska, O. Osters, and G. Wittstock, PEDOT:PSS as a functional binder for cathodes in lithium ion batteries. *Journal of the Electrochemical Society*, 2015. 162: p. A674–A678.

[21] Eliseeva, S.N., O.V. Levin, E.G. Tolstopjatova, E.V. Alekseeva, R.V. Apraksin, and V.V. Kondratiev, New functional conducting poly-3,4-ethylenedioxythiopene: polystyrene sulfonate/carboxymethylcellulose binder for improvement of capacity of $LiFePO_4$-based cathode materials. *Materials Letters*, 2015. 161: p. 117–119.

[22] Guerfi, A., M. Kaneko, M. Petitclerc, M. Mori, and K. Zaghib, $LiFePO_4$ water-soluble binder electrode for Li-ion batteries. *Journal of Power Sources*, 2007. 163: p. 1047–1052.

[23] He, J., J. Wang, H. Zhong, J. Ding, and L. Zhang, Cyanoethylated carboxymethyl chitosan as water soluble binder with enhanced adhesion capability and electrochemical performances for $LiFePO_4$ cathode. *Electrochimica Acta*, 2015. 182: p. 900–907.

[24] He, J., H. Zhong, J. Wang, and L. Zhang, Investigation on xanthan gum as novel water soluble binder for $LiFePO_4$ cathode in lithium-ion batteries. *Journal of Alloys and Compounds*, 2017. 714: p. 409–418.

[25] He, J., H. Zhong, and L. Zhang, Water-soluble binder PAALi with terpene resin emulsion as tackifier for $LiFePO_4$ cathode. *Journal of Applied Polymer Science*, 2018. 135: p. 46132.

[26] Huang, J., J. Wang, H. Zhong, and L. Zhang, N-cyanoethyl polyethylenimine as a water-soluble binder for $LiFePO_4$ cathode in lithium-ion batteries. *Journal of Materials Science*, 2018. 53: p. 9690–9700.

[27] Huang, S., J. Ren, R. Liu, *et al.*, Enhanced electrochemical properties of $LiFePO_4$ cathode using waterborne lithiated ionomer binder in Li-ion batteries with low amount. *ACS Sustainable Chemistry & Engineering*, 2018. 6: p. 12650–12657.

[28] Jeong, S.S., N. Bockenfeld, A. Balducci, and S. Passerini, Natural cellulose as binder for lithium battery electrodes. *Journal of Power Sources*, 2012. 199: p. 331–335.

[29] Kvasha, A., I. Urdampilleta, I. de Meatza, *et al.*, Towards high durable lithium ion batteries with waterborne $LiFePO_4$ electrodes. *Electrochimica Acta*, 2016. 215: p. 238–246.

[30] Lee, J.-H., J.-S. Kim, Y.C. Kim, *et al.*, Effect of carboxymethyl cellulose on aqueous processing of $LiFePO_4$ cathodes and their electrochemical performance. *Electrochemical and Solid-State Letters*, 2008. 11(10): p. A175–A178.

[31] Lee, J.-H., J.-S. Kim, Y.C. Kim, D.S. Zang, and U. Paik, Dispersion properties of aqueous-based $LiFePO_4$ pastes and their electrochemical performance for lithium batteries. *Ultramicroscopy*, 2008. 108: p. 1256–1259.

[32] Li, C.-C., C.-A. Chen, and M.-F. Chen, Gelation mechanism of organic additives with $LiFePO_4$ in the water-based cathode slurries. *Ceramics International*, 2017. 43: p. S765–S770.

[33] Li, C.-C. and Y.-S. Lin, Interactions between organic additives and active powders in water-based lithium iron phosphate electrode slurries. *Journal of Power Sources*, 2012. 220: p. 413–421.

[34] Li, C.-C., X.-W. Peng, J.-T. Lee, and F.-M. Wang, Using poly(4-styrene sulfonic acid) to improve the dispersion homogeneity of aqueous-processed $LiFePO_4$ cathodes. *Journal of the Electrochemical Society*, 2010. 157(4): p. A517–A520.

[35] Li, J., B. L. Armstrong, C. Daniel, J. Kiggans, and D. L. Wood III, Optimization of multicomponent aqueous suspensions of lithium iron phosphate ($LiFePO_4$) nanoparticles and carbon black for lithium-ion battery cathodes. *Journal of Colloid and Interface Science*, 2013. 405(405): p. 118–124.

[36] Li, J., B. L. Armstrong, J. Kiggans, C. Daniel, and D. L. Wood III, Lithium ion cell performance enhancement using aqueous $LiFePO_4$ cathode dispersions and polyethyleneimine dispersant. *Journal of the Electrochemical Society*, 2013. 160(2): p. A201–A206.

[37] Li, J., B. L. Armstrong, J. Kiggans, C. Daniel, and D. L. Wood III, Optimization of $LiFePO_4$ nanoparticle suspensions with polyethyleneimine for aqueous processing. *Langmuir*, 2012. 28: p. 3783–3790.

[38] Lux, S.F., F. Schappacher, A. Balducci, S. Passerini, and M. Winter, Low cost, environmentally benign binders for lithium-ion batteries. *Journal of the Electrochemical Society*, 2010. 157(3): p. A320–A325.

[39] Ma, X., S. Zou, A. Tang, *et al.*, Three-dimensional hierarchical walnut kernel shape conducting polymer as water soluble binder for lithium-ion battery. *Electrochimica Acta*, 2018. 269: p. 571–579.

[40] Porcher, W., B. Lestriez, S. Jouanneau, and D. Guyomard, Optimizing the surfactant for the aqueous processing of $LiFePO_4$ composite electrodes. *Journal of Power Sources*, 2010. 195: p. 2835–2843.

[41] Porcher, W., B. Lestriez, S. Jouanneau, and D. Guyomard, Design of aqueous processed thick $LiFePO_4$ composite electrodes for high-energy lithium battery. *Journal of the Electrochemical Society*, 2009. 156(3): p. A133–A144.

[42] Prasanna, K., T. Subburaj, Y.N. Jo, W.J. Lee, and C.W. Lee, Environment-friendly cathodes using biopolymer chitosan with enhanced electrochemical behavior for use in lithium ion batteries. *ACS Applied Materials & Interfaces*, 2015. 7: p. 7884–7890.

[43] Prosini, P.P., C. Cento, M. Carewska, and A. Masci, Electrochemical performance of Li-ion batteries assembled with water-processable electrodes. *Solid State Ionics*, 2015. 274: p. 34–39.

[44] Raj, H. and A. Sil, Aqueous processing based novel composite electrode for Li-ion batteries using an environmentally benign binder. *Ceramics International*, 2021. 47: p. 34639–34647.

[45] Starke, B., S. Seidlmayer, S. Jankowsky, O. Dolotko, R. Gilles, and K.-H. Pettinger, Influence of particle morphologies of $LiFePO_4$ on water- and solvent-based processing and electrochemical properties. *Sustainability*, 2017. 9: 888.

[46] Sun, M., H. Zhong, S. Jiao, H. Shao, and L. Zhang, Investigation on carboxymethyl chitosan as new water soluble binder for $LiFePO_4$ cathode in Li-ion batteries. *Electrochimica Acta*, 2014. 127: p. 239–244.

[47] Wu, C.-Y. and J.-G. Duh, Ionic network for aqueous-polymer binders to enhance the electrochemical performance of Li-ion batteries. *Electrochimica Acta*, 2019. 294: p. 22–27.

[48] Yang, S., Y. Huang, S. Su, G. Han, and J. Liu, Hybrid humics/sodium carboxymethyl cellulose water-soluble binder for enhancing the electrochemical performance of a Li-Ion battery cathode. *Powder Technology*, 2019. 351: p. 203–211.

[49] Zheng, Z., X. Gao, and Y. Luo, Influence of copolymer chain sequence on electrode latex binder for lithium-ion batteries. *Colloid and Polymer Science*, 2019. 297: p. 1287–1299.

[50] Zhong, H., A. He, J. Lu, M. Sun, J. He, and L. Zhang, Carboxymethyl chitosan/conducting polymer as water-soluble composite binder for $LiFePO_4$ cathode in lithium ion batteries. *Journal of Power Sources*, 2016. 336: p. 107–114.

[51] Bauer, W., F.A. Cetinel, M. Muller, and U. Kaufmann, Effects of pH control by acid addition at the aqueous processing of cathodes for lithium ion batteries. *Electrochimica Acta*, 2019. 317: p. 112–119.

[52] Bichon, M., D. Sotta, E. De Vito, W. Porcher, and B. Lestriez, Performance and ageing behavior of water-processed $LiNi_{0.5}Mn_{0.3}Co_{0.2}O_2$/graphite lithium-ion cells. *Journal of Power Sources*, 2021. 483: 229097.

[53] Bichon, M., D. Sotta, N. Dupre, *et al.*, Study of immersion of $LiNi_{0.5}Mn_{0.3}Co_{0.2}O_2$ material in water for aqueous processing of positive electrode for Li-ion batteries. *ACS Applied Materials & Interfaces*, 2019. 11: p. 18331–18341.

[54] Cetinel, F.A. and W. Bauer, Processing of water-based $LiNi_{1/3}Mn_{1/3}Co_{1/3}O_2$ pastes for manufacturing lithium ion battery cathodes. *Bulletin of Materials Science*, 2014. 37(7): p. 1685–1690.

[55] Chen, Z., G.-T. Kim, D. Chao, *et al.*, Toward greener lithium-ion batteries: aqueous binder-based $LiNi_{0.4}Co_{0.2}Mn_{0.4}O_2$ cathode material with superior electrochemical performance. *Journal of Power Sources*, 2017. 372: p. 180–187.
[56] Cui, Y., J. Chen, J. Zhao, *et al.*, Aqueous lithium carboxymethyl cellulose and polyacrylic acid/acrylate copolymer composite binder for the $LiNi_{0.5}Mn_{0.3}Co_{0.2}O_2$ cathode of lithium-ion batteries. *Journal of the Electrochemical Society*, 2022. 169: 010513.
[57] Doberdo, I., N. Loeffler, N. Laszczynski, *et al.*, Enabling aqueous binders for lithium battery cathodes – carbon coating of aluminum current collector. *Journal of Power Sources*, 2014. 248: p. 1000–1006.
[58] Du, Z., J. Li, M. Wood, C. Mao, C. Daniel, and D.L. Wood III, Three-dimensional conductive network formed by carbon nanotubes in aqueous processed NMC electrode. *Electrochimica Acta*, 2018. 270: p. 54–61.
[59] Du, Z., K.M. Rollag, J. Li, *et al.*, Enabling aqueous processing for crack-free thick electrodes. *Journal of Power Sources*, 2017. 354: p. 200–206.
[60] Hofmann, M., M. Kapuschinski, U. Guntow, and G.A. Giffin, Implications of aqueous processing for high energy density cathode materials: Part I. Ni-rich layered oxides. *Journal of the Electrochemical Society*, 2020. 167: 140512.
[61] Ibing, L., T. Gallasch, A. Friesen, *et al.*, The role of the pH value in water-based pastes on the processing and performance of Ni-rich $LiNi_{0.5}Mn_{0.3}Co_{0.2}O_2$ based positive electrodes. *Journal of Power Sources*, 2020. 475.
[62] Ibing, L., T. Gallasch, P. Schneider, *et al.*, Towards water based ultra-thick Li ion battery electrodes – a binder approach. *Journal of Power Sources*, 2019. 423: p. 183–191.
[63] Kuo, J.-H. and C.-C. Li, Water-based process to the preparation of nickel-rich Li $(Ni_{0.8}Co_{0.1}Mn_{0.1})O_2$ cathode. *Journal of the Electrochemical Society*, 2020. 167: 100504.
[64] Kukay, A., R. Sahore, A. Parejiya, W.B. Hawley, J. Li, and D.L. Wood III, Aqueous Ni-rich-cathode dispersions processed with phosphoric acid for lithium-ion batteries with ultra-thick electrodes. *Journal of Colloid and Interface Science*, 2021. 581: p. 635–643.
[65] Loeffler, N., G.-T. Kim, F. Mueller, *et al.*, In situ coating of Li $[Ni_{0.33}Mn_{0.33}Co_{0.33}]O_2$ particles to enable aqueous electrode processing. *ChemSusChem*, 2016. 9: p. 1112–1117.
[66] Loeffler, N., G.-T. Kim, S. Passerini, *et al.*, Performance and ageing robustness of graphite/NMC pouch prototypes manufactured through eco-friendly materials and processes. *ChemSusChem*, 2017. 10: p. 3581–3587.
[67] Loeffler, N., T. Kopel, G.-T. Kim, and S. Passerini, Polyurethane binder for aqueous processing of Li-ion battery electrodes. *Journal of the Electrochemical Society*, 2015. 162: p. A2692–A2698.

[68] Loeffler, N., J. von Zamory, N. Laszczynski, I. Doberdo, G.-T. Kim, and S. Passerini, Performance of $LiNi_{1/3}Mn_{1/3}Co_{1/3}O_2$/graphite batteries based on aqueous binder. *Journal of Power Sources*, 2014. 248: p. 915–922.

[69] Memm, M., A. Hoffmann, and M. Wohlfahrt-Mehrens, Water-based $LiNi_{1/3}Mn_{1/3}Co_{1/3}O_2$-cathodes with good electrochemical performance by use of additives. *Electrochimica Acta*, 2018. 260: p. 664–673.

[70] Pedaballi, S. and C.-C. Li, Effects of surface modification and organic binder type on cell performance of water-processed Ni-rich $Li(Ni_{0.8}Co_{0.1}Mn_{0.1})O_2$ cathodes. *Journal of Power Sources*, 2020. 472: 228552.

[71] Neidhart, L., K. Frohlich, N. Eshraghi, D. Cupid, F. Winter, and M. Jahn, Aqueous manufacturing of defect-free thick multi-layer NMC811 electrodes. *Nanomaterials*, 2022. 12: p. 317.

[72] Radloff, S., R.-G. Scurtu, M. Holzle, and M. Wohlfahrt-Mehrens, Applying established water-based binders to aqueous processing of $LiNi_{0.83}Co_{0.12}Mn_{0.05}O_2$ positive electrodes. *Journal of the Electrochemical Society*, 2021. 168: 100506.

[73] Soeda, K., M. Yamagata, and M. Ishikawa, Alginic acid as a new aqueous slurry-based binder for cathode materials of LIB. *ECS Transactions*, 2015. 64: p. 13–22.

[74] Wood, M., J. Li, R.E. Ruther, *et al.*, Chemical stability and long-term cell performance of low-cobalt, Ni-rich cathodes prepared by aqueous processing for high-energy Li-ion batteries. *Energy Storage Materials*, 2020. 24: p. 188–198.

[75] Xu, J., S.-L. Chou, Q.-F. Gu, H.-K. Liu, and S.-X. Dou, The effect of different binders on electrochemical properties of $LiNi_{1/3}Mn_{1/3}Co_{1/3}O_2$ cathode material in lithium ion batteries. *Journal of Power Sources*, 2013. 225: p. 172–178.

[76] Zhong, H., M. Sun, Y. Li, J. He, J. Yang, and L. Zhang, The polyacrylic latex: an efficient water-soluble binder for $LiNi_{1/3}Co_{1/3}Mn_{1/3}O_2$ cathode in Li-ion batteries. *Journal of Solid State Electrochemistry*, 2016. 20: p. 1–8.

[77] Hawley, W.B., H.M. Meyer III, and J. Li, Enabling aqueous processing for $LiNi_{0.80}Co_{0.15}Al_{0.05}O_2$ (NCA)-based lithium-ion battery cathodes using polyacrylic acid. *Electrochimica Acta*, 2021. 380: 138203.

[78] Hawley, W.B., A. Parejiya, Y. Bai, H.M. Meyer III, D.L. Wood III, and J. Li, Lithium and transition metal dissolution due to aqueous processing in lithium-ion battery cathode active materials. *Journal of Power Sources*, 2020. 466: 228315.

[79] Hofmann, M., F. Nagler, U. Guntow, G. Sextl, and G.A. Giffin, Long-term cycling performance of aqueous processed Ni-rich $LiNi_{0.8}Co_{0.15}Al_{0.05}O_2$ cathodes. *Journal of the Electrochemical Society*, 2021. 168: 060511.

[80] Hofmann, M., F. Nagler, M. Kapuschinski, U. Guntow, and G.A. Giffin, Surface modification of $LiNi_{0.8}Co_{0.15}Al_{0.05}O_2$ particles via Li_3PO_4 coating to enable aqueous electrode processing. *ChemSusChem*, 2020. **13**: p. 5962–5971.

[81] Kimura, K., K. Onishi, T. Sakamoto, K. Asami, and M. Yanagida, Achievement of the high-capacity retention rate for the Li $[Ni_{0.8}Co_{0.15}Al_{0.05}]O_2$ (NCA) cathode containing an aqueous binder with CO_2 gas treatment using the cavitation effect (CTCE). *Journal of the Electrochemical Society*, 2019. 166: p. A5313–A5317.

[82] Kimura, K., T. Sakamoto, T. Mukai, *et al.*, Improvement of the cyclability and Coulombic efficiency of Li-ion batteries using $Li[Ni_{0.8}Co_{0.15}Al_{0.05}]O_2$ cathode containing an aqueous binder with pressurized CO_2 gas treatment. *Journal of the Electrochemical Society*, 2018. 165(2): p. A16–A20.

[83] Tanabe, T., Y. Liu, K. Miyamoto, *et al.*, Synthesis of water-resistant thin TiO_x layer-coated high-voltage and high-capacity $LiNi_aCo_bAl_{1-a-b}O_2$ ($a > 0.85$) cathode and its cathode performance to apply a water-based hybrid polymer binder to Li-ion batteries. *Electrochimica Acta*, 2017. 258: p. 1348–1355.

[84] Watanabe, T., K. Hirai, F. Ando, *et al.*, Surface double coating of a $LiNi_aCo_bAl_{1-a-b}O_2$ ($a > 0.85$) cathode with TiOx and Li2CO3 to apply a water-based hybrid polymer binder to Li-ion batteries. *RSC Advances*, 2020. 10: p. 13642–13654.

[85] Watanabe, T., T. Yokokawa, M. Yamada, *et al.*, Surface coating of a $LiNi_xCo_yAl_{1-x-y}O_2$ ($x > 0.85$) cathode with Li_3PO_4 for applying a water-based hybrid polymer binder during Li-ion battery preparation. *RSC Advances*, 2021. 11: p. 37150–37161.

[86] Abeywardana, M.Y., N. Laszczynski, M. Kuenzel, D. Bresser, S. Passerini, and B. Lucht, Increased cycling performance of Li-ion batteries by phosphoric acid modified $LiNi_{0.5}Mn_{1.5}O_4$ cathodes in the presence of LiBOB. *International Journal of Electrochemistry*, 2019. 2019.

[87] Bigoni, F., F. De Giorgio, F. Soavi, and C. Arbizzani, Sodium alginate: a water-processable binder in high-voltage cathode formulations. *Journal of the Electrochemical Society*, 2017. 164: p. A6171–A6177.

[88] De Giorgio, F., N. Laszczynski, J. von Zamory, M. Mastragostino, C. Arbizzani, and S. Passerini, Graphite//$LiNi_{0.5}Mn_{1.5}O_4$ cells based on environmentally friendly made-in-water electrodes. *ChemSusChem*, 2017. 10: p. 379–386.

[89] Dienwiebel, I., M. Diehl, B. Heidrich, X. Yang, M. Winter, and M. Borner, Enabling aqueous processing for $LiNi_{0.5}Mn_{1.5}O_4$-based positive electrodes in lithium ion batteries by applying lithium-based processing additives. *Advanced Energy and Sustainability Research*, 2021. 2: 2100075.

[90] Hitomi, S., K. Kubota, T. Horiba, *et al.*, Application of acrylic-rubber-based latex binder to high-voltage spinel electrodes of lithium-ion batteries. *ChemElectroChem*, 2019. 6: p. 5070–5079.

[91] Kuenzel, M., D. Bresser, T. Diemant, *et al.*, Complementary strategies toward the aqueous processing of high-voltage $LiNi_{0.5}Mn_{1.5}O_4$ lithium-ion cathodes. *ChemSusChem*, 2018. 11: p. 562–573.

[92] Kuenzel, M., D. Bresser, G.-T. Kim, P. Axmann, M. Wohlfahrt-Mehrens, S. Passerini, Unveiling and amplifying the benefits of carbon-coated

aluminum current collectors for sustainable $LiNi_{0.5}Mn_{1.5}O_4$. *ACS Applied Energy Materials*, 2020. 3: p. 218–230.

[93] Kuenzel, M., H. Choi, F. Wu, *et al.*, Co-crosslinked water-soluble biopolymers as a binder for high-voltage $LiNi_{0.5}Mn_{1.5}O_4$ | graphite lithium-ion full cells. *ChemSusChem*, 2020. 13: p. 2650–2660.

[94] Kuenzel, M., R. Porhiel, D. Bresser, *et al.*, Deriving structure-performance relations of chemically modified chitosan binders for sustainable high-voltage $LiNi_{0.5}Mn_{1.5}O_4$. *Batteries & Supercaps*, 2020. 3: p. 155–164.

[95] Pieczonka, N.P.W., V. Borgel, B. Ziv, *et al.*, Lithium polyacrylate (LiPAA) as an advanced binder and a passivating agent for high-voltage Li-ion batteries. *Advanced Energy Materials*, 2015. 5: 1501008.

[96] Tanabe, T., T. Gunji, Y. Honma, *et al.*, Preparation of water-resistance surface coated high-voltage $LiNi_{0.5}Mn_{1.5}O_4$ cathode and its cathode performance to apply a water-based hybrid polymer binder to Li-ion batteries. *Electrochimica Acta*, 2017. 224: p. 429–438.

[97] Wang, Z., N. Dupre, A.-C. Gaillot, *et al.*, CMC as a binder in $LiNi_{0.4}Mn_{1.6}O_4$ 5 V cathodes and their electrochemical performance for Li-ion batteries. *Electrochimica Acta*, 2012. 62: p. 77–83.

[98] Zhong, H., J. He, and L. Zhang, Better cycle stability and rate capability of high-voltage $LiNi_{0.5}Mn_{1.5}O_4$ cathode using water soluble binder. *Materials Research Bulletin*, 2017. 93: p. 194–200.

[99] Zhong, H., J. Lu, A. He, M. Sun, J. He, and L. Zhang, Carboxymethyl chitosan/poly(ethylene oxide) water soluble binder: challenging application for 5 V $LiNi_{0.5}Mn_{1.5}O_4$ cathode. *Journal of Materials Science & Technology*, 2017. 33: p. 763–767.

[100] Ryou, M.-H., S. Hong, M. Winter, H. Lee, and J.W. Choi, Improved cycle lives of $LiMn_2O_4$ cathodes in lithium ion batteries by an alginate biopolymer from seaweed. *Journal of Materials Chemistry A*, 2013. 1: p. 15224–15229.

[101] Zhang, Z., T. Zeng, Y. Lai, M. Jia, and J. Li, A comparative study of different binders and their effects on electrochemical properties of $LiMn_2O_4$ cathode in lithium ion batteries. *Journal of Power Sources*, 2014. 247: p. 1–8.

[102] Han, Z., H. Zhan, and Y. Zhou, Preparation and performance of layered $Li[Li_{0.182}Ni_{0.182}Co_{0.091}Mn_{0.545}]O_2$ cathode with different binders. *Materials Letters*, 2014. 114: p. 48–51.

[103] Iturrondobeitia, A., A. Kvasha, J.-M. Lopez del Amo, *et al.*, A comparative study of aqueous and organic processed $Li_{1.2}Ni_{0.2}Mn_{0.6}O_2$ Li-rich cathode materials for advanced lithium-ion batteries. *Electrochimica Acta*, 2017. 247: p. 420–425.

[104] Kazzazi, A., D. Bresser, A. Birrozzi, J. von Zamory, M. Hekmatfar, and S. Passerini, Comparative analysis of aqueous binders for high-energy Li-rich NMC as a lithium-ion cathode and the impact of adding phosphoric acid. *ACS Applied Materials & Interfaces*, 2018. 10: p. 17214–17222.

[105] Laszczynski, N., J. von Zamory, J. Kalhoff, N. Loeffler, V.S.K. Chakravadhanula, and S. Passerini, Improved performance of VO_x-coated Li-rich NMC electrodes. *ChemElectroChem*, 2015. 2: p. 1768–1773.

[106] Li, J., R. Klopsch, S. Nowak, M. Kunze, M. Winter, and S. Passerini, Investigations on cellulose-based high voltage composite cathodes for lithium ion batteries. *Journal of Power Sources*, 2011. 196: p. 7687–7691.

[107] Notake, K., T. Gunji, H. Kokubun, *et al.*, The application of a water-based hybrid polymer binder to a high-voltage and high-capacity Li-rich solid-solution cathode and its performance in Li-ion batteries. *Journal of Applied Electrochemistry*, 2016. 46: p. 267–278.

[108] Wu, Q., S. Ha, J. Prakash, D.W. Dees, W. Lu, Investigations on high energy lithium-ion batteries with aqueous binder. *Electrochimica Acta*, 2013. 114: p. 1–6.

[109] Yin, Z.-W., T. Zhang, S.-J. Zhang, *et al.*, Understanding the role of water-soluble guar gum binder in reducing capacity fading and voltage decay of Li-rich cathode for Li-ion batteries. *Electrochimica Acta*, 2020. 351: 136401.

[110] Yu, M., Y. Wang, Z.-Y. Wang, *et al.*, Weakened capacity fading of Li-rich cathode via aqueous binder for advanced lithium ion batteries. *Journal of the Electrochemical Society*, 2019. 166: p. A4122–A4127.

[111] Zhang, G., B. Qiu, Y. Xia, *et al.*, Double-helix-superstructure aqueous binder to boost excellent electrochemical performance in Li-rich layered oxide cathode. *Journal of Power Sources*, 2019. 420: p. 29–37.

[112] Zhang, T., J.-T. Li, J. Liu, *et al.*, Suppressing the voltage-fading of layered lithium-rich cathode materials via an aqueous binder for Li-ion batteries. *Chemical Communications*, 2016. 52: p. 4683–4686.

[113] Zhao, T., Y. Meng, R. Ji, F. Wu, L. Li, and R. Chen, Maintaining structure and voltage stability of Li-rich cathode materials by green water-soluble binders containing Na^+ ions. *Journal of Alloys and Compounds*, 2019. 811: 152060.

[114] Bresser, D., D. Buchholz, A. Moretti, A. Varzi, and S. Passerini, Alternative binders for sustainable electrochemical energy storage – the transition to aqueous electrode processing and bio-derived polymers. *Energy & Environmental Science*, 2018. 11: p. 3096–3127.

[115] Li, C.-C., S.-J. Chang, and C.-A. Chen, Effects of sp^2- and sp^3-carbon coatings on dissolution and electrochemistry of water-based $LiFePO_4$ cathodes. *Journal of Applied Electrochemistry*, 2017. 47: p. 1065–1072.

[116] Zhang, Q., X. Xie, W. Fan, and X. Wang, Lithium polyacrylate-coated $LiMn_2O_4$ cathode materials with excellent performance for lithium ion batteries. *Ionics*, 2016. 22: p. 2273–2280.

[117] Qiu, L., Z. Shao, D. Wang, F. Wang, W. Wang, and J. Wang, Carboxymethyl cellulose lithium (CMC-Li) as a novel binder and its electrochemical performance in lithium-ion batteries. *Cellulose*, 2014. 21: p. 2789–2796.

[118] Qiu, L., Z. Shao, D. Wang, W. Wang, F. Wang, and J. Wang, Enhanced electrochemical properties of $LiFePO_4$ (LFP) cathode using the carboxymethyl cellulose lithium (CMC-Li) as novel binder in lithium-ion battery. *Carbohydrate Polymers*, 2014. 111: p. 588–591.
[119] Qiu, L., Z. Shao, D. Wang, *et al.*, Novel functional carboxymethyl cellulose lithium (CMC-Li) for enhanced performance of lithium-ion batteries. *RSC Advances*, 2014. 4: p. 24859–24862.
[120] Qiu, L., Z. Shao, W. Wang, *et al.*, Enhanced cyclability of C/lithium iron phosphate cathodes with a novel water-soluble lithium-ion binder. *Electrochimica Acta*, 2014. 145: p. 11–18.
[121] Armstrong, B.L., K.A. Hays, R.E. Ruther, *et al.*, Role of silicon-graphite homogeneity as promoted by low molecular weight dispersants. *Journal of Power Sources*, 2022. 517.
[122] Hays, K.A., R.E. Ruther, A.J. Kukay, *et al.*, What makes lithium substituted polyacrylic acid a better binder than polyacrylic acid for silicon-graphite composite anodes? *Journal of Power Sources*, 2018. 384: p. 136–144.
[123] Li, J., C. Daniel, S.J. An, and D. Wood, Evaluation of residual moisture in lithium-ion battery electrodes and its effect on electrode performance. *MRS Advances*, 2016. 1(15): p. 1029–1035.
[124] Kraytsberg, A. and Y. Ein-Eli, Conveying advanced Li-ion battery materials into practice: the impact of electrode slurry preparation skills. *Advanced Energy Materials*, 2016. 6: 1600655.
[125] Hawley, W.B., Understanding colloidal and surface phenomena to manufacture energy-dense lithium-ion and solid-state battery cathodes, in *Energy Science & Engineering*, 2021, University of Tennessee.
[126] Kumar, A. and C.K. Dixit, *Methods for characterization of nanoparticles, in* S. Nimesh, R. Chandra, and N. Gupta, Editors, (eds.) *Advances in Nanomedicine for the Delivery of Therapeutic Nucleic Acids*, Woodhead Publishing. p. 43–58, 2017.
[127] Li, J., C. Rulison, J. Kiggans, C. Daniel, and D.L. Wood, Superior performance of $LiFePO_4$ aqueous diserpsions via corona treatment and surface energy optimization. *Journal of the Electrochemical Society*, 2012. 159(8): p. A1152–A1157.
[128] Porcher, W., P. Moreau, B. Lestriez, S. Jouanneau, and D. Guyomard, Is $LiFePO_4$ stable in water? *Electrochemical and Solid-State Letters*, 2008. 11 (1): p. A4–A8.
[129] Hofmann, M., M. Kapuschinski, U. Guntow, and G.A. Giffin, Implications of aqueous processing for high energy density cathode materials: Part II. Water-induced surface species on $LiNi_{0.8}Co_{0.15}Al_{0.05}O_2$. *Journal of the Electrochemical Society*, 2020. 167: 140535.
[130] Pourbaix, M., *Atlas of Electrochemical Equilibria in Aqueous Solutions*, Houston, TX: National *Association of Corrosion Engineers,* 1966..
[131] Sahore, R., D.L. Wood III, A. Kukay, K.M. Grady, J. Li, and I. Belharouak, Towards understanding of cracking during drying of thick aqueous-

processed $LiNi_{0.8}Mn_{0.1}Co_{0.1}O_2$ cathodes. *ACS Sustainable Chemistry & Engineering*, 2020. 8: p. 3162–3169.

[132] Chen, T., F. Pan, X. Xu, and G. Zhang. Carbon-coated aluminum foil, preparation method and application. Patent no. CN106602076A. 2017.

[133] Hamam, I., N. Zhang, A. Liu, M.B. Johnson, and J.R. Dahn, Study of the reactions between Ni-rich positive electrode materials and aqueous solutions and their relation to the failure of Li-ion cells. *Journal of the Electrochemical Society*, 2020. 167: 130521.

[134] Li, J., Y. Lu, T. Yang, D. Ge, D.L. Wood III, Z. Li, Water-based electrode manufacturing and direct recycling of lithium-ion battery electrodes – a green and sustainable manufacturing system. *iScience*, 2020. 23: 101081.

[135] Wang, M., X. Dong, I.C. Escobar, and Y.-T. Cheng, Lithium ion battery electrodes made using dimethyl sulfoxide (DMSO) – a green solvent. *ACS Sustainable Chemistry & Engineering*, 2020. 8: p. 11046–11051.

[136] Zhou, H., B. Pei, Q. Fan, F. Xin, and M.S. Whittingham, Can greener Cyrene replace NMP for electrode preparation of NMC 811 cathodes? *Journal of the Electrochemical Society*, 2021. 168: 040536.

[137] Ravikumar, V.R., A. Schorder, S. Kohler, *et al.*, y-Valerolactone: an alternative solvent for manufacturing of lithium-ion battery electrode. *ACS Applied Energy Materials*, 2021. 4: p. 696–703.

[138] Hansen, C.M., *The Three Dimensional Solubility Parameter and Solvent Diffusion Coefficient: Their Importance in Surface Coating Formulation*, 1967, Technical University of Denmark.

[139] Bai, Y., W.B. Hawley, C.J. Jafta, N. Muralidharan, B.J. Polzin, and I. Belharouak, Sustainable recycling of cathode scraps via Cyrene-based separation. *Sustainable Materials and Technologies*, 2020. 25: e00202.

[140] Hawley, W.B. and J. Li, Beneficial rheological properties of lithium-ion battery cathode slurries from elevated mixing and coating temperatures. *Journal of Energy Storage*, 2019. 26: 100994.

Chapter 11
Electrolyte filling and formation of lithium-ion cells

Ali Davoodabadi[1], Jianlin Li[2] and Congrui Jin[3]

It is common practice that the liquid electrolyte is injected into the cell once other components are packed together and have formed a cell of the desired geometry. This injection marks the beginning of the electrolyte filling process, during which the electrolyte is allowed to infiltrate the porous electrodes and the separator layers and wet the originally dry components of the cell. The formation step follows the filling step as the fully wetted cell undergoes a few initial cycles to form an electronically passive layer on the electrode.

The electrolyte filling/wetting and formation of lithium-ion battery cells are two very time-consuming steps, which have direct influences on the product's final cost, performance, and safety. An improvement in this step will reduce the processing time of the cells, helping unleash their full functionality and alleviate their safety concerns. In this chapter, the importance of complete electrolyte wetting, the trade-off between cost and performance, and the latest technologies used to predict and assess the extent of electrolyte filling are discussed.

11.1 Introduction

The electrolyte filling and formation steps in the manufacturing of lithium-ion batteries are recognized as a distinctive bottleneck in the production process [1]. To achieve complete filling, the electrolyte has to wet electrodes and separators with potentially poor wettability, diffuse long distances through the micropores, and push out any gaseous species left inside. It may take several hours to a few weeks at elevated temperatures, depending on the size and chemistry of the cells. In particular, recent trends to develop thicker and more compact electrodes further hamper the electrolyte filling process, and proper measures are urgently needed to improve the filling and formation steps.

[1]Department of Chemical Engineering, University of Michigan, USA
[2]Electrification and Energy Infrastructures Division, Oak Ridge National Laboratory, USA
[3]Department of Civil and Environmental Engineering, University of Nebraska – Lincoln, USA

Incompletely wetted electrodes do not contribute fully to electrochemical reactions within the cell, and it leads to poor utilization of electrode capacity. Moreover, incomplete wetting is also related to the formation of a needle-like structure called dendrites on the electrodes that penetrate the separator and directly connect the anode with the cathode, creating internal short circuits and leading to serious safety issues. Due to the adverse effects of incomplete electrolyte wetting, it is common practice in the industry to overestimate the wetting time to ensure complete wetting, which significantly increases the overall cost, as the process requires a large number of battery cyclers and environmental chambers to be installed, large floor space, as well as a tremendous amount of low-grade heat and electricity.

Due to their importance in lithium-ion battery manufacturing, the electrolyte filling and formation steps have attracted many researchers in recent years. A variety of techniques have been employed to achieve a better understanding of the processes. In this chapter, some of the recently developed techniques and their theoretical basis have been discussed.

11.2 Surface free energy of the electrodes

One of the main parameters influencing the electrolyte filling is its interfacial affinity with different solid surfaces inside the cell. Interfacial affinity depends on the strength and nature of the molecular interactions present at the interface and manifests itself macroscopically in form of a contact angle. A low contact angle usually indicates that wetting of the surface is favorable, and the fluid will spread over a large area of the surface, whereas a high contact angle generally means that wetting of the surface is unfavorable, so the fluid will reduce contact area with the surface by producing a compact liquid droplet.

The contact angle can be predicted by the well-known Young's equation [2]:

$$\gamma_{lv}\cos\theta = \gamma_{sv} - \gamma_{sl} \tag{11.1}$$

where θ is the contact angle, γ_{sl} is the interfacial energy at the solid/liquid interface, γ_{lv} is the surface tension of the liquid, and γ_{sv} is the surface free energy of the solid. It can be seen that the contact angle can be predicted based on the information on the free surface energy. Thus, effective protocols for evaluating surface free energy are important for achieving optimal filling.

While it is a straightforward process to measure the surface tension of liquids using well-established techniques, such as bubble pressure tensiometry, Wilhelmy plate, and Du Noüy ring, the surface free energy of solids cannot be measured directly. Nonetheless, (11.1) demonstrates a clear relationship between the surface free energy of the solid γ_{sv} and the contact angle θ. Assuming that the surface tension of the probe liquid γ_{lv} and the contact angle θ can be obtained from direct measurements, there remain two unknowns, i.e., γ_{sv} and γ_{sl}, and thus the surface free energy of the solid cannot be extracted. This problem has challenged researchers to propose various techniques to modify and solve (11.1) for γ_{sv}.

The Owens–Wendt–Rabel–Kaelble (OWRK) model is one of the most widely used techniques to evaluate surface free energies of a range of solids from wooden and polymeric surfaces to two-dimensional heterostructures such as graphene and tungsten disulfide [3]. The OWRK model is established based on two main assumptions. First, the surface free energy of a solid is the sum of two independent components, i.e., a dispersive component and a polar component. In other words, the surface free energy can be written as:

$$\gamma_{sv} = \gamma_{sv}^{D} + \gamma_{sv}^{P} \tag{11.2}$$

where the superscripts D and P represent dispersive and polar components, respectively. The dispersive interactions are those caused by random fluctuations of the charge distribution in atoms and molecules. Polar interactions include coulombic interactions between permanent dipoles and between permanent and induced dipoles, such as hydrogen bonds. The second assumption is that Berthelot's combining rule holds as described in the following equation:

$$W_{sl} = \sqrt{W_{ll}^{D} + W_{ss}^{D}} + \sqrt{W_{ll}^{P} + W_{ss}^{P}} \tag{11.3}$$

where $W_{sl} = \gamma_{lv} + \gamma_{sv} - \gamma_{sl}$ is the work of adhesion and $W_{ll} = 2\gamma_{lv}$ and $W_{ss} = 2\gamma_{sv}$ represent the works of cohesion for the liquid and the solid, respectively. Combining (11.1), (11.2), and (11.3), the governing equation for OWRK model can be derived as:

$$\frac{\gamma_{lv}(1+\cos\theta)}{2\sqrt{\gamma_{lv}^{D}}} = \sqrt{\gamma_{sv}^{P}}\sqrt{\frac{\gamma_{lv}^{P}}{\gamma_{lv}^{D}}} + \sqrt{\gamma_{sv}^{D}} \tag{11.4}$$

which clearly indicates a linear relationship between $y = \gamma_{lv}(1+\cos\theta)/2\sqrt{\gamma_{lv}^{D}}$ and $x = \sqrt{\gamma_{lv}^{P}/\gamma_{lv}^{D}}$, assuming $y = ax + b$. Typically, (11.4) can be solved for a and b by substituting x and y values for at least two different liquids.

In addition to the surface tension of the probe liquids, their contact angles on composite electrodes need to be measured and substituted into the left-hand side of (11.4) to solve for the surface free energy. While it is straightforward to measure the contact angle on an ideal surface, composite electrodes are far from such a surface, as they possess roughness and heterogeneity. Deviation of solid surfaces from ideality results in an apparent contact angle θ_{app}, differing from the intrinsic Young's contact angle θ_{act}. Wenzel and Cassie-Baxter equations describe relationships between the apparent and actual contact angles and are commonly used to extract θ_{act} from θ_{app} for non-ideal surfaces [4].

Four wetting regimes between a liquid and a rough surface can be defined as follows. Wetting Cassie–Baxter regime, where there is a strong affinity between the liquid and the surface ($\theta \ll 90°$) and liquid fills out the surface pores in front of the moving contact line before the bulk of the drop has reached them. Under such a condition, a liquid drop can be considered to advance on a surface composed of solid and liquid. Wetting Wenzel regime takes place when the affinity of liquid and solid surface is relatively strong ($\theta < 90°$). In such a case, the liquid only penetrates the surface pores as the contact line passes them, so only the pores underneath the drop will be filled with liquid. Non-wetting Wenzel regime is similar to wetting Wenzel regime but applies to interfaces with lower liquid/solid affinity ($\theta > 90°$). Non-wetting Cassie–Baxter regime demonstrates the ultimate non-wetting condition ($\theta \gg 90°$) when the liquid cannot impregnate the surface pores and a liquid drop will stand on a surface composed of solid and air. For both wetting and non-wetting Wenzel states, Wenzel equation described below can be used:

$$\cos\theta_{app} = r\cos\theta_{act} \tag{11.5}$$

where r is the surface roughness ratio. For very small or very large contact angles, which fall under the Cassie–Baxter regime, the following equation holds:

$$\cos\theta_{app} = f\cos\theta_{act} \pm (1 - f) \tag{11.6}$$

where f is the solid area fraction of the surface. The plus or minus sign should be used for the wetting and non-wetting regimes, respectively. Hence, depending on the value of the contact angle and the roughness information of the surface, Cassie–Baxter or Wenzel equations can be used appropriately to obtain θ_{act} from θ_{app}.

Davoodabadi *et al.* calculated the θ_{act} of three probe liquids on common lithium-ion battery electrodes, including graphite anodes and NMC cathodes with a range of surface roughness and used that information to compare their surface free energies [5]. The results indicate higher surface free energy of NMC cathodes than graphite anodes. In terms of dispersive and polar components, the study discovers that the polar-to-dispersive ratio is higher for electrodes processed with aqueous solvents compared to their analogues processed with organic solvent. This result could be particularly important as most lithium-ion battery electrolytes possess a large polar component and preferentially spread on surfaces with higher polar values.

11.3 Electrolyte filling through the electrodes

Electrodes are the main active components in a cell, and they need to form and retain good contact with the electrolyte for the cell to operate properly. Since non-wetted parts of the electrodes do not contribute to the electrochemical reactions, the nominal capacity of the cell cannot be realized if the porous electrodes are not completely filled with electrolyte. Complete filling is also essential to form a stable solid electrolyte interphase layer, which is critical for the battery performance and safety.

There are a number of factors that play a role in electrolyte filling rate and extent in electrodes. Electrode's porosity, average pore size, pore size distribution, pore geometry, pore connectivity, and surface free energy are among the parameters that directly influence the electrolyte filling process. Electrolyte's surface tension and viscosity are also crucial players. The electrolyte imbibition process through the electrodes haven been investigated by both experimental techniques and computational methods.

Wetting balance is one of the conventional experimental techniques to characterize the filling rate of a liquid into a porous material and has been widely used to study the filling process through electrodes [6]. This technique is based on gravimetric measurements where the porous sample mass is monitored during the liquid infiltration process. The increase in mass of the electrode with respect to time indicates the amount of electrolyte that has penetrated the electrode at each moment. The infiltrated electrolyte mass information is then converted to volume, which is subsequently used to obtain the depth of electrolyte penetration into the electrode with respect to time $h(t)$:

$$h(t) = K\sqrt{t} \tag{11.7}$$

where K with the unit of $m/s^{0.5}$ quantifies the speed at which the electrode is being impregnated.

While the wetting balance technique is useful as a relatively fast test to acquire basic information about electrolyte filling rate, it may not be the most precise technique. Electrodes are porous materials with very high surface area. This high surface area leaves the infiltrated electrolyte exposed to evaporation which directly impacts the mass readings. It is also important to note that there is an underlying assumption in using this technique, where the infiltration is assumed uniform across the entire sample cross-section so that the electrolyte volume can be transferred to infiltrated depth. Additionally, the wetting balance test is executed in a vertical configuration where the gravitational force exerted on the liquid can counteract the capillary suction forces that drive the infiltration and produce results that would be valid only if the testing conditions are replicated during the actual filling process.

Another experimental scheme to analyze electrolyte filling through electrodes is to employ optical modules to directly visualize the infiltration process. An example is the technique developed by Davoodabadi *et al.* where the in-plane penetration of an electrolyte drop inside a thin electrode film was optically monitored [7]. The videos were subsequently processed to extract the filling rate. To better understand the experimental results from in-plane infiltration of the electrolyte and use that for filling rate analysis, the work also develops a theoretical framework that relates filling rate to the experimental observed variables e.g., the radius of the impregnated region.

In this study, Darcy's law for liquid flow through porous media [8] is customized for the specific configuration that the experiments are carried out i.e., considering a liquid source in the middle of a porous material film that supplies liquid to infiltrate and spread radially. The governing equation for growth of imbibition radius r with time t is derived as:

$$\frac{1}{2}r^2[2\ln(r/r_0) - 1] + \frac{1}{2}r_0^2 = Dt \tag{11.8}$$

where the left-hand side is a function of radius, $f(r)$, that varies linearly with time. The coefficient in this relationship, D, is defined as the imbibition coefficient and depends on both electrode and electrolyte properties:

$$D = \frac{2k\Delta p}{\epsilon\mu} \tag{11.9}$$

where k is the permeability of the electrode, Δp is the pressure gradient driving the flow that in this case is the capillary pressure, ϵ is the electrode porosity, and μ is the electrolyte viscosity. The imbibition coefficient introduced in this work is of great significance as it can be used to predict the required time for complete wetting for any electrode–electrolyte pair. The capillary pressure exerted on the liquid electrolyte within the porous electrode can be written as $2B\gamma\cos\theta/r$, which can be substituted for Δp in (11.9) to yield:

$$D = 4\left(\frac{\gamma}{\mu}\right)\left(\frac{kB}{\epsilon r}\right)\cos\theta \tag{11.10}$$

where B is the pore geometrical coefficient, γ is the electrolyte surface tension, r is the electrode effective pore radius, and θ is the intrinsic contact angle between electrode and electrolyte. By plotting the experimental results for $f(r)$ versus time, a linear relationship is observed that is subsequently matched against (11.8) to determine the imbibition coefficient for the electrode-electrolyte pairs.

While the imbibition coefficient contains the most precise data about the filling rate, it depends on both electrode and electrolyte properties and cannot be used to compare filling in electrodes regardless of the electrolyte. The solid permeability coefficient has been introduced as a variable that primarily depends on electrode properties and hence is useful to rate different electrodes according to their willingness to be impregnated by a given electrolyte. The experimental results demonstrated that as-deposited graphite anodes have a solid permeability coefficient that is one order of magnitude larger than as-deposited NMC532 cathodes.

Among the factors that determine the filling rate, some are the electrolyte properties. Washburn in his original work on flow through porous media [9] defined the coefficient of penetrance, $\mathrm{COP} = \gamma\cos\theta/(2\mu)$, as a variable that is primarily a function of liquid properties, i.e., surface tension γ and viscosity μ, and directly affects the imbibition rate. This is consistent with (11.10), which shows that an increase in γ/μ leads to larger imbibition coefficients, i.e., faster penetration.

Wu *et al.* studied the penetration rate of different electrolytes into lithium cobalt oxide (LCO) and meso-carbon microbead (MCMB) electrodes to identify the main electrolyte parameters that influence this process [10]. The electrolytes used in this study were made by adding different concentrations of either $LiClO_4$ or $LiPF_6$ salts to a mixture of organic solvents. Since viscosity and surface tension are the main properties affecting the filling process, they focused on the variation in these quantities.

Slower penetration was observed for higher concentrations of ethylene carbonate (EC) in solvent mixtures including EC/diethyl carbonate (DEC) and

EC/dimethyl carbonate (DMC). This was attributed to the significant increase in viscosity for higher EC concentration while the surface tension increased only slightly, leading to smaller γ/μ values. Similarly, increasing the salt concentration, either $LiClO_4$ or $LiPF_6$, in the electrolyte resulted in slower filling. Switching the electrodes from LCO to MCMB did not change the trends, but the filling rates were consistently lower for LCO, presumably due to its significantly smaller effective pore radius compared to MCMB. Overall, salt concentration and solvent mixture composition seem heavily responsible for the observed filling rates.

Recently, there has been a surge of interest in computer-aided techniques to analyze the filling process. Computational techniques have the ability to take into account and analyze all the variables that play a role in the process. Shodiev *et al.* developed a machine learning model that could be used to predict the dynamics of flow within porous electrodes and their required filling time [11]. The model based on multi-layer perceptron approach can predict the imbibition process characteristics with very low computational cost. To achieve high precision in predicting the filling process, they trained the model on a database generated using a three dimensional physics-based model developed using the Lattice Boltzmann method. Shodiev *et al.* confirmed their model's ability to predict the electrolyte filling process in an electrode with a given structure. The model also demonstrated that increasing the applied pressure during the filling process can decrease the time to saturation by orders of magnitude.

Jeon also studied the wetting process through electrodes using the Lattice Boltzmann numerical simulations [12]. While Shodiev *et al.* based their model on meso-structures of typical electrodes obtained from CT scans, Jeon's simplified model used spheres and ellipsoids to reconstruct cathodes and anodes, respectively. These particle geometries were chosen considering that cathode materials are hard ceramics, whereas anodes are made from more compliant carbon materials that can undergo plastic deformation during the manufacturing process. Joen's simulation is a two-dimensional model that further reduces the computational cost.

His results showed a non-uniform electrolyte penetration inside the cathode that is in contrast with the experimental data obtained by Davoodabadi *et al.* for organic-based cathodes [7]. Nonetheless, non-uniform electrolyte imbibition was also observed by Davoodabadi *et al.* for aqueous-processed cathodes [13]. An interesting aspect of Jeon' study is the post-penetration investigation on the saturated electrodes. For instance, the study claimed that electrolyte saturation level in the anode is lower than that in the cathode, despite the faster penetration in anode, due to the presence of entrapped gaseous phase caused by poor network connectivity. This entrapped air, however, can be disposed of during the aging process with elevated temperatures or by running repetitive charging and discharging cycles that cause anodes particles to expand and contract and free up the air.

Jeon also established a framework to investigate the effect of electrode filling degree on the cell performance [12]. It was observed that, for a fully wetted anode, the cell capacity at 1C discharge is reduced to 97% and 92% for cathode filling degrees of 95% and 90%, respectively. The capacity fade is more significant at higher rates. For example, for the same cell to undergo a 5C discharge, the capacities for 95% and 90% wetted cathodes will be 96% and 91%, respectively. On the other hand, capacity fading was not observed if anodes were not completely wetted. The results suggest that cell performance strongly depend on the state of electrolyte filling in the cathodes.

11.3.1 Effect of calendering on electrolyte filling

Electrode calendering is very common in the manufacturing process, where the as deposited electrodes with nominal porosities of about 55% are compressed to make denser and more stable. In addition to changing electrode porosity and thickness, calendering impacts the geometry and meso-structure of the electrode pores, and hence directly affects the filling process. To improve the wetting process, quantitative evaluation of the correlation between the calendering degree and the resulting wetting performance is of critical importance.

An experimental investigation of electrolyte filling in electrodes with different degrees of calendering was presented by Davoodabadi *et al.* [13]. Four types of electrodes, i.e., $LiNi_{0.5}Mn_{0.3}Co_{0.2}O_2$ (NMC532) cathodes and graphite A12 anodes that were processed with either water or organic solvent, were studied. Imbibition coefficients D as defined in (11.10) were measured, which demonstrated a consistent decrease when increasing the calendering degree. An interesting observation was that the water-processed NMC532 has a completely different pore structure with pore sizes varying with orders of magnitude that resulted in very non-uniform imbibition footprints, as shown in Figure 11.1(a)–(d). Hence, further quantification was not possible for this specific type of electrode, since (11.10) was no longer applicable.

Lee *et al.* used a two-dimensional Lattice Boltzmann model to better understand the calendering effect [14]. The model assumes that calendering changes not only the overall electrode porosity but also the particle shape due to plastic deformation, specifically in anodes with softer MCMB particles. The simulation results suggest a steady decrease in filling rate as the calendering degree increases, consistent with the experimental observations. In the simulation, because the anode particles have a through-plane compression and in-plane expansion, the through plane permeation of liquid electrolyte decreases. However, in most of the practical cases, the electrolyte is injected into the cell from top or bottom of the cell with direct contact with the electrode, the imbibition will be in-plane dominant, and thus the anode will be filled sooner.

Sheng *et al.* used the wetting balance technique to study infiltration into graphite anodes calendered to different extents, as shown in Figure 11.1(e) [6]. Interestingly, they observed an initial improvement in filling rate with calendering until reaching a maximum rate after which the filling rate plummets with additional compression. They hypothesized that the pressure applied on the electrodes during calendering can induce two different types of movement in the ellipsoidal particles. They can either move closer to one another and become more aligned or slip on top of each other and potentially move out of alignment. Since graphite particles are originally in random orientations, modest calendering initially improves filling by enhancing the alignment of the particles. Once the particles are closely packed,

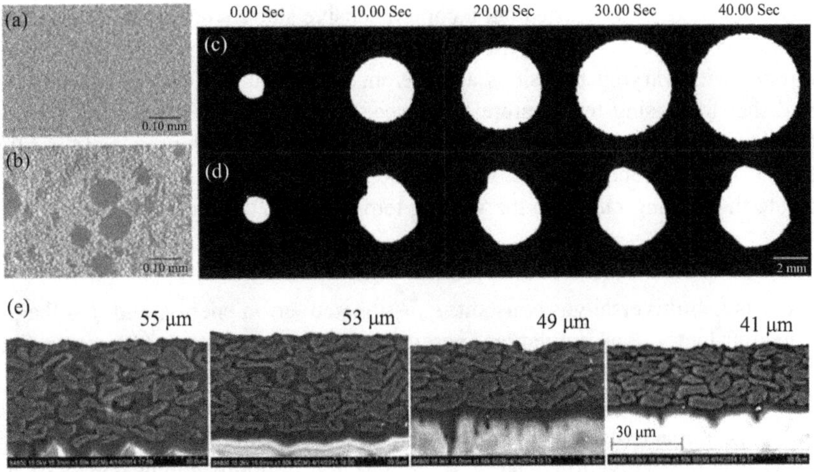

Figure 11.1 *(a) SEM microstructure of an organic solvent-based graphite anode with uniform pore size and distribution, (b) SEM microstructure of a water-based NMC cathode with irregular macro-pores, (c) isotropic electrolyte imbibition footprint in an organic solvent-based graphite anode, (d) non-uniform electrolyte imbibition footprint in a water-based NMC cathode, and (e) cross-sectional SEM images of graphite anode microstructure evolution with calendering. (a), (b), (c) and (d) from [13] with permission and (e) from [6] with permission. Copyright 2019 Elsevier.*

additional pressure induces more slipping movements among the particles that adversely impacts the filling rate.

Note that both Davoodabadi *et al.* [13] and Sheng *et al.* [6] measured the filling rate of 1.2 $LiPF_6$ in EC:EMC solvent in as-deposited graphite anodes using direct visualization and wetting balance techniques, respectively. Davoodabadi *et al.* showed a filling rate of $10.0 \pm 0.18 \times 10^{-8}$ m^2/s, whereas Sheng *et al.* demonstrated a value of $14.4 \pm 2.1 \times 10^{-8}$ m^2/s. These average imbibition rates have the same order of magnitude and the difference can be attributed to the variations in porosity and structural properties among the electrodes used in the studies. The striking point, however, is the error margins for these measurements. The superiority of the direct visualization technique compared to the conventional wetting balance test can be seen in its extremely enhanced precision where the error bars have shrunk by an order of magnitude.

11.3.2 Effect of temperature on electrolyte filling

Increasing temperature has been found to be beneficial for the wetting process and is commonly utilized in the manufacturing plants. While elevated temperature is empirically useful to promote wetting, it also brings additional costs to the production line. Thus, it is important to accurately analyze the impact of increased temperature on the filling process to determine if it is worthwhile.

Davoodabadi *et al.* carried out a comprehensive study where they measured the imbibition coefficients of different electrolytes into a range of positive and negative electrodes with varying porosities at different temperatures [13]. The results confirmed that increasing temperature helps speed up the filling process for all electrode–electrolyte pairs. The work then extends to quantify the improvement in filling rate by using the concept of activation energy. The Arrhenius equation is used to correlate the wetting rate D to the wetting temperature T as shown below [15]:

$$D = A\exp(-E_a/RT) \tag{11.11}$$

where R is the universal gas constant, E_a is the activation energy, and A is the pre-exponential factor. The activation energy is an important parameter, as it reflects the sensitivity of the electrolyte wetting rate to the change of temperature.

All the activation energies obtained in the study were positive, demonstrating the positive impact of increasing temperature on the imbibition rate. The results showed that the activation energy is similar for graphite anode and NMC532 cathode for any given electrolyte. In contrast, the activation energy varies significantly when the electrolyte is switched for any given electrode. Both surface tension and viscosity of the electrolytes were observed to decrease with increasing temperature. Equation (11.10) shows that a reduced viscosity improves the wetting rate while a lower surface tension may not be necessarily helpful to speed up the filling process. It is also important to note that the viscosity of an electrolyte is primarily determined by its salt concentration, whereas its surface tension heavily depends on its solvent composition. Thus, the faster filling dynamics for all electrolytes at higher temperatures can be attributed to the decrease in their viscosity.

The study shows that the filling rate improves for all different electrolytes at higher temperatures; however, the improvements are marginal for some electrolytes and very drastic for the others. For examples, for an as-deposited graphite anode, an electrolyte with 0.6M concentration of $LiPF_6$ in EC:EMC demonstrated that $E_a = 1.2$ kJ/mol, while an electrolyte with 1.2M concentration of $LiPF_6$ in EC:EMC showed that $E_a = 11.7$ kJ/mol. This much higher activation energy indicates that increasing temperature is most effective for electrolytes with high salt concentrations. On the other hand, the results of $E_a = 9.5$ kJ/mol was found for filling of 1.2M $LiPF_6$ in EC:DEC into graphite anodes, relatively close to the result of $E_a = 11.7$ kJ/mol for 1.2M $LiPF_6$ in EC:EMC, implying that the effect of electrolyte solvent on the activation energy is not as significant. Overall, it seems that the effectiveness of elevated temperature depends on the electrolyte type. The small activation energy observed for electrolytes with low salt concentrations indicates that heating up will not be very beneficial to improve the wetting process for such electrolytes.

11.4 Electrolyte filling through the separators

Separator is known for its role as a physical barrier between the positive and negative electrodes of a cell. However, its another role, which is equally important, is to function as an electrolyte reservoir for ion transport. Therefore, sufficient electrolyte absorption is needed for the separators to maintain their low internal resistance and provide high ionic conductivity which is crucial for optimal battery performance.

Similar to electrolyte filling through electrodes, the filling rate through separators is of great importance to the manufacturing process. Since separators are typically made from polymeric materials such as polypropylene (PP), polyethylene (PE), and ultra-high molecular weight polyethylene (UHMWPE), with low surface energies, the wetting process becomes more of a challenge for them. Ceramic coatings, such as alumina (Al_2O_3), zirconia (ZrO_2), stannic oxide (SnO_2), and silica (SiO_2), are often used to coat the separators to improve their wettability.

While most of the techniques used to study the wetting process through electrodes can be used to analyze the process through separators, anisotropic microstructure is expected for separators manufactured by the dry process, and thus the wetting rates in different directions are expected to vary significantly. The separators manufactured by the dry process are generally treated as an orthotropic material, composed of thin fibrils in the machine direction and thick lamellae arranged in the transverse direction, and therefore, both through-plane and in-plane infiltration rates need to be measured for each sample.

Direct visualization is one of the most reliable techniques to study the wetting rate through separators. Davoodabadi *et al.* have demonstrated the application of optical cameras to monitor the propagation of electrolyte inside different separators for both in-plane and through-plane analysis [16]. A drop of electrolyte is placed on a thin strip of separator to limit the penetration into a one-dimensional process and the advancement of the electrolyte front line is closely watched. The through-plane wetting rate is characterized using two synchronized high-speed cameras above and below the piece of separator and measuring the time elapsed between electrolyte placement on the top surface and the visualization of the electrolyte from the bottom camera.

Different types of separators, including single layer PP and UHMWPE separators, three-layer PP/PE/PP separators, and separators with Al_2O_3 coating, were studied in this work. Experimental results for through-plane imbibition produced different values of the imbibition coefficient D ranging from 1.1×10^{-9} m^2/s to 6.7×10^{-9} m^2/s, all of the same order of magnitude. The three-layer separator and the ceramic-coated separators showed the slowest through-plane imbibition rates. For the single layer PP separator, the through-plane imbibition coefficient showed a five-fold increase as the porosity was increased from 40% to 55%. Interestingly, no in-plane penetration was observed for any of the dry processed separators, including single layer PP and three-layer PP/PE/PP separators. This result can be explained as the pores in dry-processed separators are not well connected and thick lamellae formed during the stretching process act as barriers against electrolyte advancement, as shown in Figure 11.2(a).

Further investigation showed that the fast in-plane infiltration for the ceramic-coated separator resulted from advancement of the electrolyte through the polymer/coating interface. In fact, electrolyte penetration is slow through both UHMWPE and Al_2O_3, but the existence of a high surface energy interface between them allows the electrolyte to advance quickly, as shown in Figure 11.2(b) and (c).

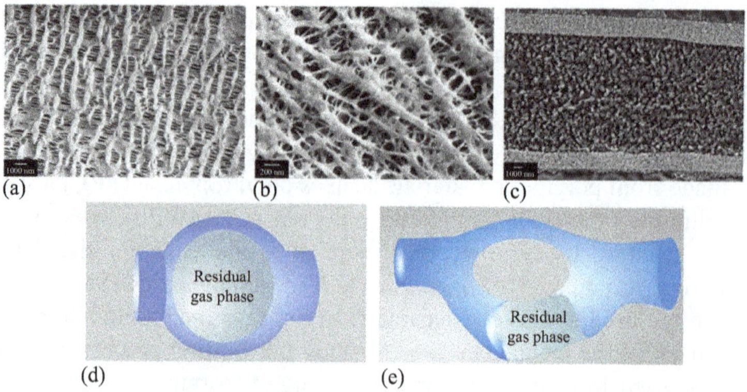

Figure 11.2 (a) SEM image showing the microstructure of a Celgard PP separator fabricated by dry stretching, (b) SEM image of a wet processed Entek EPX separator, (c) cross-sectional structure of a ceramic coated Entek EPX, (d) schematic illustration of gas entrapment through snap-off mechanism in separator pore with variable cross-section, and (e) schematic illustration of gas entrapment through pore-doublet model in a pore with two branches. (a), (b) and (c) are from [16]. (d) and (e) are from [18]. Copyright 2020 Elsevier and 2020 IOP Publishing, Ltd.

Overall, the application of a ceramic coating on polymer separators is advantageous as it increases the separator's mechanical strength and thermal stability in addition to enhancing its in-plane wetting rate, although it adversely impacts the through-plane imbibition rate.

Analyzing the ion conductivity through separators can indicate the level of electrolyte saturation in them, which can be used to analyze the final filling state of a separator. MacMullin number, defined as the ratio of the resistivity of the separator soaked in electrolyte to the resistivity of free electrolyte, is a good measure to quantify the filling state of separators [17]. Sauter *et al.* studied the effect of separator pore structure on gas entrapment inside the pores and how it results in incomplete wetting and inefficient ion transport [18]. Gas entrapment inside the pores can be seen as blockage of some of the lithium transport pathways that increases the overall path for ion transport. Electrochemical impedance spectroscopy was used to measure the effective conductivity of lithium ion through a single layer PE separator. The measured MacMullin numbers for all different electrolytes were higher than what was predicted for a completely wetted separator, indicating incomplete wetting of all electrolytes. Although the MacMullin number tended to increase with increasing electrolyte viscosity, no obvious trend between electrolyte properties and the MacMullin number was observed.

Two different potential mechanisms for gas entrapment inside the pores are commonly considered, the snap-off model and the pore doublet model. Snap-off takes place in pores with variable cross-sectional area, where the capillary pressure varies from point to point within a pore. The pore doublet model applies to a pore that branches into two different pathways and then reconnects, as shown in Figure 11.2(d) and (e). Analysis of the pore structure of the PE separator showed a pore aspect ratio of 1.3 that is small enough to rule out the possibility of gas entrapment through snap-off mechanism. This observation indicates that separator's final filling status cannot be improved solely by electrolyte properties, i.e., electrolyte properties and separator pore structures must be considered concurrently to improve filling.

Lagadec *et al.* studied the effect of separator porosity and tortuosity on effective transport coefficient to find optimized pore structure for separators [19]. Their analysis extended from hypothetical separators with simple cylindrical pores to three-dimensional separator geometries reconstructed from SEM images of real single layer PP and PE separators. It was found that, among different structural factors, connectivity of the pores is essential for higher transport efficiencies. Highly connected pores such as those found in single layer PE separators allow for ion gradient at the top of the separator to smooth out within a fraction of the separator thickness. A hypothetical separator with isolated cylindrical pores stretched from one side to the other might yield a very fast transport, but such a separator with zero connectivity density is highly susceptible to lithium plating due to the likely presence of defects.

Schilling *et al.* compared filling rate between a single layer Celgard PP separator and a ceramic-coated Separion separator [20]. Wetting balance technique

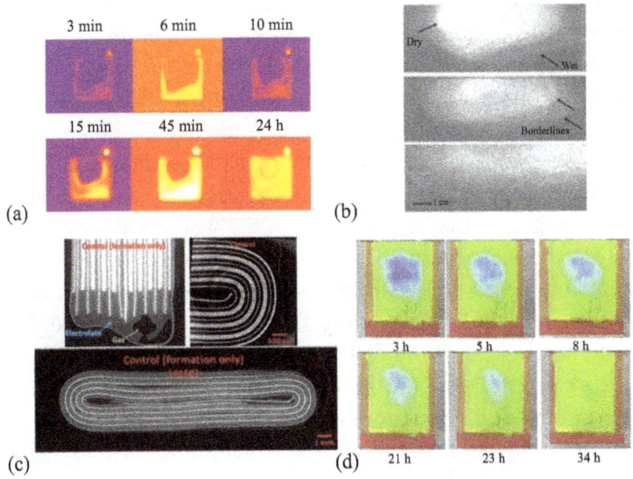

Figure 11.3 Images for studying the electrolyte filling process in lithium-ion cells obtained from (a) thermographic scan of a pouch cell [20], (b) neutron scattering of a pouch cell [21], (c) X-ray CT scan of a pouch cell [24], and (d) ultrasonic scanning of a pouch cell [27]. Copyright 2019 John Wiley and Sons, 2016 Elsevier, 2022 IOP Publishing Ltd, 2020 Elsevier.

showed that penetration rate in Separion was 60% higher than that in Celgard. In addition to wetting rate characterization in stand-alone separators, they used a thermography visualization technique to monitor electrolyte filling in pouch cells with identical components except for the separator layer to investigate if the wetting rate results still hold once the cells are packed, as shown in Figure 11.3(a). The observations indicated faster wetting in ceramic-coated separator, where almost 95% of the cell is wetted after 45 min compared to a 45% wetting in PP separator in a similar time duration. It must be mentioned, however, that the thermal images taken from the cell may not necessarily represent filling in separators. The high surface energy ceramic coating on the separator allows for fast spreading of electrolyte, thus, the electrolyte can favorably wet the interface between the separator and the electrodes first. It is not possible to determine from the thermal images whether the observed wetting is associated with electrolyte penetration through the separator or at the interfaces between the separator and electrodes.

11.5 Electrolyte filling in full cells

Direct visualization of the filling process can arguably provide the most precise results if it could be done for packed cells. However, the cell housings are not transparent to visible light. Thus, in situ neutron imaging is an important tool for this purpose. Specifically, neutrons are strongly attenuated both by lithium and by the hydrogen atoms in the organic electrolytes used in lithium-ion cells, whereas the metallic parts of the lithium-ion cell, i.e., the aluminum cell casing and the current collectors made of aluminum and copper, are rather transparent to neutrons. However, the spatial and temporal resolutions of neutron imaging techniques are lower than the optical techniques, as shown in Figure 11.3(b).

Knoche *et al.* used neutron imaging to study pouch cell stacks with four cathode sheets, five anode sheets, and a z-folded PE separator [21]. The obtained images had sufficient contrast to distinguish between dry and wet regions. It was also possible to investigate the homogeneity of the wetting process as the presence of bubbles within the wet region could be seen as a sign of heterogeneous imbibition. An interesting observation was that while the electrolyte is injected from the upper side of the pouch cell, the wetting front advancement is negligible on top compared to the other sides of the cell. In fact, most of the electrolyte seemed to flow along the pouch foils to the bottom of the cell where it is subsequently absorbed into separator and electrode layers through capillarity. Decreased pressure during the dosing improved the final electrolyte saturation degree inside the cell as more gas was removed at lower pressures. However, reduced pressure had a negative effect on the initial wetting degree i.e., before reaching the saturated level. In addition, it was shown that imposing wetting cycles on the cell prior to sealing is effective in reducing the amount of gas entrapment.

Another study performed by Weydanz *et al.* investigated the electrolyte filling in hard case prismatic cells [22]. In contrast to pouch cell, filling in prismatic cell took place isotropically where electrolyte advances from all four edges toward the center of the cell with similar speeds in all directions. Increasing the vacuum level imposed on the cell during the filling cycle from 960 mbar to 100 mbar increased the wetting coefficient from 0.53 $mm/s^{0.5}$ to 0.77 $mm/s^{0.5}$.

Imaging a full cell to understand the state of electrolyte filling is also possible with X-ray. X-ray imaging offers improved spatial resolution compared to neutron imaging and requires shorter exposure times. An example of in situ imaging of the filling process using X-ray was described by Schilling *et al.* [23]. They evaluated the filling rate by calculating the average gray scale of each frame from the recorded X-ray image sequence. Since the dry region looks brighter in the X-ray images, a higher gray scale indicates more filling. This method is useful but cannot determine the exact degree of wetting. Another way to analyze the progression of the fill level of the cell during the filling process was to monitor the location of the flow front and follow its advancement with time. It must be noted that the liquid used in this work was a contrast agent, but the setup is expected to function with common electrolytes containing conductive salts such as $LiPF_6$ with some modification.

X-ray CT imaging of the cells is yet another technique that can deliver three-dimensional views of the full cells for studying the electrolyte filling process [24], as shown in Figure 11.3(c). Wieder *et al.* used X-ray tomography to evaluate the state of lithium-ion capacitors as they were cycled [25]. They used vertical and horizontal cross-sections of the obtained three-dimensional radiography to evaluate the degree of electrolyte accumulation and electrodes delamination that can potentially lead to malfunctioning of the device. The images with a resolution of tens of microns made it possible to locate small irregularities grown during the cycling. Du *et al.* used X-ray tomography to study the evolution of gaseous phase inside a lithium-ion cell as a function of cycle numbers [26].

X-ray scans can be carried out with high spatial and temporal resolution, however, to complete a full scan of a pouch cell, long periods of time, i.e., in the order of hours, will be needed. This implies that, if used for electrolyte filling analysis, the dynamics of the process cannot be captured except for cells with very large dimensions or high compactness in which the imbibition proceeds slowly. X-ray CT scans might still be useful to compare the final electrolyte filling states, i.e., the saturation levels, among different cells. Note that X-ray CT scans with higher spatial resolution, i.e., in the order of tens of nanometers, are available but require focused X-ray bean geometries. They would also only work for samples with diameters in the order of hundreds of microns and, thus, are not suitable for studying filling in cells. Overall, while CT scans can provide images with extremely detailed information on the status of components inside a cell, their low temporal resolution limits their application for filling rate analysis.

Ultrasonic scanning is another non-destructive imaging technique that can be used to study electrolyte filling in packed cells. Deng *et al.* designed an ultrasonic scanning system with sub-millimeter spatial resolution [27]. Due to the difference in pathways for ultrasonic waves between dry and wet electrodes, the wetted region inside the cell can be distinguished, as shown in Figure 11.3(d). In fact, a dry electrode will strongly attenuate the ultrasonic waves, but they can pass through an electrode soaked with electrolyte fairly easily. They investigated the effect of electrolyte volume injected into the cell on filling rate and extent. The results indicated that as long as sufficient amount of electrolyte is dosed into the cell, no significant difference between wetting degree after 4 h and after 24 h can be observed. It would have been useful to scan the cell a few times prior to the 4-h scan to better understand the dynamics of wetting.

Besides the imaging techniques discussed above, non-direct measurements can be made to infer data about the filling process in a cell. Electrochemical impedance spectroscopy (EIS), for example, can be carried out during the filling process to extract the internal resistance of the cell and monitor its variation as the filling progresses. Such a measurement is advantageous due to its availability and ease of application for in-line cell monitoring during the production process.

Gunter *et al.* have described the exploitation of EIS for electrolyte filling analysis [28]. To verify the reliability of the results obtained from EIS, they compared the obtained wetting rate data against results from neutron radiography. It was shown that the average wetting rate measured with EIS agrees with those extracted from images taken by neutron imaging at different times. Gunter *et al.* also used EIS to evaluate the effect of electrolyte quantity on filling and performance of the cells [29]. The results indicated that increasing the dose of electrolyte injected into the cell improves the wetting rate up to a certain point after which no further improvement is observed. While EIS can be useful as a quick way to check for wetting status inside cells, it lacks information about the cause of a potential incomplete wetting that might be needed for troubleshooting the filling process.

Peter *et al.* developed a rather different technique to characterize the filling state in pouch cells, in which a voltage potential was applied to the cell at specific times and the current response from the cell was measured [30]. Once the received current response becomes constant for several consequent applied potentials, the cell can be considered as fully wetted. An interesting observation made in this study was that wetting in cells with ceramic-coated separators is faster than cells with similar separators without coating. This may imply that the interfaces between electrodes and separators are the primary pathway for electrolyte wetting. Once these pathways are filled, electrolyte starts to penetrate laterally into the electrodes and separators. This can also imply that the observations from neutron and X-ray imaging might reflect the presence of electrolyte at these interfaces and may not necessarily indicate that all the pores in the apparently wetted region are filled.

Studying the electrolyte filling process in full cells using computational techniques is very promising. Such techniques can be especially advantageous considering the shortcomings of experimental techniques on this matter. For example, EIS can identify incomplete wetting but cannot determine which portion of the cell is not filled. Optical techniques are highly precise but cannot operate with non-transparent cell housings. Neutron imaging will also require special tools and can contaminate the cells that are studied. On the other hand, computational studies do not suffer from these conditions. In addition, different parameters can be easily changed in simulations without the need to make a new cell.

Shodiev *et al.* developed a Lattice Boltzmann model to simulate the filling process in full cells using geometries reconstructed from a typical cell geometry to study the filling rate and the saturation time [31]. The full cell studies indicated the importance of the order of the cell components on the wetting rate, i.e., the filling was completed sooner in a cathode/separator/anode cell compared to the one with an anode/separator/cathode structure. Furthermore, the initial electrolyte injection point was observed to play a crucial role the overall filling time required for the cell. This framework allows to study the effect of different variables such as calendering, pore connectivity, and electrolyte viscosity, on the overall filling time.

A subsequent work from Shodiev *et al.* used this tool to identify ideal properties for fast wetting and proposed electrode architectures that facilitate electrolyte filling rate [32]. Among various properties, porosity, porosity distribution, and particle size distribution of the active material were identified as the main variables to be tuned for fast wetting. Therefore, different combinations of layered cathodes and anodes with a variety of porosities and pore sizes were put together and analyzed for wetting rate. It was seen that a heterogeneous distribution of porosity in both positive and negative electrodes is advantageous for fast filling compared to a homogeneous mesostructure with identical overall porosity. Hence, numerical simulations are a very useful tool to predict the details of the filling process in full cells as long as their overall predictions are verified experimental observations, such as EIS or imaging techniques.

Overall, the conclusion from studying full cells generally agrees with those obtained from individual component investigations. While full-cell studies typically have a limited temporal and spatial resolution, they still provide useful and practical information that cannot be achieved otherwise, such as the effect of the location of electrolyte injection into the cell.

The experimental results indicate a significant role of separators in overall filling of the cell. For high-surface energy separators, i.e., the ceramic-coated ones, the electrolyte fills out the interfaces between the electrodes and separators first, while penetrating into the electrodes and separators from the edges with a lower rate. The interfacial regions between electrodes and separators, thus, act as more local resources and through-plane electrolyte imbibition becomes the dominant filling mechanism. For low surface energy separators, however, the in-plane imbibition from the edges may remain the primary mechanism of filling that can require much longer times due to the long imbibition distances.

11.6 Summary

The processes of electrolyte filling and formation of lithium-ion battery cells pose a distinct bottleneck in the manufacturing line, especially for large-size batteries. In this chapter, the importance of complete electrolyte wetting, the trade-off between battery cost and performance, and the recently developed technologies to predict and assess the processes have been fully discussed. It can be concluded that despite the significance of achieving rapid and complete wetting, reliable scientific investigation has not yet been extensively reported. Such an important research area provides an arena for material scientists, chemical engineers, and fluid mechanicians to share knowledge and build collaborations, which will enable the battery researchers to pursue new design philosophies to achieve enhanced performance.

References

[1] D. L. Wood, J. Li, and C. Daniel, "Prospects for reducing the processing cost of lithium ion batteries," *J Power Sources*, vol. 275, pp. 234–242, 2015.

[2] T. Young, "An essay on the cohesion of fluids," *Philos Trans R Soc Lond*, vol. 95, pp. 65–87, 1805.

[3] D. K. Owens and R. C. Wendt, "Estimation of the surface free energy of polymers," *J Appl Polym Sci*, vol. 13, no. 8, pp. 1741–1747, 1969.

[4] A. B. D. Cassie and S. Baxter, "Wettability of porous surfaces," *Trans Faraday Soc*, vol. 40, no. 0, pp. 546–555, 1944.

[5] A. Davoodabadi, J. Li, Y. Liang, *et al.*, "Characterization of surface free energy of composite electrodes for lithium-ion batteries," *J Electrochem Soc*, vol. 165, no. 11, pp. A2493–A2501, 2018.

[6] Y. Sheng, C. R. Fell, Y. K. Son, B. M. Metz, J. Jiang, and B. C. Church, "Effect of calendering on electrode wettability in lithium-ion batteries," *Front Energy Res*, vol. 2, no. DEC, p. 56, 2014.

[7] A. Davoodabadi, J. Li, Y. Liang, D. L. Wood, T. J. Singler, and C. Jin, "Analysis of electrolyte imbibition through lithium-ion battery electrodes," *J Power Sources*, vol. 424, pp. 193–203, 2019.

[8] A. E. Scheidegger, "The physics of flow through porous media", Accessed: Feb. 03, 2023. https://utorontopress.com/9781487583750/the-physics-of-flow-through-porous-media-3rd-edition

[9] E. W. Washburn, "The dynamics of capillary flow," *Phys Rev*, vol. 17, no. 3, p. 273, 1921.

[10] M. S. Wu, T. L. Liao, Y. Y. Wang, and C. C. Wan, "Assessment of the wettability of porous electrodes for lithium-ion batteries," *J Appl Electrochem*, vol. 34, no. 8, pp. 797–805, 2004.

[11] A. Shodiev, M. Duquesnoy, O. Arcelus, M. Chouchane, J. Li, and A. A. Franco, "Machine learning 3D-resolved prediction of electrolyte infiltration in battery porous electrodes," *J Power Sources*, vol. 511, p. 230384, 2021.

[12] D. H. Jeon, "Wettability in electrodes and its impact on the performance of lithium-ion batteries," *Energy Storage Mater*, vol. 18, pp. 139–147, 2019.

[13] A. Davoodabadi, J. Li, H. Zhou, D. L. Wood, T. J. Singler, and C. Jin, "Effect of calendering and temperature on electrolyte wetting in lithium-ion battery electrodes," *J Energy Storage*, vol. 26, p. 101034, 2019.

[14] S. G. Lee and D. H. Jeon, "Effect of electrode compression on the wettability of lithium-ion batteries," *J Power Sources*, vol. 265, pp. 363–369, 2014.

[15] A. C. Leopold, "Temperature effects on soybean imbibition and leakage," *Plant Physiol*, vol. 65, no. 6, p. 1096, 1980.

[16] A. Davoodabadi, C. Jin, D. L. Wood, T. J. Singler, and J. Li, "On electrolyte wetting through lithium-ion battery separators," *Extreme Mech Lett*, vol. 40, p. 100960, 2020.

[17] R. Zahn, M. F. Lagadec, M. Hess, and V. Wood, "Improving ionic conductivity and lithium-ion transference number in lithium-ion battery separators," *ACS Appl Mater Interfaces*, vol. 8, no. 48, pp. 32637–32642, 2016.

[18] C. Sauter, R. Zahn, and V. Wood, "Understanding electrolyte infilling of lithium ion batteries," *J Electrochem Soc*, vol. 167, no. 10, p. 100546, 2020.

[19] M. F. Lagadec, R. Zahn, S. Müller, and V. Wood, "Topological and network analysis of lithium ion battery components: the importance of pore space connectivity for cell operation," *Energy Environ Sci*, vol. 11, no. 11, pp. 3194–3200, 2018.

[20] A. Schilling, S. Wiemers-Meyer, V. Winkler, *et al.*, "Influence of separator material on infiltration rate and wetting behavior of lithium-ion batteries," *Energy Technol*, vol. 8, no. 2, p. 1900078, 2020.

[21] T. Knoche, V. Zinth, M. Schulz, J. Schnell, R. Gilles, and G. Reinhart, "In situ visualization of the electrolyte solvent filling process by neutron radiography," *J Power Sources*, vol. 331, pp. 267–276, 2016.

[22] W. J. Weydanz, A. Gottschalk, M. Schulz, *et al.*, "Visualization of electrolyte filling process and influence of vacuum during filling for hard case prismatic lithium ion cells by neutron imaging to optimize the production process," *J Power Sources*, vol. 380, pp. 126–134, 2018.

[23] A. Schilling, P. Gümbel, M. Möller, F. Kalkan, F. Dietrich, and K. Dröder, "X-ray based visualization of the electrolyte filling process of lithium ion batteries," *J Electrochem Soc*, vol. 166, no. 3, pp. A5163–A5167, 2019.

[24] T. Bond, R. Gauthier, A. Eldesoky, J. Harlow, and J. R. Dahn, "In situ imaging of electrode thickness growth and electrolyte depletion in single-crystal vs polycrystalline LiNixMnyCozO2/graphite pouch cells using multi-scale computed tomography," *J Electrochem Soc*, vol. 169, no. 2, p. 020501, 2022.

[25] F. Wieder, Ch. Kallfaß, I. Manke, *et al.*, "Electrolyte distribution and discharge time – a combined study of X-ray tomography and electrical measurements of a commercially available lithium-ion capacitor," *ECS Trans*, vol. 53, no. 30, pp. 211–218, 2013.

[26] W. Du, R. E. Owen, A. Jnawali, *et al.*, "In-situ X-ray tomographic imaging study of gas and structural evolution in a commercial Li-ion pouch cell," *J Power Sources*, vol. 520, p. 230818, 2022.

[27] Z. Deng, Z. Huang, Y. Shen, *et al.*, "Ultrasonic scanning to observe wetting and 'unwetting' in Li-ion pouch cells," *Joule*, vol. 4, no. 9, pp. 2017–2029, 2020.

[28] F. J. Günter, J. B. Habedank, D. Schreiner, T. Neuwirth, R. Gilles, and G. Reinhart, "Introduction to electrochemical impedance spectroscopy as a measurement method for the wetting degree of lithium-ion cells," *J Electrochem Soc*, vol. 165, no. 14, pp. A3249–A3256, 2018.

[29] F. J. Günter, C. Burgstaller, F. Konwitschny, and G. Reinhart, "Influence of the electrolyte quantity on lithium-ion cells," *J Electrochem Soc*, vol. 166, no. 10, pp. A1709–A1714, 2019.
[30] C. Peter, K. Nikolowski, S. Reuber, M. Wolter, and A. Michaelis, "Chronoamperometry as an electrochemical in situ approach to investigate the electrolyte wetting process of lithium-ion cells," *J Appl Electrochem*, vol. 50, no. 3, pp. 295–309, 2020.
[31] A. Shodiev, E. Primo, O. Arcelus, *et al.*, "Insight on electrolyte infiltration of lithium ion battery electrodes by means of a new three-dimensional-resolved lattice Boltzmann model," *Energy Storage Mater*, vol. 38, pp. 80–92, 2021.
[32] A. Shodiev, F. M. Zanotto, J. Yu, M. Chouchane, J. Li, and A. A. Franco, "Designing electrode architectures to facilitate electrolyte infiltration for lithium-ion batteries," *Energy Storage Mater*, vol. 49, pp. 268–277, 2022.

Chapter 12

Simulation-assisted electrode processing: discrete and machine learning computational approaches for the simulation of the battery manufacturing process

Franco M. Zanotto[1,2], *Mohammed Alabdali*[1,2], *Diana Zapata Dominguez*[1,2] *and Alejandro A. Franco*[1,2,3,4]

12.1 Introduction

12.1.1 Context

The development of clean energy technology is one most important requirements to decrease energy-related carbon dioxide emissions. This challenge triggers the development of renewable energies to scale up in the short and medium terms because of the growth of fuel prices and fossil source depletion, extreme heatwaves [1,2], and zero-emission vehicle sales by 2035 [3]. Efforts such as electrification of transportation and electricity storage are rising steadily over the past ten years by the number of patent filings in batteries and energy accumulation [4]. The preferred battery technology is Li-ion batteries (LIBs) due to their large capacity, high power, and cyclability. Continued innovation in LIBs is needed to satisfy the electric vehicle demands, including the development of Gigafactories to decrease production costs [5,6]. To address these large-scale production efforts, it is necessary to optimize further the manufacturing parameters to fabricate LIBs. Optimization efforts across the battery production requests for improved materials, cell designs, manufacturing processes, carbon dioxide emission fingerprints, and recycling [7]. The manufacturing optimization is developed at the prototype level rather than in Gigafactories, aiming at the final product design and quality tests by a systematic trial-and-error process. This process is time and cost-consuming because the battery manufacturing process is

[1]Laboratoire de Réactivité et Chimie des Solides (LRCS), Université de Picardie Jules Verne, Hub de l'Energie, France
[2]Réseau sur le Stockage Electrochimique de l'Energie (RS2E), Hub de l'Energie, France
[3]ALISTORE-European Research Institute, Hub de l'Energie, France
[4]Institut Universitaire de France, France

multistep, which includes three major steps: electrode preparation, cell assembly, and formation and testing, affected by many experimental parameters [8].

The electrode preparation includes principally: mixing the electrode formulation materials, coating and drying the slurry, and electrode calendering. The mixing step consists of uniformly blending the active electrode material (AM), a conductive additive, and a binder in water or *N*-methyl-2-pyrrolidine (NMP) to form a slurry whose flow behaviour outputs are characterized by rheology that is impacted principally by particle size and shape, charge, solid fraction, and thermodynamic and hydrodynamic forces [9]. The electrode slurry needs to be stable and processable by avoiding agglomeration and sedimentation and being able to be coated at a relatively high speed [10].

Coating and drying of the slurry are performed subsequently. In those steps, the slurry is deposited on the current collector and dried to remove the solvent and produce a uniform electrode. The quality of the coated electrode highly depends on the stability of the slurry and drying protocol. The latter affects the ratio at which the solvent is evaporated, diffusing molecules through the coated electrode, and dragging the conductive additive and binder particles. This binder and conductive additive migration impact the electrode's internal and current collector–electrode adhesion. The typical output properties characterized after coating and drying the slurry are the electrode thickness, porosity, pore size distribution, and tortuosity. For instance, the electrode thickness determines the mass loading and capacity for the electrochemical performance of the battery electrodes. Porosity and pore size distribution characterize the contact between the electrode particles and electrolyte wetting and establish the energy and power density [11].

After the coating and drying, the electrode is pressed to decrease the coated thickness and adjust to a suitable porosity to increase the contact between the coating and the current collector, leading to more uniform electrodes, an increase in electron transport, and cycling life. After drying, the electrodes are on large rolls that need to be cut to the shape of the desired cell format. Therefore, slitting or punching is necessary for assembling prismatic or stacking pouch cells, respectively. Cell assembly must be handled in a controlled environment or a vacuum-oven drying step, emphasizing the overall positioning accuracy, where the electrodes are divided by the separator since defaults could provoke irreversible capacity losses [10].

Once the dry cells are assembled, they will be filled with the electrolyte, and the formation process will start, resulting in the longest manufacturing step, taking more than 12 h. The formation step consists of facilitating the electrolyte infiltration in the electrodes and separator at higher temperatures to simplify a stable electrolyte passivation layer formation by charging and discharging the cell at low current rates [12]. This electrolyte interface is formed on the cathode and the anode that hampers the transport of electrons and produces irreversible reducible species or transport of transition metal through the cathode AM [10]. Nevertheless, a well-formed passivation layer will limit the over-reduction of the electrolyte, the production of gaseous or soluble decomposition products, and its further growth over time [13].

The optimization of the LIB manufacturing process is based on the design of experiments that sometimes are insufficient to cover all the aspects of parameter correlation complexity [14,15]. Besides, it is presumed that battery companies generate scrap rates higher than 30% for battery optimization [16]. Computational tools such as Artificial Intelligence (AI), machine learning (ML), and physical modelling supported by mathematical descriptions are promising techniques to speed up LIB prototyping optimization as more efficient approaches that will save time and costs [7,11,17]. While the next chapter in this book covers the manufacturing process simulation from a continuum perspective, in this chapter, we discuss how discrete particle modelling approaches and machine learning can be used to simulate and optimize the LIB manufacturing process.

12.1.2 State of the art

LIB development is helped by the application of computational methodologies to manufacturing processes. The variety of physical processes that take place and in length scales at which they occur during both manufacturing and operation of these devices make this a challenging task [18]. However, due to the relevance of these processes, the last years have seen an abundance of models aimed towards this purpose [19]. The tools available for LIB manufacturing are either physics-based or data-based methods. These two paradigms are fundamentally different approaches. The former describes chemical and physical relationships between cause and effect. As a result, only fundamental parameters related to the underlying phenomena are required, and a wide array of conditions can be studied, but computational cost is high. The latter methods are used to relate external parameters with output data. In consequence, while they can be very fast, they do not allow extrapolations outside of the gathered data.

A wide array of existing physics-based models focuses on individual aspects of the operation of LIBs: e.g., materials, electrochemistry, transport processes, thermal management or mechanics. In this sense, existing modelling softwares include methodologies reported by academic groups, which remain at the forefront of innovation in the field. This difficulty lies in the need for validation of the models, which relies on exposure to high quality field data, sometimes difficult to find in literature or industrial manufacturers, due to intellectual property protection.

Outside of physics-based models, AI/ML models are gaining popularity in the LIB for a variety of purposes, including materials discovery, cell ageing prediction, or surrogate modeling [17,20,21].

As the manufacturing of LIB consists of different stages, when modelling the entire process chain, different simulation paradigms are more appropriate than others at different stages. Different approaches can be used in at least one of the steps, such as Coarse-Grained Molecular Dynamics (CGMD), Discrete Element Method (DEM), Computational Fluid Dynamics (CFD), Lattice Boltzmann Method (LBM), or analytical models. This model concatenation often involves combining specialized software, and thus can cause challenges related to the compatibility of data formats and the accumulation of computing cost. The intersection points that

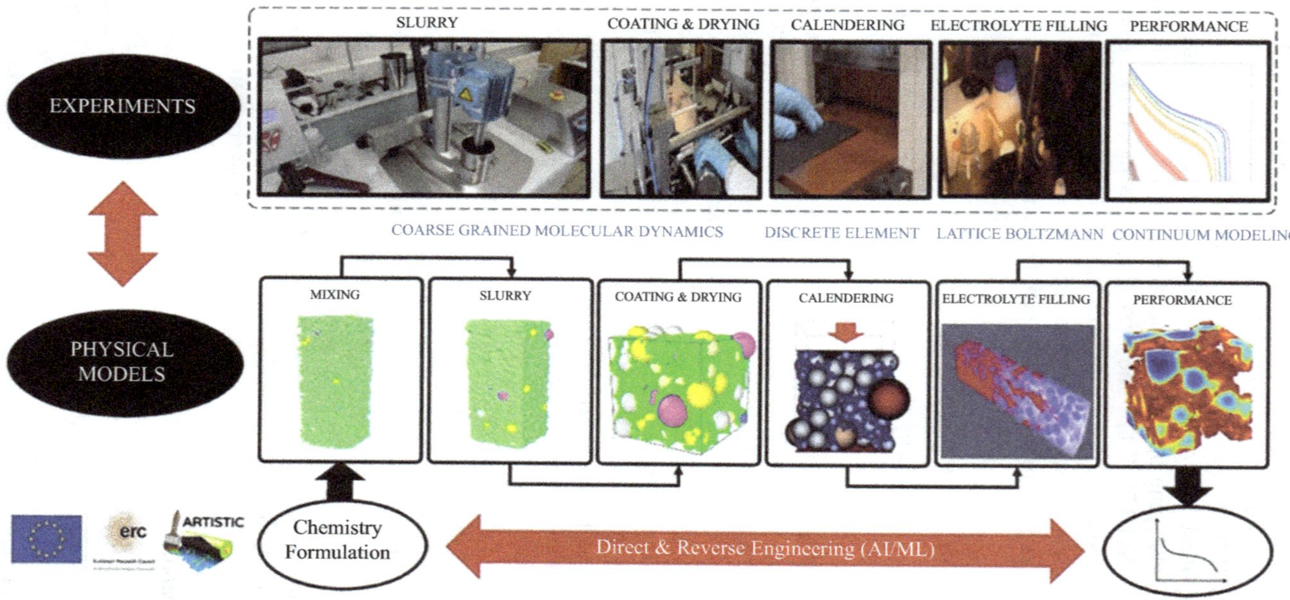

Figure 12.1 The computational workflows in the ARTISTIC project to simulate the manufacturing process of lithium ion batteries

connect each step to the following one are the structural parameters of the material as it moves from raw material to finished product. These parameters can be constant material properties (such as density of the components), can be altered (such as electrode thickness during calendering), or can be created (such as porosity during drying).

The ARTISTIC project, funded by the European Research Council and lead by Prof. Alejandro A. Franco (Université de Picardie Jules Verne, Amiens, France), started in 2018 with the goal of developing a pioneering digital twin of the entire manufacturing process of LIBs: [22] mixing, slurry coating, drying, calendering, electrolyte filling, formation, and resulting electrochemical performance (Figure 12.1). As described above, this approach sequentially couples the manufacturing stages, combining physics-based and data-driven models. It aims to predict the impact of manufacturing parameters on the final electrode architecture at the micrometer length scale, and as a function of it, the final electrochemical performance. Data driven models are used to accelerate the physics-based model's parametrization, process experimental data, and bypass more costly models through data-based surrogates. This approach can be used for the inverse design of LIB manufacturing. This consists of predicting the required initial parameters to obtain a final result with a particular set of properties, through an optimization loop (Figure 12.2).

The work done in the ARTISTIC project has recently been made available to any user from industry or academia through the ARTISTIC Online Calculator. This is a user-friendly graphical interface, accessible through any internet browser, that allows to launch slurry, drying, and calendering simulations, [23] and then download the resulting structures for analysis. Additionally, previously calculated results are available to all the users.

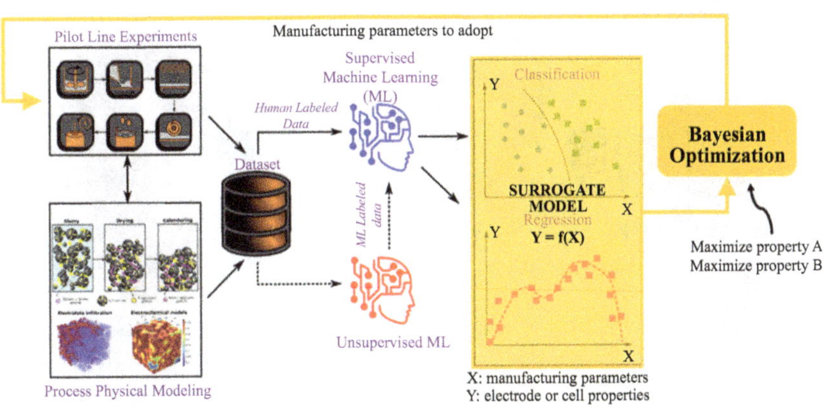

Figure 12.2 Optimization framework within the ARTISTIC project for inverse design of lithium ion battery manufacturing parameters. Adapted from Ref. [27].

The chemistry neutrality of this approach has been demonstrated by its adaptation to graphite anodes [24] organic sodium-ion batteries (SIBs), [25] and solid-state batteries cathodes [26].

12.2 Methodology

12.2.1 Slurry physics-based modeling

The CGMD method is used in the ARTISTIC project to simulate the electrode slurry in 3D using the program LAMMPS; this is the method used in this case [28]. It involves solving the conventional Newton equations numerically, which describe the link between the interaction forces (or force fields – FFs) and the acceleration of the particles in the slurry. These equations' numerical integration enables the determination of the particles' evolving velocities and positions. FFs like the Lennard–Jones and Granular–Hertzian FFs have been used to represent the interaction forces. These FFs enable the measurement of the effects of particle separation and effective radii on the magnitude of interaction forces. The AM and CBD particles, which are additives made of coarse-grained carbon, binder, and solvent, are the two types of particles taken into account. The approach depends on the formulation (weight percentages of AM and carbon-binder). The AM particle size distribution and slurry solid content are the inputs. The CBD particles' diameter is influenced by the carbon-binder ratio and the slurry's experimental density. The AM and CBD particles are randomly distributed in a simulation box with periodic boundary conditions to begin the simulation process (Figure 12.3(a)). The MD is then executed until the system reaches equilibrium (Figure 12.3(b)). This state shows the partial overlapping between the CBD particles with themselves and with

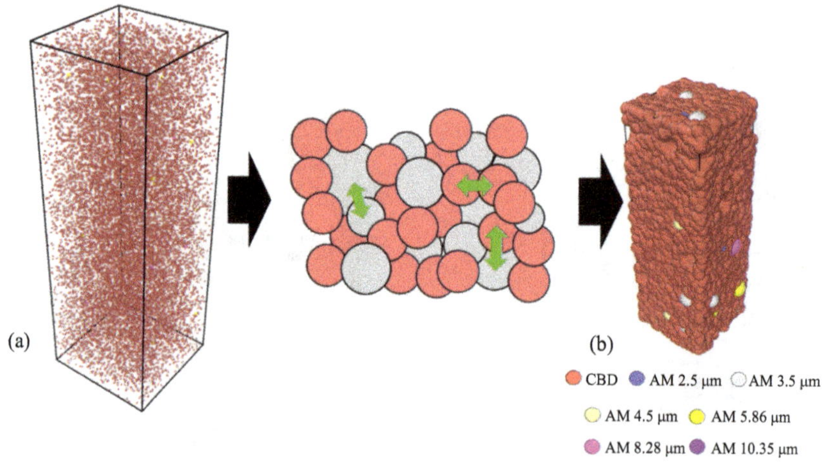

Figure 12.3 Example of slurry modelling according to ARTISTIC. (a) Particle positions are generated randomly. (b) After equilibration, a slurry microstructure is obtained. Adapted from Ref. [11].

the AM particles. The values of the FFs parameters and the anticipated size of the CBD particles rely on the corresponding 3D microstructure. Accordingly, this method requires the FFs parameter values to be fitted. This is carried out by comparing the computed viscosity-shear rate curve and the estimated slurry density to the values obtained experimentally. To calculate the viscosity-shear rate curves, the simulation box of the slurry microstructure is triclinic deformed at various shear rates during non-equilibrium MD simulations. For various slurry formulations and solid contents, the model effectively estimates the shear-thinning behavior, taking into regard the procedure of fitting the FFs parameters either manually or automatically. The second choice is given priority, obviously. We have shown that automated approaches result in a 20-fold reduction in computing costs. Particle swarm optimization (PSO), ML, and a combination of both were developed as automated procedures. PSO, or stochastic global optimization, is a swarm-based technique that attempts to iteratively improve a potential solution with respect to a specific quality metric, which is defined as the discrepancy between estimated slurry viscosities and experimental slurry densities. Additionally, since each CGMD model has a unique set of FFs parameters, ML is also used to estimate the viscosity values from the early stages of the non-equilibrium MD simulation. This allows cutting off the time of fitting, thereby faster results acquisition, and is used in optimization algorithms without the need to run completely [29]. This slurry modeling technique was shown for many AM chemistries, including NMC111, LFP, graphite, and organic AM for SIBs, as part of the ARTISTIC project [24,30].

12.2.2 Drying physics-based modeling

The CGMD method is used again in the ARTISTIC project to model the slurry drying process [31–34]. The CGMD-predicted microstructure of the slurry as a function of formulation and solid content is the input for the associated drying model. In this regard, two sub-approaches have been implemented. The initial one presumes uniform drying. Starting with the microstructured slurry, this entails reducing the CBD volume by a certain amount to simulate the evaporation of the solvent. The system is then abruptly brought back into equilibrium using an MD simulation. The dry electrode is represented by the resultant microstructure. In order to get dried electrode porosities that agree with the experiments, the CBD shrinkage factor has been adjusted. Similar to slurry simulations, Lennard–Jones and Granular–Hertzian are the FFs that are taken into consideration for the drying simulation.

The second sub-approach is based on the idea of heterogeneous drying (Figure 12.4). This sub-approach enables the explicit accounting for the displacement of the carbon-binder. It is designed in such a manner that the depth within the slurry can modulate the transition to the electrode state from the slurry. It expects that the upper part of the slurry will transition more quickly than the bottom. This is accomplished by segmenting the slurry's depth into numerous zones, each of which has a distinct pace at which the CBD particles contract (faster on the top, slower at the bottom). Although there is no specific explanation for the relationship, these

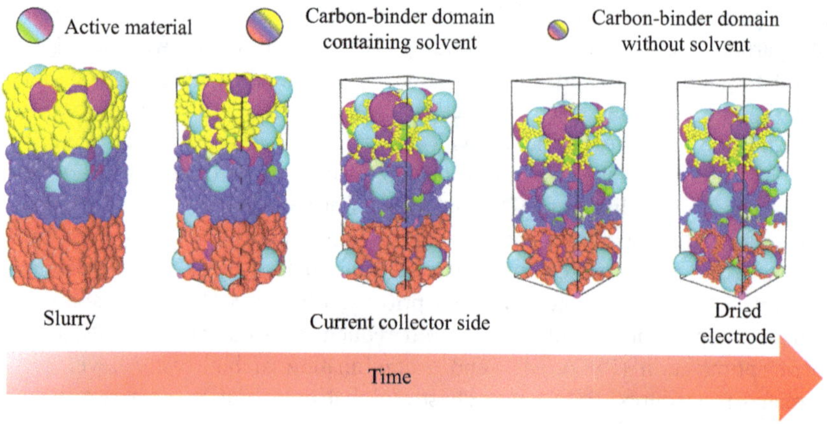

Figure 12.4 Example of heterogeneous drying model in the ARTISTIC project. Heterogeneous drying involves the continuous shrinking of CBD as the electrode dries. The system is divided into one to three layers, allowing different shrinking rates to account for the evaporation rate depending on the proximity to the current collector. Adapted from Ref. [33].

rates are inferred to be connected to experimental variables like temperature and drying rate. A correlation exists between the drying rate and the ratio of CBD shrinkage rates at various zones. This method enables us to determine that: CBD migration does not occur right away during the drying process but rather after some time. The time at which CBD movement begins depends on the drying rate; the fraction of CBD on the top area in the dried electrode at high drying rates will be larger than the fractions in the middle and bottom regions.

The first two features are consistent with what is known from experiments, whereas the third aspect is a prediction from modeling that necessitates experimental characterizations. The impact of three-stage drying techniques on the electrode microstructure may also be qualitatively captured by this CGMD-based drying model. The advantages of these methods include lower production costs. More homogenous electrodes are produced, in particular, when a high drying rate is applied at the start and end of the drying process and reduced drying rates in the middle.

12.2.3 Calendering physics-based modeling

It is essential to simulate the electrode calendering stage to enable the adjustment of the porosity and degree of compression (aiming to optimize the electronic pathways), while keeping enough space (pores) for a proper electrode wetting. By specifically accounting for both AM particles and CBD, both CGMD and DEM are employed in the ARTISTIC project to predict the effects of calendering pressure

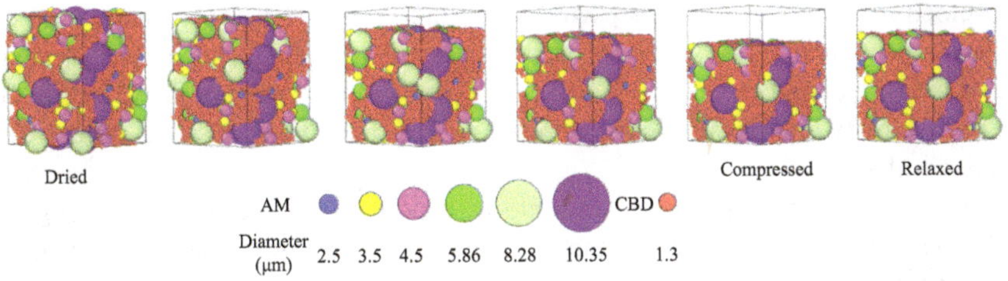

Figure 12.5 Results from the calendering simulation of a NMC111 electrode within the ARTISTIC project. AM stands for active material and CBD for carbon binder domain. Adapted from Ref. [23].

and speed on the electrode microstructure [23,35]. Such a model takes electrode microstructures from the drying simulation as an input (see section above). The simplified Johnson–Kendall–Roberts FF, which captures adhesive forces primarily caused by the binder "bridges," and the Granular Hertzian FF, which describes the mechanical characteristics of the electrodes, are used to solve Newton's equations that describe the cohesion forces between particles. The FF parameters are calculated by simultaneously fitting the experimental micro-indentation and compaction curves; the former provides information about the resistive force (hardness) a certain electrode offers when an indentator is applied. By contrasting the real the simulated microindentation curves, the mechanical model validation can be achieved. The spatial reorganization of AM particles, CBD, and pores following calendering are captured by the model (Figure 12.5). It is worth to be noticed that the model can also capture the relaxion after calendering, relaxation known as spring back phenomenon. By injecting the simulated electrode microstructures into 4D-resolved electrochemical models, it is possible to have an understanding of how this spatial reorganization affects the interfaces between the materials and pores, thereby electrochemical performance.

12.2.4 Electrolyte infiltration physics-based modeling

The next step in manufacturing is electrolyte infiltration. In practice, this process consists of two distinct stages: first, electrolyte is added to the cell. Then, it wets the internal surfaces by flowing through the pore network. The ARTISTIC project has focused on modeling the latter by using the Lattice Boltzmann Method (LBM). This is a more appropriate approach for the modelling of flow through porous media than more traditional CFD simulations [36]. This makes it ideal for the study of electrolyte wetting. It is also highly accurate, reproducible and parallelizable, which allows for short simulation times [37].

The LBM works by dividing the simulation domain into a lattice of regularly spaced points. Each point contains a given density of electrolyte and air, which evolves as virtual particles travel from each point to its neighbors. This movement is the result of two half-steps. During the first one, a local equilibrium is reached between the velocities of virtual particles due to collisions. In the second half-step, the velocity distribution dictates particle exchange between neighbor lattice points. This microscopically resolved approach results in emergent behavior that is identical to that predicted by the Navier–Stokes equations. The main input of this model are the viscosity and density of each fluid and the contact angle between the fluids and the walls.

In order to carry out these simulations, a domain geometry representing the electrode or a full cell sandwich must be generated and discretized. There is great flexibility in this process. Structures can be imported from experimental computed tomography, stochastically generated by spheres packing, or obtained from CGMD results from the calendering simulations presented in the previous section.

The ARTISTIC project has shown that during wetting, air can be trapped in small, isolated, or poorly connected pores. (Figure 12.6(a) and (b)) [38]. Trapped air impedes the complete wetting of all of the active surface, with negative effects on solid-electrolyte interphase (SEI) formation and electrochemical performance of the final product. This study allowed to identify critical variables to avoid negative effects, such as porosity, pore size distribution, and pore connectivity network. The use of structures obtained from manufacturing simulations allows to directly relate these with manufacturing parameters, particularly during calendering.

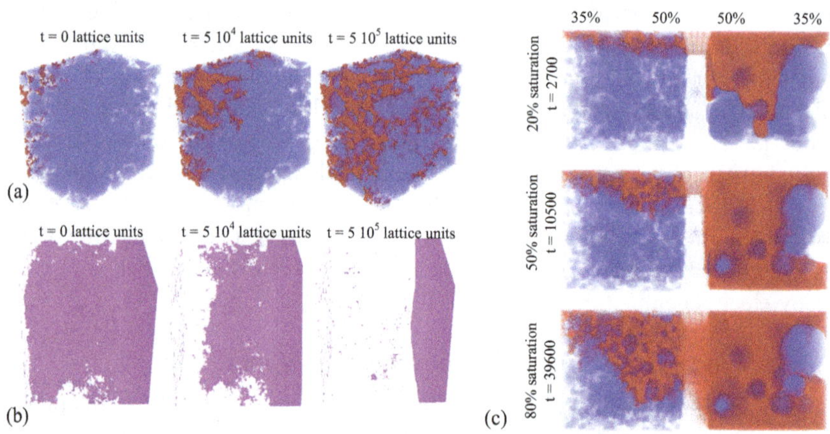

Figure 12.6 (a) Visualization of the wetting process on a tomographic reconstruction of an NMC electrode. Composition: NMC 94% – CBD 6%. Electrolyte is shown in red and solid structure in blue. (b) Visualization of the escape of air as it is displaced by the electrolyte. (c) Visualization of the wetting process on a section of a full cell with a layered architecture. The anode (right) and cathode (left) are stochastically generated following experimental particle sizes. The separator (center) is taken from tomography images. Each electrode is composed of two layers, the outer one with 35% porosity and the inner one with 50% porosity. Panels A and B adapted from Ref. [38]. Panel C adapted from Ref. [39].

A following study consisted of the use of architectures with layers of varying pore network properties as a potential approach to tune the wettability of the cell sandwich [39]. Infiltration simulations were performed based on the LBM to analyze the electrolyte saturation as a function of time. This descriptor can be used to evaluate different options for cell architectures (Figure 12.6(c)). Again porosity, porosity distribution, and particle size distribution were identified as critical variables. However, a new degree of freedom is included, related to the order in which different layers are located in the cell sandwich, in relation with the separator and current collectors. These insights allow influencing the electrolyte infiltration process for full cells in an advantageous way.

When carrying out simulations related to the electrochemical performance of electrodes generated from CGMD simulations, it is common to assume full saturation of the pores with electrolyte. In this sense, the LBM results can be directly exported to their use in such electrochemical simulations to account for incomplete wetting of the electrodes. With this approach, it was determined that structures that are not fully saturated show inhomogeneous lithium intercalation and a loss of effective ionic conductivity.

In summary, the ARTISTIC LBM modelling framework has been highly useful in producing recommendations for the reduction of energy and time of the electrolyte infiltration step.

12.2.5 Performance physics-based modeling

The last step in the manufacturing process simulation workflow is the evaluation of the generated structures in terms of their electrochemical performance. For this, they are incorporated into a three-dimensional model, which simulates lithiation and de-lithiation dynamics as a function of time. A model able to perform this task must explicitly consider the most relevant physical processes taking place in the operating electrodes. These processes include electrochemical reaction and double-layer formation at the interface between AM and electrolyte, lithium diffusion in the AM, electronic transport in both AM and CBD, electronic transfer between the AM and CBD, and lithium-ion migration in the electrolyte.

The main advantage of such a model is its independence on assumptions related to phenomenological parameters such as particle sizes, surface areas, porosity and tortuosity commonly used in Newman-type pseudo-2D electrochemical models. This is due to the explicit use of the micrometer-resolved structure as simulation domain.

This also allows a highly detailed assessment of the effect of heterogeneities on the macroscopically observed electrochemical response, since impedance can be spatially resolved to the micrometer level, and related to the location of individual particles.

This model is highly versatile, and thus has been used for a variety of applications. Some examples are the study of the effect of manufacturing parameters on NMC111 electrodes behavior [34], the proof-of-concept for an electrochemical model for Silicon–Graphite electrodes that include mechanical effects [24], and the evaluation of the advantages of battery cells with layers of varying porosity [40]. This model was also extended for the simulation of impedance spectroscopy under symmetric cell conditions [40,41].

In addition to characterization, this model can be adapted for the consideration of the SEI formation during the first electrochemical cycling (Figure 12.7(a)) [42]. Again the 3D micrometer-scale resolution represents an advantage for the study of the effect of particle morphology on the heterogeneity of the resulting SEI. Additionally, through the previous steps in modelling, these results can be traced back to the manufacturing parameters. This extended model describes the SEI through a simple equation relating SEI thickness to lithium ion concentration. The latter is directly obtained from the simulation, since it is one of the main spatially resolved variables. This study allowed to understand how the SEI is related to a local resistance for lithium intercalation and, in macroscopic terms, how the heterogeneity of the SEI affects electrode performance. By exploring different particle shapes, it was also possible to show that spherical particles, which form less tortuous electrodes, yield SEI with more homogeneous thicknesses than flat ellipsoidal particles (Figure 12.7(b)). Ultimately, this leads to improved performance.

12.2.6 Machine learning-based process modeling and optimization

Surrogate models based on machine learning (ML) are an alternative for physics-based models that promise accurate results in a fraction of the time. In the context of the ARTISTIC project, they have already been proven to predict the impact of slurry formulation, solid content, and viscosity on structural parameters of the resulting electrodes [43]. To train these ML models, large experimental data sets are required. In this example, 80 conditions, with five repetitions each were used. The resulting electrodes were manually classified into low, medium or high loading and porosity values. This dataset is then divided, as typical, into a training and testing datasets, in order to train a supervised classification model that allows to perform this sorting automatically as a function manufacturing parameter.

As a result, a surrogate model is obtained, which allows to easily visualize interdependencies between manufacturing parameters and resulting electrode properties, with an accuracy above 80%. This approach was compared with other supervised algorithms, Decision Tree and Deep Neural Network, which allows to assess predictive capabilities of each type of algorithm, identifying advantages and disadvantages for this particular application. The results were supported by additional experimental characterization of the slurry, allowing for the interpretation of results.

Application of ML tools can also be used for the evaluation of new electrode characteristics with no need of prior manual labelling, in order to identify the most relevant properties for heterogeneous electrode definition. The ARTISTIC project also demonstrated this by varying the most relevant parameters from the slurry and coating manufacturing steps for NMC-622 electrodes, and performing the subsequent experimental characterization. To do this, a dimensionality reduction and

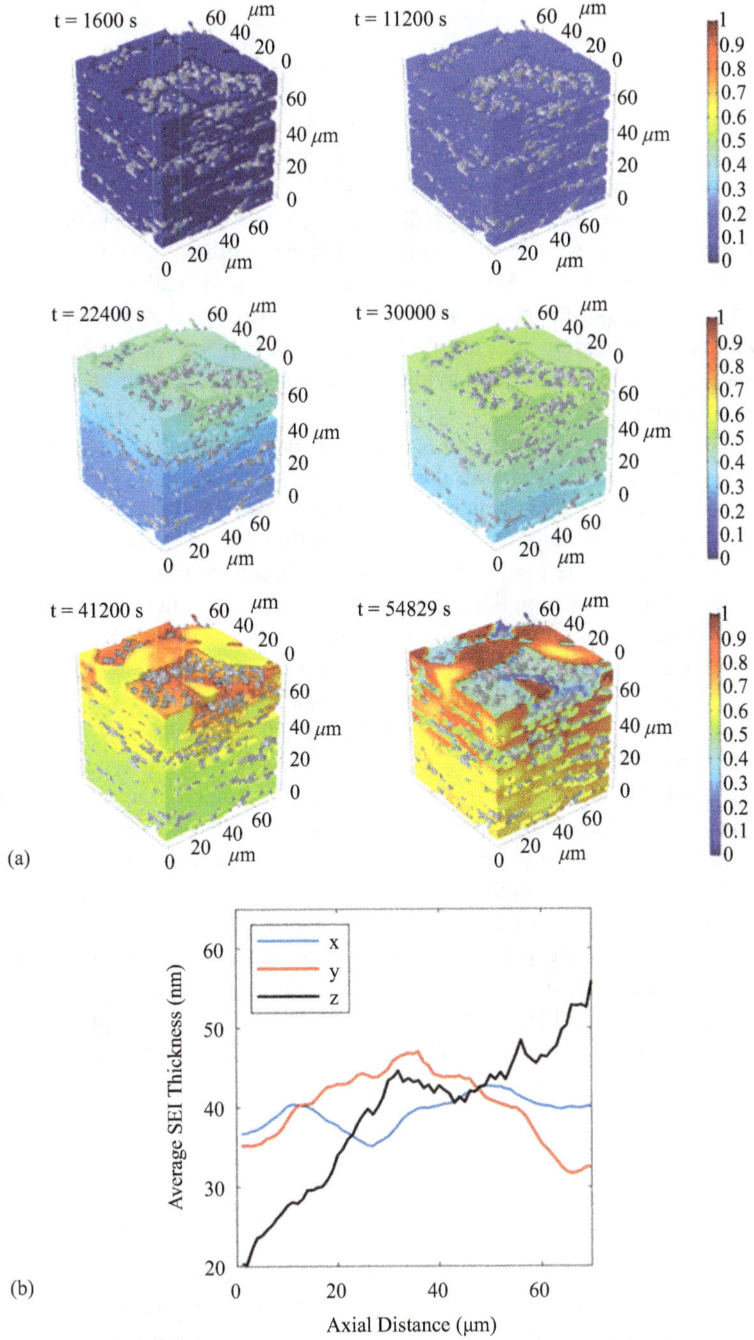

Figure 12.7 *Results of the SEI formation FEM simulations for high tortuosity graphite electrodes composed of ellipsoidal AM particles. (a) Snapshots of the half-cell at several stages during discharge at C/20, during which formation occurs. The color map represents the state of discharge of the AM. (b) Spatial evolution of the SEI thickness in a highly tortuous electrode as a function of the cartesian coordinates, showing a heterogeneous distribution of the SEI along the direction perpendicular to the current collector. Adapted from [42] and reproduced from [11].*

segmentation approach to identify groups of electrodes with common characteristics (both prior and newly defined) is performed. This strategy provides new perspectives for improving battery design [44].

ML can also be used for the prediction of contact surfaces between the electrode components, as well as the tortuosity factor in both solid and liquid phases. The mesostructures in the ML training set were stochastically generated, constrained by the experimentally defined input properties, such as porosity, formulation, and calendering pressure [45]. In this way, it was demonstrated how a combination of experimental and synthetically generated data can be relied on for the construction of training sets.

The LBM simulations used to study the electrolyte infiltration process lend themselves for the use of ML tools due to the large amount of data generated. Data-driven models can be thus generated for the prediction of saturation curves and the density of the fluid in the pores as a function of time [46]. First, LBM simulations are carried out in a wide range of conditions on structures obtained from detailed microtomography data (though it can be directly used with structures resulting from CGMD). These provide the air and electrolyte density for each lattice point in the pore network as a function of time. This information can be reduced by translating it into saturation for each individual pore as a function of time, and further condensed by selecting the saturation values for specific representative times during the simulation. These data can feed a multi-layer perceptron type ML model, resulting in a surrogate model capable of accurately predicting the saturation curve for each pore, and thus also the electrode saturation. While the one-time generation of the training data is time-consuming, this model presents an extremely low computational cost after training. This model was successfully validated by comparing it with the original LBM methodology applied to the same structures. This study concluded that the pore network connectivity is a crucial parameter for decreasing time and cost of electrolyte infiltration. Additionally, the spatially-resolved information yielded by this model allows it to be used as part of a workflow, using its results as inputs for electrochemical simulations for characterization.

12.3 Conclusions

Started in 2018, the ARTISTIC project aims at developing a digital twin of the manufacturing process of LIBs. Such digital twin is supported on a series of sequentially coupled physics-based process models able to simulate and predict the influence of manufacturing parameters on the electrode architectures in 3D and their wettability and associated electrochemical performance also in 3D. The prediction of 3D-resolved electrode architectures is possible thanks to the discrete approach underlying the different process models, representing the electrode slurry as a collection of interacting particles. Predicting such 3D-resolved architectures is important to further understand the impact of active and inactive material as well as pores heterogeneities on the wettability, the performance of the electrodes and ultimately the cells.

Such a physics-based computational workflow is complemented by a series of machine learning workflows devoted to representing in a surrogate fashion the physics-based process models. Such surrogate models are used then for optimization purposes. Machine learning is also used to speed up the parameterization of the process models and to unravel the correlations between manufacturing parameters and electrode properties. Such a computational modelling effort is supported on high quality experimental databases acquired in the pilot line of our laboratory.

Works by us have demonstrated our ARTISTIC simulation platform for different active material chemistries (e.g. NMC111, LFP, graphite [47], silicon/graphite [24]) and types of batteries (lithium ion batteries, sodium ion batteries [25], solid-state batteries [26]).

The online ARITISTIC calculator provides free online services to allow users to perform manufacturing simulations from an internet browser, and by using models developed within the Artistic project [23]. Graphical user interfaces allow to interact with parameters like in experiments and to perform simulations to investigate the influence for instance of formulation, drying rate and calendering degree on the 3D-architectures of the electrodes. The online calculator also allows to navigate and download the data already available from the simulations.

We believe that these types of computational models give the promise to accelerate battery manufacturing optimization. Ultimately, we can imagine integrating surrogate versions of these models in real pilot lines to allow them to autonomously optimize the manufacturing processes. We can reasonably expect that these autonomous characteristics can help at reducing scrap rates in battery manufacturing.

References

[1] Heating up. *Nat. Clim. Chang.* 2022, *12*, 693–693. https://doi.org/10.1038/s41558-022-01462-x

[2] A. Witze, *Nature* 2022, *608*, 464–465.

[3] E. Commission, "Zero emission vehicles: first 'Fit for 55 deal will end the sale of new CO_2 emitting cars in Europe by 2035,'" https://ec.europa.eu/commission/presscorner/detail/en/ip_22_6462, n.d.

[4] G. Gregori, S. Meini, Y. Ménière, *et al.*, *Innovation in Batteries and Electricity Storage,* Technical Report, 2020.

[5] C. Pillot, in *Battery Conference and Exhibition*, Nice, France, 2019.

[6] S. Ortiz, I. Careaga, "Gigafactories: Europe's Major Commitment to Economic Recovery Through the Development of Battery Factories | CIC energiGUNE," https://cicenergigune.com/en/blog/gigafactories-europe-commitment-economic-recovery-battery-factories, n.d.

[7] E. Ayerbe, M. Berecibar, S. Clark, A. A. Franco, and J. Ruhland, *Adv. Energy Mater.* 2021, *2102696*, 2102696.

[8] H. Bockholt, M. Indrikova, A. Netz, F. Golks, and A. Kwade, *J. Power Sources* 2016, *325*, 140–151.

[9] W. B. Hawley and J. Li, *J. Energy Storage* 2019, *25*, 100862.
[10] J. Li, J. Fleetwood, W. B. Hawley, and W. Kays, *Chem. Rev.* 2021, acs.chemrev.1c00565.
[11] F. M. Zanotto, D. Zapata Dominguez, E. Ayerbe, *et al.*, *Batter. Supercaps* 2022, *5*, doi:10.1002/batt.202200224.
[12] S. J. An, J. Li, Z. Du, C. Daniel, and D. L. Wood, *J. Power Sources* 2017, *342*, 846–852.
[13] U. S. Meda, L. Lal, S. M, and P. Garg, *J. Energy Storage* 2022, *47*, 103564.
[14] O. Rynne, M. Dubarry, C. Molson, *et al.*, *Battery.* 2019, *5*, 72.
[15] O. Rynne, M. Dubarry, C. Molson, *et al.*, *ACS Appl. Energy Mater.* 2020, *3*, 2935–2948.
[16] L. Gaines, Q. Dai, J. T. Vaughey, and S. Gillard, *Recycle.* 2021, *6*, 31.
[17] T. Lombardo, M. Duquesnoy, H. El-Bouysidy, *et al.*, *Chem. Rev.* 2022, *122*, 10899–10969.
[18] A. A. Franco, *RSC Adv.* 2013, *3*, 13027–13058.
[19] A. A. Franco, A. Rucci, D. Brandell, *et al.*, *Chem. Rev.* 2019, *119*, 4569–4627.
[20] A. Mistry, A. A. Franco, S. J. Cooper, S. A. Roberts, and V. Viswanathan, *ACS Energy Lett.* 2021, *6*, 1422–1431.
[21] A. A. Franco and A. S. Barnard, *Batter. Supercaps* 2022, e202200149.
[22] A. A. Franco, "ARTISTIC Project Website," https://www.erc-artistic.eu/, n.d.
[23] T. Lombardo, F. Caro, A. C. Ngandjong, *et al.*, *Batter. Supercaps* 2022, *5*, e202100324.
[24] C. Liu, O. Arcelus, T. Lombardo, H. Oularbi, and A. A. Franco, *J. Power Sources* 2021, *512*, 230486.
[25] T. Lombardo, F. Lambert, R. Russo, *et al.*, *Batter. Supercaps* 2022, e202200116.
[26] M. Alabdali, F. M. Zanotto, M. Duquesnoy, V. Viallet, V. Seznec, and A. A. Franco, *Three Dimensional Physical Modeling of the Wet Manufacturing Process of Solid State Battery Electrodes*, 2022.
[27] O. Arcelus and A. A. Franco, *J. Phys. Energy* 2022, *4*, 011002.
[28] T. Lombardo, J. Hoock, E. N. Primo, A. C. Ngandjong, M. Duquesnoy, and A. A. Franco, *Batter. Supercaps* 2020, *3*, 721–730.
[29] M. Duquesnoy, T. Lombardo, F. Caro, *et al.*, *arXiv Prepr. arXiv2201.04394* 2022, doi:10.48550/arxiv.2201.04394.
[30] A. C. Ngandjong, A. Rucci, M. Maiza, G. Shukla, J. Vazquez-Arenas, and A. A. Franco, *J. Phys. Chem. Lett.* 2017, *8*, 5966–5972.
[31] A. Rucci, A. C. Ngandjong, E. N. Primo, M. Maiza, and A. A. Franco, *Electrochim. Acta* 2019, *312*, 168–178.
[32] M. Chouchane, A. Rucci, T. Lombardo, A. C. Ngandjong, and A. A. Franco, *J. Power Sources* 2019, *444*, 227285.
[33] T. Lombardo, A. C. Ngandjong, A. Belhcen, and A. A. Franco, *Energy Storage Mater.* 2021, *43*, 337–347.
[34] A. C. Ngandjong, T. Lombardo, E. N. Primo, *et al.*, *J. Power Sources* 2021, *485*, 229320.

[35] C. Liu, T. Lombardo, J. Xu, A. C. Ngandjong, and A. A. Franco, *Energy Storage Mater.* 2023, *54*, 156–163.

[36] X. Shan and H. Chen, *Phys. Rev. E* 1993, *47*, 1815–1819.

[37] L. Chen, Q. Kang, Y. Mu, Y.-L. He, and W.-Q. Tao, *Int. J. Heat Mass Transf.* 2014, *76*, 210–236.

[38] A. Shodiev, E. Primo, O. Arcelus, *et al.*, *Energy Storage Mater.* 2021, *38*, 80–92.

[39] A. Shodiev, F. M. Zanotto, J. Yu, M. Chouchane, J. Li, and A. A. Franco, *Energy Storage Mater.* 2022, *49*, 268–277.

[40] A. Shodiev, M. Chouchane, M. Gaberscek, *et al.*, *Deconvoluting Benefits of Porosity Distribution in Layered Electrodes on Electrochemical Performance of Li-Ion Batteries*, Elsevier, 2022.

[41] A. Shodiev, E. N. Primo, M. Chouchane, *et al.*, *J. Power Sources* 2020, *454*, 227871.

[42] M. Chouchane, O. Arcelus, and A. A. Franco, *Batter. Supercaps* 2021, *4*, 1457–1463.

[43] R. P. Cunha, T. Lombardo, E. N. Primo, and A. A. Franco, *Batter. Supercaps* 2020, *3*, 60–67.

[44] M. Duquesnoy, T. Lombardo, M. Chouchane, E. N. Primo, and A. A. Franco, *J. Power Sources* 2020, *480*, 229103.

[45] M. Duquesnoy, I. Boyano, L. Ganborena, P. Cereijo, E. Ayerbe, and A. A. Franco, *Energy AI* 2021, *5*, 100090.

[46] A. Shodiev, M. Duquesnoy, O. Arcelus, M. Chouchane, J. Li, and A. A. Franco, *J. Power Sources* 2021, *511*, 230384.

[47] M. Duquesnoy, T. Lombardo, F. Caro, *et al.*, *npj Comput. Mater.* 2022, *8*, 161.

Chapter 13

Simulation-assisted electrode processing

Yinghui Yin[1] *and Julien Bréger*[1,2]

Li-ion battery (LIB) manufacturing, from materials to ready-to-market product, is a very complicated process that can involve more than 30 process steps. The establishment of a stable and qualified process flow is not trivial, requiring rich experiences and good understanding about the process mechanism. During the past years, all the cell components are upgraded so as to enhance battery performances and new materials emerge in an endless stream. At the same time, various novel technologies have been proposed to improve the energy efficiency and environmental friendliness of manufacturing process. As a result, there is a strong need for quick adaptation of these new materials/processes to production. Moreover, with the blooming of electrical vehicles, the demand of LIB expands at an unprecedented rate, pushing the battery production from MWh-scale to GWh-scale and even to TWh-scale. Therefore, facing the challenge of the quick materials/process adaption and fast production scale-up, in parallel to experimental trials, modeling and simulation are expected to serve as tools to enhance the understanding of process mechanism, to accelerate the process upscaling, to optimize the process parameters and to improve the quality control.

13.1 Introduction

Process modeling and simulation is a model-based representation of one or a series of process steps. While process modeling intends to describe one or more aspects of a given process step in a logical way with mathematical equations, process simulations focus on the implementation of these models. Depending on its level of details, process models can be divided into different categories:

- Phenomenological models, which are often derived directly from experimental results, are expressed in the form of mathematical formulas to show the correlation between process parameters and the product properties. Usually, phenomenological models are fast and easy to implement, but they are limited in terms of versatility.

[1]Saft Bordeaux, France
[2]ACC, Bruge, France

- Computational models, which are commonly physical-based models capturing more than one aspect of the process. As these models are often too complex to have an analytical solution, simulation techniques such as numerical approximation are needed for model implementation.
- Process chain model, which is a flow model of process chain, using intermediate product parameters as connection of different process steps for the purpose of global optimization.

Since process models at particle/pore scale have been introduced in Chapter 12, this chapter will focus on process models at larger length scale. The aim of this chapter is to give a general overview about how process modeling could potentially help in LIB manufacturing rather than provide every detail of mentioned models.

13.2 Modeling of mixing process

Battery manufacturing starts with mixing process, which aims at obtaining a stable, well-dispersed and homogenous electrode slurry. The electrode slurry is a mixture of active materials, conductive additives, polymer binder and solvent. At lab-scale, the mixing process is often performed with an agitation impeller, while at industrial level, planetary mixer and high shear mixer are popularly used. These industrial mixers have multiple shafts to ensure both dispersive mixing and distributive mixing. The dispersive mixing is to break the agglomerates of active materials and conductive additives, which is often done with a rotating dispersion disc or rotor-stator impeller. Other parts such as scrapers or 8-shape shaft are designed for a good distributive mixing by providing the motion of the slurry so as to avoid local stagnation, particularly close to the wall.

Computational fluid dynamics (CFD) is a numerical simulation technique of fluid motion, which allows to analyze in detail the slurry flow combined with mass and heat transfer. CFD models have already been used to simulate the mixing process in many industries, such as food, pharmaceutical and paper industries [1–3]. Due to the similarities of mixing process, models from these industries can be good references for that in LIB production. Thus, modeling of mixing process with CFD is discussed in Section 13.2.1. As viscosity is an important input parameter for CFD model, the rheology models of slurry are discussed in Section 13.2.2. In addition, kinetic model of carbon dispersion is discussed in Section 13.2.3.

13.2.1 CFD model of mixer

Regardless of the application, the fundamental basis of all CFD models are balance equations, which correspond to a general formulation of [4]:

$$\frac{\partial \phi}{\partial t} + v_i \frac{\partial \phi}{\partial x_i} = D \frac{\partial^2 \phi}{\partial x_i^2} + S(\phi) \tag{13.1}$$

The parameter ϕ can be a scalar (e.g. concentration), vector (e.g. velocity) or tensor (e.g. stress tensor). The terms in (13.1) represent:

rate of accumulation + transport by convection = transport by diffusion + source terms

Depending on the definition of ϕ, the balance equation is developed into partial differential equations of the balance for continuity, momentum, and energy. CFD is to solve these non-linear partial differential equations with numerical methods as it is difficult, if not impossible, to obtain the analytical solution. Many CFD packages based on finite volume method (FVM), either open-source (e.g. OpenFoam) or commercial (e.g. StarCCM + and Fluent), are available to this end. Some packages based on finite element method (FEM) provide also CFD modules, e.g. COMSOL.

As for other applications, the CFD simulation of mixing process consists of following steps:

- **Geometry modeling**: the first step is to define the domain of interest. For mixing process, it corresponds to the volume available to the slurry in the mixer, which is framed by the geometry of the vessel and mixing tools (e.g. impellers, etc.) as well as the filling level of slurry.
- **Meshing**: the domain of interest needs to be divided into small cells to implement the numerical methods. The shape and size of mesh are crucial for the accuracy of the simulation result, but a large number of cells leads to a high computational cost. It is thus important to find a good compromise.
- **Model setting**: complementary physical models are needed to simulate the turbulent flow, multiphase flow, and complex motion of the mixing tools in the mixer. More details about commonly used physical models for mixing process can be found in the review from Marshall *et al.* [5]. Besides, all physical properties of slurry should be defined, either with a simple value (e.g. density) or with a model (e.g. viscosity).
- **Results analysis**: once the simulation is done, the obtained data should be properly interpreted for meaningful results.

CFD simulation is helpful to learn about various aspects of the mixing process. Much information can be directly obtained from outputs of the simulation. The flow velocity field, which is the magnitude and direction of local flow velocity, can be used to visualize the flow pattern in the mixer and evaluate the axial/radial mixing ability of mixing tools. The shear stress distribution can indicate the intensity of dispersive mixing and locate the high shear zone. For multiphase flow simulation, the homogeneity of the slurry can be viewed in a straightforward way with the volume fraction of species. In addition to these direct information, further analysis can provide information at operational level. For example, from the pressure and shear force analysis, one can calculate the powder demand for the motor. When coupled with population balance model (PBM), the CFD model can be used to estimate the time to reach adequate carbon dispersion.

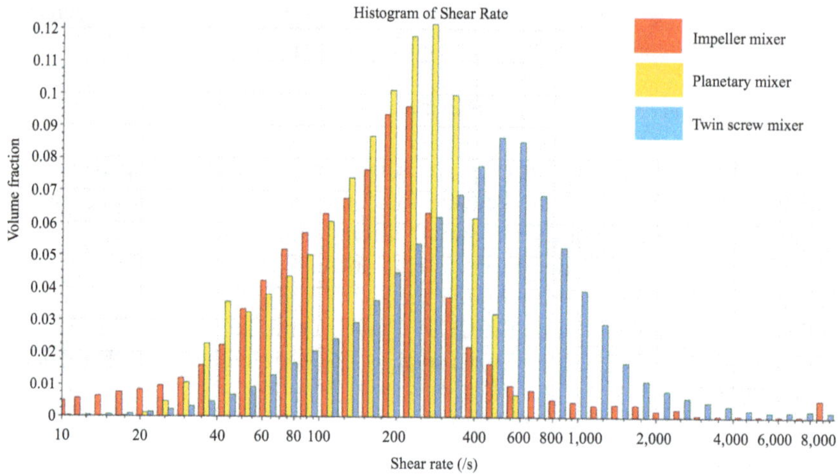

Figure 13.1 Histogram of shear rate distribution obtained from the CFD simulations of impeller mixer, blade-free planetary mixer and twin-screw mixer

As the CFD models are based on fundamental physics, its validity does not depend on the geometry of mixer. Thereby, it can be used as a tool to compare the mixing performance of distinct mixers, particularly the novel concept of mixer. At lab-scale, except for the impeller mixer, blade-free planetary mixer (BFPM) is also used for small-batch slurry mixing. In BFPM, an inclined cylinder vessel is subjected to a double-rotation movement: a rotation around its geometry axis and a revolution around an external axis. The combination of rotation and revolution ensures the motion of slurry and results in good mixing. More recently, there is a rising interest in using the counter-rotative twin-screw mixer for slurry mixing. While comparing the mixing efficiency of these three mixers via experimental trials is complicated, CFD simulations could shed light on this point. Figure 13.1 shows the shear rate distributions obtained from CFD models of a lab-scale impeller mixer, a blade-free planetary mixer and a twin-screw mixer, respectively. While the same targeted slurry composition is defined for all three models, the process parameters (e.g. rotation rates) are set accordingly to typical values for each mixer. With the same spirit, CFD simulations are also a good tool for upscaling.

13.2.2 Rheology models of electrode slurry

Slurry viscosity is not only an input parameter for modeling, but also a quality index for mixing process and a processability index for coating process. Therefore, it needs to be accurately described. The viscosity of electrode slurry depends on various factors including the shear rate, the temperature, and the slurry composition. Phenomenological models are available to address each aspect of slurry rheology.

The electrode slurries, either positive electrode or negative electrode, are non-Newtonian fluids and their viscosity decrease with the increasing shear rate. This shear-thinning behavior can be approximated by a power law model, which can be written as:

$$\eta = K\gamma^{n-1} \tag{13.2}$$

where η is the viscosity, K is the flow consistency index, and n is the flow behavior index. To take into account the yield stress (σ_0) of slurry at low shear rate, the Herschel–Bulkley model is used:

$$\sigma = \sigma_0 + K\gamma^n \tag{13.3}$$

where σ is the shear stress, which is defined as the product of shear rate and viscosity:

$$\sigma = \gamma \cdot \eta \tag{13.4}$$

A more general model is the Carreau–Yasuda model, which captures the Newtonian-like behavior of the shear thinning fluid at high shear rate. The formulation of the Carreau–Yasuda model is given by [6]:

$$\eta = \eta_\infty + (\eta_0 - \eta_\infty) \cdot \left(1 + \left(\frac{\gamma}{\gamma_c}\right)^a\right)^{\frac{n-1}{a}} \tag{13.5}$$

Here, η_0 is the zero-shear rate which refers to the low limit of viscosity, η_∞ is the infinite-shear rate which refers to the up limit of viscosity, γ_c is the critical shear rate at which the shear thinning behavior begins, and a is the Yasuda parameter.

The temperature dependence of slurry viscosity is complicated. Hawley *et al.* have shown that within the range of 20–60°C, increasing the slurry temperature leads to a rise of low shear viscosity ($<10\ \mathrm{s}^{-1}$) and a reduction of high shear viscosity ($>100\ \mathrm{s}^{-1}$) [7]. This is due to the interplay between the polymer entanglement and particle kinetic motion. Usually, shear rate applied to the slurry during the mixing process is high, the temperature dependence of slurry rheology is therefore emphasized at the high shear regime and is modeled with Arrhenius law:

$$\eta = \eta_r e^{-\frac{E_a}{R}\left(\frac{1}{T} - \frac{1}{T_r}\right)} \tag{13.6}$$

where subscript r refers to an arbitrary temperature, E_a is the activation energy for viscous flow and R is the ideal gas constant.

Moreover, when the ratio between the solid components (active materials, polymer binder, and carbon additives) remains constant, the dependence of the slurry viscosity on solid content could be approximated by the Krieger–Dougherty model [8]:

$$\eta = \eta_0 \left(1 - \frac{\phi}{\phi_m}\right)^{-[\eta]\phi_m} \tag{13.7}$$

where η_0 is the viscosity of liquid phase which refers to the solvent with dissolved binder, ϕ is the volume fraction of solid particles, ϕ_m is volume fraction of solid particles at maximum packing and $[\eta]$ is the intrinsic viscosity which depends on the particle shape [9]. A more detailed model taking into account various interparticle forces (e.g. van der Waals and steric) is proposed by Ma *et al.*, which can predict the impact of the particle size and molecular weight of the polymer binder on slurry rheology [10].

13.2.3 Kinetic model of carbon black dispersion

Carbon black (CB) is a common conductive additive in electrode slurry. For the sake of an adequate electronic conductivity of the electrode, the CB in slurry has to be deagglomerated to a sufficient degree. Two common means to improve the carbon dispersion are increasing mixing intensity and extending the mixing time. However, overmixing can cause damage to polymer binder and active materials, leading to a negative influence on electrode quality and on cell performance. Therefore, it is of importance to well define the mixing time.

Schilde *et al.* developed a model to describe the dispersion kinetic of nanosized particles, which is formulated as follows [11]:

$$r(t) = r_0 + (r_{end} - r_0) \cdot \frac{t}{t + K_t} \tag{13.8}$$

where r_0 is the initial particle size, x_{end} is the minimum achievable particle size, t is the time and K_t is a measure for the stress frequency. For a given mixer, K_t and r_{end} can be correlated to the circumferential speed u of the dissolver disc with empirical parameters a_1, a_2, b_1 and b_2 by:

$$r_{end} = a_1 \cdot u^{b_1} \tag{13.9}$$

$$K_t = a_2 \cdot u^{b_2} \tag{13.10}$$

The above model has been applied to carbon dispersion in an anode slurry of LIB and the simulation results fits well with the experimental measurements [12]. Therefore, this model can be used to predict overall mixing time from the early stage results of mixing process. However, as the parameter Kt depends on the design of mixer as well as the slurry properties, the model parameters should be refitted when the mixer or slurry formulation changes. The lack of transferability is an intrinsic limitation of model.

Another approach to model the CB dispersion is to couple the CFD simulation with population balance model (PBM). The PBM is widely used in various engineering areas, such as crystallization, granulation, milling, etc. [13]. In PBM, particles are grouped into different size classes and the variation of particles size are considered as the change of number density in size classes. A general size discrete PBM is given by:

$$\frac{dN_i}{dt} = -k_i N_i + \sum b_{i,j} k_j N_j \tag{13.11}$$

where the subscript i refers to the ith size class of particles, N is the number density of particle in size class i, k is the breaking kinetics and b_{ij} is the breakage distribution function. While the first term in the right side of (13.11) refers to the amount of breakage in size class i, the second term represents the amount of size class i particles generated from particle breakage of other size classes. Considering the small size of CB particles and the high viscosity of slurry, the shear-induced erosion can be considered as the dominant size reduction mechanism. Therefore, the size class can set in a way that a particle of class i always breaks into a particle of class $i-1$ and primary particles (N_0). As a result, except for N_0, (13.11) can be thus rewritten as:

$$\frac{dN_i}{dt} = -k_i N_i + b_{i,i+1} k_{i+1} N_{i+1} \tag{13.12}$$

Consider, as an approximation, that the particle breakage has a first-order kinetic to the excess of hydrodynamic force (F_H) to cohesion force (F_{COH}), k_i can be expressed as follows:

$$k_i = k_{i,0}(F_H - F_{COH}) = k_{i,0}(k_s \tau r_i^2 - F_{COH}) \tag{13.13}$$

where τ is the shear stress and r_i is the particle radius of size class i. In terms of the CFD-PBM coupling, the CB size classes can be implemented into CFD models as passive scalars, and the breaking kinetics of each size class can be solved by (13.12) and (13.13) with the local shear stress obtained from CFD simulations. As the PBM model is independent of mixer design, the parameters (e.g. $k_{i,0}$, k_s and F_{COH}) fitted for one mixer can be applied to the simulation of another one. Therefore, the CFD-PBM model allows to compare the mixing time between different mixers. It is also a promising tool for process upscaling.

13.3 Modeling of coating process

After mixing process, the electrode slurry is coated on a current collector to form a wet thin film. The quality of the coated wet film determines the geometrical uniformness of dry electrode, which directly impacts the downstream calendering and assembly process [14]. Besides, the presence of coating defects risks to result in deteriorated electrochemical performances of cells. Thus, it is of high importance to obtain uniform and defect-free wet electrode films.

With its advantage of precision and flexibility, slot-die coating is commonly used in industrial production of LIB. Since its first application for fabricating photographic film in 1950s, the slot-die coating has become a critical process for manufacturing of organic electronics, thin film solar cell and polymer electrolyte membrane fuel cells. Many efforts from both academia and industry have been made to enhance the understanding of slot-die coating by either theory development or field application. A thorough review of slot-die coating as well as other premetered coating methods can be found in the book of Schweizer [15]. Modeling is employed to simulate the slurry distribution in the slot-die head and to analyze the coating window, which can be further applied to slot-die coater design, coating uniformness prediction and process parameter optimization [16–19].

13.3.1 Modeling of slurry distribution in die

During the slot-die coating process, electrode slurry is pumped into slot-die head through the inlet and is then distributed to the cross-web direction in the manifold. Then it flows through the slot and coats on the web (Figure 13.2). The nominal thickness of the coating film (h_{film}) is determined by slurry flow rate (Q), coating width(W) and coating line speed (u_w) by the law of mass conservation:

$$h_{film} = \frac{Q}{Wu_w} \tag{13.14}$$

However, the uniformness of the film thickness in the cross-web direction depends on the local flow velocity of the electrode slurry at the end of the slot. As the local flow rate depends on the geometry design of the slot-die head, the slurry rheology and the slurry pumping pressure, it is difficult to predict electrode uniformness solely by empirical knowledge.

CFD model can be used to simulate the slurry flow inside the slot-die head [16,20,21]. The domain of interest consists of the slurry inlet pipe, the manifold and the slot gap framed by the shim plate and the slot die body. It is sufficient to consider the slurry as a single-phase fluid in this model and the rheology model for mixing process can also be used to describe the rheological property of the slurry. The slurry flow velocity can then be correlated to the local film thickness by:

$$h(y) = \int_{x=0}^{l} \frac{v(x,y)dx}{u_w} \tag{13.15}$$

where x, y refer to the direction of slot gap and cross-web, respectively.

For slot-die coater supplier, the above CFD model can be used to evaluate design of slot-die head, particularly the position of the inlet, the shape and depth of the manifold as well as the geometry of shim plates.

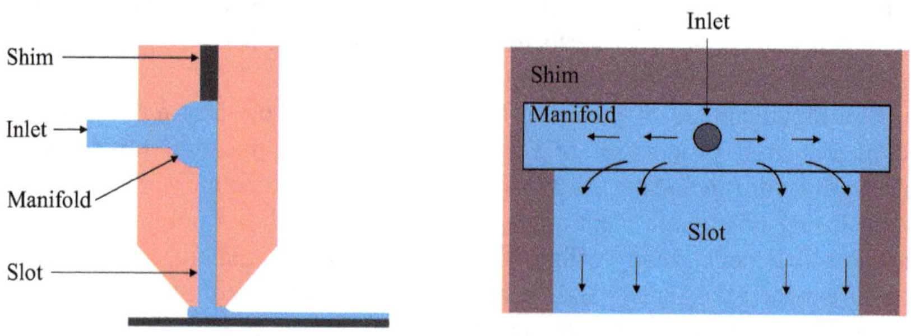

Figure 13.2 Graphic illustration of slot-die head (side view) and the slurry distribution inside

For battery manufacturer, this CFD model can serve as a tool to define process parameters such as the shim thickness and slurry pumping rate for a given electrode slurry to ensure the cross-web uniformness of the coating thickness.

13.3.2 Modeling of coating window

The slot-die coating process can produce a stable coating only within a certain range of operating parameters, beyond which various coating defects will appear. Plotting the operating boundaries with respect to relevant process parameters gives rise to a diagram of coating window.

The operating limits are closely related to the stability of coating bead, which is determined by the capillary force, viscous force, inertial forces and applied vacuum pressure [22]. A coating window predicted from the visco-capillary model is showed in Figure 13.3, in which the pressure drop across the coating bead is plotted versus the ratio of coating gap to film thickness. For a stable coating bead, the upstream meniscus is pinned between the upstream lip and the web. When the vacuum pressure is too high, the upstream meniscus leaks into the vacuum box leading to the failure known as swelling. To the other extreme, when the vacuum pressure so low that the upstream meniscus moves beyond the inner edge of the upstream lip, the air entrainment occurs due to the failure of dynamic wetting. Another process limit refers to the low flow limit, which happens when the web speed is too high, or the target film thickness is too small. Beyond the low flow limit, the downstream meniscus is too curved, and it can no longer bridge the downstream lip and the coating. The low flow limit is responsible for the ribbing defect which is a sinusoidal pattern in the web moving direction.

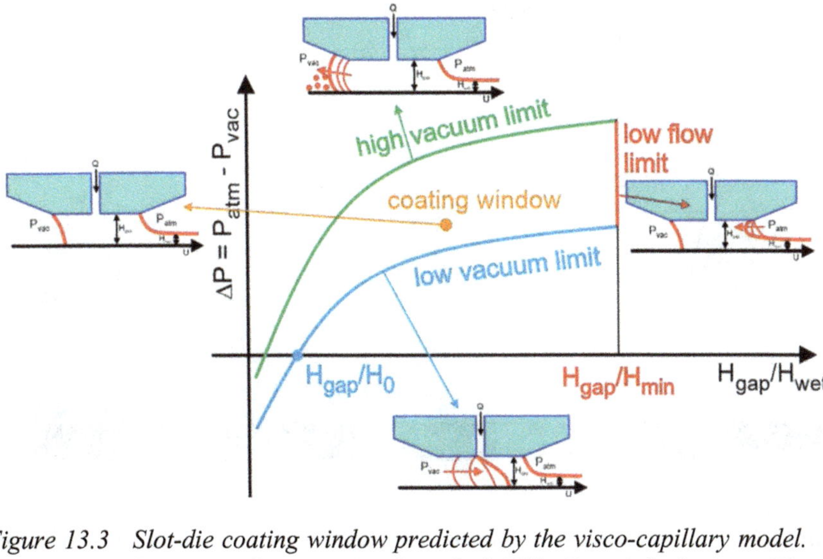

Figure 13.3 Slot-die coating window predicted by the visco-capillary model. Reprint with permission from Ref. [15]. Copyright Springer Nature.

The above-discussed process limits can be modeled with analytical equations. Details about the evolution of equations can be found in the review paper from Ding *et al.* [22]. More specifically, Schmitt has applied this analytical model to the slot-die coating process of LIB production, with considering the slurry as power-law non-Newtonian fluid (13.2) [23]. The equations of the process limits are formulated in (13.16)–(13.18). All relevant parameters are illustrated in Figure 13.4. The analytical model is easy to implement with Python code or Excel spreadsheet, thus it can serve as a fast engineering tool to estimate the feasibility of process parameters for target coating film design.

- High vacuum limit:

$$\Delta P_U = 1.34\left(K\frac{u_w^n}{\sigma h_G^{n-1}}\right)^{\frac{2}{3}}\frac{\sigma}{h} + l_D K\left(\frac{(n+1)(2n+1)}{n}\right)^n \frac{(2h-h_G)^n}{h_G^{2n+1}} u_w^n + l_U K\left(\frac{(n+1)(2n+1)}{n}\right)^n h_g^{-(n+1)} u_W^n + \frac{2\sigma}{h_G} \tag{13.16}$$

- Low vacuum limit:

$$\Delta P_U = 1.34\left(K\frac{u_w^n}{\sigma h_G^{n-1}}\right)^{\frac{2}{3}}\frac{\sigma}{h} + l_D K\left(\frac{(n+1)(2n+1)}{n}\right)^n \frac{(2h-h_G)^n}{h_G^{2n+1}} u_w^n + \frac{2\sigma}{h_G} \tag{13.17}$$

- Low flow limit:

$$Ca_{crit} = K\frac{u_w^n}{\sigma h_G^{n-1}} = 0.6\left(\frac{2}{\frac{h_g}{h} - 1}\right)^{\frac{3}{2}} \tag{13.18}$$

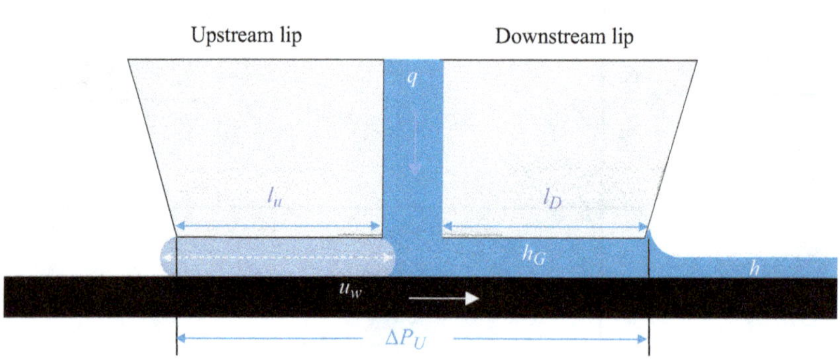

Figure 13.4 Illustration of a stable coating bead showing all parameters relevant for bead stability

CFD models can also be used to simulate the slot-die coating on the web [19,24,25]. The domain of interest, in either 2D or 3D, consists of the slot end, the coating gap, part of coated film and surrounding volume occupied by air. Both slurry and air must be considered in the model to simulate the capillarity. Thus, the volume of fluid (VOF) method is used to allow the tracking of slurry/air interface. Comparing to analytical model, the CFD model could include more details into the simulation, such as the contact angle between slurry and web, the slip of the slurry to the web and the local flow velocity at the end of the slot gap. For example, Hoffmann *et al.* have found that CFD model with the local-slip condition has the closest results to the experimental data, in comparison with the CFD model with no-slip condition and the 1D analytical model [19].

Moreover, both the 1D analytical model and 2D or 3D CFD model can be extended to simulate multilayer coating [26,27]. Besides, the CFD model can also be applied to simulate intermittent coating [28].

13.4 Modeling of drying process

Right after the coating process, the wet electrode films are sent to the dryer to remove the solvent. At industrial scale, multi-compartment convection dryers are used for electrode drying. The temperature and the air flow rate in each compartment can be set independently. To meet the requirement of high throughout production, the drying speed can be as high as 100 m/min, in accordance with the upstream coating process. As a result, the industrial dryer can be more than 100 m. The current electrode drying is an energy-intensive process, which accounts for more than 30% of energy consumption of LIB production [29]. Moreover, during the drying process, the solvent is evaporated and leaves the porosity in the electrode. The solvent flow in wet electrode may lead to a redistribution of the inactive component, e.g. CB and polymer binder, which are prone to accumulate on the electrode surface. This phenomenon is known as the "binder migration" and described in many publications [30,31]. As the CB and polymer binder provide the electrical contact and mechanical strength, this segregation will result in bad power performances and deteriorated adhesion of the coating layer. Therefore, it is of high importance to optimize the drying process parameters for the purpose of improving the energy efficiency without compromising the electrode quality.

The drying process can be modeled at different length scales for different applications. At the machine level, CFD models are used to simulate the flow field of hot air in dryer, which helps to optimize the dryer design, e.g. the size and position of the air nozzle [32,33]. At the electrode film level, the solvent removing rate electrode can be modeled as a function of temperature profile in the dryer and slurry composition [34]. Then, at a more local scale, the binder migration can be correlated to the drying condition by mathematical models [35]. In this section, modeling of the solvent removing and of binder migration will be discussed as they are more process-oriented.

13.4.1 Modeling of solvent removing

In convection dryers, the hot air flows over the top of the wet electrode film, transferring heat to the wet film and bringing away the solvent vapor at the same time. As it involves the mass and heat transfers in all three phases (solid, liquid and gas), modeling of the drying process is inherently complex. The electrode drying process can be divided into two stages. At the first stage, the evaporation front is always pinned at the electrode surface and the thin film thickness keeps decreasing. The second stage starts when the solid particles in the electrode film are drawn together, and the film stays at a minimum thickness. Thereby, the evaporation occurs inside the porous structure of electrode and the solvent vapor has to move to the electrode surface to be removed. Due to the characteristics of these two stages, different mathematical descriptions are needed.

Inspired by a drying model from Mesbah *et al.* for pharmaceutical manufacturing [36], Oppegård *et al.* derived a model for the first stage of the drying process for LIB production [37]. At this stage, it is reasonable to consider that the solvent removing depends solely on the solvent evaporation. The relationship between the film thickness h and the solvent evaporation rate ($\dot{m}$) is then described as follows:

$$\dot{m} = \frac{\partial(h\rho_f)}{\partial t} = \rho_f\frac{\partial h}{\partial t} + h\frac{\partial \rho_f}{\partial t} \tag{13.19}$$

where ρ_f refers to the density of film. The dependence of evaporation rate on the partial pressure of solvent vapor in the air (p_v) is given by:

$$\dot{m} = \frac{k_m M_v}{R}\left(\frac{p_{eq}}{T} - \frac{p_v}{T_a}\right) \tag{13.20}$$

where k_m is the mass transfer coefficient, M_v is the molecular weight of the solvent, R is the ideal gas constant, p_{eq} is the equilibrium vapor pressure of solvent, and T_a is the temperature of the hot air. Moreover, the energy balance of the film is given by:

$$\frac{dT}{dt} = -\frac{T}{h}\frac{dh}{dt} - \frac{T}{\rho_f}\frac{d\rho_f}{dt} - \frac{k_m M_v \lambda}{\rho_f C_p R}\cdot\left(\frac{p_v}{T} - \frac{p_a}{T_a}\right) + \frac{\dot{q}_{in}}{\rho_f C_p} \tag{13.21}$$

where λ is the latent heat of vaporization of the solvent, C_p is the specific heat of the film, and q_{in} represents the heat transferred to the film which can be described as following for convection heat transfer:

$$\dot{q}_{in} = k_c(T_a - T) \tag{13.22}$$

where k_c is the heat transfer coefficient. It is worth noting that with drying process going on, there is less and less solvent in the wet film. As a result, the density and the heat capacity of the wet film change accordingly. Combining the above equations and solving them numerically could obtain the evolution of film thickness and temperature at the first stage. However, this model is not valid for the second stage since the solvent removing is determined not only by the solvent evaporation kinetics but also by the mass transfer of solvent.

At the second stage of drying process, both liquid and gas phases exist in the electrode porous structure. To facilitate the model description, a parameter "saturation" (S) is defined as the ratio of solvent volume (V_L) to the total volume of void space (V_p):

$$S = \frac{V_L}{Vp} \tag{13.23}$$

The mass conservation of liquid solvent and solvent vapor along the direction of film thickness (z) are given in (13.24) and (13.25):

$$\frac{\partial}{\partial t}(\rho_l \varepsilon S) + \frac{\partial}{\partial z}(\rho_l \varepsilon S u_l) = \dot{m} \tag{13.24}$$

$$\frac{\partial}{\partial t}(\rho_v \varepsilon(1 - S)) + \frac{\partial}{\partial z}(\rho_v \varepsilon(1 - S)u_v) = \dot{m} \tag{13.25}$$

where ρ_l and ρ_v are the viscosities of solvent in liquid and vapor phase, u_l and u_v are their flux velocities. As the gas phase in the pore consists of solvent vapor and air, the mass balance equation of dry air can be written as:

$$\frac{\partial}{\partial t}(\rho_a \varepsilon(1 - S)) + \frac{\partial}{\partial z}(\rho_a \varepsilon(1 - S)u_a) = 0 \tag{13.26}$$

The velocity of gas phase and liquid phase can be solved with Darcy's equations as follows:

$$u_g = -\frac{K_g}{\eta_g \varepsilon(1 - S)}\left(\frac{\partial p_g}{\partial z} - \rho_g g\right) \tag{13.27}$$

$$u_l = -\frac{K_l}{\eta_l \varepsilon(S)}\left(\frac{\partial p_l}{\partial z} - \rho_l g\right) \tag{13.28}$$

where η is the viscosity, K is the permeability and p_l is the pressure of liquid solvent. For solvent vapor and air in gas phase, the diffusion flux should be considered, resulting in:

$$u_v = u_g - \frac{\rho_g M_a M_v D}{\rho_v M_g^2}\frac{\partial}{\partial z}\left(\frac{p_v}{p_g}\right) \tag{13.29}$$

$$u_a = u_g - \frac{\rho_g M_a M_v D}{\rho_v M_a^2}\frac{\partial}{\partial z}\left(\frac{p_a}{p_a}\right) \tag{13.30}$$

where ρ_g is the density of the gas phase, M_a, M_v, M_g are molecular weights for air, solvent, and solvent–air mixture, respectively, D is the diffusivity of solvent vapor in air. The evaporate rate of solvent at this stage is given by:

$$\dot{m} = \frac{k_m M_v}{RT} a(S - S_f)(p_{v_{eq}} - p_v) \tag{13.31}$$

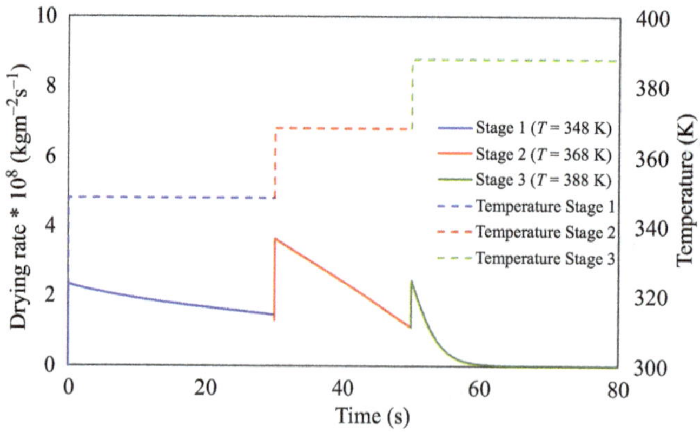

Figure 13.5 Drying rate and temperature profile in a 3-stage dryer with stage 1 at 348 K, stage 2 at 368 K, and stage 3 at 388 K. Reprint with permission from Ref. [34]. Copyright 2018 Elsevier.

where S_f is the final irreducible solvent content of the porous electrode and a is the evaporating area per unit volume. Moreover, the equation of energy conservation is given by:

$$(\rho_f C_{pf})\frac{\partial T}{\partial t} + \lambda \dot{m} + (\rho_l \varepsilon S u_l C_{pl} + \rho_v \varepsilon (1-S) u_v C_{pv} + \rho_a \varepsilon (1-S) u_a C_{pa})\frac{\partial T}{\partial z} = \frac{\partial}{\partial x}\left(\epsilon_f \frac{\partial T}{\partial z}\right) \tag{13.32}$$

where ϵ is the film thermal conductivity.

Based on (13.22)–(13.32), Susarla *et al.* have simulated the drying rate evolution of a NMC532 based cathode electrode in a three-zone dryer with temperature set at 348, 368, and 388 K, respectively [34]. As shown in Figure 13.5, while the drying rate increases spontaneously when the temperature rises, it keeps decreasing at the same zone. Moreover, the drying rate drops to close to zero at the middle of the third zone, which indicates that the electrode is dried from about 70 s. Consequently, this drying process can be optimized by decreasing the temperature at the third zone or increasing the web speed to avoid the overheating the electrode and to improve the energy efficiency at the same time.

13.4.2 Modeling of binder migration

It is widely observed that during the drying process, the polymer binder migrates towards the electrode surface. The depletion of binder at the bottom leads to bad electrode adhesion and cohesion. At the same time, the accumulation of binder at the electrode surface results in an elevating internal resistance. This phenomenon becomes more obvious for high drying temperatures or thicker electrodes [30,31,38].

Font *et al.* have developed a model to simulate the transport of binder in solvent by an advection-diffusion equation [35]. As the volume fraction of binder in electrode slurry is very small (usually 1–5%), the impact of binder on the mass transport of solvent and solid phase can be considered as negligible. Therefore, the advection–diffusion equation of the binder is solved with the solvent flow. Like the previous drying model, the model of binder migration is also solved only in the direction cross the film thickness (*z* direction). The volume-averaged continuity equation for dissolved binder is given by:

$$\frac{\partial}{\partial t}(\phi_l c) + \frac{\partial J}{\partial z} = 0 \tag{13.33}$$

where ϕ_l refers to the volume fraction of solvent, c is the concentration of dissolved binder in solvent, and J is the mass flux of dissolved binder which is calculated by:

$$J = F_l c - D_{eff}\frac{\partial c}{\partial z} \tag{13.34}$$

where F_l is the volume fraction flux of solvent and D_{eff} is the effective diffusion coefficient. In (13.34), the term of $F_l c$ represents the advection flow which is directly linked to the evaporation rate of solvent by

$$F_l = \frac{\dot{m}\phi_s^0 H_0}{H^2} z \tag{13.35}$$

where H_0 is the wet film thickness at the beginning of drying. Solving the above equations numerically gives rise to the evolution of binder distribution across the thickness of electrode film. At low drying rate, the quantity of binder brought to the electrode surface is limited due to a small convection flow of solvent. At the same time, there is enough time for these binders to diffuse back to the bottom. Consequently, the binder distribution across the thickness of electrode is quite homogeneous. In contrast, at high drying rate, as the solvent is removed rapidly from the wet electrode, leaving limited time for binder diffusion and thus resulting in a large gradient of binder quantity from the surface to the bottom of electrode.

The binder transport model provides a method to evaluate the impact of drying rate on the binder distribution across the electrode thickness. However, it is still challenging to characterize parameters, such as the drying rate and the effective diffusion coefficient, as inputs for the model.

13.5 Modeling of calendering process

After the drying process, the electrodes are calendered to reduce the pore volume, to reach the targeted thickness, and to enhance the energy density. During the calendering process, the electrode is forced to pass through a gap between two counter-rotating rollers. As the gap distance is smaller than the electrode thickness, a compaction force is applied to the electrode by the rolls. Either the gap distance or the applied force is controlled by the calendering machine. However, defining the right applied force or gap distance to achieve the target electrode porosity is so far done by trial-and-error, which risks at having a high scrap rate as well as causing damages to electrode and calendering machine.

Discrete element model (DEM) has been used by many researchers to simulate the calendering process at particle level, which has been discussed in the previous chapter. In this section, the mechanistical model and computational mechanics (CM) model of calendering process will be discussed.

13.5.1 Mechanistical model of calendering process

The force applied to the electrode by the calendering machine is expressed with a parameter of line loading (q_L). The relationship between the porosity of calendered electrode (ε_c) and line loading can be described by Heckel equation:

$$\varepsilon_c = \varepsilon_{c,\min} + \left(\varepsilon_{c,0} - \varepsilon_{c,\min}\right) \cdot \exp\left(-\frac{q_L}{\gamma_c}\right) \tag{13.36}$$

where $\varepsilon_{c,0}$ is the initial porosity of electrode, $\varepsilon_{c,\min}$ is the minimum achievable porosity of electrode, γ_c is the compaction resistance. The value of $\varepsilon_{c,\min}$ depends on the geometrical shape and size distribution of the solid particles in electrode. By comparing the compaction behavior of 15 different electrodes, Meyer *et al.* found the ratio of $\varepsilon_{c,\min}$ to $\varepsilon_{c,0}$, which is noted as p, varies between 0.3 and 0.6, with 0.4 being a good approximation for most cases [39]. The compaction resistance can be correlated to mass loading (M_c) by:

$$\gamma_c = \mu \cdot M_c \tag{13.37}$$

where μ is the mass loading dependency that depends on the electrode materials and temperature (T_R). The influence of temperature on calendering process can be further included in the model by implementing its impacts on p and μ [40]. The temperature dependence of these two parameters is given by:

$$p_T = p_0 - m_T T_R \tag{13.38}$$

$$\mu_T = \mu_0 - \xi T_R \tag{13.39}$$

where m_T and ξ are parameters fitted from experimental data.

This mechanistic model can be used as a fast tool to estimate the processability of a given electrode and to define the specifications of machine. Moreover, due to its simplicity, this model can be easily implemented to a process chain model.

13.5.2 CSM model of calendering process

Computational solid mechanics (CSM) is to use computational method to study the motion and deformation of solid materials. The most popular numerical technique for CSM is the finite element method (FEM), which has been widely used in machine design and structure analysis, etc.

In either DEM model or mechanistic model, rollers are considered as rigid body without deformation. However, driven by the pursuit of high energy density, the electrode porosity is designed to be very low, resulting in a high calendering pressure. Thus, the rollers in calendering machine may deform under such a high pressure, which may lead to a non-uniform electrode thickness. With FEM, Wang *et al.* studied the influence of roller parameters on the uniformity of the electrode thickness [41]. It is found that increasing the roller diameter could improve the uniformness of electrode thickness, which is ascribed to a smaller flexure deformation. This model can be helpful for equipment selection. Moreover, Ma *et al.* have obtained the relationship between gap-distance and applied force for a $LiMn_2O_4$ electrode with FEM simulations [42]. Given the mechanical property of electrode is properly characterized, similar model can provide guidance for setting the gap distance in production.

13.6 Modeling of electrolyte filling process

The aim of filling process is to fill the pore volume of electrodes and separator with electrolyte which provides transport path for Li^+ ions. There are two sub-steps in electrolyte filling process: dosing and impregnation. The dosing is to inject the electrolyte into the LIB cells housing, either a soft pouch or a hard can. The impregnation refers to the process that the electrolyte gradually wets and fills in the pores of electrodes and separators. Moreover, in order to accelerate the filling process, a vacuum step is often performed before the dosing step [43].

The direct numerical simulation (DNS) of electrolyte filling process, which is usually based on Lattice Boltzmann method (LBM), is introduced in Chapter 11. In this section, models at electrode level are discussed.

At macroscale, the Lucas–Washburn equation (LWE) is widely applied to simulate the liquid uptake of porous materials. By considering the pores as a bundle of parallel cylinders with an effective radius r_{eff}, the relationship between capillary rise (h) and time (t) is described by:

$$h = \sqrt{\frac{\sigma \cos \theta r_{eff}}{2\eta}} \cdot t^{0.5} = K \cdot t^{0.5} \tag{13.40}$$

where σ is the surface tension of the electrolyte, θ is the contact angle, and η is the viscosity of electrolyte. The LWE is often used to evaluate the dynamic of electrolyte impregnation in electrode [44]. Moreover, Knoche *et al.* and Weydanz *et al.* have used a modified version of LWE to fit the experimental wetting curve from neutron diffraction and satisfying results have been obtained [43,45]. However, LWE is too simplified to simulate the electrolyte filling process.

Günter *et al.* have applied a general model of liquid transport in porous materials [46] to electrolyte filling process [47]. In this model, the electrode/separator assembly is considered as a closed tortuous capillary and the microstructure of the assembly is characterized by its tortuosity (τ) and effective pore radius (r_{eff}). In addition to the capillary force, the viscous loss, the electrolyte temperature, and the pressure drop between internal residual gas and external gas are considered. Therefore, this model allows to simulate the evolution of electrolyte filling level under a given filling protocol that refers to the pressure and temperature profile during the filling process. As the model is developed in 1D, it is mainly valid for wound cells, either in cylindrical format or in prismatic format. In these cells, the electrolyte wetting occurs from the top and bottom and the electrolyte transport in radial direction is blocked by solid current collector. However, for stacking cells where the electrolyte can enter the cell from directions parallel to stackings, 2D model is needed.

13.7 Challenges and opportunities of process modeling of LIB

One of the challenges of the process modeling of LIB is the model conception, which involves defining the problem, making relevant hypothesis, and choosing modeling techniques. All these require not only a sufficient understanding of the process, but also a good knowledge of modeling methods. Fortunately, as the process in LIB shares a lot of similarities with that in other industries, the process models developed in those industries can be good references. For instance, the mixing process in LIB is similar to that in paint industry, and the drying process is just like that of fruit drying in food industry. Another challenge is the parameter characterization and model validation. For many process models, in particular the physical models, it is difficult to obtain input parameters and to validate model with currently available means. Therefore, it is necessary to develop new methods and new tools to this end. For example, the validation of electrolyte filling model relies on the experiments of neutron diffraction or X-ray diffraction, which allow the visualization of this process [43,48].

Despite these challenges, there are vast opportunities for process modeling. It can play an important role in the entire production chain. It can be used to have a deep understanding about the process, to help achieve smart design, to accelerate the upscaling and to improve the quality control. All these will contribute to a more agile production and a better product quality.

References

[1] Sahu AK, Kumar P, Patwardhan AW, and Joshi JB. CFD modelling and mixing in stirred tanks. *Chem Eng Sci.* 1999;54(13):2285–93.

[2] Xia B and Sun DW. Applications of computational fluid dynamics (cfd) in the food industry: a review. *Comput Electron Agric.* 2002;34(1):5–24.

[3] Pohar A. A review of computational fluid dynamics (CFD) simulations of mixing in the pharmaceutical industry. *Biomed J Sci Tech Res.* 2020; 27(3):20732–6. [cited 2022 Nov 17]. https://biomedres.us/fulltexts/BJSTR.MS.ID.004494.php

[4] Andersson B, Andersson R, Håkansson L, Mortensen M, Sudiyo R, and van Wachem B. *Computational Fluid Dynamics for Engineers.* Cambridge University Press; 2011. 203 p.

[5] Marshall EM and Bakker A. Computational fluid mixing. In: *Handbook of Industrial Mixing*. John Wiley & Sons, Ltd; 2003. [cited 2022 Nov 17]. p. 257–343. https://onlinelibrary.wiley.com/doi/abs/10.1002/0471451452.ch5

[6] Macosko CW and Larson RG. *Rheology: Principles, Measurements, and Applications*. VCH; 1994. 576 p.

[7] Hawley WB and Li J. Beneficial rheological properties of lithium-ion battery cathode slurries from elevated mixing and coating temperatures. *J Energy Storage*. 2019;26:100994.

[8] A discrete model for the apparent viscosity of polydisperse suspensions including maximum packing fraction. *J Rheol*. 2022;57(3):743–65. [cited2022 Nov 19]. https://sor.scitation.org/doi/abs/10.1122/1.4795746

[9] Pabst W, Gregorová E, and Berthold C. Particle shape and suspension rheology of short-fiber systems. *J Eur Ceram Soc*. 2006;26(1):149–60.

[10] Ma F, Fu Y, Battaglia V, and Prasher R. Microrheological modeling of lithium ion battery anode slurry. *J Power Sources*. 2019;438:226994.

[11] Schilde C, Kampen I, and Kwade A. Dispersion kinetics of nano-sized particles for different dispersing machines. *Chem Eng Sci*. 2010;65(11):3518–27.

[12] Grießl D, Huber K, Scherbauer R, and Kwade A. Dispersion kinetics of carbon black for the application in lithium-ion batteries. *Adv Powder Technol*. 2021;32(7):2280–8.

[13] Ramkrishna D and Singh MR. Population balance modeling: current status and future prospects. *Annu Rev Chem Biomol Eng*. 2014;5(1):123–46.

[14] Günther T, Schreiner D, Metkar A, Meyer C, Kwade A, and Reinhart G. Classification of calendering-induced electrode defects and their influence on subsequent processes of lithium-ion battery production. *Energy Technol*. 2020;8(2):1900026.

[15] Schweizer PM. *Premetered Coating Methods: Attractiveness and Limitations*. New York, NY: Springer Nature; 2022. 691 p.

[16] Jin GL, Ahn WG, Kim SJ, Nam J, Jung HW, and Hyun JC. Effect of shim configuration on internal die flows for non-Newtonian coating liquids in slot coating process. *Korea-Aust Rheol J*. 2016;28(2):159–64.

[17] Oh SH, Kim SI, Park KH, and Yu BH. A Numerical Study on Improvement of Coating Uniformity by Controlling the Pressure at the Exit of the Slot Die Nozzle. In: AJKFluids2019. Volume 3A: Fluid Applications and Systems; 2019 [cited 2022 Nov 20]. https://doi.org/10.1115/AJKFluids2019-5139

[18] Kim S, Lee J, and Lee C. Computational fluid dynamics model for thickness and uniformity prediction of coating layer in slot-die process. *Int J Adv Manuf Technol*. 2019;104(5):2991–7.

[19] Hoffmann A, Spiegel S, Heckmann T, Scharfer P, and Schabel W. CFD model of slot die coating for lithium-ion battery electrodes in 2D and 3D with load balanced dynamic mesh refinement enabled with a local-slip boundary condition in OpenFOAM. *J Coat Technol Res*. 2022;20: 3–14. [cited 2022 Nov 20]; https://doi.org/10.1007/s11998-022-00660-8

[20] Jung H and Nam J. Numerical analysis of pulsatile flows in a slot-die manifold. *J Coat Technol Res*. 2019;16(4):1141–51.

[21] Lee SH, Koh HJ, Shim SH, Jung HW, and Hyun JC. An optimal die design for the coating uniformity of non-Newtonian liquids in slot coating process. *Korean Chem Eng Res.* 2011;49(3):314–9.

[22] Ding X, Liu J, and Harris TAL. A review of the operating limits in slot die coating processes. *AIChE J.* 2016;62(7):2508–24.

[23] Marcel S. *Slot Die Coating of Lithium-Ion Battery Electrodes*. KIT Scientific Publishing; 2016. 202 p.

[24] Lin CF, Wang BK, Tiu C, and Liu TJ. On the pinning of downstream meniscus for slot die coating. *Adv Polym Technol.* 2013;32(S1): E249–57.

[25] Jang I and Song S. A model for prediction of minimum coating thickness in high speed slot coating. *Int J Heat Fluid Flow.* 2013;40:180–5.

[26] Schmitt M, Raupp S, Wagner D, Scharfer P, and Schabel W. Analytical determination of process windows for bilayer slot die coating. *J Coat Technol Res.* 2015;12(5):877–87.

[27] Taylor SD and Hrymak AN. Visualization and flow simulation of a two-layer slot coater. *Chem Eng Sci.* 1999;54(7):909–18.

[28] Tan P, Diao S, Huang T, Yang Z, Zhou H, and Zhang Y. Mechanism and control of the trailing edge in intermittent slot die coating. *Ind Eng Chem Res.* 2020;59(35):15758–67.

[29] Thomitzek M, von Drachenfels N, Cerdas F, Herrmann C, and Thiede S. Simulation-based assessment of the energy demand in battery cell manufacturing. *Procedia CIRP.* 2019;80:126–31.

[30] Baunach M, Jaiser S, Schmelzle S, Nirschl H, Scharfer P, and Schabel W. Delamination behavior of lithium-ion battery anodes: Influence of drying temperature during electrode processing. *Dry Technol.* 2016;34(4):462–73.

[31] Jaiser S, Kumberg J, Klaver J, *et al.* Microstructure formation of lithium-ion battery electrodes during drying – an ex-situ study using cryogenic broad ion beam slope-cutting and scanning electron microscopy (Cryo-BIB-SEM). *J Power Sources.* 2017;345:97–107.

[32] Sun DW (eds.) *Computational Fluid Dynamics in Food Processing.* Boca Raton, FL: CRC Press; 2007. 760 p.

[33] Li J, Fleetwood J, Hawley WB, and Kays W. From materials to cell: state-of-the-art and prospective technologies for lithium-ion battery electrode processing. *Chem Rev.* 2021;122:903–56. [cited 2021 Nov 8]. https://doi.org/10.1021/acs.chemrev.1c00565

[34] Susarla N, Ahmed S, and Dees DW. Modeling and analysis of solvent removal during Li-ion battery electrode drying. *J Power Sources.* 2018; 378:660–70.

[35] Font F, Protas B, Richardson G, and Foster JM. Binder migration during drying of lithium-ion battery electrodes: Modelling and comparison to experiment. *J Power Sources.* 2018;393:177–85.

[36] Mesbah A, Ford Versypt AN, Zhu X, and Braatz RD. Nonlinear model-based control of thin-film drying for continuous pharmaceutical manufacturing. *Ind Eng Chem Res.* 2014;53(18):7447–60.

[37] Oppegård E. Optimization and Control of Battery Electrode Drying Process [Master thesis]. NTNU; 2021 [cited 2022 Nov 21]. https://ntnuopen.ntnu.no/ntnu-xmlui/handle/11250/2781024

[38] Müller M, Pfaffmann L, Jaiser S, *et al.* Investigation of binder distribution in graphite anodes for lithium-ion batteries. *J Power Sources.* 2017;340:1–5.

[39] Meyer C, Kosfeld M, Haselrieder W, and Kwade A. Process modeling of the electrode calendering of lithium-ion batteries regarding variation of cathode active materials and mass loadings. *J Energy Storage.* 2018;18:371–9.

[40] Meyer C, Weyhe M, Haselrieder W, and Kwade A. Heated calendering of cathodes for lithium-ion batteries with varied carbon black and binder contents. *Energy Technol.* 2020;8(2):1900175.

[41] Wang R, Du X, Liu B, Yan S, and Wan J. Research on influence of roll parameters on thickness uniformity of lithium ion battery electrodes. *J Plast Eng.* 2018;25(1):46–51.

[42] Ma S and Tian L. Stiffness analysis and structure optimization of rolling mill for lithium-ion battery electrode manufacturing. *Chin Mech Eng.* 2015; 26(6):803.

[43] Weydanz WJ, Reisenweber H, Gottschalk A, *et al.* Visualization of electrolyte filling process and influence of vacuum during filling for hard case prismatic lithium ion cells by neutron imaging to optimize the production process. *J Power Sources.* 2018;380:126–34.

[44] Davoodabadi A, Li J, Liang Y, Wood DL, Singler TJ, and Jin C. Analysis of electrolyte imbibition through lithium-ion battery electrodes. *J Power Sources.* 2019;424:193–203.

[45] Knoche T and Reinhart G. Electrolyte filling of large-scale lithium-ion batteries: challenges for production technology and possible approaches. *Appl Mech Mater.* 2015;794:11–8.

[46] Fries N. *Capillary Transport Processes in Porous Materials – Experiment and Model.* Cuvillier Verlag, 2010. 190 p.

[47] Günter FJ, Keilhofer J, Rauch C, *et al.* Influence of pressure and temperature on the electrolyte filling of lithium-ion cells: experiment, model and method. *J Power Sources.* 2022;517:230668.

[48] Schilling A, Gümbel P, Möller M, Kalkan F, Dietrich F, and Dröder K. X-ray based visualization of the electrolyte filling process of lithium ion batteries. *J Electrochem Soc.* 2018;166(3):A5163.

Chapter 14
Methodology in quality control for electrode processing

Carl D. Reynolds[1,2] *and Emma Kendrick*[1,2]

14.1 Introduction

Lithium-ion electrode manufacture is a complex multi-stage process, and so quality control is vital at each stage to maintain production speed and reduce wastage. Current electrode manufacture is based on a slurry casting process, where the dry ingredients are suspended in a minimum of solvent to form a slurry, which is coated onto a current collector, dried to remove the solvent, and the coating calendared (compressed) to a target porosity (Figure 14.1) [1,2]. However future technologies may include dry processing or alternative deposition methods. Quality control is performed at several key stages of the process; on the raw material components, the slurry mix before coating, the coating during deposition, after drying, and after calendaring.

The slurry casting process is represented in Figure 14.1. The materials, which include the active components for the negative and positive electrodes, binder, solvent and conductive additive, are formulated and mixed into a slurry. The conductive additive, often carbon black, forms a network throughout the electrode enhancing its electronic conductivity. The binder, often a polymer, helps to suspend the other components in the solvent to enable efficient mixing, as well as enhance the flexibility, adhesion, and mechanical properties of the final electrode. The binder selection is closely linked to the solvent being used, as it must dissolve in the solvent to enable suspension and efficient mixing (although sometimes a combination of binders is used, one for suspension and another for the final mechanical properties, which can itself be in suspension).

The materials are mixed into a slurry, where final distribution of components in the coating is optimised to achieve a well dispersed but connected conductive network of carbon black. This requires intensive mixing, however if mixed too much the carbon network breaks down and the conductivity of the final coating is reduced [3,4]. There are a variety of mixers that can be used for electrode slurries,

[1]School of Metallurgy and Materials, University of Birmingham, UK
[2]Faraday Institution, Harwell Campus, UK

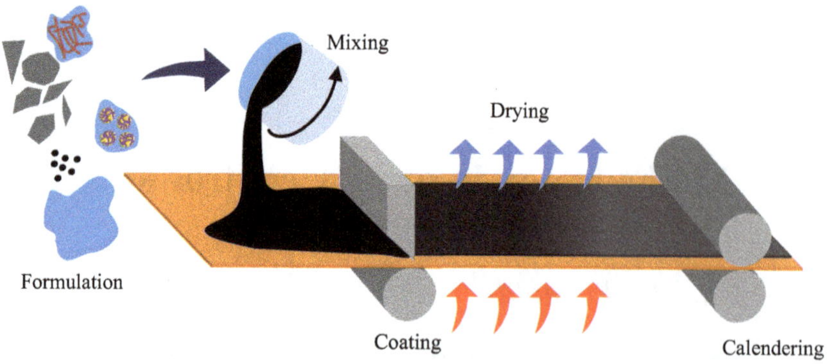

Figure 14.1 Illustration of the current slurry casting electrode manufacturing process

in research, centrifugal mixers which move the vessel, are common to conserve sample and for ease of cleaning, but these are not scalable to the large vessels required in industry, where mixers with rotary paddles are more common. There are various designs including different design of paddles, vessels that spin on their axis and scrapers to prevent material accumulating on the sides. More description on slurry preparation can be found in Chapter 3.

The electrode slurry is then coated onto a current collector. In research settings, this is often a draw down coater, to make a single sheet of coating by moving a geometry (e.g., blade or rod) with a fixed gap over excess slurry applied to the surface. However industrially, a continuous reel to reel process is used, where slurry is applied to rolls of current collector continuously, sometimes using a blade or comma bar geometry but most commonly from a slot die head. The slot die uses a pumped flow of material, allowing the flow rate to be adjusted to give the target coating thickness without moving the position of the head. This deposition step creates the coating microstructure and leads to the final mechanical and electrochemical properties. If the slurry is coated too fast, instabilities can occur leading to non-uniform coatings [5], if the slurry is not viscous enough, spreading of the coating can occur which leads to ill-defined edges and non-uniform thickness [6], and if air bubbles are present, pinhole defects can form when they burst leaving a void in the coating [7]. More information on electrode fabrication can be found in Chapter 4.

After deposition, the coating will be dried. Reel to reel coatings are drawn through specially designed ovens to ensure sufficient drying before it is rolled back up. The speed of the drying process is highly important to the final microstructure, as while removing the solvent the components can migrate. Binder migration is where the binder moves away from the current collector/coating interface during drying, leaving that interface with poor adhesion and the final electrode prone to delamination [8,9]. The speed of drying is controlled to prevent this, usually by limiting the temperature of drying and by slowing the web speed through the dryer.

Other potential defects that can occur during the drying process are cracks which form due to stress build-up between the dry and wet parts of the coatings. These can be controlled through formulation and optimisation of the current collector-coating interfacial properties, however drying parameters are also important. Fast drying will increase the speed at which internal stresses develop, and thus make it more likely they do not have time to relax before they are relieved through cracking of the coating [10,11].

The final process for the coating is calendaring, where the coated current collector is compressed by passing through two heated rollers to reduce the thickness and porosity of the coating, and compact the particles. This process results in an improved electronic conductivity and significantly improves the volumetric energy density of the electrodes, but can increase the pore network tortuosity which can influence the ionic transport of the electrode [12]. Excessive calendaring closes the pore network, and isolates parts of the electrode to the electrolyte resulting in inactive areas. It can also cause delamination, if calendared too harshly [13]. Hence, the optimum is a balance, between the electronic conductivity from the conductive particles and the active materials and the ionic conductivity, which is determined by the pore volume, shape and size for the electrolyte absorption and transportation.

As well as slurry casting, there is significant interest in moving to novel processing routes, such as extrusion and potentially additive manufacturing. Extrusion requires significantly greater viscosity pastes, which means less or no solvent is required for the processing, this reduces or even removes the drying requirement, which is currently a major energy and financial cost to electrode production (Figure 14.2) [14–16]. This also requires a transition from batch to continuous slurry mixing, as the materials can be continuously fed into the mixer, e.g. twin screw extruders [15]. This enables further opportunities for in-line control. Dosing equipment can be controlled to adjust amounts of each component fed in to achieve the required extrusion properties. When the materials are extruded as a paste, rather than a slurry, they can be directly calendared onto the current collector. Although low solvent processing often results in lower porosity, calendaring is still usually required and allows the paste to be attached to the current collector, which is required for mechanical stability and fast electron transport to all areas of the electrode.

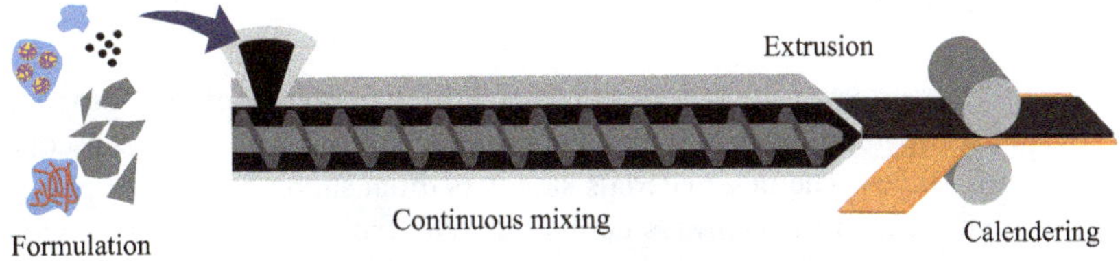

Figure 14.2 Illustration of a continuous mixing extrusion process, which can have little or no drying stage

Additive manufacturing is another potential option for electrode manufacture, direct ink writing (deposition of slurry from a position-controlled syringe) or aerosol/spray coating are methods that have been trialled for electrodes. These allow better control over the electrode microstructure which can give advantageous final properties (e.g. thick electrodes with low tortuosity can be manufactured, having favourable power and capacity characteristics) [17,18]. However, these methods are currently less scalable, often requiring higher solvent quantities and more time than conventional processing. Nevertheless, they may be applicable to speciality electrodes where precise properties are required.

In both standard and novel electrode manufacture, there are myriad input variables in the process, many of which change with time (e.g., materials properties batch to batch, external temperatures, flow through pipework that may erode or become clogged). Hence, it is vital that the electrode manufacturing process be responsive, able to adapt process parameters to keep the output at a consistent quality. The key to this is metrology, the measurements at each stage needed so that these changes can be detected early and responded to, minimising lost time and materials.

This chapter is divided into sections, each of which represents a stage in the current slurry casting process where there is an opportunity for metrology. Although the current standard process is used to discuss, much of the quality control is also applicable to novel processing and opportunities where quality control (QC) may diverge in future processing and will be highlighted.

- Materials – Characterisation of the raw materials used in electrode manufacture
- Slurry – Measurement of the physical properties of the mixed electrode slurry and characterisation of the efficiency of mixing
- Coating – Evaluation of the electrode coating structure, wet and dry, and the identification of coating defects
- Cell testing and electrochemistry – Measurement of the electrical properties in a cell, and the impact of the previous manufacturing parameters on cell performance
- Input–property relationships – Combining metrology with modelling approaches to link data from each stage of the process and develop predictive understanding

14.2 Materials

The physical properties of the raw materials and the consistency of their properties are key to ensure reproducibility and no negative impact upon the subsequent manufacturing stages. The raw materials suppliers often supply a range of physical properties, but these are not always comprehensive and subtle differences within the target range can impact processing. It is therefore vital that raw materials undergo QC to test their quality and adjust later processing parameters to compensate.

The active materials typically are shipped in powder form, there are various parameters that can be measured to gauge the quality of these. Specifically, the chemical composition, particle size, shape and size distribution, surface area and water or moisture content.

The physical properties of the powders such as the size and shape of the particles can have a significant impact on processing and formulation. The active materials typically have the largest particle size of the components within an electrode and their geometry can have a significant impact on the packing and ultimate structure formed in the slurry and during coating stages [18]. Active materials commonly have a bimodal size distribution to enable efficient packing in the final electrode. Larger particle sized materials can cause problems during coating, and typically the particle size needs to be an order of magnitude less than the size of the coating gap required, to avoid jams and uneven coating. The shape of active particles is also very important. Flake like materials can form networks more efficiently than spherical particles which means they are more prone to gelling in the slurry [19]. To avoid these issues and enable more efficient packing, spherical particles are usually preferred. The particle size and shape distributions can be directly measured through imaging techniques e.g. scanning electron microscopy (SEM) [20], whereas the particle size distribution can be measured using light scattering after dispersion in a solvent [21,22].

The chemical composition is often difficult to quantify and the supplied materials may have low levels of dopants or surface treatments to stabilise and improve the materials properties. This information is not always supplied by the raw materials supplier and can significantly impact the manufacturing process. For example the ratio of Nickel, manganese and cobalt in NMC cathode material can vastly change the final electrochemistry and stability of the ink [23,24]. Therefore, measuring that the materials have the expected chemistry is important (e.g. using Inductively coupled plasma (ICP) to measure the ratios of elements) [25].

Often materials have a shelf life, and particularly the nickel rich layered oxides such as NMC811 should be stored with care. The surface of the materials can be altered over time through reaction with the air, and residuals such as lithium carbonate (Li_2CO_3) or lithium hydroxides (LiOH) can form, removing the lithium from the layered oxide structure, particularly in moist environments [26]. The surface residuals also cause problems in subsequent processing such as gelation of the inks and slurries if not removed. Detecting these residuals is not simple, as standard structure methods such as X-ray diffraction do not pick up the small levels of lithium compounds. These species can be detected with FTIR, but the changes are subtle and require fitting to extract [23]. If these are detected before use, surface species can be removed by processing (e.g. through washing in ethanol), which can improve the performance of the cell, reducing off gassing during formation [27,28].

As storage time and environment can significantly impact the properties of battery materials, careful control of these parameters is required, by sampling to ensure the quality has not degraded during storage. As water is a common driver of degradation of these materials, it can be useful to track the moisture content of the powders before use, even when kept in moisture-controlled storage. This is commonly measured by Karl–Fisher titration, where the moisture is removed from the sample by heating, and then used to oxidise sulphur dioxide with iodine. This consumes one mole of iodine per mole of water added and so iodine is gradually added until it is in excess, at which point all the water has been consumed. Modern apparatus can detect trace amounts of water in the range 10–10,000 ppm by weight [29]. Thermal gravimetric analysis (TGA) can also be used to measure water content, which works on a similar principle of heating the sample until water is removed, and analysing the change in weight. However, there can be other components released on heating e.g. adsorbed CO_2, which may mean analysis of the removed gases is required, mass spectrometry is one method commonly coupled with TGA [30].

Conductive additives, such as carbon black, nanotubes or fibres, can be characterised in much the same way. They are typically comprised of much smaller particles, which can be dispersed in the electrode to form a conductive network. Their precise size and shape usually has less impact on the overall slurry/coating structure, but can influence if or how a conducting network is formed. Carbon black, a common conductive additive, is usually orders of magnitude smaller than the micron sized actives. It typically agglomerates, which means in order to get a true particle size that represents the particles after dispersion in the mixing stage, the agglomerates must be dispersed [31]. Some particle sizers have dispersion units that can efficiently disperse agglomerates before using light scattering to assess the particle size.

Binder materials are polymeric, and are characterised through molecular weight, which is typically measured by size exclusion chromatography. Molecular weight can change sometimes change between batches of materials and can impact the slurry rheology and resulting coating structure, higher molecular weights are typically more viscous in a solvent [32]. Typically high molecular weight (100,000–1000,000 g/mol) linear polymers are used as battery electrode binders, because they can increase the viscosity of the slurry more significantly at lower weight percentages, meaning less binder is necessary overall [33].

Moisture content is again important, for example carboxy methyl cellulose (CMC), a common anode binder, is hygroscopic. As it absorbs water it will have different moisture content at different humidity levels, which may have an impact on the mass added to formulations (as a differing amount of the mass will be water).

Often stock solutions of binders are prepared in advance, as dispersion of binder can be time-consuming or require additional heat. It is important to monitor the quality of these as well as the dry powders, as polymers are more likely to degrade when stored in solution due to increased mobility and reaction rates [34]. One method to inspect quality is to monitor the rheology of the solution, e.g. replace the stock solution when the viscosity at a specific shear rate is outside acceptable limits.

As well as the molecular weight, the particle size of the polymers can also be measured, which is related to the molecular weight and the polymer structure. Similar to the active materials, this is typically measured using light scattering and is another way to study the distribution of polymer chain sizes, alongside their interaction with the solvent. Some binders, usually part of a multi-binder system, are suspended in solvent, e.g., styrene-butadiene rubber (SBR), which is added alongside CMC for adhesion and flexibility in the final electrode. Suspended polymers need to be low in molecular weight (~100 g/mol) and collapse to small sizes (~0.1 μm) to stay suspended [35,36].

Quality and impurity levels in other electrode and cell components are also important. The required purity of the current collector (usually Al or Cu foil) is extremely high >99.9%. The reason for this is that other metals alloying with the aluminium or copper may be electrochemically active, and therefore be removed or deposited on the electrodes during use. The thickness of the foil needs to be high enough to give mechanical stability during the coating, but excess thickness will add excess size/weight and decrease the volumetric and gravimetric capacity of the cell [37]. During coating, thickness variations of the current collector can lead to inhomogeneous coatings and inaccuracies in the measured mass loadings and cell capacities. Depending upon how the foil is made (i.e., rolling, electrodeposition), there also may be oily surface residues that need removed before coating, this affects the surface properties, and wettability of the current collector which can also be measured. The surface properties are key to obtaining good coating adhesion and electrical transport between the coating and current collector. The precise morphology of the surface as well as any surface species present can impact all these properties. Contact angle is one method to quantify the wetting of the surface which can change with morphology and surface species e.g. graphene/nanotube coating on the current collector reduces the contact angle and increases adhesion [38]. The surface morphology can be characterised by microscopy (e.g. optical, atomic force microscopy). Roughened current collector surfaces have been shown to enhance adhesion and electrochemical performance of electrodes [39].

When assembled into a cell, the anode and cathode are sandwiched with a layer of separator between, which allows ion but not electron transport to prevent a short circuit. This is soaked in an electrolyte to allow ion transport between electrodes. For a coin cell this is enclosed in a cell casing, and a spacer and a spring is used to maintain pressure on the electrodes. In pouch cells, this is sealed in pouch material (usually aluminium foil sandwiched between layers of polymer), and pressure is applied externally to the cell. Electrolytes have a shelf-life and degrade over time especially in the presence of air and moisture [40]. Consistency in the cell components is also required, separator, usually a polymeric membrane, needs to be a consistent thickness to maintain desirable transport properties without bringing excess weight. The size/shape of coin cell components will be key to the pressure that results on the electrodes, which impacts their performance. The pouch cell material also needs to efficiently seal and prevent electrolyte escape and air ingress.

For all the materials used, quality control measures (summarised in Table 14.1) should be established to check that materials coming in are consistent and to understand any changes batch to batch or after storage and how they may impact subsequent stages.

Table 14.1 Measurable parameters for raw battery materials and measurement methods

Property	Technique	Notes
Materials		
Particle size distribution	Light scattering [41,42]	• Requires dispersion in solvent • Can inform whether further processing (e.g., ball milling) of materials is required before use
Chemistry	Fourier transform infrared (FTIR) spectroscopy [23]	• Evolution of chemistry and particle surface • E.g., stretching peaks typical of adsorbed water, CO_2 or carbonates or hydroxy species from surface reactions
	X-ray diffraction (XRD) [40]	• Bulk crystal structure and phase changes • Can detect lithium removal from the structure, or hydration of active material
	Energy dispersive X-ray spectroscopy (EDS) [40]	• Combined with SEM, can give elemental composition of very specific areas in SEM image, useful for distinguishing different components
	X-ray photoelectron spectroscopy (XPS) [40]	• Characterising surface chemistry, e.g., evolution in surface species over time
Polymer molecular weight distribution	Gel permeation chromatography [33]	• Molecular weight – key to slurry viscosity and coating adhesion • Usually not sensitive to branching (rheology is useful to distinguish this)
Moisture content	Karl–Fisher titration [29]	• Highly sensitive to low moisture content
	Thermal analysis [30]	• Can study compounds other than water released on heating, but needs to be coupled with other techniques e.g., mass spec to identify the mass coming from water alone
Foil surface properties	Contact angle [43,44]	• Standard liquid (e.g., water, NMP) can be used to study surface changes • Slurry contact angle can be measured, higher values reduce slumping after coating but also typically decreases adhesion
	Scanning electron microscopy (SEM) [38,43]	• Direct visualisation of surface, small sample areas
Density	Tapped density [45]	• Measures density of well-packed powder, insight into particle size distribution, larger, more monodisperse particles will give lower values

14.3 Slurry

The materials are typically mixed into a slurry for coating, which contains the conductive additive, active material, binder and a solvent. In this section, the physical properties of the slurry are discussed, which can be measured during and after mixing. This gives opportunities to gauge the effectiveness of mixing and the quality of the electrode slurry.

The flow properties, or rheology of the mixed slurry are vital to its processing in the later coating step, and will dictate the microstructure created in coating. There are several key rheological properties required. First, a high viscosity at low shear rates, analogous to flow under gravity, which is required to maintain a stable dispersion as well as maintain the coating structure created and avoid flow of the wet coating, spreading, which leads to thin and uneven coating. Second, a low viscosity at high shear rates, which is required to allow flow in the coater, without excess build-up of pressure which can limit coating speed. This response also needs to be fast, the material must respond rapidly to each of these shears so as not to delay coating or allow spreading before the viscosity has returned after the coating shear (see Figure 14.3). Finally, the viscoelasticity of the slurry is important, if the slurry is in an elastic region at the coating shear rate, coating instability may occur causing uneven coating [31,46].

As these slurries are complex mixtures, their rheology will also be complex, and the active material, conductive additive, binder and solvent all have a part to play in the rheological properties. The impact of binder is perhaps most clear, as its purpose is to thicken the solvent so that the particles can stay in stable suspension. Therefore, it increases the viscosity of the solvent, but as most binders are polymeric, they also impart viscoelastic properties to the slurry. The conductive additive can also significantly impact the rheology. For example, carbon black can form a network in the slurry. This can give the slurry weak gel like properties, which

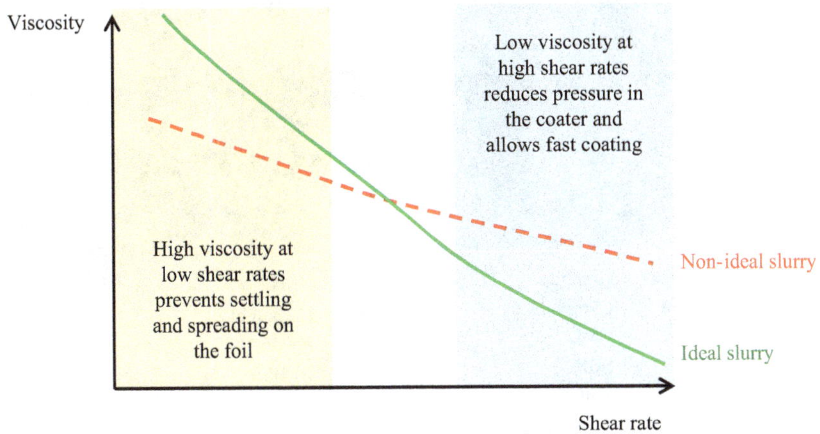

Figure 14.3 Illustration of ideal rheology in steady shear

easily break up with shear. This is usually favourable as the structure can easily be broken up to allow flow in the coater, but it returns quickly, preventing spreading of the wet coating. The network also imparts favourable conductivity later in the process. Whether this occurs depend on the solvent (carbon black tends to agglomerate in water but less so in NMP, so a dispersed network structure is more likely to form in NMP [31]), as well as the efficiency of mixing (must be enough to wet and deagglomerate the powder), and ratio of components in the slurry (both carbon black and binder preferentially line the surface of active particles so this must be balanced to allow enough free carbon black to form the network). The active material usually serves to increase viscosity, as it is a large bulk of particles added to the slurry, but can impact other components (such as the carbon black above). Any additives can also alter rheology, such as dispersants, which help to disperse carbon black and active. While dispersion of agglomerates may be expected to reduce the viscosity, it can also cause an increase, as more interface for interaction and network formation is created [47]. Figure 14.4 shows a comparison of slurry structures in NMP and water.

Hence, it is important to measure the rheology, which can be done offline on a rheometer but there is also high value in measuring it in-line such as using pressure across a geometry during pumping of the slurry between mixer and coater to extract viscosity. This can be used to check whether the slurry viscosity is within expected ranges to monitor quality, and detect variations over time. Rheological properties can also be used to monitor the efficiency of mixing, for example the power input into the mixer is related to the mixer geometry and the slurry viscosity, so this can be measured and monitored over time, to study the dispersion processes and determine when to end the mixing for the optimum microstructure. E.g. Wenzel *et al.* [48] monitored motor current with number of mixing revolutions and found a sharp decrease as carbon black agglomerate network began to fragment, but a

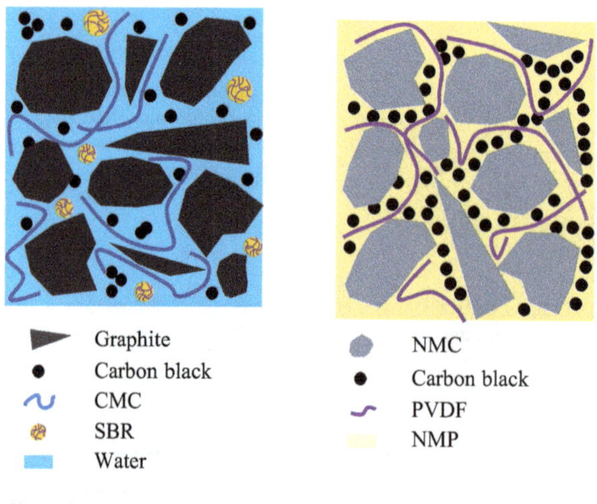

Figure 14.4 Illustration of slurry structure for a graphite slurry in water and an NMC slurry in NMP

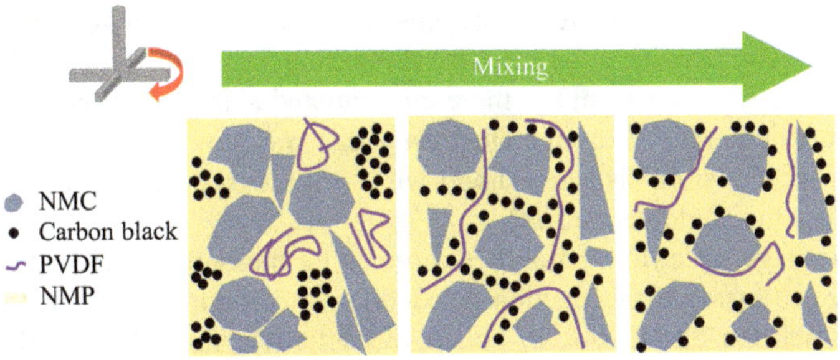

Figure 14.5 *Illustration of structure evolution with mixing, showing how agglomerates can be dispersed, but over-mixing can lead to excessive adsorption to active materials and loss of the conductive network*

subsequent increase as the branched CB agglomerates adsorbed to active material particles, resisting flow. This then decreased again as these agglomerates on the surface fracture and compact, reaching an equilibrium. Figure 14.5 shows illustrations of structure with mixing, as networks form and then break, as adsorption to the particle surfaces is promoted with over-mixing, leading to a reduction in conductivity and rate performance in the final cell [49].

Surface properties of the slurry help to measure the wettability and slumping of the slurry after coating, which can be characterised using surface tension of the slurry and the interfacial tension between the slurry and the current collector (measured through contact angle) [50]. High contact angles will reduce spreading of the slurry after coating, but also may promote poor adhesion, so the contact angle must be optimised to balance these effects. The surface tension can be a useful indicator of the slurry structure. For example, badly mixed slurries may have an upper layer of solvent and so present surface tensions identical to the solvent alone.

The density of the slurry is another useful physical property. This can be predicted from the density and ratios of added components, and so comparing a measured value (usually on a small sample) to this prediction is a useful measure of uniformity of the slurry. Badly mixed slurries will have density variations across the batch and so samples will likely not match the predicted density. There also can be interactions formed on mixing which can alter the density of the slurry from that of the combined components.

Agglomerate size in the slurry is a very important measure. Similar to the initial particle size, this can cause coating issues where agglomerates are on the same order of magnitude as the coating gap. It is also a useful measure of the efficiency of mixing. Usually the target is to disperse the active material to individual particles, so if the agglomerate size is several times the particle size of the active, more intensive mixing is likely needed. The opaque slurries with high weights solids typically pose a problem for most particle sizing methods, and must be diluted for light scattering measurements [21,22]. The particle size distributions obtained from light scattering can be highly useful to ascertain whether the active material has been well dispersed, and whether carbon black is free in suspension or has excessively coated the active material. However, care must be taken in interpreting the results because the dilution can change the structure from the more concentrated slurry. Agglomerate size can also be simply ascertained using a Hegman gauge, a channel with decreasing depth through which the slurry is scraped and will not pass through uniformly once the depth reaches the agglomerate size.

At this stage, electrical properties can also be measured. Impedance analysis can be used to measure the fraction of conductive components across the mixing vessel to track the distribution of conductive additive, to target a high, uniform conductivity in the slurry and end the mixing when this is achieved [51]. Electrical resistance tomography can also be used to measure the fraction of particles throughout the mixing vessel and determine the efficiency of mixing [52].

All of the properties discussed can also be measured for the dry pastes involved in low solvent processing, although the consistency may pose a challenge for their measurement. Density, agglomerate size and electrical properties can be measured in the same ways. Rheology will be equally important, as whatever the consistency of the mix, it will need to flow to form the final coating shape, and so the flow properties (rheology) will be key to the processing speeds that can be achieved and the final electrode structure. It is likely the rheological properties of these pastes will dominate over surface interactions when forming the interface between the mix and the current collector. Thus, surface properties may become less important, but could still be relevant to the adhesion and coating structure. Current methods for measuring surface tension will struggle with pastes but contact angle of known liquids on the surface of both the paste and the current collector can be used to extract surface energies.

These measurements, summarised in Table 14.2, all provide the option to detect and remedy problems at a relatively early stage in manufacturing, and catching issues here can reduce wastage of time, energy and materials (e.g., current collector, cell components) in the later steps. When the QC at this stage indicates a high-quality slurry with optimised rheology, high conductivity, uniformity and low agglomerate size, this can be taken forward to the coating stage.

Table 14.2 Measurable parameters for battery electrode slurries and measurement methods

Property	Technique	Notes
Slurry		
Rheology	Shear rheology [31,53]	• Steady shear gives viscosity at relevant shear rates • Multi-step tests can give insights into relaxation times • Oscillation gives viscoelastic properties, can be used to probe structure and predict processing behaviour
	Extensional rheology [54]	• Can give structural information e.g. presence of polymer branching and insights into processing e.g. swell at slot die

(Continues)

Table 14.2 (*Continued*)

Property	Technique	Notes
Density	Pycnometry	• Direct measurement for a sample • Low cost glass pycnometer can be used
	Calculation from formulation	• Assumes no density change on mixing
Agglomerate size	Hegman/fines gauge [55,56]	• Fast, low cost measure of maximum agglomerate size in a slurry sample, no information on distribution
	Light scattering [41]	• Can be used for agglomerated structures as well as individual particles • Requires dispersion and dilution which may change structure
	Imaging (optical/SEM) [57]	• SEM/optical can be used depending on agglomerate size, gives visualisation of agglomerate size/structure but on a very localised sample

14.4 Coating

Moving to the coating stage, the electrode slurry is deposited onto the current collector, which is then dried, and calendared to form the final electrode. There are opportunities for QC on the wet and dry coating, as well as after calendaring. These are equally applicable to novel processing routes as once the coating on the current collector is made, it should be comparable in each processing route. The aim of the QC at the coating stage is to measure the electrode properties and enable the relationships with processing parameters to be extracted, so electrodes can be produced without defects and with maximum and controlled energy density which can be balanced against a counter electrode.

The thickness and coatweight of the coating are interlinked properties which are both important to measure. The coatweight, mass of material per unit area of electrode, determines the amount of active material and thus the capacity of the electrode produced. The thickness, when combined with the coatweight, allows extraction of density of the coating, and for the dry coating, its porosity. Calculation of the porosity is vital as the dry coating must be calendared to a fixed porosity to give increased conductivity and volumetric capacity without hindering ion transport.

Thickness can be measured offline on the wet coating with a wet film comb, a series of tabs at different heights which can be pressed onto the coating, leaving residue on the tabs that extend enough to make contact. This only gives a range for the wet thickness and is destructive as it leaves marks in the small area of coating it is applied to, however is a low cost and fast measurement for small scales. Dry thickness can be measured using a dial gauge or micrometer, where the gauge tip makes contact with the coating surface to measure its thickness. This is again low cost and easy to measure on a lab scale, but measures a very localised area, and many measurements are needed to understand the thickness distribution across the coating. Coatweight is commonly measured in the lab by sampling the electrode, taking a known area of coating, weighing it on a balance, and subtracting the mass of an equivalent area of foil.

Moving towards industrial scales, and looking towards in-line measurements which can facilitate the use of feedback loops for in-line control, there are other options for thickness and coatweight measurement.

Thickness is commonly measured with laser sensors. Laser triangulation uses single sided sensors which can accurately determine the distance of the coating from the sensor. This can be measured over a large area, enabling thickness maps across the coating to be produced. This does require the coating to be stable, so must be done as the coating passes over a fixed roller for example. It also requires the bare foil to be measured as a reference to extract the coating thickness. For non-opaque samples, laser techniques can detect the thickness by detection of scattered light from interfaces, without zeroing against the substrate, however this is unfortunately not possible for the highly opaque battery slurries [2,5,58].

Laser callipers are an alternative, double sided sensor technique. They are a highly accurate single point measurement. It is difficult to traverse them across the coating to produce thickness maps as the two sensors need very precise alignment. For this reason, a combination of a laser calliper for thickness calibration and laser triangulation for mapping of the surface has been suggested to extract the best possible thickness measurement across the entire coating [59].

There are various techniques for coatweight measurement. X-ray or beta sensors can be used, relying on the transmittance of the radiation between two sensors, which is reduced when a mass of material is moved between them. In this way, the coatweight can be extracted. These techniques work well and can be traversed across the coating to produce maps, but the main drawback is the need for radiation safety controls [60,61]. These sensors can also be used in a single sided configuration for backscattering measurements, but like single sided thickness measurement, this makes them more sensitive to movement and vibrations [62]. Ultrasound is an alternative to radiation-based measurements, relying on a similar measurement principle for transmission of ultrasound waves between two sensors, requiring fewer safety controls [63].

With both thickness and coatweight in-line, density maps can be created which could enable in-line control of the calendaring pressure or gap height to achieve a target porosity. Currently, offline calculations are used to determine these parameters, but in-line control would enable new formulations to be calendared to consistent porosity without intervention in the process.

Porosity can also be studied using imaging, which allows visualisation of the coating structure, e.g., optical microscopy or SEM can be used to inspect the coating. Cross-sectional SEM and image analysis can be used to calculate porosity of the electrode. This gives a direct measure, and distributions of pore sizes can be extracted. This is useful for comparison to that calculated from thickness and coatweight, but SEM images will represent a very localised sample, so may not be representative of the bulk porosity. As well as porosity, imaging also allows the extraction of particle distributions and the location of active, conductive additive and binder particles, to evaluate if the mixing has efficiently dispersed the components, and the coating and drying process has given a uniform distribution without migration of components during drying.

There are also porosity measurements that rely on intrusion of a fluid into the sample. Helium pycnometry is one method that gives a full skeletal porosity (due to helium's small size), that includes the pores inside the active material as well as in between the particles. Alternatively, Mercury intrusion can be used, where the larger mercury does not penetrate into the particles structure. Like imaging, these techniques have the advantage that can give a pore size distribution, as the pressure is increased gradually the fluid explores smaller pores, allowing distributions to be extracted [41,64].

Conductivity of the coating can also be measured, this can be measured in plane (along the coating surface) with techniques such as 4 point probe testing, or through plane (from coating surface to current collector) with specialist setups [65]. While in-plane techniques are useful to probe the coating, through plane is most representative of the conductivity in the final cell.

Adhesion is also a useful measure, as it is predicative of how well the coating will handle the subsequent calendaring and cell assembly steps without delamination. As the coating will undergo volume expansion and contraction during cycling in the cell, delamination is also a cause of premature cell failure, so ensuring good adhesion at this stage can help avoid failures in the cell.

This can be measured in different modes, peel testing, which involves peeling away the coating and measuring the force to do so or pull off testing, where the coating is removed from the current collector vertically upwards. Peel is perhaps most useful to predict adhesion when handling and calendaring, as the coating is likely to undergo some degree of bending where any delamination that occurs would be analogous to peeling the coating. Pull off however gives a more direct measure of adhesion which is easier to use in models [47,66,67].

14.4.1 Coating defects

When the coating is created, dried and calendared, there are various defects, or non-uniformities that can occur (see Figure 14.6). Detection of these defects is an additional goal of the QC process, and efficient detection can enable adjustment of the processing parameters to return to uniform coating to reduce wastage. Hence a discussion of the types of defects that can form and their causes is relevant to QC in manufacturing.

Defects can first occur in the wet coating. If the shear rate encountered during coating, exceeds the inverse of the material relaxation time, instabilities will form. This can be the shear rate in the coating gap:

$$\dot{\gamma} = \frac{v}{h}$$

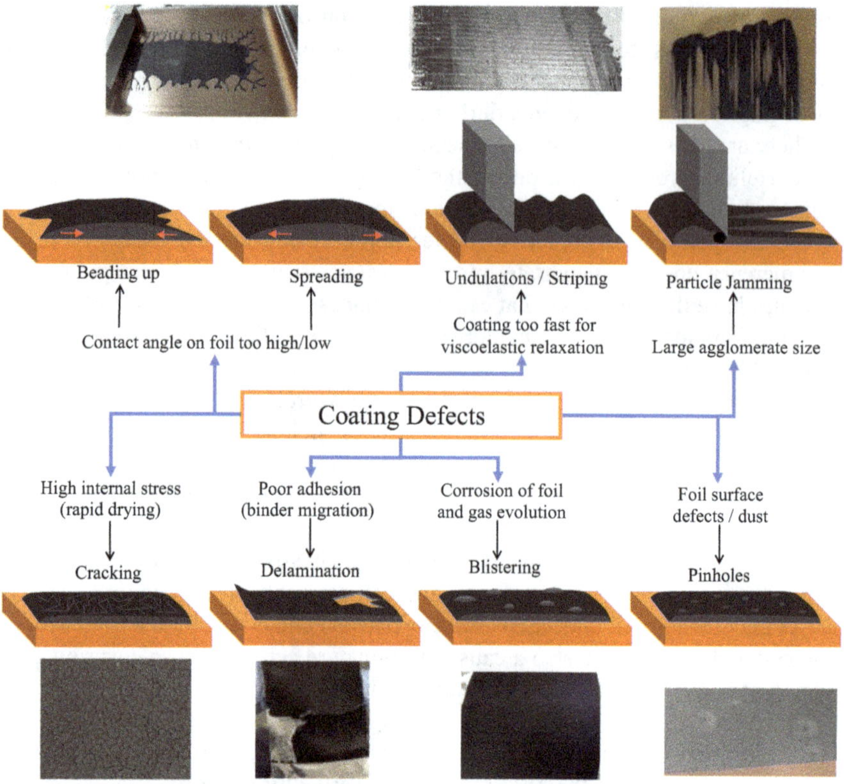

Figure 14.6 Illustration and pictures of coating defects and their causes

where v is the coating speed and h is the gap between the blade and the foil. For slot die coaters there is also a shear rate in the slot, created by the shear along the slot walls:

$$\dot{\gamma} = \frac{6Q}{bh^2}$$

where Q is the volume flow rate, b is the coating width and h is the slot gap. This value is typically larger than the shear rate in the coating gap so important to measure for slot-die coating.

These instabilities typically manifest as undulations or in extreme cases stripes in the coating. As well as the flow rate being too high, it can also be too low. As slurries typically exhibit a yield stress and are steeply shear thinning, if the flow rate is too low this can produce little or no flow through the coating head and thus a thin, uneven coating. This is particularly a problem when alternative binders and complex additives are used that strengthen the network throughout the slurry, increasing the yield stress. Stripes can also form in slot-die coatings, when the flow rate is too low for the coating speed, and thus there is not enough material to form a uniform coating, and thus repeating gaps are left in the coating.

Pinholes are another defect that can occur during coating, where small holes in the coating form. These can be caused by surface effects, where the contact angle is high and so slurry beads up around any defects or external particles (e.g., dust) on the current collector surface. In extreme cases, when the slurry exhibits very high contact angle on the current collector, often because of low weight solids slurry and oxidised current collectors, the entire coating can bead up, leading to unusable coatings.

Pinholes can also be caused by reaction with the current collector surface, in particular aluminium is known to react with water-based slurries when they are basic, evolving hydrogen, which can leave gaps in the coating where the gas bubbles form [68].

As discussed previously, if the particle size is too large during coating, particle jams can occur, where the particles become stuck in the coating gap. This typically leaves streaks in the coating parallel to the coating gap, where slurry could not flow past the stuck particles or agglomerates.

Finally, the coating can have ill-defined edges, which can be caused again by surface effects. If the slurry contact angle is too low, it can spread at the edges leading to thin and uneven coating near the edges. Edge effects can also be caused by material drying at the edges of the coating gap, when the coating is stopped and started, this causes uneven flow around these surfaces and badly defined coating edges.

Further defects can be caused by the drying process. Delamination has been previously discussed, it is at the drying stage that this manifests, and is caused by high contact angle leading to poor contact between the coating and current collector, or by fast drying, causing migration of the adhesive binder away from the current collector. It is particularly a problem in thicker coatings, where greater internal stresses build up and mean peeling is more likely.

Cracks are another defect that occurs during drying, also due to internal stresses building up within the coating. If adhesion is sufficient that these cannot be relieved by delamination, cracks will form in the coating. Therefore, again they are most common in thicker coatings, and can be reduced with slower drying rates, to allow the coating to relax internal stress. They can also be influenced by the drying mode, as if the outer coating dries to form a skin around the wet inner material, the stresses formed during the rest of the drying process are likely to form cracks in this outer skin. So heating from the current collector side with airflow over wet coating can promote drying near the current collector first, which reduces cracking.

Finally blistering can be formed during drying, which is analogous to pinholes caused by evolving gases – if the coating is dried too fast, before this gas can escape, it will form a blister under the dried surface, rather than a pinhole in the coating. Optimisation of the slurry (e.g. adjusting the pH) or the surface (protective coatings or changing the current collector) can reduce this gas evolution and prevent this [11].

Figure 14.6 summarises the types of defects that can form with examples and their potential causes. Coating defects can be detected using many of the coating metrology options discussed above. Thickness and coatweight are useful, particularly when mapped across the coating, as the defects will be highlighted by the variation. Defects are also detectable by optical inspection, which can be done manually, or by in-line sensors and image analysis algorithms, which provides opportunities for in-line control. As well as optical imaging, IR provides a potential way to easily identify defects and make the processing simpler, as bare foil absorbs and releases IR much faster than the electrode coating, so uncoated areas are easily highlighted with IR tomography after exiting the heated drying ovens [59].

Table 14.3 summarises the measurable parameters for electrode coatings, during coating, drying and calendaring. These measurements are highly important as they can be used to study the electrode structure during its creation, and evolution in manufacturing. There are multiple easily adjustable parameters during coating, drying and calendaring (e.g. coating speed, slurry flow rates, drying temperature, calendaring pressure), which can be used for in-line control if deviations in quality are detected, and so coating metrology offers significant opportunities for cyber-physical systems.

14.5 Cell testing and electrochemistry

Often, the electrochemical testing is the final check for a coating run to check whether the quality of the coating. The electrochemical testing looks at the gravimetric capacity of the active material content (to check whether the full capacity is being achieved for the materials) and the areal capacity of the coating (so that the electrodes can be balanced when assembled into a full cell). It should be highlighted that materials testing usually investigates the specific capacity (mAh/g), whereas for electrode formulation and cell development the areal capacity is more important (mAh/cm^2). For material testing, lower coatweight electrodes are made for testing, with high quantities of carbon and binder. Further optimisation and control of formulation, thickness, and porosity is required to develop electrodes and cells for maximum energy or power.

Quality control of the electrodes can be performed in half cell tests, where the coated electrode is tested as the working electrode against a metallic counter electrode such as lithium for lithium-ion batteries, and sodium for sodium-ion batteries. A positive constant low current test, where the alkali metal is deintercalated or removed from the active material component in the positive electrode (cathode) and then deposited upon the metallic counter electrode is performed. The quantity of electrons or charge that is removed from the electrode and then returned is measured and the Columbic efficiency calculated. A formation process is first undertaken where the cell undergoes a few low-rate cycles to ensure good interphase formation on the counter electrode before further testing and cycling is performed. This process gives us a first cycle loss or coulombic efficiency for the working electrode, and a subsequent rated capacity for further testing. For a

Table 14.3 Measurable parameters for battery electrode slurries and measurement methods

Property	Technique	Notes
Coating		
Coatweight	Radiation (X-ray, Beta) [60,61]	• High cost, double sided sensor, radiation safety required
	Backscattering [62]	• Lower cost, single sided sensor, sensitive to movement and vibrations
	Ultrasound [55,63,69]	• Higher cost, double sided sensor, radiation free
Thickness	Wet thickness comb [47]	• Low cost, destructive, manual offline measurement of wet thickness
	Laser triangulation [5,58]	• In-line, single sided and easy to traverse, sensitive to movement of foil
	Laser callipers [59]	• In-line, high accuracy but difficult to traverse, can be combined with triangulation to address this
	Dial gauge/micrometer [70,71]	• Offline, low cost, manual, non-destructive measurement of dry thickness
Defect detection	Coatweight/thickness tomography	• Defects can be highlighted by variations in thickness/coatweight measured using techniques above
	Optical inspection [72,73]	• Simple experimental setup (camera) but needs manual inspection or image analysis code to identify defects
	IR thermography [59]	• Uncoated foil absorbs more IR during drying, defects can be quantitatively identified directly after oven, can be automated, more costly than optical
Adhesion	Peel [47,66]	• Peeling usually removes majority of coating • Measure adhesion without cohesion
	Pull-off [67]	• Greater forces required to remove coating • Results commonly a combination of adhesion and cohesion
Porosity	Mercury intrusion [41]	• Measures porosity accessible to Mercury, excluding very small pore throats an internal porosity in particles
	Helium pycnometer [64]	• Measures porosity inside particles giving full skeletal porosity
	Calculation from coatweight + thickness	• Assumes uniform density, simple to execute, can be mapped across electrode
	Cross-sectional SEM + image analysis [64]	• Can build up 3D tomographs, can be complex
Conductivity	4 Point Probe [74]	• In-plane measurement • Simple, non-destructive, low-cost setups available
	Through plane setups [65]	• Require specialist setups • More representative of conductivity in the final cell

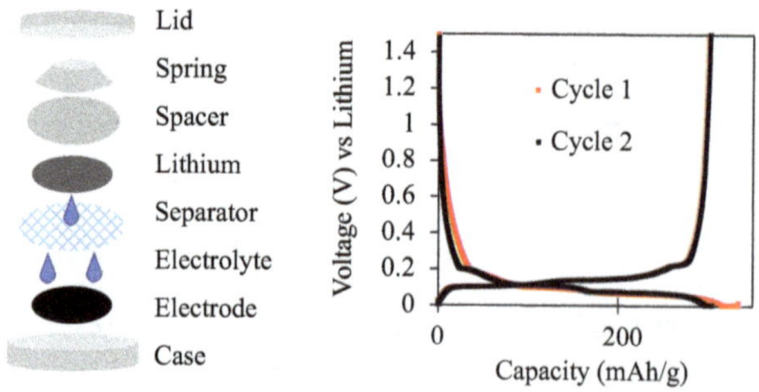

Figure 14.7 (Left) illustration of the construction of a half coin cell, (right) example formation data, showing a first cycle loss in capacity between cycles 1 and 2

negative electrode, the lithium is first inserted into the active materials component rather than removed, and this is performed with a negative current between the working and the counter electrode. Figure 14.7 shows a half cell set up in a coin cell and an example of formation.

Subsequent rate, capacity and cycling testing probe the properties of the electrode further. This final electrochemistry can be a useful QC measure that gives insights into the electrode manufacturing. Poor capacity could result from insufficient dispersal of the active and conductive additive. More intensive mixing of the initial slurry can improve the initial capacity and reduce the internal resistance in the cell. Poor stability, drop off in capacity over time, could result from coating defects, causing delamination during the volume expansion and contraction of the coatings. This can typically be improved with thinner coating, slower drying, or additives to lower surface tension or improve adhesion. Tortuosity can also be measured from electrochemistry, using symmetrical cells with pore blocking electrolyte. This is a useful output as it gives insight into how efficiently the microstructure created in the previous stages allows ion transport, and thus can be related to the processing parameters [75,76]. All the specific properties of the electrode such as the adhesion, conductivity, porosity, lead to the final cell performance parameters and these relationships are discussed in the next section.

The parameters involved in cell making can also influence the results. Inhomogeneous coatings, with poor electronic conductivities and poorly wetted areas can also promote lithium plating which can lead to dendrite formation, short circuits, and premature failure of the cell [77]. For this reason, the amount of electrolyte is important and the cell is left to soak before testing to ensure good wetting.

The pressure applied to the cell is also key to the final electrochemistry. For some cells, this is determined by the components. For coin cells where the casing is solid, the thickness of components determines the pressure applied, and spacers are used to increase this [78]. For cells with flexible casings, like pouch cells, an external pressure must be applied, which can have a significant impact on the final properties [79]. Pressure is advantageous to promote electrode wetting with electrolyte, stabilise the electrodes and slow cracking and delamination. However, excess pressure can reduce the reaction surface area available and lead to undesirable side reactions, as well as cause stresses within the electrode which may accelerate mechanical degradation.

A highly useful guide to coin cell testing, including discussion of the different cell parameters is provided by Murray, Hall and Dahn [78].

14.6 Input–property relationships

Measurement at each interval of the process can help to break down the complex manufacturing process and understand the impact of each stage. However, it is still very difficult to fully separate any of the stages of manufacture, because of the complex interdependencies of the physical and processing parameters, that carry through the entire process. Only with full characterisation of every output that could be relevant to later processing, would isolation of the individual steps be possible.

Figure 14.8 shows a diagram of the parameters involved at each stage, to highlight the level of complexity, but is by no means exhaustive.

One way to help unify the parameters in the process and understand the underlying relationships is through modelling. Accurate models of the manufacturing process will be required to speed up optimisation of new lines and new materials, and develop digital twins for prediction and control of the process without user input, which will accelerate the progress of battery manufacturing towards Industry 4.0 [80].

A strategy is to model the underlying relationships with physical models. For example, computational fluid dynamics has been used to capture the coating process and predict conditions for the onset of air entrapment in the coating, experimentally validated with the slurry rheology and contact angle [81]. Discrete element modelling has been used to predict the change in electrode structure during calendaring, validated with structures from synchrotron radiation X-ray tomography [82]. However, despite the large body of work on physical modelling, there are still gaps in our understanding of the processes occurring in manufacturing. These models all require experimental data for validation, and with the vast amount of parameter space to explore, there is a significant role for in depth QC in development of these models. There also has been progress in combining physical models into a full model of the process [83]. However, the complexity becomes extreme when combining all the manufacturing steps, and cases of different materials and equipment, and so the models involved must be simplified and often require experimental data at the intermediate stages of manufacture.

Machine learning is an alternate strategy to modelling the process, which does not require a full understanding of the physical process in each step. There has been a significant amount of work on applying data-driven approaches to battery manufacturing, which have demonstrated prediction of optimum formulations and processing parameters [6,47,84]. Once trained on sufficient data, these models can also be used to independently vary parameters (many of which are co-dependant so

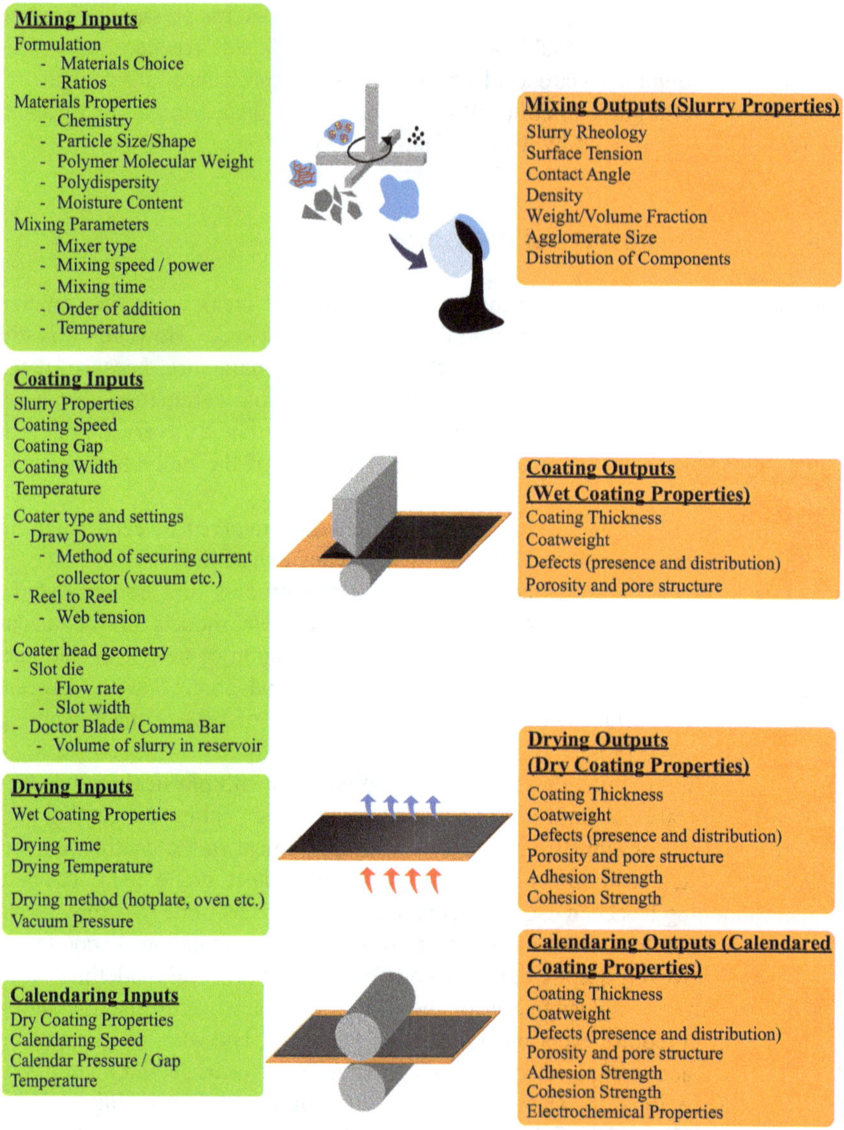

Figure 14.8 Table of parameters involved at each stage of electrode manufacture

cannot be independently varied in reality) to extract relationships and validate physical models. This combination of physical and data-driven modelling has the potential to be powerful for prediction of manufacturing outputs. One of the limitations of machine learning currently is the availability of sufficient quantity and quality of data which captures all the relevant manufacturing variables, another reason why advanced QC is necessary.

Currently, no model can capture the full complexity of the process. The complexity (both computational and for our scientific understanding) rises too fast in physics-based models, and in statistical models there is a need for more understanding of the physics to guide feature selection (e.g., there is no way to differentiate if a dataset is not large enough to detect the importance of a feature, or if it is not relevant to the process). Therefore, to develop digital twins that fully capture industrial manufacturing and allow prediction of the results and rapid optimisation, a combination of physical, data-driven modelling, and advanced QC methods to fully characterise the process is required.

14.7 Conclusions

There is a large variety of variables involved in battery manufacture, as illustrated in Figure 14.8. This means there is huge scope for metrology to monitor and control manufacturing quality, but also a huge complexity in understanding the relationships between these variables and the process outputs and the inter-dependencies between variables. Future manufacturing and Industry 4.0 will require a thorough understanding of these relationships for the development of models and cyber-physical systems [85]. However, there is still a way to go to fully understand the physics throughout the process. Quality control methods have a significant role to play in providing the data to elucidate the underlying physical relationships, and in providing inputs to feedback loops to enable in-line control.

Metrology is important throughout the process, but certain properties stand out. The slurry rheology is highly useful as it can be captured very early in the process giving insights into the properties early on. It gives insights into the slurry structure, which can assess the efficiency of the mixing process, and the slurry can be sent for further mixing if required, removing problems at this stage. It also gives insights into the processing behaviour and can help determine the pumping and coating rates that are possible for that slurry, allowing high speed processing without defects.

Coatweight is also a key property, as it determines the final capacity of the cell, and must be tightly controlled so that the anode and the cathode are correctly balanced in the final cell. This is perhaps the single property that determines most heavily whether the final electrode will be usable. As whatever the quality, if the anode and the cathode coating are not correctly balanced, one must be repeated. Therefore, there is a large opportunity for reduction of waste in coatweight control. While there are established methods for measuring coatweight in-line, there is room for further development of feedback loops and in-line control to maintain specifications and reduce wastage.

Coating quality and the presence of defects are also important, and again, while there are techniques for detecting defects in coating lines, there is more work to be done on understanding their impact, determining the tolerance for defects in an electrode, and elucidating methods for in-line control of processing parameters to reduce them.

Overall, metrology is well advanced for electrode manufacture, and while there are still improvements to be made, the key outstanding challenge is the collation and combination of all the physical parameters with models, statistical or physical, to enable in-line control. The coordination of all this data and the generation of datasets large enough to extract subtle trends in the myriad physical properties is still a daunting task, but steady progress is being made across academia and industry to develop digital twins of battery manufacture and enable faster optimisation and control.

References

[1] Kendrick, E. Advancements in manufacturing. In *Future Lithium-ion Batteries*, London: RSC, 2019, pp. 262–289 (Chapter 11).

[2] Reynolds, C. D., Slater, P. R., Hare, S. D., Simmons, M. J. H., and Kendrick, E. A Review of Metrology in Lithium-Ion Electrode Coating Processes. *Mater. Des.* 2021, 209, 109971.

[3] Dominko, R., Gaberscek, M., Drofenik, J., Bele, M., Pejovnik, S., and Jamnik, J. The Role of Carbon Black Distribution in Cathodes for Li Ion Batteries. *J. Power Sources* 2003, 119–121, 770–773.

[4] Grießl, D., Huber, K., Scherbauer, R., and Kwade, A. Dispersion Kinetics of Carbon Black for the Application in Lithium-Ion Batteries. *Adv. Powder Technol.* 2021, 32 (7), 2280–2288.

[5] Schmitt, M., Baunach, M., Wengeler, L., *et al.* Slot-Die Processing of Lithium-Ion Battery Electrodes—Coating Window Characterization. *Chem. Eng. Process. Process Intensif.* 2013, 68, 32–37.

[6] Faraji Niri, M., Reynolds, C., Román Ramírez, L. A., Kendrick, E., and Marco, J. Systematic Analysis of the Impact of Slurry Coating on Manufacture of Li-Ion Battery Electrodes via Explainable Machine Learning. *Energy Storage Mater.* 2022, 51, 223–238.

[7] David, L., Ruther, R. E., Mohanty, D., *et al.* Identifying Degradation Mechanisms in Lithium-Ion Batteries with Coating Defects at the Cathode. *Appl. Energy* 2018, 231, 446–455.

[8] Zhang, Y., Bailey, J., Sun, Y., *et al.* Applications of Advanced Metrology for Understanding the Effects of Drying Temperature in Lithium-Ion Battery Electrodes Manufacturing Process. *J. Mater. Chem. A* 2022, 10, 10593–10603.

[9] Font, F., Protas, B., Richardson, G., and Foster, J. M. Binder Migration during Drying of Lithium-Ion Battery Electrodes: Modelling and Comparison to Experiment. *J. Power Sources* 2018, 393, 177–185.

[10] Rollag, K., Juarez-Robles, D., Du, Z., Wood, D. L. I., and Mukherjee, P. P. Drying Temperature and Capillarity-Driven Crack Formation in Aqueous Processing of Li-Ion Battery Electrodes. *ACS Appl. Energy Mater.* 2019, 2 (6), 4464–4476.

[11] Sahore, R., Wood, D. L. I., Kukay, A., Grady, K. M., Li, J., and Belharouak, I. Towards Understanding of Cracking during Drying of Thick Aqueous-Processed $LiNi_{0.8}Mn_{0.1}Co_{0.1}O_2$ Cathodes. *ACS Sustain. Chem. Eng.* 2020, 8 (8), 3162–3169.

[12] Parikh, D., Christensen, T., and Li, J. Correlating the Influence of Porosity, Tortuosity, and Mass Loading on the Energy Density of

$LiN_{i0.}6M_{n0.}2C_{o0.}2_O2$ Cathodes under Extreme Fast Charging (XFC) Conditions. *J. Power Sources* 2020, 474, 228601.

[13] Lu, X., Daemi, S. R., Bertei, A., *et al.* Microstructural Evolution of Battery Electrodes During Calendering. *Joule* 2020, 4 (12), 2746–2768.

[14] Seeba, J., Reuber, S., Heubner, C., Müller-Köhn, A., Wolter, M., and Michaelis, A. Extrusion-Based Fabrication of Electrodes for High-Energy Li-Ion Batteries. *Chem. Eng. J.* 2020, 402, 125551.

[15] Haarmann, M., Haselrieder, W., and Kwade, A. Extrusion-Based Processing of Cathodes: Influence of Solid Content on Suspension and Electrode Properties. *Energy Technol.* 2020, 8 (2), 1801169.

[16] Wiegmann, E., Kwade, A., and Haselrieder, W. Solvent Reduced Extrusion-Based Anode Production Process Integrating Granulate Coating, Drying, and Calendering. *Energy Technol.* 2022, 10 (6), 2200020.

[17] Soares, D. M., Ren, Z., Mujib, S. B., *et al.* Additive Manufacturing of Electrochemical Energy Storage Systems Electrodes. *Adv. Energy Sustain. Res.* 2021, 2 (5), 2000111.

[18] Gastol, D., Capener, M., Reynolds, C., Constable, C., and Kendrick, E. Microstructural Design of Printed Graphite Electrodes for Lithium-Ion Batteries. *Mater. Des.* 2021, 205, 109720.

[19] Jeschull, F., Surace, Y., Zürcher, S., *et al.* Graphite Particle-Size Induced Morphological and Performance Changes of Graphite–Silicon Electrodes. *J. Electrochem. Soc.* 2020, 167 (10), 100535.

[20] Hietaniemi, M., Hu, T., Välikangas, J., Niittykoski, J., and Lassi, U. Effect of Precursor Particle Size and Morphology on Lithiation of $Ni_{0.6}Mn_{0.2}Co_{0.2}(OH)_2$. *J. Appl. Electrochem.* 2021, 51 (11), 1545–1557.

[21] 14:00-17:00. *ISO 21501-2:2019*. ISO. https://www.iso.org/standard/75048.html (accessed 2022-12-08).

[22] 14:00-17:00. *ISO 22412:2017*. ISO. https://www.iso.org/standard/65410.html (accessed 2022-12-08).

[23] Roberts, S., Chen, L., Kishore, B., Dancer, C. E. J., Simmons, M. J. H., and Kendrick, E. Mechanism of Gelation in High Nickel Content Cathode Slurries for Sodium-Ion Batteries. *J. Colloid Interface Sci.* 2022, 627, 427–437.

[24] Choi, K. H., Liu, X., Ding, X., Xia, Y., and Li, Q. Impact of Gelation in Nickel-Rich Ternary Lithium-Ion Batteries. *Ionics* 2021, 27 (12), 5159–5166.

[25] Xuan, W., Otsuki, A., and Chagnes, A. Investigation of the Leaching Mechanism of NMC 811 ($LiNi_{0.8}Mn_{0.1}Co_{0.1}O_2$) by Hydrochloric Acid for Recycling Lithium Ion Battery Cathodes. *RSC Adv.* 2019, 9 (66), 38612–38618.

[26] Li, T., Yuan, X.-Z., Zhang, L., Song, D., Shi, K., and Bock, C. Degradation Mechanisms and Mitigation Strategies of Nickel-Rich NMC-Based Lithium-Ion Batteries. *Electrochem. Energy Rev.* 2020, 3 (1), 43–80.

[27] Renfrew, S. E. and McCloskey, B. D. Residual Lithium Carbonate Predominantly Accounts for First Cycle CO_2 and CO Outgassing of Li-Stoichiometric and Li-Rich Layered Transition-Metal Oxides. *J. Am. Chem. Soc.* 2017, 139 (49), 17853–17860.

[28] Su, Y., Li, L., Chen, G., *et al.* Strategies of Removing Residual Lithium Compounds on the Surface of Ni-Rich Cathode Materials. *Chin. J. Chem.* 2021, 39 (1), 189–198.

[29] Kosfeld, M., Westphal, B., and Kwade, A. Correct Water Content Measuring of Lithium-Ion Battery Components and the Impact of Calendering via Karl-Fischer Titration. *J. Energy Storage* 2022, 51, 104398.

[30] Sicklinger, J., Metzger, M., Beyer, H., Pritzl, D., and Gasteiger, H. A. Ambient Storage Derived Surface Contamination of NCM811 and NCM111: Performance Implications and Mitigation Strategies. *J. Electrochem. Soc.* 2019, 166 (12), A2322.

[31] Reynolds, C. D., Hare, S. D., Slater, P. R., Simmons, M. J. H., and Kendrick, E. Rheology and Structure of Lithium-Ion Battery Electrode Slurries. *Energy Technol.* 2022, 10 (10), 2200545.

[32] Ishii, M. and Nakamura, H. Influence of Molecular Weight and Concentration of Carboxymethyl Cellulose on Rheological Properties of Concentrated Anode Slurries for Lithium-Ion Batteries. *JCIS Open* 2022, 6, 100048.

[33] Lee, B.-R. and Oh, E.-S. Effect of Molecular Weight and Degree of Substitution of a Sodium-Carboxymethyl Cellulose Binder on $Li_4Ti_5O_{12}$ Anodic Performance. *J. Phys. Chem. C* 2013, 117 (9), 4404–4409.

[34] Lyu, S. and Untereker, D. Degradability of Polymers for Implantable Biomedical Devices. *Int. J. Mol. Sci.* 2009, 10 (9), 4033–4065.

[35] Gordon, R., Orias, R., and Willenbacher, N. Effect of Carboxymethyl Cellulose on the Flow Behavior of Lithium-Ion Battery Anode Slurries and the Electrical as Well as Mechanical Properties of Corresponding Dry Layers. *J. Mater. Sci.* 2020, 55 (33), 15867–15881.

[36] Park, J., Willenbacher, N., and Ahn, K. H. How the Interaction between Styrene-Butadiene-Rubber (SBR) Binder and a Secondary Fluid Affects the Rheology, Microstructure and Adhesive Properties of Capillary-Suspension-Type Graphite Slurries Used for Li-Ion Battery Anodes. *Colloids Surf. Physicochem. Eng. Asp.* 2019, 579, 123692.

[37] Zhu, P., Gastol, D., Marshall, J., Sommerville, R., Goodship, V., and Kendrick, E. A Review of Current Collectors for Lithium-Ion Batteries. *J. Power Sources* 2021, 485, 229321.

[38] Wang, R., Li, W., Liu, L., *et al.* Carbon Black/Graphene-Modified Aluminum Foil Cathode Current Collectors for Lithium Ion Batteries with Enhanced Electrochemical Performances. *J. Electroanal. Chem.* 2019, 833, 63–69.

[39] Chen, J., Wang, X., Gao, H., *et al.* Rolled Electrodeposited Copper Foil with Modified Surface Morphology as Anode Current Collector for High Performance Lithium-Ion Batteries. *Surf. Coat. Technol.* 2021, 410, 126881.

[40] Gorman, S. F., Pathan, T. S., and Kendrick, E. The 'Use-by Date' for Lithium-Ion Battery Components. *Philos. Transact. A Math. Phys. Eng. Sci.* 2019, 377 (2152), 20180299.

[41] Konda, K., Moodakare, S. B., Kumar, P. L., *et al.* Comprehensive Effort on Electrode Slurry Preparation for Better Electrochemical Performance of LiFePO4 Battery. *J. Power Sources* 2020, 480, 228837.

[42] Capone, I., Hurlbutt, K., Naylor, A. J., Xiao, A. W., and Pasta, M. Effect of the Particle-Size Distribution on the Electrochemical Performance of a Red Phosphorus–Carbon Composite Anode for Sodium-Ion Batteries. *Energy Fuels* 2019, 33 (5), 4651–4658.

[43] Xiao, Z., Rao, X., Chen, J., Potapenko, H., Zhang, Q., and Zhong, S. Organic Carbonized Copper Foil Facilitates the Performance of the Current Collector for Lithium-Ion Batteries. *Mater. Chem. Front.* 2022, 6 (17), 2478–2490.

[44] Wen, L., Liang, J., Liu, C.-M., *et al.* Li4Ti5O12 on Graphene for High Rate Lithium Ion Batteries. *J. Electrochem. Soc.* 2016, 163 (14), A2951.

[45] Ying, J., Wan, C., Jiang, C., and Li, Y. Preparation and Characterization of High-Density Spherical $LiNi_{0.8}Co_{0.2}O_2$ Cathode Material for Lithium Secondary Batteries. *J. Power Sources* 2001, 99 (1), 78–84.

[46] Hawley, W. B. and Li, J. Electrode Manufacturing for Lithium-Ion Batteries—Analysis of Current and next Generation Processing. *J. Energy Storage* 2019, 25, 100862.

[47] Drakopoulos, S. X., Gholamipour-Shirazi, A., MacDonald, P., *et al.* Formulation and Manufacturing Optimization of Lithium-Ion Graphite-Based Electrodes via Machine Learning. *Cell Rep. Phys. Sci.* 2021, 2 (12), 100683.

[48] Wenzel, V., Moeller, R. S., and Nirschl, H. Influence of Mixing Technology and the Potential to Modify the Morphological Properties of Materials Used in the Manufacture of Lithium-Ion Batteries. *Energy Technol.* 2014, 2 (2), 176–182.

[49] Wang, M., Dang, D., Meyer, A., Arsenault, R., and Cheng, Y.-T. Effects of the Mixing Sequence on Making Lithium Ion Battery Electrodes. *J. Electrochem. Soc.* 2020, 167 (10), 100518.

[50] Li, J., Rulison, C., Kiggans, J., Daniel, C., and Wood, D. L. Superior Performance of $LiFePO_4$ Aqueous Dispersions via Corona Treatment and Surface Energy Optimization. *J. Electrochem. Soc.* 2012, 159 (8), A1152.

[51] Karatsu T., Wang Z., Zhao T., Kawashima D., and Takei M. Dynamic Fields Visualization of Carbon-Black (CB) Volume Fraction Distribution in Lithium-Ion Battery (LIB) Cathode Slurry by Electrical Resistance Tomography (ERT). *J. Soc. Powder Technol. Jpn.* 2021, 58 (3), 119–126.

[52] Williams, R. A., Jia, X., and McKee, S. L. Development of Slurry Mixing Models Using Resistance Tomography. *Powder Technol.* 1996, 87 (1), 21–27.

[53] Ouyang, L., Wu, Z., Wang, J., *et al.* The Effect of Solid Content on the Rheological Properties and Microstructures of a Li-Ion Battery Cathode Slurry. *RSC Adv.* 2020, 10 (33), 19360–19370.

[54] Reynolds, C. D., Lam, J., Yang, L., and Kendrick, E. Extensional Rheology of Battery Electrode Slurries with Water-Based Binders. *Mater. Des.* 2022, 222, 111104.

[55] Román-Ramírez, L. A., Apachitei, G., Faraji-Niri, M., Lain, M., Widanage, D., and Marco, J. Experimental Data of Cathodes Manufactured in a

Convective Dryer at the Pilot-Plant Scale, and Charge and Discharge Capacities of Half-Coin Lithium-Ion Cells. *Data Brief* 2022, 40, 107720.

[56] *Paint and Coating Testing Manual: Fourteenth Edition of the Gardner – Sward Handbook|Titel.* https://library.wur.nl/WebQuery/titel/940131 (accessed 2022-12-06).

[57] Ma, F., Fu, Y., Battaglia, V., and Prasher, R. Microrheological Modeling of Lithium Ion Battery Anode Slurry. *J. Power Sources* 2019, 438, 226994.

[58] Schmitt, M., Scharfer, P., and Schabel, W. Slot Die Coating of Lithium-Ion Battery Electrodes: Investigations on Edge Effect Issues for Stripe and Pattern Coatings. *J. Coat. Technol. Res.* 2014, 1 (11), 57–63.

[59] Mohanty, D., Li, J., Born, R., *et al.* Non-Destructive Evaluation of Slot-Die-Coated Lithium Secondary Battery Electrodes by In-line Laser Caliper and IR Thermography Methods. *Anal. Methods* 2014, 6 (3), 674–683.

[60] Azimi, Y., Kortschot, M., and Farnood, R. A Non-Destructive Method for Obtaining a Local Coat Weight Map Using X-Ray Imaging. *J. Pulp Pap. Sci.* 2009, 35, 11–16.

[61] Yeager, B. New Detection Advancement Enhances Basis Weight, Coat Weight Measurement. *Pulp Pap.* 1996, 70 (1), 45–58.

[62] MacKenzie, I. K. Method and Apparatus for Measuring Thickness of Coating Layers on Substrates Using Backscattering of X-Rays and Gamma Rays. US6252930B1, June 26, 2001.

[63] Knorr, H. Ultrasonic Transmitting and Receiving Device for Thickness and/or Grammage Measurement. US8904874B2, December 9, 2014.

[64] Beuse, T., Fingerle, M., Wagner, C., Winter, M., and Börner, M. Comprehensive Insights into the Porosity of Lithium-Ion Battery Electrodes: A Comparative Study on Positive Electrodes Based on $LiNi_{0.6}Mn_{0.2}Co_{0.2}O_2$ (NMC622). *Batteries* 2021, 7 (4), 70.

[65] Wei, Q., Suga, H., Ikeda, I., *et al.* An Accurate Method to Determine the Through-Plane Electrical Conductivity and to Study Transport Properties in Film Samples. *Org. Electron.* 2016, 38, 264–270.

[66] Gaikwad, A. M. and Arias, A. C. Understanding the Effects of Electrode Formulation on the Mechanical Strength of Composite Electrodes for Flexible Batteries. *ACS Appl. Mater. Interfaces* 2017, 9 (7), 6390–6400.

[67] Haselrieder, W., Westphal, B., Bockholt, H., Diener, A., Höft, S., and Kwade, A. Measuring the Coating Adhesion Strength of Electrodes for Lithium-Ion Batteries. *Int. J. Adhes. Adhes.* 2015, 60, 1–8.

[68] Church, B. C., Kaminski, D. T., and Jiang, J. Corrosion of Aluminum Electrodes in Aqueous Slurries for Lithium-Ion Batteries. *J. Mater. Sci.* 2014, 49 (8), 3234–3241.

[69] Knorr, H. Method and Apparatus for Non-Contact Determination of the Weight per Unit Area of Thin Materials. US5621173A, April 15, 1997.

[70] Kumagai, S., Mukaiyachi, K., and Tashima, D. Rate and Cycle Performances of Supercapacitors with Different Electrode Thickness Using Non-Aqueous Electrolyte. *J. Energy Storage* 2015, 3, 10–17.

[71] Mayur, M., Yagci, M. C., Carelli, S., Margulies, P., Velten, D., and Bessler, W. G. Identification of Stoichiometric and Microstructural Parameters of a Lithium-Ion Cell with Blend Electrode. *Phys. Chem. Chem. Phys.* 2019, 21 (42), 23672–23684.

[72] Khandavalli, S. and Rothstein, J. P. The Effect of Shear-Thickening on the Stability of Slot-Die Coating. *AIChE J.* 2016, 62 (12), 4536–4547.

[73] Gutoff, E. B. and Cohen, E. D. Water- and Solvent-based Coating Technology. In Wagner, J. R., (ed.), *Multilayer Flexible Packaging* (Second Edition), Plastics Design Library, William Andrew Publishing, 2016, pp. 205–234 (14).

[74] Bommel, A. and van Divigalpitiya, R. Effect of Calendering $LiFePO_4$ Electrodes. *J. Electrochem. Soc.* 2012, 159 (11), A1791.

[75] Malifarge, S., Delobel, B., and Delacourt, C. Determination of Tortuosity Using Impedance Spectra Analysis of Symmetric Cell. *J. Electrochem. Soc.* 2017, 164 (11), E3329.

[76] Fu, J., Thomas, H. R., and Li, C. Tortuosity of Porous Media: Image Analysis and Physical Simulation. *Earth-Sci. Rev.* 2021, 212, 103439.

[77] Jeon, D. H. Wettability in Electrodes and Its Impact on the Performance of Lithium-Ion Batteries. *Energy Storage Mater.* 2019, 18, 139–147.

[78] Murray, V., Hall, D. S., and Dahn, J. R. A Guide to Full Coin Cell Making for Academic Researchers. *J. Electrochem. Soc.* 2019, 166 (2), A329.

[79] Mussa, A. S., Klett, M., Lindbergh, G., and Lindström, R. W. Effects of External Pressure on the Performance and Ageing of Single-Layer Lithium-Ion Pouch Cells. *J. Power Sources* 2018, 385, 18–26.

[80] Ayerbe, E., Berecibar, M., Clark, S., Franco, A. A., and Ruhland, J. Digitalization of Battery Manufacturing: Current Status, Challenges, and Opportunities. *Adv. Energy Mater.* 2022, 12 (17), 2102696.

[81] Hoffmann, A., Spiegel, S., Heckmann, T., Scharfer, P., and Schabel, W. CFD Model of Slot Die Coating for Lithium-Ion Battery Electrodes in 2D and 3D with Load Balanced Dynamic Mesh Refinement Enabled with a Local-Slip Boundary Condition in OpenFOAM. *J. Coat. Technol. Res.* 2022, 20, 3–14.

[82] Ge, R., Cumming, D. J., and Smith, R. M. Discrete Element Method (DEM) Analysis of Lithium Ion Battery Electrode Structures from X-Ray Tomography - The Effect of Calendering Conditions. *Powder Technol.* 2022, 403, 117366.

[83] Lombardo, T., Lambert, F., Russo, R., *et al.* Experimentally Validated Three-Dimensional Modeling of Organic-Based Sodium-Ion Battery Electrode Manufacturing. *Batter. Supercaps* 2022, 5 (8), e202200116.

[84] Franco, A. A. and Barnard, A. S. Artificial Intelligence in Electrochemical Energy Storage. *Batter. Supercaps* 2022, 5 (5), e202200149.

[85] Cañas, H., Mula, J., Díaz-Madroñero, M., and Campuzano-Bolarín, F. Implementing Industry 4.0 Principles. *Comput. Ind. Eng.* 2021, 158, 107379.

Chapter 15
Industrial perspective of electrode manufacturing

Yangping Sheng[1], Nan Zhang[1], Chengwei Yi[1], Wenqiang Cheng[1] and Yuhao Lu[1,2]

15.1 Introduction

The world is transferring from fossil fuels to renewable energy, targeting a carbon neutral society. Batteries with various chemistries are playing critical roles in storing and releasing electricity generated intermittently by renewable sources, including wind and solar power, so as to supply stable power and maintain grid operation [1]. Relied on electrochemical redox reactions, the electrode structures are enabling ionic and electronic exchange in the interface between active materials and electrolytes, which contributes to a faradic process for electric energy storage and conversion. Herein, we need to point out the difference between a battery and a cell according to the terminology of power sources. The battery is an assembly of cells, battery management system (BMS) and other connecting parts; and the cell is a basic unit of electrochemical storage enabled by anode, cathode, electrolyte, and other auxiliary materials. Active materials, binders and electronically conductive additives which are called electrode compositions or recipes are mixed together and then coated onto a current collector to configure an electrode of the battery. Aside from recipes, the electrode qualities, for example uniformity, porosity, defects, etc., impact its electrochemical performance to a great extent. In this chapter, the electrode manufacturing is described with its state of the art, especially from the industrial point of view.

15.2 The electrode manufacturing and cell production

The manufacturing process for a cell includes steps roughly that are electrode preparation, cell assembly, electrochemical activation, and cell qualification. In the electrode preparation, a recipe is coated onto a current collector. Every recipe is elaborately formulated by means of an adjustment among active materials loading,

[1]Amperex Technology Limited, ATL, Hong Kong, China
[2]Key Laboratory of Consumer Lithium-Ion Battery in Fujian, Fujian, China

conductive additive content and binder amount, ensuring the electrode quality (e.g., electrode adhesive force and inherent strength), and satisfying the required electrochemical performance for a specific application. In general, the loading for active materials is expected to be as high as possible in order to obtain the high energy density. Although the minimal conductive additives and binders in the recipe are expected, they are still able to provide good electronic network and maintain acceptable mechanical quality for the electrode, respectively.

The electrode quality, from manufacturing perspective, is an important and decisive factor in the battery quality [2] that dedicates to its stable performance and safety during operation. The defects of the electrodes, discussed in later sections, definitely result in various battery performance failures, e.g. lithium plating, cell swelling, extra heating, and low columbic efficiency. Therefore, it is very necessary to understand how an electrode is fabricated in industry as well as inspire researchers in the battery community to explore new materials, develop new binders, design novel cell structures, and develop adaptable manufacturing processes.

Compared with other three steps, the step of electrode preparation has easy means to detect imperfections, e.g., cracks and scratches, to avoid potential failures in cells and batteries. Unfortunately, detections are performed by a lot of laborers rather than automatic equipment because of two points below.

1. The step underlies many complicated mechanisms. From particles to electrode, it covers interactions between solid particles and binders, particles dynamic movement in solvents, vaporization of solvents, and pore structure formation along with the solidification of binders. Currently the step overly relies on Edisonian approach, trial-error empirical experiments and observation, and hence, the human inspection is necessary.
2. In the electrode preparation, trial runs, roughly optimizing parameters prior to the standard of procedure, are required for coating, calendering and slitting. Under the circumstance of immature Industry 4.0 [3], to find pre-determined parameters is difficult with automation and human beings are mandatory.

In order to accelerate the process of Industry 4.0 on the electrode preparation [4], process engineering and equipment teams are dedicating to reducing the human intervention and realizing full automation.

15.3 The configuration of electrode in industry

It is noteworthy to note that cell performance is subject to its structures in which electrode configurations are determined. Meanwhile, processes for the electrode preparation need to be modified in according with those configurations. In pouch cells, thanks to different tab positions and the number of tabs, the cell structures, as widely used in industry, have M6S (mainstream 6th square, single tab), MMT (Middle to Middle tab), MTW (Multi-tab Winding), stacking cells and so on. In order to reach the requirement (e.g., energy and power densities) of a specific application, a cell structure as well as the corresponding electrode configuration are selected. In the charging/discharging processes, the tabs from electrodes bridge outside circuits and internal electrodes of cells. The properly positioned and distributed tabs lead to uniform current distribution in the cells. In the competitive market, it is forecasted that the cell structure evolves quickly along with the urgent demand of cell performance since the properly designed structure can bring better performance, less manufacturing cost, and more suitable for end devices. In industry, a typical configuration of electrode can be seen from Figure 15.1, from the views of top and side. Overhang is designed for anode in case of its misalignment with cathode that results in capacity mismatch between them to avoid lithium plating on anode, which could cause a potential safety issue during the operation of cells with such misaligned electrodes inside. In consumer electronics, according to IEEE Standard 1725-2021, a standard for rechargeable batteries for mobile phones, the anode active area shall extend at least 0.1 mm (plus process margin) beyond the edges of the cathode area unless process capability/stability is demonstrated to be less than 0.1 mm.

Electrode configurations complying with cell structures are multiform where head/tail positions, edges, tab positions, A/B side alignment need to be adjusted or redesigned carefully. Vivid views of cell structures and electrode configurations are demonstrated in Figure 15.2. The winding structures for M6S, MMT, and MTW jelly rolls are shown in the top, and related electrodes with top and side views. The blue circle is marked for the cathode tabs position and the underlying light grey line stands for dried cathode recipes. The red circle indicates the anode tabs, and the connecting light purple line stands for the dried anode recipes. The dash line stands for a layer of separator between the anode and the cathode. Green parts are the tapes used to cover the tabs and exposed heads/tails to prevent internal shorts that are caused by welding or cutting burrs around edges of current collectors. Differences in the electrode configurations are obvious such as the spaces of bare foil left for tab welding, A/B side misalignment for anode and cathode and the matching ways of anode and cathode. It is very effective to optimize energy and power densities if the use of bare foils and electrode are minimized for a certain capacity cell, which contributes to the less weight of cell and lower process/material cost.

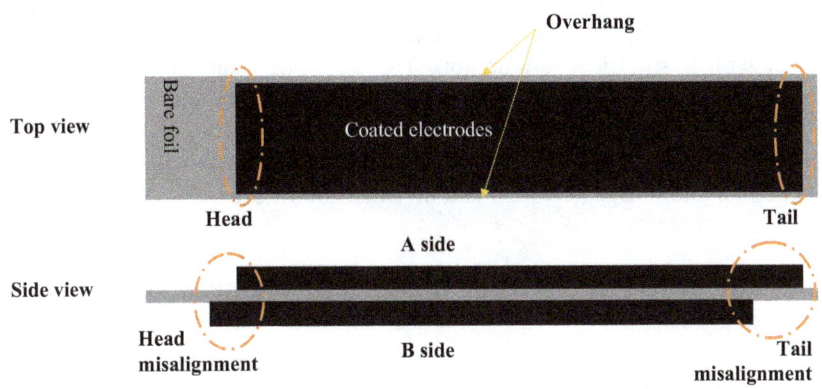

Figure 15.1 Schematic structures for an industry electrode, from top view and side view

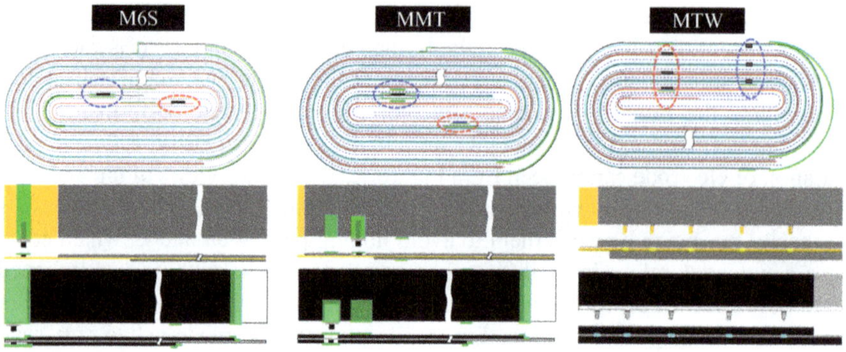

Figure 15.2 Cell structures and electrode configurations

The manufacturing processes and quality standards are very different for each cell structure. The slitting process for multi-tab configuration is more complicated than that of the single tab configuration because the protruded edge cutting is more difficult for multiple-tab positions than the straight line cutting for the single tab positions. More details will be discussion in later sections. In the quality standards side, for example, the electrode adhesive force required by the winding cell is much higher than that by stacking cell type since the inside loops of the winding structure generate quite high tension to layers of dried recipes that may cause those layers stripping off the current collectors.

Principles in industry to select proper cell structures are taken into accounts of many factors such as the electrochemical performance, safety and manufacturing complication of the cell. For example, the 4680 cylindrical structure is proposed by Tesla [5] to achieve high gravimetric energy density, power density but cheap manufacturing process at the large scale for EV applications. Once the cell structure is selected, the electrode manufacturing processes need to be deliberated for the required electrode configurations accordingly.

In some cases, novel manufacturing processes need to be developed if the current SOP (standard of practice) could not be satisfied for the electrode preparation with a specific configuration. The electrode manufacturing processes, from raw materials to final cell structures, have material storage → slurry mixing → coating → calendering→ slitting & cutting. With in-depth learning of the mechanism in each process, the generated knowledge would help to shorten the R&D period, and intrigue creation and invention in the area.

15.4 The practical electrode manufacturing processes and controls

15.4.1 Material storage

Before discussion the manufacturing processes, it is worth discussing about how the raw materials, auxiliary materials and WIP (work in process) are stored properly. The WIP storage condition and time are not only determined by all relevant materials' properties but also required by manufacturing processes. For example, the storage time for a slurry is an important parameter to design recipes and their mixing process because the slurry could not be processed for the next process in time at industrial environment. Sedimentation or agglomeration in the slurry can easily happen if the waiting time is too long [6], especially when the recipe composition is not appropriately designed, or the active materials are not stable enough. As a result, the sequential coating process would be greatly impacted, or even not able to be proceeded due to severe jams in the pumping systems caused by the unstable slurry. Some binders used in the recipes are also sensitive to humidity, and thus cause the thickness rebound of the dried slurry in the electrodes. The typical storage condition and time of relevant materials are listed below for the electrode preparation.

1. Storage time for anode slurries should be less than 48 h, and for cathode slurries should be less than 60 h. Overtime storage would lead to sedimentation, agglomeration and gelation in those slurries. During the storage time, the slurry should be under slow stirring.
2. Pre-coated current collectors (e.g., conductive layer coated) should be stored in normal environment by 15 days, and the time can be extended for another 15 days under some conditions. It must be noticed that the binders used for the pre-coated current collectors may swell and lose adhesion by absorbing moisture from the storage environment.
3. Prepared anode electrodes should be stored in an ambient environment for less than 7 days and it can be extended to 30 days in a low humidity environment. Prepared cathode electrodes should not be stored in normal environment because in general cathode materials are humidity sensitive [7] and thus their storage environment must be controlled at low humidity and their storage time should be less than 10 days.

15.4.2 Slurry mixing

The purpose of slurry mixing is to disperse all components of a recipe uniformly, with minimal time and energy consumption. Non-uniformities in the slurry, for example, including the non-ideal dispersed large particles and the residual bubbles after de-gassing process, lead to coating defects in electrodes (i.e., bright spots and needle holes even to appear metal current collectors). The slurry is non-Newtonian fluid and usually has shear-thinning behavior. In much more complicated ways, some rheological problems such as abnormal elasticity or viscosity influence on the coating quality in vicinity of edges or heads/tails of electrodes. The anode slurry is usually aqueous based, and cathode is N-methylpyrrolidone (NMP) based. In the mixing, active materials, conductive additives and binders are added into a tank of planetary mixer with a specific composition [8]. With shearing and kneading, those components form a solid suspension with good uniformity and stability. The capacity of mixing tank is usually 600–1,500 l, the mixing time is from 4 to 8 h, and the solid content is controlled in the range of 40–70%. The viscosity of produced slurries is controlled for different coating methods:

1. Gravure coating: anode slurry, 3,000–5,000 mPa s, cathode slurry, 2,000–4,000 mPa s
2. Slot die coating: anode slurry, 4,000–8,000 mPa s, cathode slurry, 4,000–8,000 mPa s

The typical mixing procedures can be seen in Figure 15.3. In the anode mixing process, graphite undergoes a kneading process by mixing with a small portion of carboxymethyl cellulose (CMC) and distilled (DI) water. In the kneading processes, the anode recipe forms a dough with a smooth and non-sticky surface. Then, more DI water is added to dilute the dough and make it flowable. In the last, the styrene butadiene rubber (SBR) emulsion is added with a mild mixing speed. After the viscosity and agglomerate distribution inspection, the slurry is transferred to the storage tank for the next process. In the cathode slurry mixing process, carbon black (a kind of conductive additive) is dispersed into NMP solvent at the beginning to avoid its agglomeration, followed by the addition of polyvinylidene fluoride (PVdF) binder solution in which PVdF is pre-mixed with NMP by several hours to form a uniform solution. At last, active material, for example lithium cobalt oxide (LCO), is disperse into the solution to form a stable slurry with uniform distribution of LCO, carbon black and PVdF. After the viscosity and agglomerate distribution inspection, the slurry is transferred to the storage tank for the next process. In the mixing process, the temperature for the mixing tank is controlled by water bath heating since the internal friction of the slurry results in heat-up during its preparation, which may cause the binder gelation and abnormal viscosity. In the planetary mixing process, the slurry is susceptible to residual air in the tank and then form bubbles. Therefore, vacuum degassing is introduced for both anode and

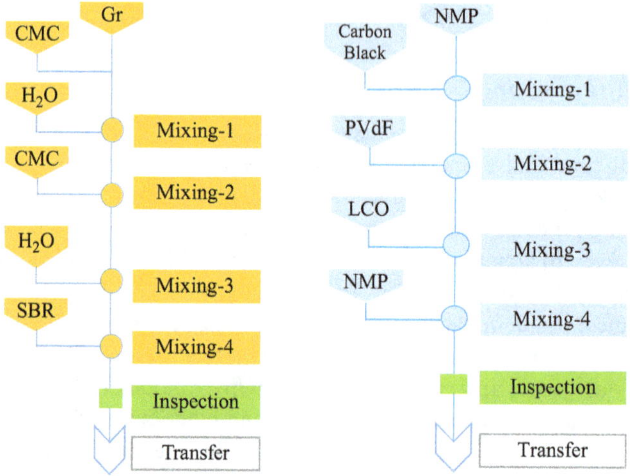

Figure 15.3 Typical mixing procedures for anode and cathode slurries

cathode slurry processing. After the agglomerate distribution inspection, the anode slurry is screened by the 100–175 mesh filter, and the cathode slurry is passed through the 150–300 mesh filter, which avoids generating high pressure in the slurry pumping as well as form poor quality electrodes in the subsequent coating process.

15.4.3 Coating

The coating machine mainly consists of four parts, namely a winding/unwinding system, a coating system, a drying system and an outlet system, as shown in Figure 15.4. Taking the slot die coating that is mostly used in industry as an example, the slurry is filtered first and then transferred by a screw pump and a metering system from the external buffer tank to the manifold (or cavity) within a slot die head through the middle inlet. The manifold is the distribution system to spread the slurry evenly across the current collectors with a specific width. In practical design, the manifold is composed of the slurry storage and flow buffering parts. As the storage part is full, the slurry overflows into the buffering part (the buffering part is used to adjust the slurry flow rate); finally, the slurry in the buffering part is "extruded" from the die lip and then uniformly coated onto the current collectors [2,9]. Subsequently, the solvent in the slurry is evaporated in the drying oven so that the solid materials are adhered onto the current collectors strongly. The winding/unwinding system drives the roll of the current collectors, complied with the coating speed. In order to reduce cost, the wider current collectors e.g. 1,400 mm, are used and the coating speed has been increased faster than 100 m/min. However, the wider current collector and faster coating speed bring huge challenges on the quality of electrodes such as uniformity (especially transverse uniformity) and consistency (e.g., how to mediate the fluctuation caused by large width).

Compared with the common coating methods, some improvements have been made in industry for cell safety and novel cell structures. For instance, in order to reduce the slitting burrs that may lead to the short circuit between the cathode and the anode, ceramic layers are introduced onto the edges of the cathode electrode. Such ceramic layers are done by adding two coating outlets of the ceramic slurry on both sides of the main outlet. In order to enable tabs welded onto the current

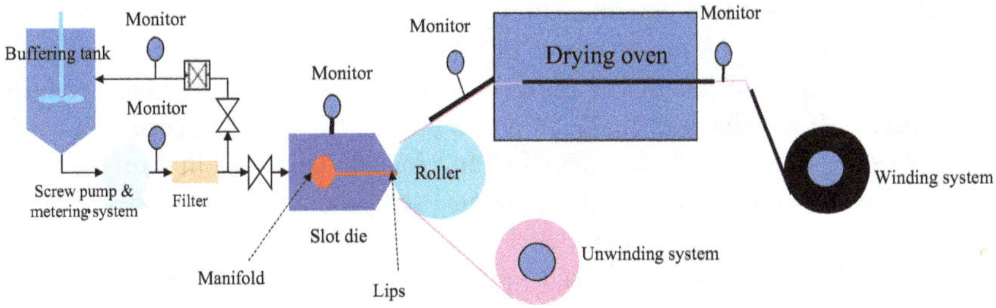

Figure 15.4 Schematic of the coating system in industry

collectors, an intermittent coating method is used to reserve bare spaces for welding tabs, or a peelable method is introduced to remove some coated parts that appear the surface of current collectors for welding tabs.

In the drying process, suspending drying with double-sided air supply is generally used. Upon suspending drying, specially designed air nozzles are settled on both sides of the coated electrode to provide high-speed air flow. The air flow with under the Coanda effect [10] moves along with the wet-coated electrode film, and the slurry on the current collectors is gradually dried. The volatilized solvent gases are processed by a condenser and then recycled, but the residual air re-enters into the air intake duct through a circulation system and then is heated again. The whole process does not have any cooling step thereby reducing the energy consumption. The drying dynamics are complicated [11], but it is valuable to understand the underlying mechanisms of the slurry consolidation & shrinkage, pores and adhesion formation.

Common defects of coating and the causes are listed below:

1. **Scratches/dark marks**: Most of them are caused by the scraper blocked by large particles or the slurry agglomeration. It may also be caused by the failure in filter compartments not to effectively filter large particles or contamination of the drive roller or idler by the solidified slurry.
2. **Air bubbles**: In some cases, the slurry is not stable to produce gases, and meanwhile they cannot be exhausted thoroughly during stirring. Bubbles may also be caused by the leakage from the buffering system of the coating machine and/or the pipe clamps that are not locked tightly, resulting in the air bubbles being brought into the electrode film during coating.
3. **Bright spot**: The slurry is not uniformly dispersed; or there are impurities in the powders; or agglomeration occurs during coating. Normally, the density at the bright spot is higher than the normal areas.
4. **Concave points**: The equidistant concave points are usually caused by the contamination of the drive roller/floating roller that is stained with dry solid slurries. The equidistant concave points are formed by pressing the coating layers during the transmission process of slurries. The distance between the concave points is usually equal to the perimeter of the roller that has been stained.
5. **Peeling off**: The dried slurries peel off from the current collectors, and sometimes the bare current collectors could be seen from the damaged parts. The peeled slurries go into the electrolyte and short internal circuits of the cell, which leads to a potential safety issue.
6. **Crease/wrinkle**: As the levelness of the roller is poor and the parallelism is not good, the tension of the current collectors is easy to produce creases/wrinkles. If the slurry is not dried thoroughly in the coating process, it easily causes the wrinkling during the winding process. The wrinkling on the pinch roller is generally ascribed to the inconsistent clamping degree between the left and right or the serious wear of the pinch roller.
7. **Wavy edges**: The kind of defects appears as the viscosity of the slurry is too low, the gasket is not installed properly, the chamfering of the gasket is not good, and uneven control of unwinding/winding tension cause wavy edges

8. **Trailing edges**: Trailing edges are formed due to the low viscosity and good fluidity of the slurry. In addition, too large gap between the die head and the current collectors is another reason to form trailing edges.
9. **Poor drying**: If the oven temperature is not enough high and/or the air flow is too slow, the slurry coated on current collectors would not be dried well, making it easy to stick to the roller, which causes wrinkling upon winding.
10. **Excessive drying**: The drying process of the electrode happens from its surface to inside. If the oven temperature is too high and the airflow is too fast, the surface of coated slurry is dried too fast, blocking the evaporation of solvents inside so as to slow down the drying process. Meanwhile, a great deal of cracks on the surface appear in such drying process and thus result in the dried slurry peeling off during calendering or winding. Tiny cracks are often difficult to be detected, and it is necessary to repeatedly wipe the coating area, and then observe the layers of dried slurry under backlight.
11. **Breakage of electrodes**: The main reasons for the defect are excessive tension setting, the scratches in the extrusion head, mis- and matched speeds among rollers. In addition, the cutting burrs of the current collectors can also easily cause band breakage, especially for the cathode current collectors.
12. **Exposed metal**: The current collectors could not be fully covered because of underweight coating or bubbles in the slurries.

15.4.4 Calendering

Calendering is a process that the coated electrode passes through the gap between a pair of reverse rotating pressing rollers [12]. Under the compression of the rollers, the electrode decreases in the height direction and increases in the length and width directions, and thus its density is increased [13]. According to the operating temperature, the mainstream calendering methods can be called as the hot pressing and cold pressing, respectively.

The typical calendering machine, as shown in Figure 15.5, is composed with the roller system, hydraulic system and gap wedges. The roller system consists of the roll body, roll neck and shaft head. The roll body is the middle rolling parts of the roller system. The parts have smooth surfaces for the anode calendering

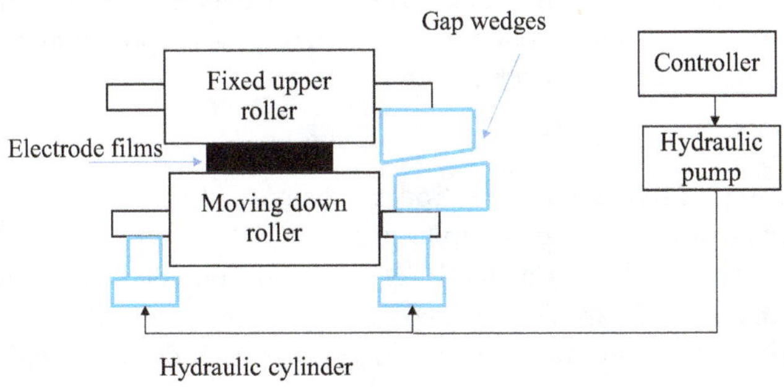

Figure 15.5 Schematic of an industrial calendering system

because of softness of graphite anode. For the cathode calendering, the surfaces of rolling parts are grooved to deal with the hard ceramic particles of cathodes. As indicated in Figure 15.5, two rollers are connected by a pair of gap wedges. When the down moving roller is lifted towards the fixed upper roller by the hydraulic system, the rolling force is generated between gap wedges. Meanwhile, the gap can also be adjusted by the interaction between the upper and lower wedges. Therefore, the pressure and gap applied to the calendering process can be controlled with certain independent levels.

In the industrial manufacturing process, the roller cleaning/pinch systems are necessary attachments. Since the graphite is soft and electrostatic attractive, the roller cleaning system is attached inevitably onto the roller during anode calendering. The cathode materials are not attachable to the roller, the cleaning system is usually not needed. Therefore, it is mostly termed as anode cleaning system in industry. The pinch system is applied to solve the elongation difference between coated area and bare foil area. It is mostly applied for zebra coating which has uniform distribution of coated and bare foil areas. For example, the bare Al foil of the cathode is more ductile than the coated parts, and thus the two parts are uneven elongated at pressing. As a result, the bare foil part is wrinkled, which is suitably for later slitting, tab welding, etc. processes. Pinch system is used to resolve these kinds of problems. The details of two systems are discussed below.

Cleaning system for anode calendering: The system consists of a water spray nozzle, a wet wiper and a dry wiper. A non-woven fabric makes of the wipers that are pressed down onto the surface of the anode calendering roller by double rubber rollers. The water spray nozzle is spraying water evenly on the wet wiper simultaneously. There are a speed reduction motor and a servo motor to control the unwinding speed and tightening tension of the wipers, respectively. The graphite particles attached onto the calendering roller surface are removed by the wet wiper first. Then the dry wiper system is keeping the roller dry and clean which is ready for roll-to-roll calendering process.

Pinch system for calendering: The system is a pinch roller with wrapped Teflon that fits for the bare foil parts of the electrodes. Due to the existence of Teflon strips, the strain between coated and bare parts on the electrodes is minimized, which retains the elongation of electrodes during calendering and ensures the quality of calendering. The thickness of Teflon, its alignment with bare foil parts, pressure of the pinch roller and calendering speed are important factors affecting the function of the pinch system.

Common defects of calendering:

1. **Consistency of thickness** of electrodes is controlled by the roller straightness, roller jump, roller bending and roller wear.
2. **Poor compaction density control**: As the compaction density of anode is low, for example, its electronic conductivity would be poor, which restrains the electronic contribution to the retard kinetics of Li^+-ion intercalation. On the other hand, as the anode is over-calendered to form high compaction density, the graphite particles are crashed, which creates narrow Li^+-ion channels and

thus hinders the Li^+ intercalation and results in accumulation of Li^+ on the anode surface [14] to form lithium plating further. Moreover, the over-calendered electrode makes it difficult to be wetted by electrolytes [15], which leads to incomplete Li^+-ions intercalation into anode.

3. **Electrode wrinkle**: Wrinkling electrodes are usually caused by the improper levelness and parallelism of the guide roller, uneven tension, pinch tension, winding tension and uneven tension of the incoming as-coated electrodes.
4. **The elastic recovery of the electrode film**: There is an elastic recovery of the anode electrode film after calendering which causes unstable electrode thickness. Baking can accelerate this process to achieve a stable thickness value. The elastic recovery of the electrode film is mainly caused by the elastic recovery of SBR. The mechanism is that the Hookean spring (the elastic part of SBR) drives the Newtonian glue pot (the viscous part of SBR) to return to the initial position when the pressure is removed [16]. Increasing the pressing time such as pressing at relative low speed or multiple pressing, will improve the deformation of viscous part of the SBR and reduce the elastic recovery of the electrode film.
5. **Broken electrodes**: The large particles or their agglomeration can be embedded into the current collectors under calendering, which may break the electrode if a tension is applied. Meanwhile, poor levelness of the cold roller produces the abnormal tension that breaks the electrode starting from the weaken parts, e.g. burrs on the edges of current collectors.
6. **Disconnect/exposed metal**: Layers of dried slurries disconnected with the current collectors are easily caused by large particles, under or over dried parts of the electrode, especially when over-calendering is applied. Most severely, the disconnected parts strip off the electrode and the bare parts of current collectors are exposed.
7. **Black spots/shine spots**: Black spots or shine spots are caused by the uneven distribution of slurries. As a result, the areas with denser or less slurries form shine spots or black spots after calendering, respectively.
8. **Rippled edges**: Uneven tension under calendering produces ripples readily in the vicinity of edges of the electrode. Wear of the roller also exacerbates the rippling phenomenon because of non-uniform fiction between the roller and electrode films.

15.4.5 Slitting and cutting

To improve the production efficiency, an approach is to produce the electrode with a large width. According to the size requirements of electrodes in cells, the prepared electrodes with the large width need slitting to narrow pieces. The disc cutter, as shown in Figure 15.6, is mainly used to slit the electrodes. The cutter is equipped with a dust suction pipe for the upper and lower blades, respectively. Two de-dusting brushes are used to clean up dusts generated during cutting. In addition, the high resolution camera is used to capture images during slitting with which the pixel calculation and OFFSET value correction are used to monitor the difference

Figure 15.6 A typical industrial disc cutter

between the real width and it is designed one during cutting. Therefore, the cutting positions could be adjusted as long as the difference deviates the tolerances for the designed cells.

To make electrodes with MTW configuration is more complex and the challenge of slitting is also greater based on zebra coating. The welding positions of tabs in the electrodes are zig-zag non-continuous style. Slitting based on disc cutter is not applicable, the laser die-cutting techniques with programmable laser cutting paths are developed. The basic principle of laser die-cutting is to use high-power laser beams to irradiate the electrode films, then the electrode film can be heated to a very high temperature, quickly melted, vaporized, burned through or reached the ignition point to form holes. With the movement of the laser beams, the holes are continuously formed and generated an extremely narrow slit to complete the cutting. The laser intensity determines the effect of laser die-cutting. If the laser intensity is too high, it can cause the exfoliation of coated films at cutting edges, the explosion of the substrate foils, the breakage of the tabs, and producing foreign matters and molten metal beads. Otherwise, if the laser intensity is too low, it may not shear completely, and generate sheared burrs, broken sheets and wrinkled tap and other abnormalities.

In the slitting and cutting processes, it is also feasible to create the welding position inside the coated area of the electrodes. Similar to the peelable method in the coating process, laser cleaning is another method to produce welding position for embedded tab, which enable more choses of electrode configurations and possibilities of cell structures. Laser cleaning is a process that uses the laser to quickly burn out and vaporizes the coating layer at a specified position on both front and back of electrode sheets, and then the metal current collectors are exposed and meet the requirements of the tab welding. It is necessary to cooperate with local sensors to measure the positions and distances between the heads of electrode film and the specified cutting areas accurately, so as to clean the welding positions that meet the size requirements.

Common defects of slitting & cutting processes:

1. **Inaccurate slitting width**: The slitting width is determined by the width of the cutting dies. The insufficient capacity may happen in a cell if the real slitting widths of electrodes are smaller than the designed ones.
2. **Slitting burr**: Burrs may penetrate through the separator and cause internal short circuits in cells. Burrs may be attributed to die wear, misalignment between the upper and lower dies, improper slitting speeds and so on.
3. **Wavy edges**: Wavy edges may give rise to the deviation of tape positions during welding process because of wrinkles on electrodes and high reject ratio in X-ray detection. Wavy edges can be caused by several reasons: (a) inconsistent thicknesses on both sides of electrodes, (b) uneven tension during slitting, (c) insufficient weight of the winding rollers, (d) deviation of the position of the pressure wheels, (e) inappropriate size and hardness of the sponge gasket of the cutting die which change the tension of electrode film before and after the cutting die.

15.5 Novel routes for industrial electrode manufacturing

The current industrialized electrode manufacturing processes have been achieved tremendous success in promoting lithium ion batteries business. The manufacturing processes are constantly motivated by lower cost, higher efficiency, and better quality control. In this section, novel routes for industrial electrode manufacturing focuses on emerging mixing, drying, and novel coating technologies. Those could provide a new picture of how the electrode could be manufactured in the future.

15.5.1 Mixing

The current double planetary mixer [17] mainly uses rotational mechanical disperse force to prepare slurries. Disperse efficiency, uniformity, batch to batch reliability are the bottleneck of the route. It also occupies a large layout space, is hard to clean for next batch, and has insufficient degree of automation. Some new technologies of mixing [18] are emerging to prepare slurries according to different ways of dispersing, and they can be put into three categories: high speed mixing, ultra-shear dispersion, and twin screw extrusion dispersion. A summative comparison of these three routes is shown in Table 15.1.

(a) High speed mixing
High speed mixing is a discontinuous process so-called the "batch" mixing method. It uses a porous dispersion wheel, as shown in Figure 15.7, in a high speed rotation to enable particles dispersed well in a very short time. The slurry

Table 15.1 Comparison of different mixing routes in different aspects

	Planetary mixer	**High speed mixing**	**Super-high-shear dispersion**	**Twin-screw extrusion**
Production type	Batch to batch	Batch to batch	Batch to batch	Continuous
Core parts	Planet and disper	Porous disperse wheel	Stator and rotor	Twin screw rods
Mixing method	Kettle blade-based revolution and rotation	Recirculated shearing force	High speed spin of impeller and rotor	Twin screw rods rotation in shape
Production capacity	~50 kg/h	~280 kg/h	>200 kg/h	~300 kg/h
Disperse uniformity	★★★	★★★☆	★★★★	★★★☆
Slurry stability	★★★☆	★★★★	★★★☆	★★★★

Figure 15.7 Porous dispersion wheel

in the mixing tank is sheared and collided circularly with negative pressure generated by the high speed rotation of the wheel, and experiences a repetition of "pull→spin→impact→pull" cycle. The agglomerated particles are dispersed to form a stabilized slurry. The structure design of porous dispersion wheel is critical to the dispersion efficiency. The advantages of high speed mixing include the higher mixing efficiency, less damage to the particle surface, smaller equipment size, and lower energy consumption.

(b) Super-high-shear mixing

Super-high-shear mixing is also a batch mixing method. With the high speed rotation of the inner impeller, the slurry is sucked into the space between the stator and rotor as shown in Figure 15.8. Undispersed particles in the slurry bear complex forces of shearing, friction and collision which are generated by laminar, turbulent fluids, and cavitation effect, respectively. The cavitation effect is an important factor to uniform dispersion of particles in the slurry. It is due to perturbation from the blast wave that is generated by the blowing of cavitation bubbles. According to this mechanism, some techniques like high pressure homogenizer and ultrasonic disperser have the similar effect. The core parts of super-high-shear mixer are the stator and the rotor. To improve its dispersion efficiency, it must be deliberate to design the shapes of the stator and rotor, optimize the gap between them, and adjust the ratio and width of slot inside them.

The advantages of super-high-shear mixing have the high mixing efficiency, smaller layout space, and low equipment, operation and energy cost. The effectiveness of dispersion can be improved by increasing the number of stators and rotors. However, the high shear force in the dispersion process could cause damages to the particles, which may deteriorate electrochemical performance of assembled cells. Consistent behavior and degree of automation are two challenges to employ the route making cells for the industrial application.

(c) Twin-screw extrusion

Twin-screw extrusion method has been successfully applied in plastic, food and paper industries. The application in the battery industry is still under developing [19]. Unlike the other two batch mixing routes, the twin screw extrusion is a continuous mixing process that combines material weighting, transferring, kneading and mixing processes together. As shown in Figure 15.9, the two parallel aligned screw rods generate the mixing force by reversing or parallel rotation. In the reversing rotation process, the two screw rods are engaged to

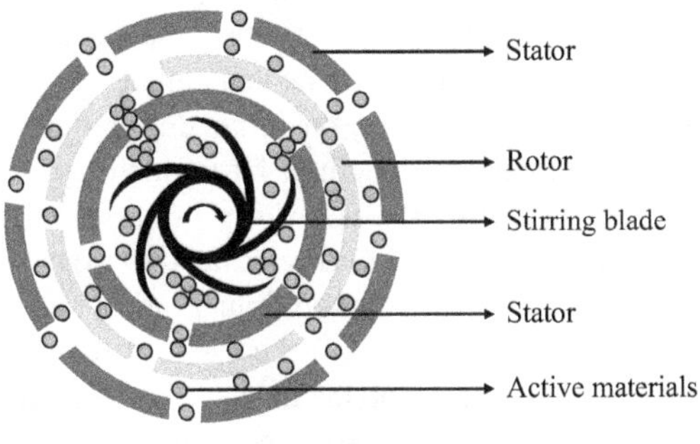

Figure 15.8 Schematic diagram of stator and rotor

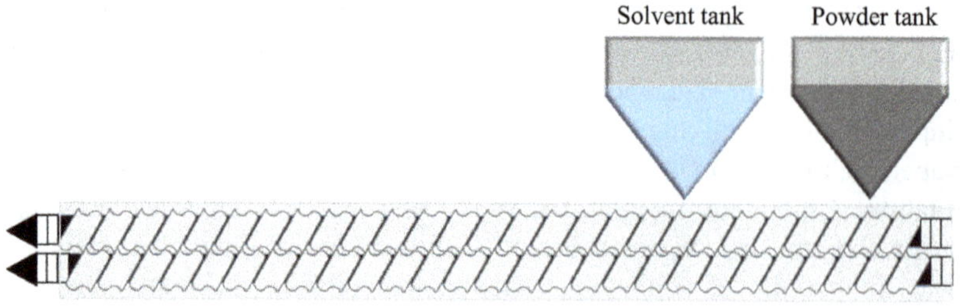

Figure 15.9 Schematic diagram of twin screw mixing equipment

form a "C" shape space. With the rotation of the two rods, the slurry is pushed forward vertically. The shearing force in the "C" space makes the particles dispersed. In the parallel rotation process, the slurry goes "∞" shape movement, and the shearing efficiency is higher than that in the "C" shape space.

The advantages of twin-screw mixing have high dispersing efficiency, continuous processing, and good consistency. Meanwhile, such mixing process can be realized with high degree of automation because relevant equipment can be made with modules of kneading, dispersion, and flow control functions. To achieve best mixing performance, the twin-screw-rods structure needs to be optimized. In addition, twin screw mixing is the best route suitable for mixing one recipe of slurry in large scale.

15.5.2 New drying methods

New drying methods such as infrared drying and microwave drying are attempted to improve drying efficiency and solvent volatilization compared to blast drying [20]. Other methods like UV curing [21] and electron beam (EB) [22] curing are not under discussion due to the binder-types changed.

Currently infrared drying has been widely studied in the fields of food, paper and spray paint [23]. For infrared drying, radiation heat transfer occurs via infrared light which is an electromagnetic wave with a wavelength between 0.75 and 1,000 μm. The radiation energy emitted by the infrared heat source is absorbed by the wet slurry coated on the current collectors. When the infrared wavelength is equal to the natural vibration wavelength of the molecular motion of the components (solvent, binder etc.), resonance is induced, and the molecules vibrate and collide violently, generating heat energy and resulting in drying. In industry, far-infrared drying devices are commonly used for heating. Due to the long wavelength of far-infrared rays, infrared radiation is not only absorbed on the surface of the wet slurry but also reaches into its inside. Therefore, the whole coated slurry can be heated at the same time, improving the heating efficiency and uniformity. The infrared drying rate is affected by a number of factors such as the thickness and initial solvent content of the wet coated slurry, the radiation power, the radiation distance, and the radiation temperature. There are several advantages of low equipment cost, low energy consumption, high drying efficiency, less heat loss and continuous operation for infrared drying. However, infrared drying is not suitable for drying the ultra-thick electrodes. Additionally, the volatilization process produces a thick solvent atmosphere above the electrodes, which alleviates the infrared interaction with solvents inside the wet coated slurry and thus reduces the drying efficiency. For the improvement, convective circulation by hot air is combined with the infrared drying to exsufflate vapors above electrodes in time. However, its practical application is still restrained by the drying depth currently.

Microwave drying technology dated back to 1940s and got significant development in the 1960s and 1970s [24]. At present, microwave drying plays a vital role in many areas such as chemical industry, food and paper industry, and is gradually expanding into the battery field. Microwaves are defined as electromagnetic waves with wavelengths ranging from about 1 m to 1 mm corresponding to frequencies between 300 MHz and 300 GHz, respectively. The commonly used frequency is 915–2,450 MHz. Microwave drying is a heating method that uses electromagnetic waves as a heating source to make the drying material self-heating. The core component of the microwave drying equipment is the microwave generator. In the high frequency changing electromagnetic field via the microwave generator, the polar molecules inside the wet coated slurry continuously rotate as they try to align themselves with the alternating electromagnetic field. The rotating molecules collide other molecules violently, and friction occurs among molecules, which thereby converts microwave energy into molecular kinetic energy and then into heat energy to achieve the wet coated slurry drying. The microwave drying efficiency is related to the microwave frequency, the material thickness and the microwave absorption capability of materials in the slurry. Microwave drying demonstrates the advantages of high heating efficiency, uniform heating, high energy utilization and easy operation of equipment. Compared with infrared drying, microwave drying has higher energy-efficiency with longer wavelengths and better penetration capability. However, its cost is relatively high, and microwave leakage would endanger human health. Meanwhile, the industrial application and development of microwave drying technology are imperfect, such as equipment specifications with dimensions, structure design of equipment, processing research, and reliability control.

15.5.3 Dry coating technologies

Dry coating, without any solvent during mixing and coating, is one of the most attractive technologies in the industry. The benefits of excluding solvents can be seen in several aspects, (a) no drying process can save major energy consumption; (b) the efficiency in the mixing process is greatly improved; and (c) the equipment occupies smaller floor space. The dry coating process can be divided into two categories, extruding calendering and electrostatic spray.

Extrusion coating is one of the most mature technologies for dry coating manufacturing and has been industrialized in the field of super capacitors [25]. The characteristic of extrusion coating lies in the use of polymers that can be fiberized (e.g. polytetrafluoroethylene, PTFE) as stirred and hot rolled at high temperatures, Figure 15.10. The fiberized binders connect the active materials and conductive

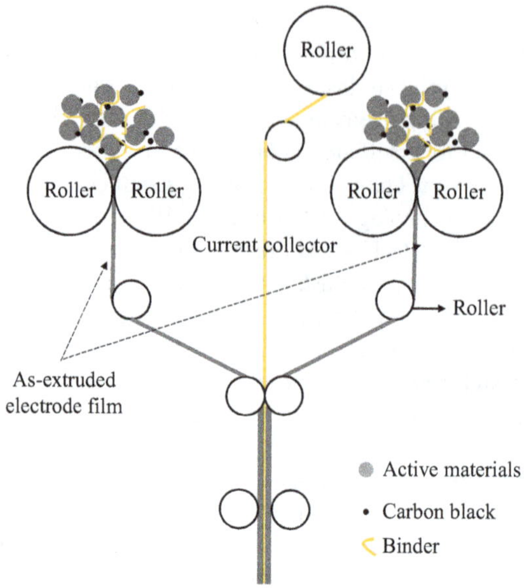

Figure 15.10 Schematic diagram of "solvent-free" dry electrode fabrication: extrusion coating

additives into the fibrous network, and then can be applied onto the current collectors to form the electrodes.

Extrusion coating is suitable for electrode preparation with the thickness ranging from 50 to 1,000 μm. Compared with the classical wet coating, extrusion coating is easier to achieve active layers simultaneously on both sides of metal foils. In addition, the extrusion coating has good compatibility with new manufacturing and new materials such as solid-state electrolytes, pre-lithiation technology, and porous current collectors. However, the electrochemical performance of dry-process cells still needs to be further optimized. Currently, the extrusion coating speed is relatively low, and the industrialized manufacturing capacity is being accelerated.

With the development since 1953, the electrostatic spray nowadays has been widely used in automobile paint technology. In the field of lithium ion batteries, it is still mainly under academia research. Unlike the extrusion coating, electrostatic spraying usually uses PVdF as a binder [26]. Electrostatic spraying equipment mainly includes a high-speed mixer, an electrostatic spray gun/electrostatic screen, and hot rollers, etc. The mixed powder is charged by an electrostatic spray chamber or electrostatic screen, and then sprayed and deposited onto the surface of the grounded current collector. At last, PVdF is melted by hot pressing to form the electrode films, as shown in Figure 15.11. Compared with extrusion coating, electrostatic spraying has lower energy consumption and cost. Electrostatic spray is a process in which the electrode thickness depends on the spray time and be in a range of 30–500 μm. The electrostatic spray gun is highly flexible, and multiple

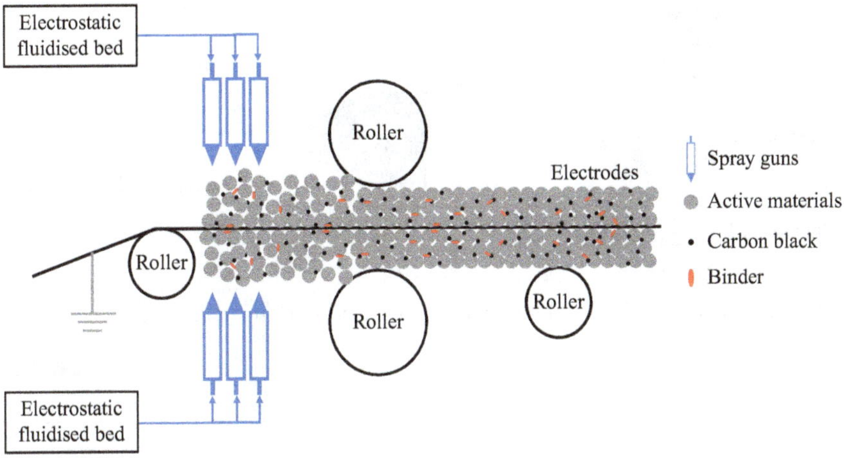

Figure 15.11 Schematic diagram of "solvent-free" dry electrode fabrication: electrostatic spraying

independent gun heads allow multiple materials being sprayed together to achieve customized design of electrode structure. The challenge of electrostatic spray technology is to improve the active material loading as well as the uniformity of electrode film and technical maturity.

15.5.4 Other coating technologies

Increasing the thickness of electrodes can reduce the use of current collectors, separators, packaging and other materials, which thereby improves the energy density and lowers the cost of the battery. However, their electrons and ions pathways are elongated and thus enhance the resistance in thicker electrodes, which make it difficult to meet the requirements for some high power applications. Therefore, the configurations of electrodes need to be optimized or customized. For example, the electrochemical kinetics can be improved by the porous electrode designs to decrease the tortuosity of ion diffusion pathways, and thus reduces the ionic impedance. From the perspective of coating technology, the current possible methods include multilayer coating, laser ablation, magnetic coating, etc.

The most effective way in industry to improve the kinetics of thick electrodes is to design the electrodes with gradient porosity structure by multi-layer coating technology [27], as shown in Figure 15.12. The core components of the multi-layer coating technology are the multi-chamber slot die and the feeding module. Inside the die, there are several different parallel chambers. When different slurries flow into these chambers and out of the lips in parallel, the multi-layer coating can be achieved. The porosity of each layer can be controlled by adjusting the particle size and the solid content of the slurries. The key to the multi-layer coating is to adjust the properties of the slurries to avoid the interlayer mixing. During the drying process, the surface cracking of the multilayer thick electrodes can be alleviated by

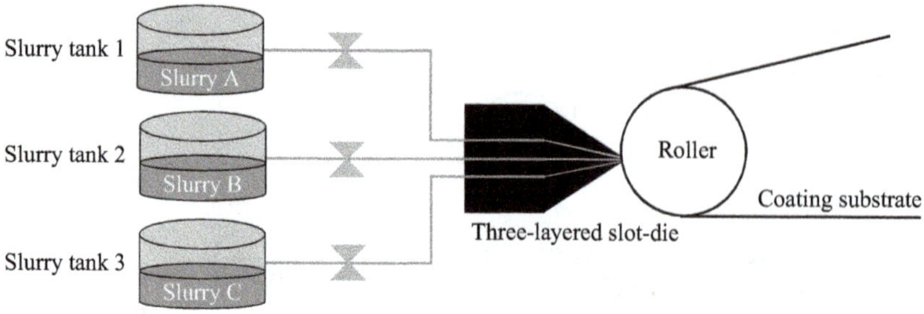

Figure 15.12 Schematic diagram of multilayer coating

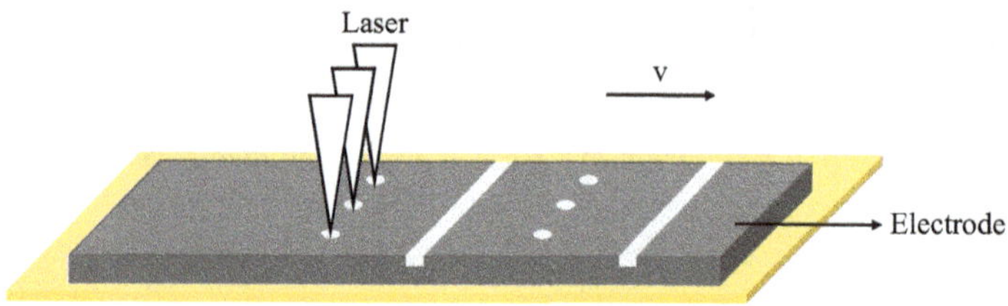

Figure 15.13 Schematic diagram of laser ablation

gradient porosity design. In addition to the porous electrodes, the multi-layer coating machine is applied to process various functional layers such as adhesive coatings, safety coatings and high-capacity coatings so as to meet the different requirements of battery applications.

Laser ablation [28] is a transient cleaning technology developed in the 1970s and is one of the most important applications in laser industry. The laser emits a high-power density laser beam onto the surface of electrodes so that those irradiated electrodes absorb energy, melt and vaporize. As a result of laser irradiation, as depicted in Figure 15.13, the materials can be removed to form holes or lines patterns in the electrodes, which reduces the tortuosity, enhances capillary transmission, improves electrolyte wettability and reduces ion diffusion impedance.

The effect of laser ablation is taken by the thermal conductivity, reflectivity, and absorptivity of materials. In addition, there is also a certain correlation with the laser types and technical parameters (e.g. laser power, wavelength, pulse number, gas type, incident angle, etc.). Currently, laser etching is developing towards short pulses and high energy. According to the different laser pulse widths [29], it can be categorized as (1) nanoseconds (10^{-9}~10^{-10} s), (2) picoseconds (10^{-11}~10^{-12} s), and (3) femtoseconds (10^{-15} s). The advantages of the laser etching mentioned above include high processing accuracy, no scraps or burrs, wide size range (0.2 μm–1,500 μm) and strong controllability. Besides, the non-contact processing can avoid the contamination of materials and thus minimize damages to surrounding materials. Although the speed of laser ablation seems to be quick already, the efficiency of industrialization is still low due to the limitation of the drilling depth and spacing requirements. The efficiency of laser ablation can be improved by the assistance of electromagnetic fields. Holes and lines generated by the laser help the ion transportation and thus improve the kinetics of the electrode pronouncedly [30]. However, in consideration of the fact that laser etching removes active material from the electrode and reduces its capacity, the parameters for laser etching should be controlled well to balance capacity and kinetics. Although multilayer coating technology shows more advantages in terms of technical impacts and technical maturity, laser etching is still significant for further exploration.

15.6 Future perspective of electrode manufacturing

With the burst demands of batteries, the electrode manufacturing, from powders to electrodes, has been paid great attention and made substantial improvement. The demands still grow exponentially, with the expanding markets of EV and renewable energy storage. However, many challenges for electrode manufacturing are being confronted such as production speed acceleration, energy consumption reduction, and quality improvement. Battery manufacturers are eager for the exploration of new manufacturing technologies to satisfy the demands of prosperous power markets.

There is another important aspect to achieve intelligentized and digitalized in the manufacturing processes. These processes should be highly rooted in the deep mechanisms of how the electrode configurations are formed. In the process of making slurry, for example, no matter which method is adopted, dried or wetted, it must be elucidated how morphologies of particles lead to a practical slurry with proper rheology and how the pore structures of electrodes form and change during the coating, drying and calendering processes. With the simulation [31], advanced investigation technologies, machine learning [32], AI [33], etc., manufacturing processes in the future could be dramatically different from the current ones.

References

[1] IEA. (2022). World Energy Outlook 2022, IEA, Paris. https://www.iea.org/reports/world-energy-outlook-2022, License: CC BY 4.0 (report); CC BY NC SA 4.0 (Annex A).

[2] Li, J., Fleetwood, J., Hawley, W. B., and Kays, W. (2021). From materials to cell: state-of-the-art and prospective technologies for lithium-ion battery electrode processing. *Chemical Reviews*, 122(1), 903–956.

[3] Haghi, S., Summer, A., Bauerschmidt, P., and Daub, R. (2022). Tailored digitalization in electrode manufacturing: the backbone of smart lithium-ion battery cell production. *Energy Technology*, 10(10), 2200657.

[4] Schmied, J., Puchta, A., Scharmann, T., *et al.* (2022). Framework for the application of industry 4.0 in lithium-ion battery cell production. In *Proceedings of the Conference on Production Systems and Logistics: CPSL 2022* (pp. 151–160). Hannover: Publish-Ing.

[5] 2020 Battery Day Presentation. https://tesla-share.thron.com/content/?id=96ea71cf-8fda-4648-a62c-753af436c3b6&pkey=S1dbei4 (accessed Jan 6, 2023).

[6] Wenzel, V., Nirschl, H., and Nötzel, D. (2015). Challenges in lithium-ion-battery slurry preparation and potential of modifying electrode structures by different mixing processes. *Energy Technology*, 3(7), 692–698.

[7] Mijung, N., Lee, Y., and Cho, J. (2006). Water adsorption and storage characteristics of optimized $LiCoO_2$ and $LiNi_{1/3}Co_{1/3}Mn_{1/3}O_2$ composite cathode material for Li-ion cells. *Journal of the Electrochemical Society*, 153(5), A935.

[8] Wang, M., Dang, D., Meyer, A., Arsenault, R., and Cheng, Y. T. (2020). Effects of the mixing sequence on making lithium ion battery electrodes. *Journal of the Electrochemical Society*, 167(10), 100518.

[9] Ding, X., Liu, J., and Harris, T. A. (2016). A review of the operating limits in slot die coating processes. *AIChE Journal*, 62(7), 2508–2524.

[10] Trancossi, M. (2007). An overview of scientific and technical literature on Coanda effect applied to nozzles. *SAE Technical Paper* (2011-01-2591).

[11] Zhang, Y. S., Courtier, N. E., Zhang, Z., *et al.* (2022). A review of lithium-ion battery electrode drying: mechanisms and metrology. *Advanced Energy Materials*, 12(2), 2102233.

[12] Gunther, T., Schreiner, D., Metkar, A., Meyer, C., Kwade, A., and Reinhart, G. Classification of calendering-induced electrode defects and their influence on subsequent processes of lithium-ion battery production. *Energy Technology*. 2020, 8, 1900026.

[13] Meyer, C., Bockholt, H., Haselrieder, W., and Kwade, A. (2017). Characterization of the calendering process for compaction of electrodes for lithium-ion batteries. *Journal of Materials Processing Technology*, 249, 172–178.

[14] Stephenson, D. E., Walker, B. C., Skelton, C. B., Gorzkowski, E. P., Rowenhorst, D. J., and Wheeler, D. R. (2011). Modeling 3D microstructure and ion transport in porous Li-ion battery electrodes. *Journal of the Electrochemical Society*, 158(7), A781.

[15] Sheng, Y., Fell, C. R., Son, Y. K., Metz, B. M., Jiang, J., and Church, B. C. Effect of calendering on electrode wettability in lithium-ion batteries. *Frontiers in Energy Research* 2014, 2, 56.

[16] Jeschull, F., Lacey, M. J., and Brandell, D. (2015). Functional binders as graphite exfoliation suppressants in aggressive electrolytes for lithium-ion batteries. *Electrochimica Acta*, 175, 141–150.

[17] Keppeler, M., Roessler, S., and Braunwarth, W. Production research as key factor for successful establishment of battery production on the example of large-scale automotive cells containing nickel-rich LiNi0.8Mn0.1Co0.1O2 electrodes. *Energy and Technology* 2020, 8, 2000183.

[18] Liu, Y., Zhang, R., Wang, J., and Wang, Y. (2021). Current and future lithium-ion battery manufacturing. *IScience*, 24(4), 102332.

[19] Martin, C. (2016). Twin screw extruders as continuous mixers for thermal processing: a technical and historical perspective. *AAPS PharmSciTech*, 17 (1), 3–19.

[20] von Horstig, M. W., Schoo, A., Loellhoeffel, T., Mayer, J. K., and Kwade, A. (2022). A perspective on innovative drying methods for energy-efficient solvent-based production of lithium-ion battery electrodes. *Energy and Technology*, 10, 2200689.

[21] Shukla, V., Bajpai, M., Singh, D. K., Singh, M., and Shukla, R. (2004). Review of basic chemistry of UV-curing technology. *Pigment & Resin Technology*, 33, 272–279.

[22] Du, Z., Janke, C. J., Li, J., and Wood III, D. L. (2019). High-speed electron beam curing of thick electrode for high energy density Li-ion batteries. *Green Energy & Environment*, 4(4), 375–381.

[23] Escobedo, R., Miranda, R., and Martínez, J. (2016). Infrared irradiation: toward green chemistry, a review. *International Journal of Molecular Sciences*, 17(4), 453.

[24] Rattanadecho, P. and Makul, N. (2016). Microwave-assisted drying: a review of the state-of-the-art. *Drying Technology*, 34(1), 1–38.

[25] Zhang, Y., Huld, F., Lu, S., Jektvik, C., Lou, F., and Yu, Z. (2022). Revisiting polytetrafluorethylene binder for solvent-free lithium-ion battery anode fabrication. *Batteries*, 8(6), 57.

[26] Ludwig, B., Zheng, Z., Shou, W., Wang, Y., and Pan, H. (2016). Solvent-free manufacturing of electrodes for lithium-ion batteries. *Scientific Reports*, 6(1), 1–10.

[27] Demiryürek, R., Gürbüz, N., Hatipoglu, G., *et al.* (2021). Roll-to-roll manufacturing method of aqueous-processed thick LiNi0.5Mn0.3Co0.2O2 electrodes for lithium-ion batteries. *International Journal of Energy Research*, 45(15), 21182–21194.

[28] Pfleging, W. (2018). A review of laser electrode processing for development and manufacturing of lithium-ion batteries. *Nanophotonics*, 7(3), 549–573.

[29] Ion, J. (2005). *Laser Processing of Engineering Materials: Principles, Procedure and Industrial Application*. Elsevier.

[30] Dunlap, N., Sulas-Kern, D. B., Weddle, P. J., *et al.* (2022). Laser ablation of Li-ion electrodes for fast charging: material properties, rate capability, Li plating, and wetting. *Journal of Power Sources*, 537, 231464.

[31] Ramadesigan, V., Northrop, P. W., De, S., Santhanagopalan, S., Braatz, R. D., and Subramanian, V. R. (2012). Modeling and simulation of lithium-ion batteries from a systems engineering perspective. *Journal of the Electrochemical Society*, 159(3), R31.

[32] Niri, M. F., Liu, K., Apachitei, G., *et al.* (2021). Machine learning for optimised and clean Li-ion battery manufacturing: revealing the dependency between electrode and cell characteristics. *Journal of Cleaner Production*, 324, 129272.

[33] Niri, M. F., Liu, K., Apachitei, G., *et al.* (2022). Quantifying key factors for optimised manufacturing of Li-ion battery anode and cathode via artificial intelligence. *Energy and AI*, 7, 100129.

Index

www.ingramcontent.com/pod-product-compliance
Lightning Source LLC
LaVergne TN
LVHW020750170326
834039LV00003B/133
* 9 7 8 1 8 3 9 5 3 6 6 9 4 *